Non-destructive Testing of Materials in Civil Engineering

Non-destructive Testing of Materials in Civil Engineering

Special Issue Editor

Krzysztof Schabowicz

MDPI • Basel • Beijing • Wuhan • Barcelona • Belgrade

MDPI

Special Issue Editor
Krzysztof Schabowicz
Wrocław University of Science and Technology
Poland

Editorial Office
MDPI
St. Alban-Anlage 66
4052 Basel, Switzerland

This is a reprint of articles from the Special Issue published online in the open access journal *Materials* (ISSN 1996-1944) from 2018 to 2019 (available at: https://www.mdpi.com/journal/materials/special issues/non-destructive testing).

‑   ‑

For citation purposes, cite each article independently as indicated on the article page online and as indicated below:

LastName, A.A.; LastName, B.B.; LastName, C.C. Article Title. *Journal Name* **Year**, *Article Number*, Page Range.

ISBN 978-3-03921-690-1 (Pbk)
ISBN 978-3-03921-691-8 (PDF)

Contents

About the Special Issue Editor

Krzysztof Schabowicz is the Author and co-author of 5 books and over 200 scientific papers, including 40 articles indexed in the database Web of Science. These publications have >500 citations in Web of Science. He serves as the Guest Editor of *Materials* (MDPI) and Editorial Board member of *Civil Engineering and Architecture* (HRPUB), and *Nondestructive Testing and Diagnostics* (SIMP). He has been involved in the development of more than 200 reviews of journal and conference articles. He is co-author of 5 patents and 5 patent applications. His memberships include the Polish Association of Civil Engineers and Technicians (PZITB) as well as the Polish Association of Building Mycology (PSMB). His research interests include concrete, fiber cement, ultrasonic tomography, impact-echo, impulse-response, GPR and other nondestructive tests, as well as artificial intelligence.

materials MDPI

Editorial

Non-Destructive Testing of Materials in Civil Engineering

Krzysztof Schabowicz

Faculty of Civil Engineering, Wrocław University of Science and Technology, Wybrzeże Wyspiańskiego 27, 50-370 Wrocław, Poland; krzysztof.schabowicz@pwr.edu.pl

Received: 23 September 2019; Accepted: 29 September 2019; Published: 3 October 2019

check for updates

Abstract: This issue was proposed and organized as a means to present recent developments in the field of non-destructive testing of materials in civil engineering. For this reason, the articles highlighted in this editorial relate to different aspects of non-destructive testing of different materials in civil engineering, from building materials to building structures. The current trend in the development of non-destructive testing of materials in civil engineering is mainly concerned with the detection of flaws and defects in concrete elements and structures, and acoustic methods predominate in this field. As in medicine, the trend is towards designing test equipment that allows one to obtain a picture of the inside of the tested element and materials. Interesting results with significance for building practices were obtained.

Keywords: non-destructive testing; diagnostic; acoustic methods; ultrasound; building materials; defects

---

1. Introduction

The current trend in the development of non-destructive testing of materials in civil engineering is mainly concerned with the detection of flaws and defects in concrete elements and structures, and acoustic methods predominate in this field. As mentioned in [1,2] much attention has been devoted to acoustic techniques because they have been greatly developed in recent years and there is a clear trend towards acquiring information about a tested element or structure from acoustic signals processed by proper software using complex data analysis algorithms. Another trend in the development of nondestructive techniques is the assessment of characteristics other than strength in elements or structures, particularly those made of concrete or reinforced concrete. As in medicine, the trend is towards designing test equipment that allows one to obtain a picture of the inside of the tested element. Increasingly, the offered apparatus is equipped with software based on sophisticated mathematical algorithms and artificial intelligence, which makes advanced analysis of the test results possible [2].

2. Non-Destructive Testing

In construction, modern diagnostic methods are applied to building structural members and structures. Another major diagnostic field is the non-destructive testing of building materials. Such materials as wood, masonry units, concrete, fiber-cement and steel are subjected to tests for various reasons and at different times, e.g., during construction, but mainly during the service life. Many investigative methods are used for this purpose. Depending on the degree of their invasiveness, they can be divided into destructive, semi-destructive and non-destructive methods. A general classification of methods that are useful for diagnosing buildings and building materials [2] is presented in Figure 1.

Non-destructive methods are mainly used to test strength and investigate its changes over time. Usually samples taken from the structure, and sometimes whole members or structures, are tested in this way. Also load tests, which rather rarely are applied to buildings, but more often to bridges and roads, can be put into this category.

Semi-destructive and destructive methods are used to test samples and members. They can also be used to test whole structures. Strength and its changes over time are tested, but mainly other properties are tested in this way.

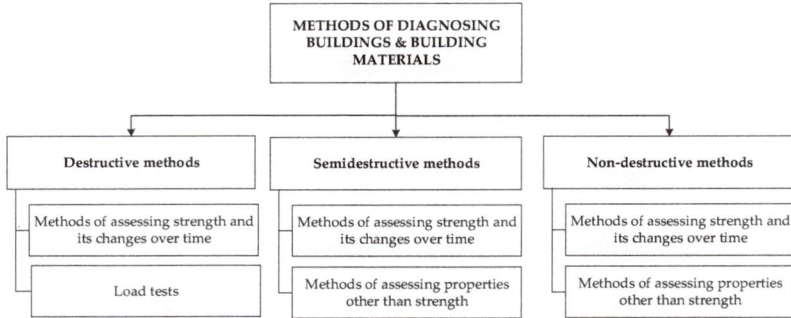

Figure 1. General classification of investigative methods useful for diagnosing buildings and building materials [2].

The difference between semi-destructive and non-destructive methods is that in the case of the former, the material is usually locally and superficially damaged when tested. No such damage occurs in the case of non-destructive methods. This is one of the reasons why they are suitable for testing large surfaces down to a considerable depth, and in general construction. Moreover, in the case of non-destructive methods, measurements can be repeated, whereby the test results can be verified and validated.

A general classification of non-destructive methods that are useful for diagnosing buildings and building materials [3] is presented in Figure 2.

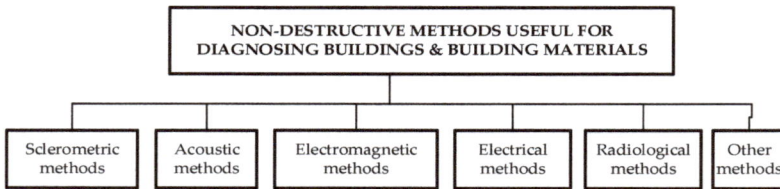

Figure 2. General classification of non-destructive methods useful for diagnosing buildings and building materials [3].

Figure 3 shows a detailed classification of non-destructive methods useful for diagnosing buildings and building materials [2,3].

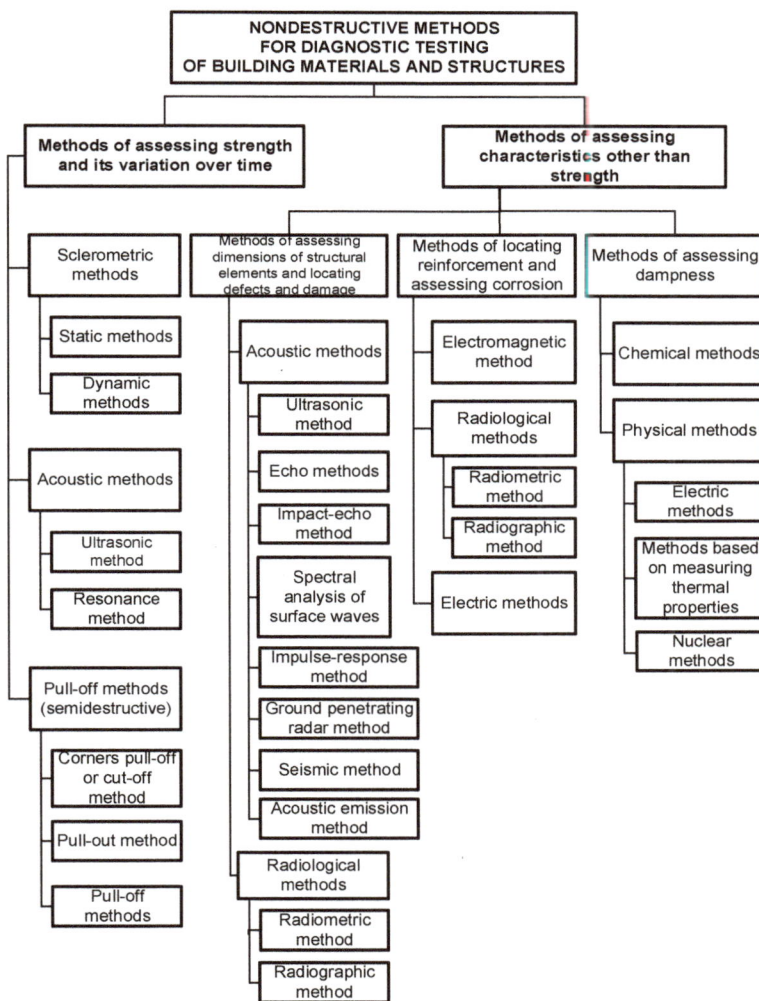

Figure 3. Non-destructive methods useful for diagnosing building structures and building materials [2,3].

Figure 4 shows a slightly modified classification (based on [2]) of the acoustic methods suitable for testing concrete materials and structures. This classification was presented in [1]. Figure 5 shows (on the basis of [1]) the suitability of the particular state-of-the-art non-destructive acoustic methods, used individually or combined for testing concrete and materials structures.

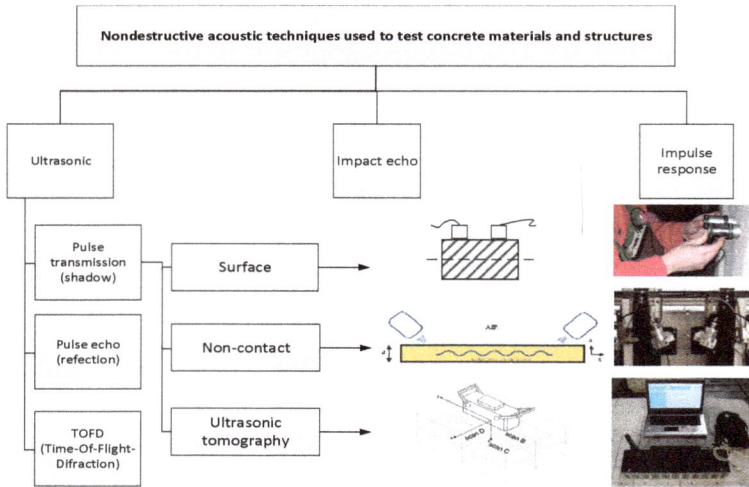

Figure 4. Classification of non-destructive acoustic methods suitable for testing concrete materials and structures [1,2].

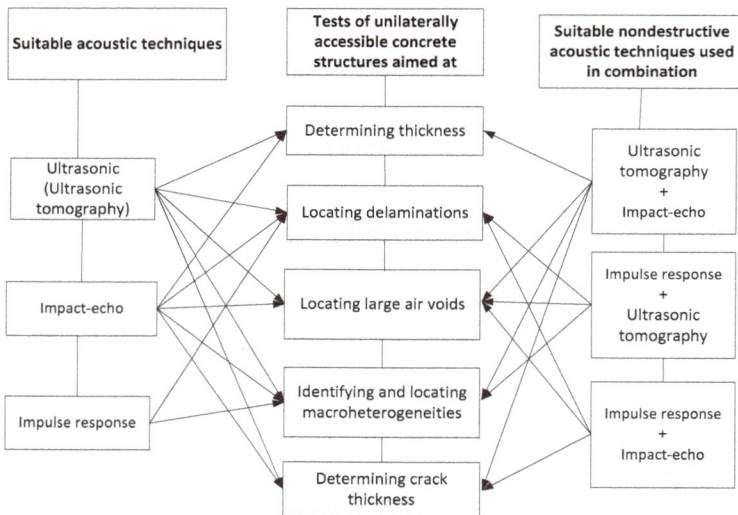

Figure 5. Suitability of state-of-the-art non-destructive acoustic methods, used individually or combined for testing concrete materials and structures [1].

Table 1 lists, according to [1], the above (geometric and material) imperfections occurring in concrete structures, together with the terms proposed for their description and the assigned (as mentioned earlier) state-of-the-art non-destructive acoustic methods suitable for testing such structures [1].

The terms, i.e., identification, location, extent and intensity that are proposed in Table 1 for describing the imperfections are explained, using as an example the geometric imperfection in Table 2, and materials imperfection—delamination in Table 3. The proposed terms, based on tests carried out by the state-of-the-art non-destructive acoustic methods, make the description of the imperfections in concrete materials and structures more precise and, in my opinion, represent a scientific and research achievement [4–14].

Table 1. Selected imperfections occurring in concrete structures, together with terms proposed for their description and assigned state-of-the-art non-destructive acoustic methods suitable for testing such structures [1].

Type/Description of Imperfection		Test Method					
		Ultrasonic Tomography Method	Impact-Echo Method	Impulse Response Method	Ultrasonic Tomography + Impact Echo	Impulse Response + Ultrasonic Tomography	Impulse Response + Impact-Echo
Incorrect thickness of member	Identification	●	●	-	●	-	-
	Location	●	●	-	●	-	-
	Extent	●	●	-	●	-	-
	Intensity	○	○	-	○	-	-
Delamination	identification	●	●	●	●	●	●
	Location	●	●	○	●	●	●
	Extent	●	●	●	●	●	●
	Intensity	-	○	-	○	-	○
Large air voids	Identification	●	●	-	●	●	●
	Location	●	●	-	●	●	●
	Extent	●	●	-	●	●	●
	Intensity	N.A.	N.A.	N.A.	N.A.	N.A.	N.A.
Zones of concrete macroheterogeneities	Identification	●	●	●	●	●	●
	Location	●	●	●	●	●	●
	Extent	●	●	○	●	●	●
	Intensity	-	○	-	○	-	○
Cracks	Identification	●	●	-	●	-	●
	Location	●	-	-	●	-	-
	Extent	-	●	-	●	-	●
	Intensity	-	●	-	●	-	●

Symbols: ●—suitable method, ○—partially suitable method, -—unsuitable method, N.A.—not applicable.

Table 2. Explanation of proposed terms for describing geometric imperfection in concrete structures tested by non-destructive acoustic methods.

Description of Imperfection	Illustration of Imperfection Description	
	View	Cross Section
Identification An imperfection (incorrect thickness of the structure) is found to be present.		a - a
Location The place of occurrence of the imperfection in the cross section of the structure is determined with accuracy dependent on the equipment used.		a - a
Extent The size (surface area and volume) of the place where the imperfection occurs is determined with accuracy dependent on the equipment used.		a - a
Intensity The degree of advancement of the imperfection (the distribution of locally incorrect thickness in the whole place of its occurrence).		a - a

Table 3. Explanation of proposed terms for describing materials imperfection—delamination in concrete structures tested by non-destructive acoustic methods.

Description of Imperfection	Illustration of Imperfection Description	
	View	Cross Section
Identification An imperfection (incorrect thickness of the structure) is found to be present.		a - a
Location The place of occurrence of the imperfection in the cross section of the structure is determined with accuracy dependent on the equipment used.		a - a
Extent The size (surface area and volume) of the place where the imperfection occurs is determined with accuracy dependent on the equipment used.		a - a
Intensity The degree of advancement of the imperfection (the distribution of locally incorrect thickness in the whole place of its occurrence).		a - a

3. Description of the Articles Presented in the Issue

In paper [15], a condition assessment of masonry pillars is presented. Non-destructive tests were performed on an intact pillar as well as three pillars with internal inclusions in the form of a hole, a steel bar grouted by gypsum mortar, and a steel bar grouted by cement mortar. The inspection utilized ultrasonic stress waves and the reconstruction of the velocity distribution was performed by means of computed tomography. The results showed the potential of tomographic imaging in characterizing the internal structure of pillars. Particular attention was paid to the assessment of the adhesive connection between a steel reinforcing bar embedded inside pillars, and the surrounding pillar body [15].

Paper [16] describes the validation of the following methods: semi-non-destructive, non-destructive, and ultrasonic technique for autoclaved aerated concrete (AAC). This study covers the compressive strength of AAC test elements with various density classes of 400, 500, 600, and 700 (kg/m^3), at various moisture levels. Empirical data including the shape and size of specimens were established from tests on 494 cylindrical and cuboid specimens, and standard cube specimens using the general relationship for ordinary concrete (Neville's curve). The developed methods turned out to be statistically significant and can be successfully applied during in-situ tests [16].

In paper [17], non-destructive tests of gantry cranes by means of the residual magnetic field (RMF) method were carried out for a duration of 7 years. The distributions of the residual magnetic field tangential and the normal components of their gradients were determined. A database of magnetograms was created. The results show that the gradients of tangential components can be used to identify and localize stress concentration zones in gantry crane beams. Special attention was given to the unsymmetrical distribution of the tangential component gradient on the surface of the crane beam. The anomaly was the effect of a slight torsional deflection of the beam as it was loaded. Numerical simulations with the finite element method (FEM) were used to explain this phenomenon. The displacement boundary conditions introduced into the simulations were established experimentally. Validation was carried out using the X-ray diffraction method, which confirmed the location of strain concentration zones (SCZs) identified by means of RMF testing [17].

Paper [18] presents the results of research aimed at identifying the degree of degradation of fiber-cement boards exposed to fire. The fiber-cement board samples were initially exposed to fire at various durations in the range of 1–15 min. The samples were then subjected to three-point bending and were investigated using the acoustic emission method. Artificial neural networks (ANNs) were employed to analyse the results yielded by the acoustic emission method. Fire was found to have a degrading effect on the fibers contained in the boards. As the length of exposure to fire increased, the fibers underwent gradual degradation, which was reflected in a decrease in the number of acoustic emission (AE) events recognized by the artificial neural networks as accompanying the breaking of the fibers during the three-point bending of the sample. It was shown that it is not sufficient to determine the degree of degradation of fiber-cement boards solely on the basis of bending strength (MOR) [18].

In paper [19] the aim of the experiment was to explore the possibility of using active thermography for testing large-sized building units (with high heat capacity) in order to locate material inclusions. As part of the experiment, two building partition models—one made of gypsum board (GB) and another made of oriented strand board (OSB)—were built. Three material inclusions (styrofoam, granite, and steel) with considerably different thermal parameters, were placed in each of the partitions. The distribution of the temperature field was studied on both sides of the partition for a few hours. The results showed that using the proposed investigative method, one can detect defects in building partitions under at least 22 mm of thick cladding. Active thermography can be used in construction for non-destructive materials testing. When the recording of thermograms is conducted for an appropriate length of time, inverse contrasts can be observed (on the same front surface) [19].

The aim of the study presented in [20] was to investigate the degradation of the microstructure and mechanical properties of fiber-cement board (FCB), which was exposed to environmental hazards, resulting in thermal impact on the microstructure of the board. The process of structural degradation was conducted under laboratory conditions by storing the FCB specimens in a dry, electric oven for 3 h at a temperature of 230 °C. Five sets of specimens, that differed in cement and fiber content, were tested. The fiber reinforcement morphology and the mechanical properties of the investigated compositions were identified both before, and after their carbonization. Visual light and scanning electron microscopy, X-ray micro tomography, flexural strength, and work of flexural test Wf measurements were used. The results obtained suggest a possible application of the UT method for an on-site assessment of the degradation processes occurring in fiber-cement boards [20].

Paper [21] presents procedures for investigating spun concrete and interpreting the results of such investigations, which make it possible to characterize the microstructure of the concrete. Three investigative methods were used to assess the distribution of the constituents in the cross section of the element: micro-computed tomography, 2D imaging (using an optical scanner) and nanoindentation. A procedure for interpreting and analysing the results is proposed. The procedure enables one to quantitatively characterize the following features of the microstructure of spun concrete: the mechanical parameters of the mortar, the aggregate content, the pore content, the cement paste content, the aggregate grading and the size (dimensions) of the pores. The proposed procedures constitute a valuable tool for evaluating the process of manufacturing spun concrete elements [21].

An enhanced singular value truncation method is proposed in paper [22] to evaluate structural damage more effectively by using a few lower order natural frequencies. The main innovations of the enhanced singular value truncation method lie in two aspects. The first is the normalization of linear systems of equations; the second is the multiple computations based on feedback evaluation. The proposed method is very concise in theory and simple to implement. Two numerical examples and an experimental example are employed to verify the proposed method. In the numerical examples, it was found that the proposed method can successively obtain more accurate damage evaluation results compared with the traditional singular value truncation method. In the experimental example, it was shown that the proposed method possesses more precise and fewer calculations compared with the existing optimization algorithms [22].

Article [23] presents results from non-destructive testing (NDT) that referred to the location and diameter or rebars in beam and slab members. The aim of this paper was to demonstrate that the accuracy and deviations of the NDT methods could be higher than the allowable execution or standard deviations. Tests were conducted on autoclaved aerated concrete beam and nine specimens that were specially prepared from lightweight concrete. The testing equipment was used to analyse how the rebar (cover) location affected the detection of their diameters and how their mutual spacing influenced the detected quantity of rebars. The considerations included the impact of rebar depth on cover measurements and the spread of obtained results. Tests indicated that the measurement error was clearly greater when the rebars were located at very low or high depths. Electromagnetic and radar devices were unreliable when detecting the reinforcement of small (8 and 10 mm) diameters at close spacing (up to 20 mm) and of large (20 mm) diameters at a close spacing and greater depths. Recommendations for practical applications were developed to facilitate the evaluation of a structure [23].

As mentioned in [24], predictable compressive strength of concrete is essential for concrete structure utilization and is the main feature of its safety and durability. Recently, machine learning has gained significant attention and future predictions for this technology are even more promising. Data mining on large sets of data has attracted attention because machine learning algorithms have achieved a level whereby they can recognize patterns that are difficult to recognize using human cognitive skills. In this paper, state-of-the-art achievements in machine learning techniques were utilized for concrete mix design. The authors prepared an extensive database of concrete recipes along with the relevant destructive laboratory tests, which was used to feed the selected optimal architecture of an artificial neural network. They translated the architecture of the artificial neural network into a mathematical equation that can be used in practical applications [24].

Paper [25] presents an assessment of the condition of wood from a wharf timber sheet wall after 70 years of service in a (sea)water environment. Samples taken from the structure's different zones, i.e., the zone impacted by waves and characterized by variable water-air conditions, the zone immersed in water and the zone embedded in the ground were subjected to non-destructive or semi-destructive tests. Moreover, ultrasonic, stress wave and drilling resistance methods were used. Then, an X-ray microtomographic analysis was carried out. The results provided information about the structure of the material on the micro and macroscale and the condition of the material was assessed on that basis. Also, correlations between the particular parameters were determined. Moreover, the methods themselves were evaluated with regard to their usefulness for the in situ testing of timber and to estimate the mechanical parameters needed for the static load analysis of the whole structure [25].

In paper [26], the adsorption properties of waste brick dust (WBD) were studied by removal of PbII and CsI from an aqueous system. For adsorption experiments, 0.1 M and 0.5 M aqueous solutions of Cs^+ and Pb^{2+} and two WBD (Libochovice (LB) and Tyn nad Vltavou (TN)) in the fraction below 125 m were used. The structural and surface properties of WBD were characterized by X-ray diffraction (XRD) in combination with solid-state nuclear magnetic resonance (NMR), supplemented by scanning electron microscopy (SEM), specific surface area (SBET), total pore volume and zero point of charge (pHZPC). LB was a more amorphous material showing better adsorption conditions than that of TN. The adsorption process indicated better results for Pb^{2+} due to the inner-sphere surface complexation in all Pb^{2+} systems, supported by the formation of insoluble $Pb(OH)_2$ precipitation on the sorbent surface. A weak adsorption of Cs^+ on WBD corresponded to the non-Langmuir adsorption run followed by the outer-sphere surface complexation. The leachability of Pb^{2+} from saturated WBDs varied from 0.001% to 0.3%, while in the case of Cs^+, 4–12% of the initial amount was leached. Both LB and TN met the standards for PbII adsorption, yet completely failed for any CsI removal from water systems [26].

The aim of the study presented in [27] was to experimentally and numerically research the effects of the wave frequency on damage identification in a single-lap adhesive joint of steel plates. The ultrasonic waves were excited at one point of an analyzed specimen and then measured in a certain area of the joint. The recorded wave velocity signals were processed by way of a root mean square (RMS) calculation, giving the actual position and geometry of defects. In addition to the visual assessment of damage maps, a statistical analysis was conducted. The influence of an excitation frequency value on the obtained visualizations was considered in the wide range for a single defect. Supplementary finite element method (FEM) calculations were performed for three additional damage variants. The results revealed some limitations of the proposed method. The main conclusion was that the effectiveness of measurements strongly depends on the chosen wave frequency value [27].

As mentioned in [28], the nondestructive testing of reinforced concrete chimneys, especially the high ones is an important element in the assessment of their condition, making it possible to forecast their safe service lifespan. Industrial chimneys are often exposed to the strong action of acidic substances—they are adversely affected by the flue gas condensate on the inside and by acid precipitation on the outside. During the service life of such chimneys, their condition should be monitored in order to prevent structural failures and indicate the most endangered parts of the structure. The methods for the interpretation of results from thermovision studies to determine the safety and durability of industrial chimneys are shown in [28].

In paper [29], the authors present two methods used in the identification of viscoelastic parameters of asphalt mixtures used in pavements. The static creep test and the dynamic test, with a frequency of 10 Hz, were carried out based on the four-point bending beam (4BP). In the method for identifying viscoelastic parameters for the Brugers' model, they included the course of a creeping curve (for static creep) and fatigue hysteresis (for dynamic test). It was shown that these parameters depend significantly on the load time, method used, and temperature and asphalt content. A similar variation in the parameters depending on temperature was found for the two tests, but different absolute values were obtained. The authors also found that the parameters should be determined using the creep curve for the static analyses with persistent load, whereas in the case of the dynamic studies, the hysteresis is more appropriate. In the 4BP dynamic test, the authors determined the relationships between damping and viscosity coefficients, showing that material variability depends on the test temperature [29].

Article [30] presents the application of the acoustic emission technique (AE) in order to detect the initiation and examination of the crack growth process in steel used in engineering construction. The tests were carried out on specimens with a single edge notch in bending (SENB) made of 40CrMo steel. Crack opening displacement, force parameter and potential drop signal were measured in the tests. During loading of the specimens, the fracture mechanism was classified as brittle. Accurate investigations of the cracking process by the acoustic emission (AE) method and observation of fracture surfaces by scanning electron microscopy (SEM) have made it possible to determine that the cracking process is more complex than the classically understood brittle fracture. The work focuses on the comparison of selected parameters of the acoustic emission signals in the phase before the initiation and development of brittle fracture cracking [30].

Paper [31] examines the repair of a stand of a motorbike speedway stadium. The synchronized dancing of fans cheering during a meeting brought the stand into excessive resonance. The main goal of this research was to propose a method for the structural tuning of stadium stands. Non-destructive testing by vibration methods was conducted on a selected stand segment, the structure of which recurred in the remaining stadium segments. Through experiments, the authors determined the vibration forms throughout the stand, taking into account the dynamic impact of fans. Numerical analyses were performed on the 3-D finite element method (FEM) stadium model to identify the dynamic jump load function. The results obtained on the basis of sensitivity tests using the finite element method allowed the tuning of the stadium structure to successfully meet the requirements of the serviceability limit state [31].

In paper [32] the author asked a question: is the variation in the compressive strength of concrete across the thickness of horizontally cast elements negligibly small, or rather, does it need to be taken into account at the design stage? In order to determine if the compressive strength of concrete varies across the thickness of horizontally cast elements, ultrasonic tests and destructive tests were carried out on core samples taken from a 350 mm thick slab made of class C25/30 concrete. Special point-contact probes were used to measure the time taken for the longitudinal ultrasonic wave to pass through the tested sample. The correlation between the velocity of the longitudinal ultrasonic wave and the compressive strength of the concrete in the slab was determined It was found that the destructively determined compressive strength varied only slightly (by 3%) across the thickness of the placed layer of concrete, whereas the averaged ultrasonically determined strength of the concrete in the same samples does not vary across the thickness of the analyzed slab. Therefore, it was concluded that the slight increase in concrete compressive strength with depth below the top surface is a natural thing and need not be taken into account in the assessment of the strength of concrete in the structure [32].

Paper [33] presents the results of investigations into the effect of freeze-thaw cycling on the failure of fiber-cement boards and on the changes taking place in their structure. An artificial neural network was employed to analyze the results yielded by the acoustic emission method. The investigations conclusively proved that freeze-thaw cycling has an effect on the failure of fiber-cement boards, as indicated mainly by the fall in the number of AE events recognized as accompanying the breaking of fibers during the three-point bending of the specimens. SEM examinations were carried out to get a better insight into the changes taking place in the structure of the tested boards. Interesting results with significance for building practice were obtained [33].

The findings of the study in article [34] deepen the understandings of pore structure damage in cement-based materials (CBMs) by mercury intrusion, and provide methodological insights into the microstructure characterization of CBMs by XCT. Mercury intrusion porosimetry (MIP) is questioned for possibly damaging the micro structure of CBMs, but this theme still has a lack of quantitative evidence. By using X-ray computed tomography (XCT), this study reported an experimental investigation of the pore structure damage in paste and mortar samples after a standard MIP test. XCT scans were performed on the samples before and after mercury intrusion. Because of its very high mass attenuation coefficient, mercury can greatly enhance the contrast of XCT images, providing a path to probe the same pores with and without mercury fillings. The paste and mortar showed different MIP pore size distributions but similar intrusion processes. A grey value inverse for the pores and material skeletons before and after MIP was found. With the features of excellent data reliability and robustness verified by a threshold analysis, the XCT results characterized the surface structure of voids, and diagnosed the pore structure damages in terms of pore volume and size of the paste and mortar samples [34].

Paper [35] presents the identification of the destruction process in a quasi-brittle composite based on acoustic emission and the sound spectrum. The tests were conducted on a quasi-brittle composite. The sample was made from ordinary concrete with dispersed polypropylene fibers. The possibility of identifying the destruction process based on the acoustic emission and sound spectrum was confirmed and the ability to identify the destruction process was demonstrated. Three- and two-dimensional spectra were used to identify the destruction process. The three-dimensional spectrum provides additional information, enabling a better recognition of changes in the structure of the samples on the basis of the analysis of sound intensity, amplitudes, and frequencies. The paper shows the possibility of constructing quasi-brittle composites to limit the risk of catastrophic destruction processes and the possibility of identifying those processes with the use of acoustic emission at different stages of destruction [35].

The aim of the study in article [36] is to investigate two moisture monitoring techniques to promote moisture safety in wood-based buildings (i.e., new structures, as well as renovated and protected buildings). The study is focused on the comparison of two electrical methods that can be employed for the non-destructive moisture monitoring of wood components integrated in the structure of buildings. The main principle of the two presented methods of measuring the moisture by electric resistance is based on a simple resistor–capacitor (RC) circuit system improved with an ICM7555 chip and integrator circuit using a TLC71 amplifier. The RC-circuit is easier to implement thanks to the digital signals of the used chip, whilst the newly presented integration method allows faster measurement at lower moisture contents. A comparative experimental campaign utilizing spruce wood samples was conducted in this regard. Based on the results obtained, both methods can be successfully applied to wood components in buildings for moisture contents above 8% [36].

The methodology of multi-scale structural assessment of the different cellulose fiber-cement boards subjected to high temperature treatment was proposed in article [37]. Two specimens were investigated: Board A (air-dry reference specimen) and Board B (exposed to a temperature of 230 °C for 3 h). At macroscale all considered samples were subjected to the three-point bending test. Next, two methodologically different microscopic techniques were used to identify evolution (caused by temperature treatment) of geometrical and mechanical morphology of boards. For that purpose, SEM imaging with EDS analysis and nanoindentation tests were utilized. High temperature was found to have a degrading effect on the fibers contained in the boards. Most of the fibers in the board were burnt-out, or melted into the matrix, leaving cavities and grooves which were visible in all of the tested boards. Nanoindentation tests revealed significant changes of mechanical properties caused by high temperature treatment: "global" decrease of the stiffness (characterized by nanoindentation modulus) and "local" decrease of hardness. The results observed at microscale are in a very good agreement with macroscale behavior of considered composite. It was shown that it is not sufficient to determine the degree of degradation of fiber-cement boards solely on the basis of bending strength; advanced, microscale laboratory techniques can reveal intrinsic structural changes [37].

In paper [38], the ultrasonic method using exponential heads with spot surface of contact with the material was chosen for the measurements of concrete strength in close cross sections parallel to the corroded surface. The test was performed on samples taken from compartments of a reinforced concrete tank after five years of operation in a corrosive environment. Test measurements showed heterogeneity of strength across the entire thickness of the tested elements. It was determined that the strength of the elements in internal cross sections of the structure was up to 80% higher than the initial strength. A drop in the mechanical properties of concrete was observed only in the close zone near the exposed surface. The dependence of the compressive strength of standard cubic samples on the duration of their exposure in the sulphate corrosion environment has been described [38].

The ultrasonic pulse velocity test, the rebound hammer test is the most common NDT method currently used for this purpose. However, estimating compressive strength using general regression models can often yield inaccurate results. The experiment results in paper [39] show that the compressive strength of any concrete can be estimated using one's own newly created regression model. A traditionally constructed regression model can predict the strength value with 50% reliability, or when two-sided confidence bands are used, with 95% reliability. However, civil engineers usually work with the so-called characteristic value defined as a 5% quantile. Therefore, it seems appropriate to adjust conventional methods in order to achieve a regression model with 95% one-sided reliability. This paper describes a simple construction of such a characteristic curve. The results show that the characteristic curve created for the concrete in question could be a useful tool, even outside of practical application [39].

4. Conclusions

As mentioned at the beginning, this issue was proposed and organized as a means of presenting recent developments in the field of non-destructive testing of materials in civil engineering. For this reason, the articles highlighted in this editorial relate to different aspects of non-destructive testing of materials in civil engineering, from building materials to building structures. Interesting results with significance for the materials were obtained and all of the papers have been precisely described.

Funding: This research received no external funding.

Conflicts of Interest: The authors declare no conflict of interest.

References

1. Schabowicz, K. Modern acoustic techniques for testing concrete structures accessible from one side only. *Arch. Civ. Mech. Eng.* **2015**, *15*, 1149–1159. [CrossRef]
2. Hoła, J.; Schabowicz, K. State-of-the-art non-destructive methods for diagnostic testing of building structures—Anticipated development trends. *Arch. Civ. Mech. Eng.* **2010**, *10*, 5–18. [CrossRef]
3. Hoła, J.; Schabowicz, K. Non-destructive diagnostics for building structures: Survey of selected state-of-the-art methods with application examples. In Proceedings of the 56th Scientific Conference of PAN Civil Engineering Committee and PZITB Science Committee, Krynica, Poland, 19–24 September 2010. (In Polish).
4. Schabowicz, K.; Gorzelańczyk, T. Fabrication of fibre cement boards. In *The Fabrication, Testing and Application of Fibre Cement Boards*, 1st ed.; Ranachowski, Z., Schabowicz, K., Eds.; Cambridge Scholars Publishing: Newcastle upon Tyne, UK, 2018; pp. 7–39. ISBN 978-1-5276-6.
5. Drelich, R.; Gorzelanczyk, T.; Pakuła, M.; Schabowicz, K. Automated control of cellulose fiber cement boards with a non-contact ultrasound scanner. *Autom. Constr.* **2015**, *57*, 55–63. [CrossRef]
6. Chady, T.; Schabowicz, K.; Szymków, M. Automated multisource electromagnetic inspection of fibre-cement boards. *Autom. Constr.* **2018**, *94*, 383–394. [CrossRef]
7. Schabowicz, K.; Jóźwiak-Niedźwiedzka, D.; Ranachowski, Z.; Kudela, S.; Dvorak, T. Microstructural characterization of cellulose fibres in reinforced cement boards. *Arch. Civ. Mech. Eng.* **2018**, *4*, 1068–1078. [CrossRef]
8. Schabowicz, K.; Gorzelańczyk, T.; Szymków, M. Identification of the degree of fibre-cement boards degradation under the influence of high temperature. *Autom. Constr.* **2019**, *101*, 190–198. [CrossRef]
9. Schabowicz, K.; Gorzelańczyk, T. A non-destructive methodology for the testing of fibre cement boards by means of a non-contact ultrasound scanner. *Constr. Build. Mater.* **2016**, *102*, 200–207. [CrossRef]
10. Schabowicz, K.; Ranachowski, Z.; Jóźwiak-Niedźwiedzka, D.; Radzik, Ł.; Kudela, S.; Dvorak, T. Application of X-ray microtomography to quality assessment of fibre cement boards. *Constr. Build. Mater.* **2016**, *110*, 182–188. [CrossRef]
11. Ranachowski, Z.; Schabowicz, K. The contribution of fibre reinforcement system to the overall toughness of cellulose fibre concrete panels. *Constr. Build. Mater.* **2017**, *156*, 1028–1034. [CrossRef]
12. Rucka, M.; Wilde, K. Experimental study on ultrasonic monitoring of splitting failure in reinforced concrete. *J. Nondestruct. Eval.* **2013**, *32*, 372–383. [CrossRef]
13. Rucka, M.; Wilde, K. Ultrasound monitoring for evaluation of damage in reinforced concrete. *Bull. Pol. Acad. Sci. Tech. Sci.* **2015**, *63*, 65–75. [CrossRef]
14. EN 12467—Cellulose Fibre Cement Flat Sheets. Product Specification and Test Methods. 2018. Available online: https://standards.cen.eu/dyn/www/f?p=204:110:0::::FSP_PROJECT,FSP_ORG_ID:66671,6110&cs=1151E39EDCD9EF75E3C2D401EB5818ACD (accessed on 15 September 2019).
15. Zielińska, M.; Rucka, M. Non-Destructive Assessment of Masonry Pillars Using Ultrasonic Tomography. *Materials* **2018**, *11*, 2543. [CrossRef] [PubMed]
16. Jasiński, R.; Drobiec, Ł.; Mazur, W. Validation of Selected Non-Destructive Methods for Determining the Compressive Strength of Masonry Units Made of Autoclaved Aerated Concrete. *Materials* **2019**, *12*, 389. [CrossRef] [PubMed]
17. Juraszek, J. Residual Magnetic Field Non-Destructive Testing of Gantry Cranes. *Materials* **2019**, *12*, 564. [CrossRef] [PubMed]
18. Schabowicz, K.; Gorzelańczyk, T.; Szymków, M. Identification of the degree of degradation of fibre-cement boards exposed to fire by means of the acoustic emission method and artificial neural networks. *Materials* **2019**, *12*, 656. [CrossRef] [PubMed]
19. Noszczyk, P.; Nowak, H. Inverse Contrast in Non-Destructive Materials Research by Using Active Thermography. *Materials* **2019**, *12*, 835. [CrossRef] [PubMed]
20. Ranachowski, Z.; Ranachowski, P.; Dębowski, T.; Gorzelańczyk, T.; Schabowicz, K. Investigation of structural degradation of fiber cement boards due to thermal impact. *Materials* **2019**, *12*, 944. [CrossRef]
21. Michałek, J.; Pachnicz, M.; Sobótka, M. Application of Nanoindentation and 2D and 3D Imaging to Characterise Selected Features of the Internal Microstructure of Spun Concrete. *Materials* **2019**, *12*, 1016. [CrossRef]

22. Yang, Q.; Wang, C.; Li, N.; Wang, W.; Liu, Y. Enhanced Singular Value Truncation Method for Non-Destructive Evaluation of Structural Damage Using Natural Frequencies. *Materials* **2019**, *12*, 1021. [CrossRef]
23. Drobiec, Ł.; Jasiński, R.; Mazur, W. Accuracy of Eddy-Current and Radar Methods Used in Reinforcement Detection. *Materials* **2019**, *12*, 1168. [CrossRef]
24. Ziolkowski, P.; Niedostatkiewicz, M. Machine Learning Techniques in Concrete Mix Design. *Materials* **2019**, *12*, 1256. [CrossRef] [PubMed]
25. Nowak, T.; Karolak, A.; Sobótka, M.; Wyjadłowski, M. Assessment of the Condition of Wharf Timber Sheet Wall Material by Means of Selected Non-Destructive Methods. *Materials* **2019**, *12*, 1532. [CrossRef] [PubMed]
26. Doušová, B.; Koloušek, D.; Lhotka, M.; Keppert, M.; Urbanová, M.; Kobera, L.; Brus, J. Waste Brick Dust as Potential Sorbent of Lead and Cesium from Contaminated Water. *Materials* **2019**, *12*, 1647. [CrossRef] [PubMed]
27. Wojtczak, E.; Rucka, M. Wave Frequency Effects on Damage Imaging in Adhesive Joints Using Lamb Waves and RMS. *Materials* **2019**, *12*, 1842. [CrossRef] [PubMed]
28. Maj, M.; Ubysz, A.; Hammadeh, H.; Askifi, F. Non-Destructive Testing of Technical Conditions of RC Industrial Tall Chimneys Subjected to High Temperature. *Materials* **2019**, *12*, 2027. [CrossRef]
29. Mackiewicz, P.; Szydło, A. Viscoelastic Parameters of Asphalt Mixtures Identified in Static and Dynamic Tests. *Materials* **2019**, *12*, 2084. [CrossRef]
30. Krampikowska, A.; Pała, R.; Dzioba, I.; Świt, G. The Use of the Acoustic Emission Method to Identify Crack Growth in 40CrMo Steel. *Materials* **2019**, *12*, 2140. [CrossRef]
31. Grębowski, K.; Rucka, M.; Wilde, K. Non-Destructive Testing of a Sport Tribune under Synchronized Crowd-Induced Excitation Using Vibration Analysis. *Materials* **2019**, *12*, 2148. [CrossRef]
32. Michałek, J. Variation in Compressive Strength of Concrete aross Thickness of Placed Layer. *Materials* **2019**, *12*, 2162. [CrossRef]
33. Gorzelańczyk, T.; Schabowicz, K. Effect of Freeze–Thaw Cycling on the Failure of Fibre-Cement Boards, Assessed Using Acoustic Emission Method and Artificial Neural Network. *Materials* **2019**, *12*, 2181. [CrossRef]
34. Wang, X.; Peng, Y.; Wang, J.; Zeng, Q. Pore Structure Damages in Cement-Based Materials by Mercury Intrusion: A Non-Destructive Assessment by X-ray Computed Tomography. *Materials* **2019**, *12*, 2220. [CrossRef] [PubMed]
35. Logoń, D. Identification of the destruction process in quasi brittle concrete with dispersed fibres based on acoustic emission and sound spectrum. *Materials* **2019**, *12*, 2266. [CrossRef] [PubMed]
36. Slávik, R.; Čekon, M.; Štefaňák, J. A Nondestructive Indirect Approach to Long-Term Wood Moisture Monitoring Based on Electrical Methods. *Materials* **2019**, *12*, 2373. [CrossRef] [PubMed]
37. Gorzelańczyk, T.; Pachnicz, M.; Różański, A.; Schabowicz, K. Multi-Scale Structural Assessment of Cellulose Fibres Cement Boards Subjected to High Temperature Treatment. *Materials* **2019**, *12*, 2449. [CrossRef] [PubMed]
38. Stawiski, B.; Kania, T. Examining the distribution of strength across the thickness of reinforced concrete elements subject to sulphate corrosion using the ultrasonic method. *Materials* **2019**, *12*, 2519. [CrossRef]
39. Kocáb, D.; Misák, P.; Cikrle, P. Characteristic Curve and Its Use in Determining the Compressive Strength of Concrete by the Rebound Hammer Test. *Materials* **2019**, *12*, 2705. [CrossRef] [PubMed]

materials

MDPI

Article

A Nondestructive Indirect Approach to Long-Term Wood Moisture Monitoring Based on Electrical Methods

Richard Slávik [1], Miroslav Čekon [2,3,*] and Jan Štefaňák [3]

1   Department of Wood Science and Technology, Faculty of Forestry and Wood Technology,
    613 00 Brno, Czech Republic
2   Department of Physics, Faculty of Civil Engineering, Slovak University of Technology,
    810 05 Bratislava, Slovakia
3   AdMaS Centre, Faculty of Civil Engineering, Brno University of Technology, 602 00 Brno, Czech Republic
*   Correspondence: miroslav.cekon@stuba.sk or mcekon@gmail.com; Tel.: +421-(2)-59-274-489

Received: 24 June 2019; Accepted: 24 July 2019; Published: 25 July 2019

check for updates

Abstract: Wood has a long tradition of use as a building material due its properties and availability. However, it is very sensitive to moisture. Wood components of building structures basically require a certain level of moisture protection, and thus moisture monitoring to ensure the serviceability of such components during their whole lifespan while integrated within buildings is relevant to this area. The aim of this study is to investigate two moisture monitoring techniques promoting moisture safety in wood-based buildings (i.e., new structures, as well as renovated and protected buildings). The study is focused on the comparison of two electrical methods that can be employed for the nondestructive moisture monitoring of wood components integrated in the structures of buildings. The main principle of the two presented methods of the moisture measurement by electric resistance is based on a simple resistor–capacitor (RC) circuit system improved with ICM7555 chip and integrator circuit using TLC71 amplifier. The RC-circuit is easier to implement thanks to the digital signals of the used chip, whilst the newly presented integration method allows faster measurement at lower moisture contents. A comparative experimental campaign utilizing spruce wood samples is conducted in this relation. Based on the results obtained, both methods can be successfully applied to wood components in buildings for moisture contents above 8%.

Keywords: resistance measurement; wood moisture sensing; non-destructive testing; moisture safety

---

1. Introduction

The properties and lifespan of wood strongly depend on many aspects, and therefore the permanent moisture monitoring of wood has a specific application in a wide range of technical fields. Primarily, wood-based wall technologies are widely considered to be suitable building materials for low-environmental impact composites in the building engineering field. Wood framed and composite wood wall technologies that utilize advanced insulation techniques have been well-integrated for several decades in the building sector, and specifically meet thermal and environmental requirements [1]. Thermal properties, such as thermal conductivity and specific heat, are given significant consideration when designing buildings, especially low-energy or passive ones. Building components that are exposed to the weather outdoors are mainly affected by moisture and temperature-related factors. It has been two decades since exterior building surfaces began to be designed to be monitored by methods for the continuous monitoring of temperature and moisture in the micro-environment of structures, and within the wood itself. A system was introduced that maps the climate index for the decay of wood at various geographical levels via the use of existing climatic data, standards, and moisture content measurements [2]. Moisture and temperature can also play an important role indoors and within the structure of a building, i.e., throughout the whole building envelope. The existence of high moisture content can initiate decay or the growth of fungi. Particularly for bio-based building materials such as wood, the use of biological agents should be considered in order to predict service life, particularly with regard to fungal decay and mold growth risks. The control and reduction of wood moisture content is therefore a key instrument for wood protection [3,4]. However, the permanent monitoring of wood components may also have a relevant role to play in this field, especially when such components are incorporated in the building structure and their continuous physical monitoring is impossible. The correct estimation of timber moisture content and the subsequent initiation of potentially necessary measures are therefore essential tasks during the planning, execution, and maintenance of buildings built with wood or wood-based products. This fact has contributed to a recent and considerable rise in interest concerning the in-situ monitoring of the moisture content of structural timber elements [5]. In this connection, the modeling of the outdoor performance of wood products is also attracting specific attention [6]. Furthermore, moisture content measurement has a lot of potential for use in testing the durability of timber products [7].

A technique for the nondestructive evaluation of moisture content distribution during drying using a newly developed soft X-ray digital microscope and absortimetry was investigated by Tanaka et al. [8] and Tanaka and Kawai [9]. X-ray-based methods and diagnostics have already been successfully developed and used in many applications to identify aspects of wood decay [10,11]. In particular, wood temperature and moisture content have a direct impact on fungus and its ability to metabolize and degrade wood cell wall material over time [12]. Moisture requirements for the growth and decay of different fungi and wood species have already been determined, though relationships between wood moisture content, wood temperature, and fungal decay play an important part when applying the method in specific climates [13]. In addition, the relationship between microclimate, material environment, and decay is being studied in order to achieve a better understanding of issues concerned with the service life prediction of wood and wood-based products. Dietsch et al. [5] describes common methods of determining wood moisture content and evaluates them with respect to their applicability in monitoring concepts. Continuous moisture measurements using calibrated load cells and a data logger coupled with a weather station are an efficient way to record moisture in all kinds of material [14]. Unfortunately, the most accurate direct methods, which use oven drying and distillation, are time-consuming and cause the destruction of specimens.

Many indirect methods have been developed based on electric conductance or dielectric properties which allow results to be obtained fast and with satisfactory accuracy. In this relation, wood moisture measurement has a long history. Indirect methods are often applied for in-situ measurements and monitoring. They are based on determining a different property, which is correlated with the water content. Thus, resistivity measurements require material-specific characteristics and a temperature compensation, since both parameters have a significant effect on electrical conductivity [15]. Dunlap [16] discusses twelve commercial electrical moisture meters. Most of them are based on resistance (conductance) measurement. The Wood Handbook [17] states that the resistance of wood ranges from a few petaohms for oven-dry wood to a few kiloohms for wood with fiber saturation. In the range between fiber saturation (appr. 30%) to complete saturation, the change of resistance is not so significant. James [18] mentioned that the conductance and dielectric properties of wood vary consistently with moisture content when it is less than 30%, with a roughly linear relationship between the logarithm of conductance and the logarithm of moisture content. Thanks to this relationship, the moisture content can be determined. The measurement principle is based on the application of direct or alternating current, but higher resistances than a few hundred megaohms are not so easy to measure. Electric-current through such huge resistance is often very small, and direct current measurement in a simple electrical circuit with an appropriate error is not possible. Measuring very high resistance values is a difficult task, since low voltage or currents are present and thus, noise and amplification must be carefully done, especially when low resistance values are required to be measured using the same circuit, too [19]. Moisture content is not the only phenomenon which has an effect on the conductance of wood material; another significant factor is temperature dependency. James [18] mentioned that an increase of 10 °C causes an approximately two-fold increase in conductance in regions with more than 10% moisture content. Another problem is the anisotropy of wood in the direction of the grain. Conductance measured parallel to the grain is approximately double that of perpendicular conductance.

Many methods, complex electrical circuits, and devices have been developed in past decades. The most common and most easily applicable methods are electrical resistance measurements [15]. Typical devices are equipped with probes and a display for showing calculated values. They are portable, and very useful for taking technical measurements or conducting on-site inspections of materials. A long-term moisture measuring and data logging method for wood exposed to weather conditions was developed by Brischke et al. [20]. The method involves measuring the electrical resistance with glued electrodes for a sustainable connection. The measuring points at the tips of the electrodes are glued conductively into the wood while the remaining outer parts of the electrodes are glued with insulating adhesive. For this purpose, special conductive and insulating glues and electrodes were developed and comparatively evaluated in laboratory tests. In a recent study, the comparison of accuracy and ease of operation between voltammetry and digital bridge method for electrical resistance measurement in wood specimens and the factors influencing voltammetry were examined in detail [21]. Another useful approach is based on interdigital capacitance sensing of moisture content in rubber wood where the electrode contact is non-penetrating. It has linear sensitivity and better accuracy at very low moisture content. This represents faster measurement and is more convenient to be used in the industry for the production line quality control monitoring of moisture content of rubber wood [22].

Nowadays, with innovative approaches in many fields, we often use technologies which allow us to monitor the service life of all building elements permanently. Portable independent devices often cannot be integrated into structures for a long time or connected to networks. For that reason, there are potential future applications for sensors which could be connected to a network and placed into a composite structure containing wood for a long period of time (Figure 1). The sophisticated circuits used in commercial portable devices are often trade secrets of their producers and are not so easy to adapt for use in automatized systems. If the resistance measurement approach is used, there are two methods which are applicable: They are subjected to further analysis here. Both of the analyzed methods use the capacitor charge principle in a different way.

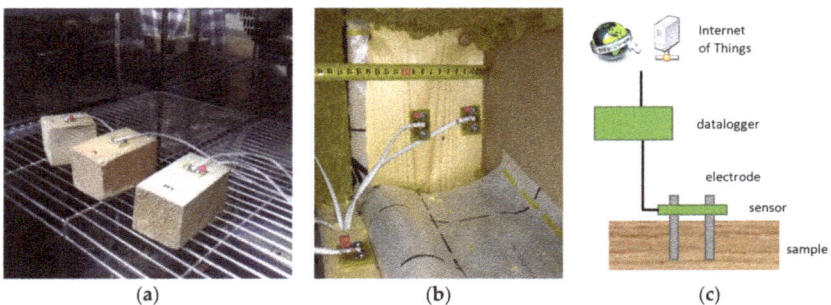

Figure 1. Implementation of small sensor module at real-world scale; (**a**) sensor testing in experimental chamber; (**b**) real wall structure integration; (**c**) functional scheme.

This study describes experimental work which aimed to obtain wood moisture monitoring data indirectly using two electrical methods. The main aims of this research work are first, to introduce the theoretical principles of both measurement methods used, and second, to evaluate their applicability and finally to verify them based on the experimental results obtained. Both methods are applied to samples made of spruce wood. The experimental results were evaluated via a statistical approach based on Bland-Altman plots. Originally, the Bland-Altman procedure [23] was used in medicine to compare two clinical measurements that each produce some error in the data they measure [24]. Bland-Altman plots are now extensively used to evaluate the agreement between two different instruments or two measurement techniques. They allow the identification of any systematic difference between measurements, or possible outliers. The results of this statistical analysis method determine the applicability of both, monitoring methods and the overall consistency between them. The statistical analysis also identifies the moisture content range for which the methods should obtain acceptable results if properly applied.

2. Description of the Circuits, Theoretical Principles, and Applications

As mentioned above, there are basically two methods which can be used for long-term wood moisture monitoring using electrical resistance and direct current. The first method uses a resistor–capacitor (RC) circuit which has been improved by transforming it into a digital device using a 555 timer chip [25,26]. The second applicable method uses capacitor charging via an operational amplifier connected as an integrator circuit. This method was originally applied to other problems requiring the measurement of high resistance by Aguilar et al. [27]. Both methods can be usefully implemented in small sensor packages (Figure 1). It shows small sensor prototypes which were assembled based on this research work and directly attached to wood samples in a climatic chamber during a test involving long-term monitoring.

The application of circuits was tested on spruce wood samples. Several wood samples with electrodes attached in various orientations relative to the fibers of the wood were measured with both methods. The samples were fitted with electrodes and placed into a desiccator where humidity and temperature were controlled. The measurements were obtained via a digital oscilloscope at intervals during conditioning, and the weights of the samples were measured. The results from both circuits were compared to each other using the obtained data.

2.1. The RC Circuit Method

This type of circuit is also known as an RC network or RC filter. It combines a capacitor with a resistor in series and is driven by voltage or a source of current. The main principle of RC circuits is based on the relationship between capacitor and resistor values and the voltage level in the capacitor over time. In a simple example with direct current and a rise in voltage from zero to a particular voltage level at the input rail of the resistor, the time of capacitor charging is proportional to the time constant of the circuit. The time constant is defined by the value of the resistor and capacitor.

At zero time the capacitor C is discharged. If the input voltage is applied at resistor R, the voltage charge of capacitor C starts rising over time until it reaches the maximum value (Figure 2). The time constant (1) represents the time taken by the capacitor to charge to approximately 63.2% of its final voltage value. At a known and stable capacitance and a known time of charging the resistance can be estimated, and vice versa. The relation between the voltage of capacitor V_C and the stable input voltage V_{IN} on the left resistor rail can be described by Equation (2).

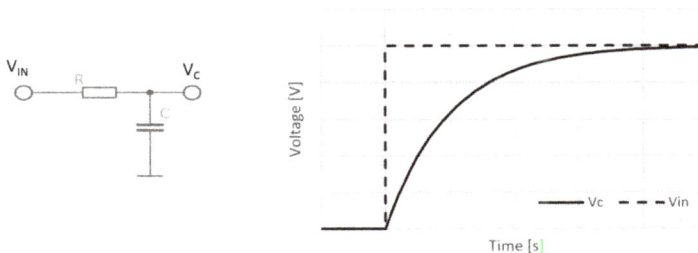

Figure 2. Resistor–capacitor (RC) circuit diagram and working principle.

The described idea is applied in the first method. The charging and discharging of the circuit are managed by a timer-integrated circuit. The most well-known timer, the 555 timer chip, has many applications, from circuits with a blinking LED (Light-Emitting Diode) to circuits which use it to generate an AC (Alternating Current) signal. The use of this chip for wood moisture measurement was already mentioned by Vodicka [25]. It is a logical application in which the chip is used as a monostable timer. This simply means that if a trigger input is pulled to a logical zero, the capacitor is discharged, and then charging starts with the output of the chip set to a high logical level. If the voltage of the capacitor reaches 63% of the input voltage level, the output is set to the zero logical level.

The time of the output pulse is proportional to the RC constant and can be calculated with Equation (3). Testing was performed with an oscilloscope; the circuit (Figure 3) was triggered by a push button and time of charging was captured by oscilloscope (Figure 4). Triggering and time counting could be handled by a small microcontroller in a real world sensor. The problem standard timer chips have with leakage current, which is often higher than the current through a wood sample with a moisture content of under 10%, has been solved by the use of the ICM7555 chip. Its leakage current is within the range of a few picoamperes according the producer's specifications.

$$\tau = RC \; [-] \tag{1}$$

$$V_C = V_{in}\left(1 - e^{\frac{-t}{\tau}}\right) \; [V] \tag{2}$$

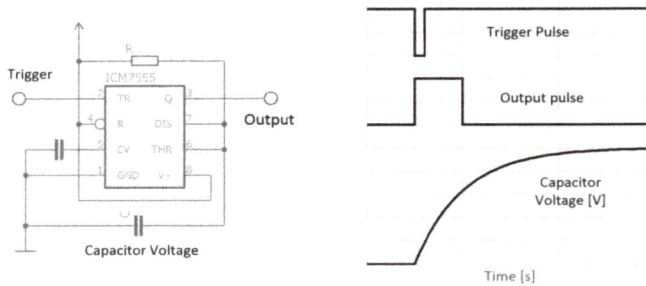

Figure 3. RC circuit improved by a 555 timer chip.

$$t = 1.1RC \; [s] \tag{3}$$

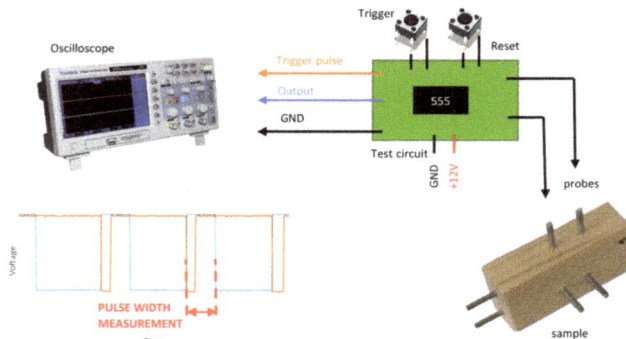

Figure 4. The complete principle of the RC circuit for the moisture content measurement.

2.2. The Integration Method

Operational amplifiers have various uses in electronics. The most common functions are as comparators, amplifiers, differentiators, and integrators. The last of these is the most relevant to high resistance measurement, and the second method implements operational amplifiers in that mode. A TLC71 amplifier in integrator mode has already been used to measure high resistance at the level of hundreds of megaohms [27]. Most operational amplifiers need a symmetric power supply, which is the main disadvantage of these devices in digital electronics.

The integrator works on the principle of capacitor charging. The voltage at the capacitor changes over time according to the voltage value applied to the negative rail of the amplifier (Figure 5).

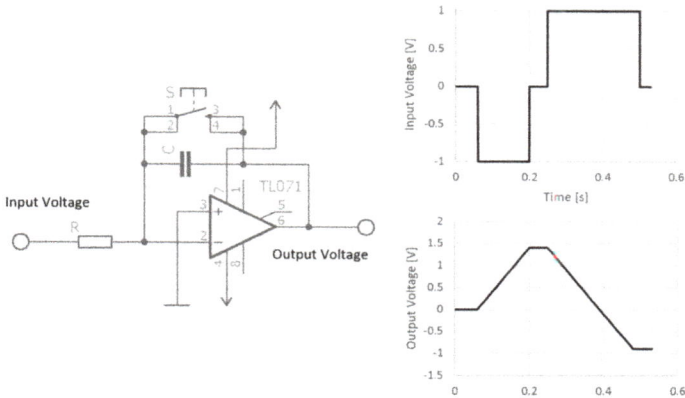

Figure 5. Integration method circuit diagram and working principle.

The positive rail of the amplifier is grounded. If a negative voltage level is reached at input through the resistor, which represents measured resistance, the capacitor voltage grows to a positive supply voltage and vice versa. The slope of voltage change is proportional to the resistance and capacity. The output voltage is measured via an oscilloscope and the resetting of the integrator is triggered by a switch (Figure 6). In the case of a real-world sensor, this could be handled by a transistor managed by a microcontroller which is equipped with an analog to digital convertor for output voltage measurement.

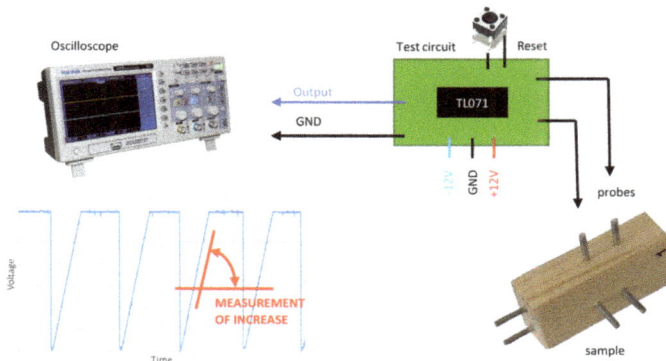

Figure 6. The complete principle of integration method for the moisture content measurement.

3. Materials and Methods

An experimental investigation was carried out using the two abovementioned electrical resistance-based methods, and the measurement results were statistically evaluated. In the initial part of this research, the spruce wood samples were prepared, and test circuits were developed for their calibration using resistor fields. Then, the samples were conditioned, measured, and weighed on a continuous basis over three months. Finally, the obtained data were analyzed and statistically evaluated. Through this approach, a complex characterization of resistance could be comprehensively obtained from the measurements to verify the applicability of both methods used. All of the information on time-dependent moisture parameters measured from the samples was expected to show a clear picture of the evolution of moisture content can be provided if these test circuits are integrated within wood-based material or a real-world structure.

3.1. Sample Preparation Procedure

Three samples were prepared in the form of blocks with a cross section of 47.0 × 27.5 mm and a length of 90.0 mm (Figure 7). They were made of planed spruce wood, dried by electric oven at 105 °C and drilled with diameter 3 mm with deep 20 mm (Figure 7c).

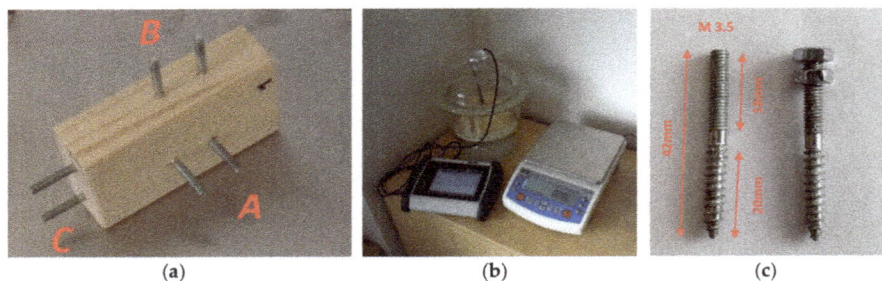

Figure 7. (a) Prepared test sample; (b) test conditioning; (c) detail of probes.

3.2. Circuit Setup and Calibration Procedure

Test circuits were built on test printed circuit boards (PCBs) with through hole technology (THT) and surface-mount device (SMD) electronic parts. The trigger pulses for measurements were initialized by push buttons. The voltage response of the trigger and the output of the circuits were measured by a digital oscilloscope with a 25 MHz bandwidth. The RC circuit with a 555 timer chip used a 12 V single power supply, while the integrating circuit had a dual power supply with −12 V and +12 V rails. A polypropylene capacitor was chosen with the value 1 nF.

Both circuits were calibrated with two resistor fields. The first test field consisted of 15 resistors with 10 MΩ resistance, and the second test field had 50 resistors with 20 MΩ resistance. Each resistor was measured separately with two digital Ohmmeters and the results were averaged. The total resistance of the first field was set to 148.87 MΩ with a maximum error of 1.5 MΩ, and 993.4 MΩ with a maximum error of 9.9 MΩ. Several combinations of resistors were tested by both experimental circuits. The proportional constants of the circuits were identified according to the obtained data using statistical analysis (Figure 8).

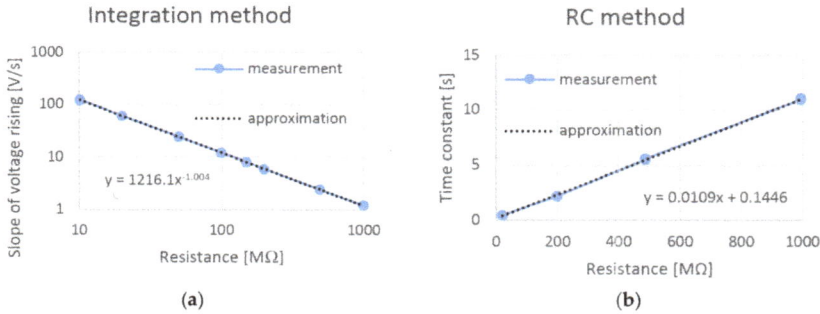

Figure 8. Calibration of (**a**) integrator and (**b**) RC circuit.

3.3. Sample Conditioning and Measurement Procedure

The samples were conditioned in a desiccator above a solution of salt in accordance with technical standard EN 12751. Sodium chloride and potassium chloride were used in desiccator. Relative humidity at laboratory temperature (21 °C) reached approximately 76% with the first solution and about 86% with the second. The environment in the desiccator was monitored by a temperature and relative humidity probe (Figure 7b). The resistance and weight of the samples were measured at specified intervals by the circuits and a laboratory balance with a resolution of 0.01 g and an accuracy of 0.05 g. Each measurement was taken three times for each electrode location and the results were averaged. The whole conditioning period lasted for roughly three months.

The evolution of sample moisture content was identified by weighing, and the relative humidity in the desiccator was measured (Figure 9). Dry samples were placed into desiccator with solution of sodium chloride for first 1300 h. A logarithmic fast increase of moisture content at the beginning of conditioning can be identified, later continuing more slowly until finally resulting in its equilibrium. In the next step, after 200 h of equilibrium conditions, the samples were placed into another desiccator with higher partial pressure of water vapor. The vapor pressure slowly rises again to reach equilibrium, approaching 2500 h of measurement.

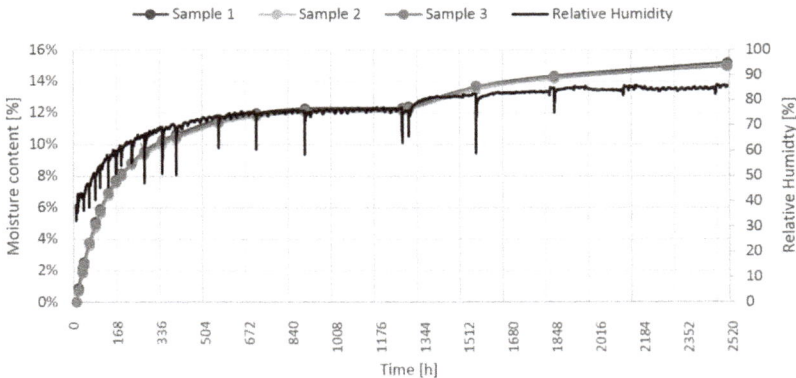

Figure 9. Evolution of moisture content over time.

Resistance value of samples could be distorted by nonlinear distribution of moisture in depth form surface of a sample. The conditioning process takes a long time, however the expectation of quasi linear distribution of moisture in the sample is more realistic closer to states where equilibrium states are reached in desiccator.

3.4. Statistical Analysis

In the real world, every value measured at a given assembly or sensor often differs slightly, thanks to variations in the repeatability and accuracy of methods. The methods used in this research are no exception, and they can be expected to provide slightly varied results, sometimes with higher and sometimes with lower differences. The comparison of both methods requires statistical analysis, which was attempted via the application of the Bland-Altman procedure, which allows the identification of any systematic difference between measurements, as well as possible outliers. The mean difference is the estimated bias, and the standard deviation (SD) of the differences measures the random fluctuations around this mean. The 95% limits of agreement, i.e., ±1.96 SD of the difference, are computed to determine the most likely difference between two measurements conducted using two methods. If the differences within the ±1.96 SD are not physically important, the two methods may be used interchangeably. The 95% limits of agreement are often unreliable estimates of population parameters, especially for small sample sizes. For small sets of data, like those in the presented study, it is appropriate to use a two-sided $1 - (\alpha/2)$ value of Student's t-distribution with $(n - 1)$ degree of freedom as a constant, which multiplies the standard deviation when calculating the limits of agreement.

4. Results

The resistance of the wood samples was obtained from the integration method circuit based on the slope of the rise in voltage. The resistance in megaohms was calculated based on data obtained from the calibration procedure. The obtained results for each electrode position differ slightly. Electrodes A and B obtained very similar values, but the values for electrode C were roughly double those measured by A and B (Table 1). This is in accordance with James [18]. Based on the proportional constant identified during the calibration procedure performed for the RC circuit, the resistance of samples was calculated. Positions A and B provided quite similar values, whilst those gained by position C were nearly two times higher (Table 2).

The relation between moisture content and calculated resistance has approximately linear behavior with resistance in logarithm scale (Figures 10–12). Power regression lines have been added to the plots of individual measurements. The values obtained for each probe orientation have quite similar tendencies and sizes. Both methods reach very close values to each other in whole scale range of measurement. The influence of the probe orientation and the direction of the wood grain is about 2% (Figure 13). The curves obtained by the A and B probes are close to each other and lie below the curves obtained by the C probes, which are shifted horizontally by roughly around 2%. The measurement in perpendicular direction of fibers reached approximately two times higher value than those in parallel directions. It practically confirms general assumptions that are well covered in the literature. Relative conductivity values in the longitudinal, radial, and tangential directions are related by the approximate ratio of 1.0:0.55:0.50 [18].

Table 1. Calculated resistances from the RC method in megaohms (MOhm).

	Sample 1												
	7.73%	8.14%	8.85%	9.57%	10.26%	10.61%	11.56%	11.96%	12.25%	12.27%	13.63%	14.34%	15.19%
A	26,550	14,936	6835	2775	1326	612	282	186	170	130	89	49	27
B	27,101	17,174	7220	3083	1461	665	313	201	191	147	102	58	29
C	46,009	28,257	12,046	5502	2601	1253	599	379	355	268	188	101	61
	Sample 2												
	7.56%	7.98%	8.71%	9.33%	9.98%	10.36%	11.29%	11.71%	12.02%	12.10%	13.45%	14.15%	15.00%
A	27,009	17,358	8752	3372	1735	777	345	224	209	162	105	61	32
B	23,725	14,569	6716	2989	1462	672	299	191	185	140	95	54	29
C	30,183	24,550	11,514	5754	3009	1454	650	428	415	282	183	102	52
	Sample 3												
	7.66%	8.09%	8.83%	9.50%	10.20%	10.51%	11.43%	11.86%	12.13%	12.19%	13.66%	14.28%	14.95%
A	24,349	16,239	6550	3453	1473	697	331	222	207	147	95	55	36
B	27,853	17,780	7321	3428	1427	669	302	193	183	140	94	55	33
C	42,294	24,936	10,945	6396	2599	1286	561	369	346	261	173	102	61

Table 2. Calculated resistances from the integration method in MOhm.

						Sample 1							
	7.73%	8.14%	8.85%	9.57%	10.26%	10.61%	11.56%	11.96%	12.25%	12.27%	13.63%	14.34%	15.19%
A	21,565	14,593	6723	2622	1371	621	281	183	174	135	86	45	26
B	22,452	15,499	7151	2918	1496	673	312	198	196	148	101	56	29
C	33,257	24,589	11,950	5109	2687	1299	575	385	367	279	184	99	60
						Sample 2							
	7.56%	7.98%	8.71%	9.33%	9.98%	10.36%	11.29%	11.71%	12.02%	12.10%	13.45%	14.15%	15.00%
A	24,570	17,251	7513	3756	1675	728	353	225	203	149	103	58	30
B	21,639	14,227	6246	3183	1419	645	308	186	180	137	94	58	27
C	31,784	23,127	10,383	6034	2853	1339	654	430	366	276	184	100	50
						Sample 3							
	7.66%	8.09%	8.83%	9.50%	10.20%	10.51%	11.43%	11.86%	12.13%	12.19%	13.66%	14.28%	14.95%
A	19,681	13,700	6558	2833	1496	747	340	213	209	156	97	57	30
B	23,737	15,293	6945	3019	1437	695	304	194	184	156	92	54	31
C	30,981	21,213	11,066	5170	2612	1328	559	379	359	271	171	99	58

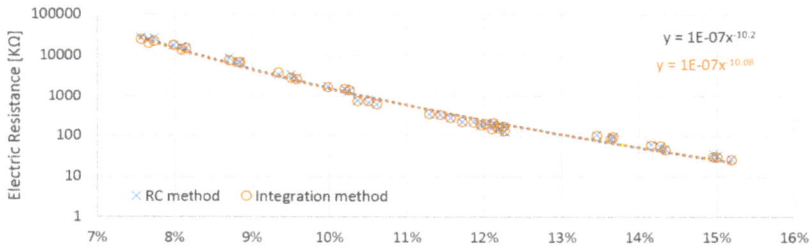

Figure 10. Obtained resistances from the integration and RC methods, the results calculated for the tops of the samples at the "A" electrodes.

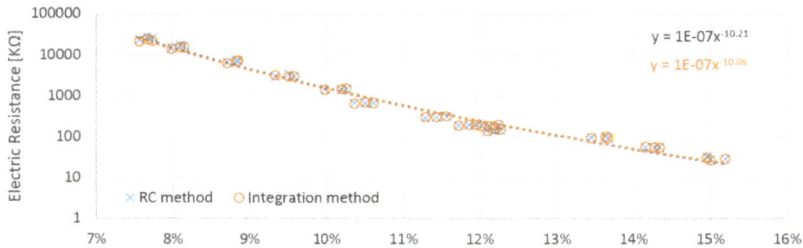

Figure 11. Obtained resistances from the integration and RC methods for the "B" probes placed on the right sides of the samples.

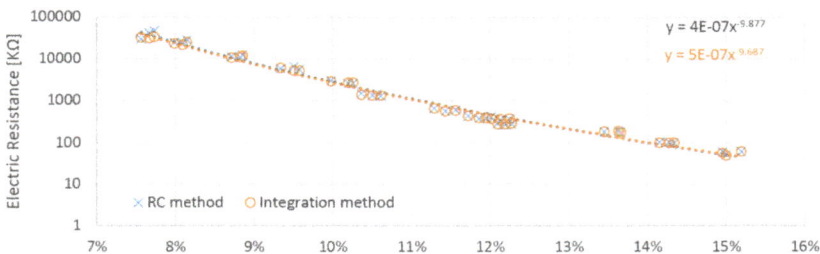

Figure 12. Obtained resistances from the integration and RC methods, the values calculated with the "C" probes placed in the cross sections.

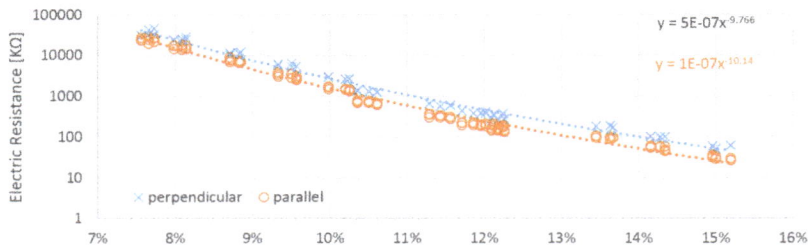

Figure 13. Obtained resistances from the integration and RC methods, the results calculated for the samples at the "A", "B" and "C" electrodes together for parallel and perpendicular direction.

5. Statistical Evaluation of Results

A study by the authors Bland and Altman [28] revealed that any two methods designed to measure the same parameter should show good correlation when a set of samples are chosen for which the parameter to be determined varies considerably. Therefore, a high correlation coefficient obtained for any two methods designed to measure the same property is just a sign that the sample chosen for measurement has a parameter which varies widely. It does not necessarily imply that there is a good agreement between the two methods. Hence, the analysis below was conducted to provide deeper insight into the differences between the two sets of measurements obtained by the two presented methods.

5.1. Construction of Bland-Altman Plots

The x axis represents the mean of the values measured by the RC method and the integration method. The y axis represents the differences between the values obtained by the two described methods. If there is agreement between two methods, the values in Figure 12 are expected to cluster around the mean of the differences (called the bias), and certainly within the limits of agreement. The dashed lines in the plots represent the lower and upper limit of agreement. In the presented case, only two points lie outside the limits of agreement.

5.2. Proportional Measurement Bias

Proportional bias can also be investigated via Bland-Altman plots, which indicate that the methods do not agree equally through the range of measurements. The limits of agreement are then dependent on the actual measurements. When the relationship between differences is identified, e.g., via regression analysis, the regression-based 95% limits of agreement should be provided, or proper transformation of the differences should be conducted. There is dependence in analyzed data between the mean values and differences. The Pearson's correlation coefficient $r = 0.85$ is high. The found regression curve with regression coefficient R2 (Figure 14). In other words, the limits of agreement are underestimated (too wide) for low values and overestimated (too narrow) for high values. In such cases, logarithmic data transformation can be used. The aim of transformation is to determine the limits of agreement that are valid for the entire range of values. Logarithmically transformed data (resistances measured at position of electrodes A) are derived from the results (Figure 15). The x axis represents the logarithm of mean of the values measured by the RC method and the integration method. The y axis represents the logarithm of differences between the values obtained by both methods.

The regression coefficient decreased ($R^2 = 0.34$) for the transformed data, and the points are more evenly distributed between the new limits of agreement. Nevertheless, three points still lie outside the limits of agreement. Logarithmic transformation was performed also for the two other data sets (resistances measured at position of electrodes B (Figure 16) and at position C (Figure 17), respectively).

Two points corresponding to the moisture 7.33% and 7.66% are outside the limits of agreement in both these cases.

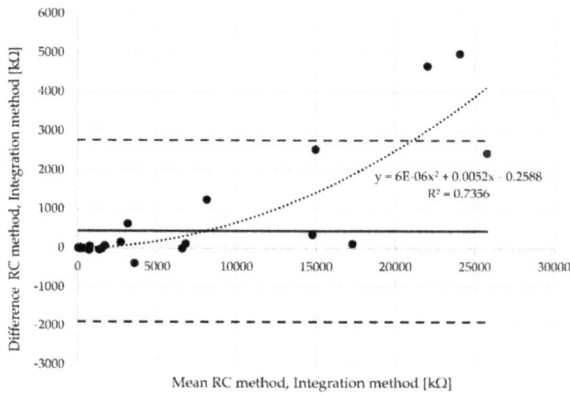

Figure 14. Bland-Altman plot for all samples measured at probe "A".

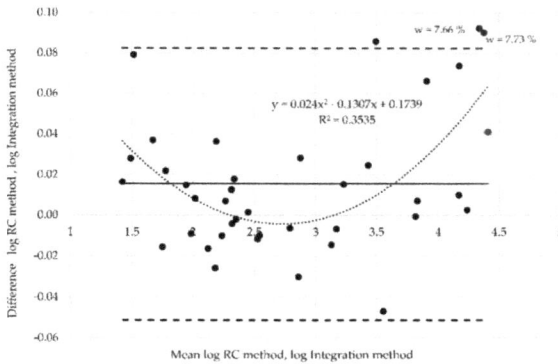

Figure 15. Altman plot for all samples measured at probe "A"—logarithmic transformation.

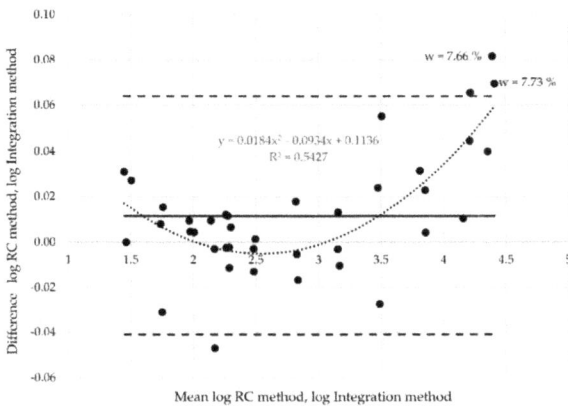

Figure 16. Altman plot for all samples measured at probe "B"—logarithmic transformation.

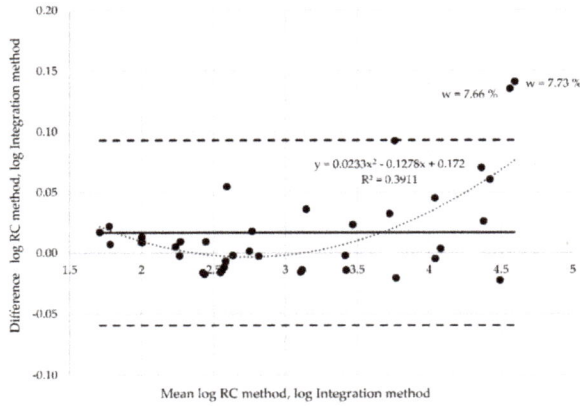

Figure 17. Altman plot for all samples measured at probe "C"—logarithmic transformation.

When the position of the electrode is ignored, there are still a few points out of the limits of agreements. Two of them are of importance, as they are even out of the outlier detector limits (dotted lines) calculated as mean ± 3 standard deviation (SD) (Figure 18).

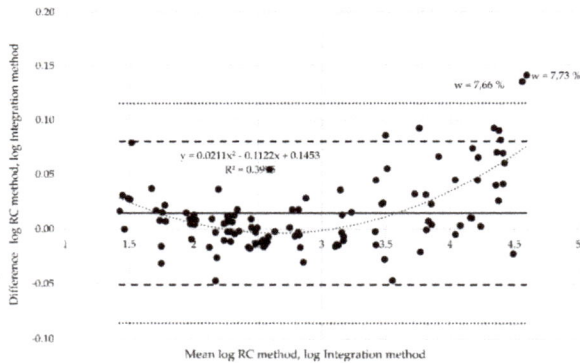

Figure 18. Bland-Altman plot for all samples and all probes—logarithmic transformation.

6. Discussion

The main advantage of RC circuits is the simplicity of adapting timer circuits to emit digital signals. It is easier to integrate this kind of circuit in the sensors utilized by today's digital electronic devices. The key limitation of such circuits is their long capacitor charge period at lower moisture contents. Sometimes, it may be necessary for a capacitor to charge for a relatively huge time; this would not be recorded by the controller, which evaluates the signal from the 555 timer chip at low moisture contents. The second problem is the resolution of charging time, which could be very small at higher moisture contents and, thus, could cause higher repeatability errors to affect measurements.

The integration method shows voltage changes, which allows the identification of lower moisture content much faster than an RC circuit. The main disadvantage of the integration method is the need for a symmetric power supply, as well as other circuits for the post-processing of analog signals. Post-processing circuits for analog signals are expensive and sensitive. The sensitivity of these circuits could lead to worse measurement repeatability and accuracy at lower moisture contents.

The range of differences between the two proposed methods is statistically important for small moisture contents of approximately below 8% (which correspond to high resistances), regardless of the fibers' direction. At moisture contents under 8%, wood has very high resistance and these methods of measurement are affected by significant errors. Based on the results of the statistical analysis, it is recommended that the threshold moisture content for the use of the proposed methods be set at 8%. Both of the proposed methods show good agreement for a range of moisture contents above this threshold.

The threshold value of moisture content 8%, revealed by statistical analysis, is in line with findings of other authors. Papez et al. [29] performed moisture measurement by local resistivity sensors with a measurement range of 7–30%. The calibration curves were developed by Fernandez-Golfin et al. [30] for the estimation of ten hardwoods by means of electrical resistance measurement. All resistance measurements by Fernandez-Golfin et al. were taken by advanced laboratory tool AGILENT 4339B high resistance meter. They reported 8.0% as the lowest value of measurable moisture content.

The direct comparison with laboratory methods was not performed in this study. However, the uncertainty of the presented method can be deduced from comparison with other authors. Some of them, e.g., Fernandez-Golfin et al. [30] and Moron et al. [31], developed similar circuits and calibrated them with laboratory methods. Fernandez-Golfin et al. [30] predicted the moisture content of wood with an error ±1.0%. They also revealed that the measuring direction has negligible influence (error of estimation <±0.5%) on moisture content estimates, below the moisture content of about 15%. Moron et al. [31] compared results of moisture content measurement conducted by their capacitance meters with those based on the variation of electromagnetic transmittance of timber. They reported similar accuracy (below 1%) between both methods in measurement range 1 MΩ to 100 GΩ. Their results show how low power and low-cost circuits can be similar to high precision, cost, and size instruments.

7. Conclusions

Based on the data obtained, both tested methods—RC and integration method—can successfully identify moisture changes within the range of 8–15%, with a resolution and high accuracy of about 0.5% of moisture content. Furthermore, the advantages and disadvantages of the methods have been analyzed. The RC circuit is easier to implement thanks to the digital signals of the 555 chip, whilst the integration method allows faster measurement at lower moisture contents. The RC method seems to be more suitable for intelligent sensors integrated within the structure of a building to perform its long-term monitoring. Low moisture content significantly limits the effectiveness of the two methods due to the high resistance of the wood, and both methods can fail in this situation. At moisture contents higher than 8%, both methods seem to be adequately suitable. Based on an analysis of the results of both methods, moisture values ranging from 8–15% also reach similar levels from a statistical point of view (the differences between both methods are within the limits of agreement for this moisture content range). When two methods are compared to determine whether they can be used interchangeably, the 'true' value of the measured quantity is unknown. Hence, the comparability differs from calibration, which is the case when a true measure is compared with measurement by a new method. Linear regression is sometimes used inaccurately for comparison of two methods. A high correlation coefficient is just a sign that a widespread sample has been chosen to measure; however it does not imply that there is a good agreement between the two methods. The Bland-Altman technique was adopted for assessment of the magnitude of disagreement here, and based on the statistical analysis, the conformity of the measurement with the integrator and the RC circuit was provided. The accuracy of measurements with RC resistance is then supported by comparison with RC resistance that is already introduced by other authors [25,26]. Statistically, the relation between moisture content and calculated resistance has approximately linear behavior with resistance in logarithm scale. Furthermore, it was clearly demonstrated that the influence of the probe orientation and the direction of the wood grain is about 2%.

It is planned that further research will focus on the improvement of circuits and signal processing from an electrical point of view. Further measurements with higher moisture contents and various kinds of wood need to be analyzed using the two methods to expand the options for their application in buildings.

Author Contributions: R.S. conceived and designed the main concept of the presented research. J.Š. analyzed and described the statistical analysis of the presented study. M.Č. compiled the texts and prepared the draft of the manuscript. R.S. proposed and performed the experimental analysis and described it in the manuscript.

Funding: This research was conducted within the strengthening and development of creative activities at the Mendel University in Brno by creating post-doc positions as the project part 7.1 in in the MENDELU institutional plan, supported by the Slovak Research and Development Agency, Project No. APVV-15-0681, project VEGA 1/0682/19 and co-funded by the European Regional Development Fund and the project No. LO1408 "AdMaS UP —Advanced Materials, Structures and Technologies", supported by Ministry of Education, Youth and Sports under the "National Sustainability Programme I" of the Czech Republic.

Conflicts of Interest: The authors declare that no conflict of interest exists in connection with this paper. The funding sponsors had no role in the design of the study, nor in the collection, analysis, or interpretation of data, nor in the writing of the manuscript or the decision to publish the results.

References

1. Kosny, J.; Asiz, A.; Smith, I.; Shrestha, S.; Fallahi, A. A review of high-value wood framed and composite wood wall technologies using advanced insulation techniques. *Energy Build.* **2014**, *72*, 441–456. [CrossRef]
2. Norberg, P. Monitoring wood moisture content using the WETCORR method Part 1: Background and theoretical considerations. *Holz Als Roh Werkstoff* **1999**, *57*, 448–453. [CrossRef]
3. Isaksson, T.; Thelandersson, S. Experimental investigation on the effect of detail design on wood moisture content in outdoor above ground applications. *Build. Environ.* **2013**, *59*, 239–249. [CrossRef]
4. Németh, R.; Tsalagkas, D.; Bak, M. Effect of soil contact on the modulus of elasticity of beeswax-impregnated wood. *BioResources* **2015**, *10*, 1574–1586. [CrossRef]
5. Dietsch, P.; Franke, S.; Franke, B.; Gamper, A.; Winter, S. Methods to determine wood moisture content and their applicability in monitoring concepts. *J. Civ. Struct. Health Monit.* **2015**, *5*, 115–127. [CrossRef]
6. Brischke, C.; Thelandersson, S. Modelling the outdoor performance of wood products—A review on existing approaches. *Constr. Build. Mater.* **2014**, *66*, 384–397. [CrossRef]
7. Brischke, C.; Meyer, L.; Bornemann, T. The potential of moisture content measurements for testing the durability of timber products. *Wood Sci. Technol.* **2013**, *43*, 869–886. [CrossRef]
8. Tanaka, T.; Avramidis, S.; Shida, S. Evaluation of moisture content distribution in wood by soft X-ray imaging. J Wood Sci. Modelling the outdoor performance of wood products—A review on existing approaches. *Constr. Build. Mater.* **2009**, *55*, 69–73.
9. Tanaka, T.; Kawai, Y. A new method for nondestructive evaluation of solid wood moisture content based on dual-energy X-ray absorptiometry. *Wood Sci. Technol.* **2013**, *47*, 1213–1229. [CrossRef]
10. Maeda, K.; Ohta, M.; Momohara, I. Relationship between the mass profile and the strength property profile of decayed wood. *Wood Sci. Technol.* **2015**, *49*, 331–344. [CrossRef]
11. Hultnas, M.; Fernandez-Cano, V. Determination of the moisture content in wood chips of Scots pine and Norway spruce using Mantex desktop scanner based on dual energy X-ray absorptiometry. *J. Wood Sci.* **2012**, *58*, 309–314. [CrossRef]
12. Meyer, L.; Brischke, C. Fungal decay at different moisture levels of selected European-grown wood species. *Int. Biodeter. Biodeger.* **2015**, *103*, 23–29. [CrossRef]
13. Brischke, C.; Rapp, O.A. Influence of wood moisture content and wood temperature on fungal decay in the field: Observations in different micro-climates. *Wood Sci. Technol.* **2008**, *42*, 663–677. [CrossRef]
14. Van den Bulcke, J.; Van Acker, J.; De Smet, J. An experimental set-up for real-time continuous moisture measurements of plywood exposed to outdoor climate. *Build. Environ.* **2009**, *44*, 2368–2377. [CrossRef]
15. Otten, K.A.; Brischke, C.; Meyer, C. Material moisture content of wood and cement mortars—Electrical resistance-based measurements in the high ohmic range. *Constr. Build. Mater.* **2017**, *153*, 640–646. [CrossRef]
16. Dunlap, M.E. *Electrical Moisture Meters for Wood*; United States Department of Agriculture, Forest Service, Forest Product Laboratory: Madison, WI, USA, 1944.

17. *Wood Handbook—Wood as an Engineering Material*; General Technical Report FPL-GTR-190; Department of Agriculture, Forest Service, Forest Products Laboratory: Madison, WI, USA, 2010; p. 508.

18. James, W.L. *Electric Moisture Meters for Wood*; General Technical Report FPL-GTR-6; United States Department of Agriculture, Forest Service, Forest Products Laboratory: Madison, WI, USA, 1988; Available online: http://www.fpl.fs.fed.us/documnts/fplgtr/fplgtr06.pdf (accessed on 1 March 2019).

19. Casans, S.; Iakymchuk, T.; Rosado-Muñoz, A. High resistance measurement circuit for fiber materials: Application to moisture content estimation. *Measurement* **2018**, *119*, 167–174. [CrossRef]

20. Brischke, C.; Rapp, O.A.; Bayerbach, R. Measurement system for long-term recording of wood moisture content with internal conductively glued electrodes. *Build. Environ.* **2008**, *43*, 1566–1574. [CrossRef]

21. Gao, S.; Bao, Z.; Wang, L.; Yue, X. Comparison of voltammetry and digital bridge methods for electrical resistance measurements in wood. *Comput. Electron. Agr.* **2018**, *145*, 161–168. [CrossRef]

22. Chetpattananondh, P.; Thongpull, K.; Chetpattananondh, K. Interdigital capacitance sensing of moisture content in rubber wood. *Comput. Electron. Agric.* **2017**, *142*, 545–551. [CrossRef]

23. Bland, J.M.; Altman, D.G. Statistical methods for assessing agreement between two methods of clinical measurement. *Lancet* **1986**, *327*, 307–310. [CrossRef]

24. Hanneman, S.K. Design, analysis and interpretation of method-comparison studies. *AACN Adv. Critic. Care* **2009**, *19*, 19–234.

25. Vodicka, A. Increased Moisture Detection and Early Warning System for Buildings. Ph.D. Thesis, Czech Technical University, Praha, Czech Republic, 2015. Available online: https://dspace.cvut.cz/ (accessed on 1 April 2019).

26. Mlejnek, P.; Vodička, A.; Včelák, J. Zařízení pro detekci vlhkosti dřeva. Available online: https://stavba.tzb-info.cz/drevene-konstrukce/12437-zarizeni-pro-detekci-vlhkosti-dreva (accessed on 19 September 2018).

27. Aguilar, H.M.; Landín, R.O. A simple technique for high resistance measurement. *Phys. Edu.* **2012**, *47*, 599–602. [CrossRef]

28. Bland, J.M.; Altman, D.G. Measuring agreement in method comparison studies. *Stat. Methods Med. Res.* **1999**, *8*, 135–160. [CrossRef] [PubMed]

29. Papez, J.; Kic, P. Wood moisture of rural timber construction. *Agron. Res.* **2013**, *11*, 505–512.

30. Fernandez-Golfin, J.I.; Conde, M.; Calvo, R.; Baonza, M.V.; De Palacios, A.P.; Palacios, Y. Curves for the estimation of the moisture content of ten hardwoods by means of electrical resistance measurements. *Forest Syst.* **2012**, *21*, 121–127. [CrossRef]

31. Moron, C.; Garcia-Fuentevilla, L.; Garcia, A.; Moron, A. Measurement of moisture in wood for application in the restoration of old buildings. *Sensors* **2016**, *16*, 697. [CrossRef]

materials

MDPI

Article

Accuracy of Eddy-Current and Radar Methods Used in Reinforcement Detection

Łukasz Drobiec *[iD]**, Radosław Jasiński**[iD] **and Wojciech Mazur**[iD]

Department of Building Structures, Silesian University of Technology; ul. Akademicka 5, 44-100 Gliwice, Poland; radoslaw.jasinski@polsl.pl (R.J.); wojciech.mazur@polsl.pl (W.M.)
* Correspondence: lukasz.drobiec@polsl.pl; Tel.: +48-32-237-11-27

Received: 4 March 2019; Accepted: 5 April 2019; Published: 10 April 2019

check for
updates

Abstract: This article presents results from non-destructive testing (NDT) that referred to the location and diameter or rebars in beam and slab members. The aim of paper was to demonstrate that the accuracy and deviations of the NDT methods could be higher than the allowable execution or standard deviations. Tests were conducted on autoclaved aerated concrete beam and nine specimens that were specially prepared from lightweight concrete. The most advanced instruments that were available on the market were used to perform tests. They included two electromagnetic scanners and one ground penetrating radar (GPR). The testing equipment was used to analyse how the rebar (cover) location affected the detection of their diameters and how their mutual spacing influenced the detected quantity of rebars. The considerations included the impact of rebar depth on cover measurements and the spread of obtained results. Tests indicated that the measurement error was clearly greater when the rebars were located at very low or high depths. It could lead to the improper interpretation of test results, and consequently to the incorrect estimation of the structure safety based on the design resistance analysis. Electromagnetic and radar devices were unreliable while detecting the reinforcement of small (8 and 10 mm) diameters at close spacing (up to 20 mm) and of large (20 mm) diameters at a close spacing and greater depths. Recommendations for practical applications were developed to facilitate the evaluation of a structure.

Keywords: NDT methods; rebar location; eddy-current method; GPR method

---

1. Introduction

Non-destructive testing that refers to the location and geometry of reinforcement in the structure is currently quite common while using different types of measuring equipment. The obtained test results are often used to make verifying calculations to determine the resistance of elements and decide whether they should not be further used or whether they should be reinforced. Rebar diameters and the precise location of reinforcement in the structure is required to calculate its resistance. For elements subjected to bending (slabs, beams) and compression and bending (columns), the reinforcement location affects internal forces, and thus the element strength. Even minor errors in detecting its location significantly change the result of resistance calculations and affect the structure safety. Moreover, non-destructive testing is employed at the acceptance of the performed facilities. It is particularly important that the accuracy of measurements is higher than the allowable execution deviations specified in standards. Therefore, tests on the structure reinforcement require information on the precision and limitations of the applied method and test equipment. All the above are useful in developing the programme for estimating the reliable building conditions.

Manufacturers are improving their products, and the measurement accuracy is increasing. It is very important to specify the measurement accuracy and to do this before the beginning of evaluating works. Non-destructive testing (NDT) methods are particularly useful for testing a wide surface or many elements. Electromagnetic and radar methods are currently the most often used NDT methods in detecting reinforcement in the structure [1–4]. Both of the methods have their advantages and disadvantages. It can be assumed that advantages of the electromagnetic method are: the accuracy of measurements and the possibility for determining the diameter of reinforcement. Its disadvantages include a short range of measurements and some errors that are triggered by resolution, which are significant when the rebars are closely arranged (lap splices, bundles of rebars) [5,6]. The advantage of the radar method is the possibility of localising the reinforcement at great depths; and its disadvantages cover difficulties in measuring diameters measurements and some measurement errors of damp structures [7,8].

Despite the continuous development, electromagnetic and radar methods still have present limitations. Many papers in the literature [3–5,8] present advantages and possibilities for using electromagnetic and radar methods. Information regarding their limitations and measurement accuracy is rarely published. According to the definition [9], the accuracy is a compatibility level between the obtained result of a single measurement and the expected value that is related to the systematic and accidental errors. However, design standards and standards specifying structures reinforced with rebars permit some execution deviations. Thus, it is interesting whether the real accuracy of methods is within standard limits of execution deviations. When the measurement accuracy is higher than the allowable execution deviations, then the non-destructive testing methods should not be used.

The accuracy of non-destructive tests on the reinforcement geometry can be analysed when considering the measurements of rebar diameter and their position in perpendicular and parallel direction to the plane the tested element [3]. The measurement accuracy mainly depends on the dimension of concrete cover and the mutual spacing of adjacent rebars. This article describes results from testing the measurement accuracy of reinforcement diameter and geometry while using the most advanced instruments that are available on the market. The main aim of paper was to demonstrate whether the measurement accuracy and related measurement uncertainties could be higher than the allowable execution deviation. The tests were conducted on reinforced precast beam made of autoclaved aerated concrete beam and nine specimens specially prepared from lightweight concrete. The analysis involved the effect of the cover thickness and horizontal spacing of rebars on determining the number of rebars and measurement results for their diameters. Additionally, the aspect of measurement accuracy of concrete cover thickness and the effect of cover dimensions on the measurement accuracy were analysed. The obtained results were compared to deviations accepted by standards. The measurement accuracy of diameters and covers was analysed in accordance with the uncertainty of results when assuming suitable estimators of uncertainty. The comparison of tests results for the same elements that were obtained by different methods and the direct analysis of obtained results had a significant contribution to the development of methods of NDT.

2. Accuracy and Possibilities of Non-Destructive Testing

2.1. Electromagnetic Method

The accuracy of electromagnetic tests mainly depends on the depth, spacing, and arrangement of rebars toward the direction of scanning, the type of reinforcement, and the quality of concrete surface [10,11]. A type of spectral analysis and compensation procedures of systematic measuring errors also affect the accuracy of the tests.

The depth of rebars location has a significant impact on the accuracy of diameter measurements, and even on the accuracy of rebars location. The maximum depth, at which reinforcement can be detected, depends on the diameter of rebars. For typical rebar diameters of 6–25 mm, depending on the employed device, the biggest depth at which the reinforcement can be detected is 100–200 mm. The closer location of rebars to the scanned surface evidently ensures a greater accuracy of the measured and real diameter. The acceptable measurement accuracy (5%) of the reinforcement diameter is obtained when the rebars are located at a depth of ca. 60 mm. Unfortunately, electromagnetic methods are not effective in detecting the non-metallic reinforcement, which is becoming increasingly common [12,13].

Drobiec et al. [11] describe tests to detect plain and ribbed rebars of 12 diameter. The tests were done with four different devices. Variations of the measured thickness of the concrete cover were within 4 mm, and the measured reinforcement diameter differed by not more than one gradation. Sivasubramanian et al. [14] presented the analysis of measurement errors of the electromagnetic device in slabs having dimensions of 400 × 400 × 250 mm, which were made of high compressive strength concrete (average 71.3 N/mm^2). One rebar was placed in each slab. Rebars of 12, 16, 20, 25, and 32 mm diameter and 70 and 85 mm covers were used (in total 10 of test models). Figure 1 illustrates the graph of the relative measuring error of the reinforcement diameter as a function of the concrete cover size. The acceptable error (<10%) was obtained for the cover thickness of up to ca. 40 mm. The measurement error of diameter of the order of 100% was obtained in tests regarding reinforcement at the depth of over 70 mm. It should be emphasized that, although the paper [15] is relatively recent, the used device is not a modern one. As no tests on models of concrete with lower strength were conducted, it is difficult to estimate the impact of the high compressive strength concrete. In typical reinforced concrete structures (beams, slabs, columns) with standard covers (c ~ 20–45 mm), electromagnetic devices that are available in practice can be used to perform tests with an accuracy of the measurement that is less than 10%. The gradation of rebar diameters causes difficulties in diameter differentiation.

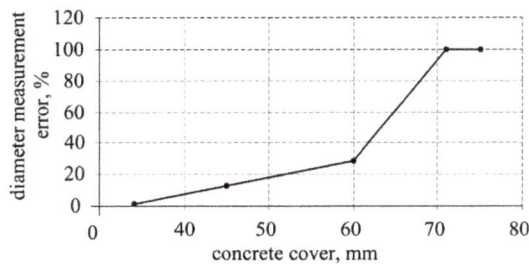

Figure 1. The error of the reinforcement diameter measured with an electromagnetic device, depending on the cover size according to [14].

2.2. Radar Method

In the radar method, the measuring range of the structure depth depends on concrete structure, a type of an antenna, and the frequencies of the excited impulse [15–17]. In typical devices, this range is up to 750 mm. The image contrast that was obtained from radar tests depends on the relative difference between dielectric constant values at the contact area between materials. There are no difficulties in interpreting the obtained image due to considerable differences in constant values for concrete and steel. Tests on reinforcement location that were conducted with radar method generate radargrams, that is, the record of all reflected signals registered during the passage of a measuring probe on the element surface. The reinforcement image on radargram is a distortion of course contour lines in the shape of hyperbola arms directed the radargram down.

The modern measurement systems perform automatic analyses of radargrams, convert radargrams taken close to each other in one image, and visualize reinforcement in the construction in a legible way for users. The additional software can be used to take a spatial image of the structure with the reinforcement. The radar devices do not provide direct information regarding reinforcement diameters. Perhaps measurements in a three-dimensional (3D) model can be a solution for this problem. However, some published articles present some mathematical correlations between the shape of a hyperbola that is shown on the radargram and the reinforcement diameter. Devices with the option of determining the reinforcement diameter are soon likely to appear on the market. However, the accuracy of the measurement method might be still a problem.

Currently, the accuracy of the cover measurements with the radar method performed with the best devices is defined as ±5–10 mm.

The papers [18–24] describe that the shape of hyperbola illustrating the reinforcement on the radargram depends on the wave propagation velocity v and passage time t_0 of measuring the probe above the tested reinforcement. The time of the measuring probe passage is the time passing from the moment when the device records the reinforcement for the first time until the moment when it stops to record it. The velocity of the wave propagation affected the hyperbola curve, whereas the time of the probe passage affected the range of its arms. Knowing the parameter values a and b of the hyperbola b (Figure 2), their relation with the reinforcement diameter ϕ, wave propagation velocity v, and time of probe passage t_0, on the tested reinforcement can be expressed, as following:

$$a = t_0 + \frac{\phi}{v},$$ (1)

$$b = \frac{v}{2}\left(t_0 + \frac{\phi}{v}\right)$$ (2)

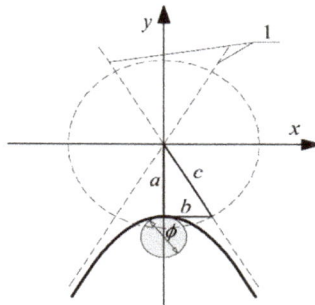

Figure 2. Assumptions to derive Equations (5) and (6), 1—hyperbola asymptotes.

The wave propagation velocity v and passage time t_0 of the measuring probe were recorded by a measuring device; and, parameters a and b were read from the radargram. Therefore, there were no obstacles to determine the reinforcement diameter ϕ from Equations (5) and (6). The paper [25] confirmed the possible use of that method for defining the reinforcement, as under laboratory conditions, the obtained measuring error of diameters of steel pipes and cables was in the order of 1.7–5.3%.

Another way of defining the diameter was suggested in the paper [26]. It was assumed that the impulse distribution angle from the transmitting antenna was 90° (Figure 3). Such an assumption was useful in determining the rebar diameter without taking into account the wave propagation velocity v and passage time t_0 of the measuring probe above the tested reinforcement. The reinforcement diameter ϕ can be determined according to:

$$\phi = 2\sqrt{2}L - 2X - 2Y$$ (3)

where: L, X, Y—geometric data according to Figure 3.

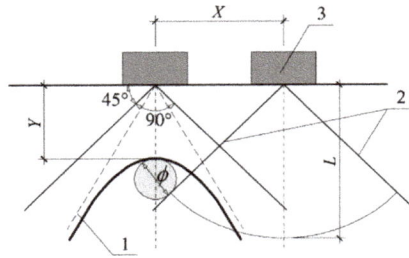

Figure 3. Assumptions to derive the Equation (7), 1—hyperbola asymptotes, 2—the angle of impulse distribution from the antenna, 3—a measuring probe.

2.3. Combination of Electromagnetic and Radar Methods

Wiwatrojanagul et al. [27] and Wiwatrojanagul et al. [28] suggested connecting the advantages of the electromagnetic and radar methods. Tests were conducted on specimens that were placed in a pen and in concrete. Rebar diameters were tested in rebar laboratories. The electromagnetic method was found to not give correct test results, because the reinforcement was placed too close. On the basis of conducted tests, the empirical equations were developed to estimate the diameter of contact rebars in the rebar laboratory. The accuracy of the determined diameters was obtained at the level of 2.35%. Empirical equations were found to have some application limitations regarding the tested reinforcement and concrete.

To sum up, the advantage of devices working according to the electromagnetic method was their high accuracy at small depths of measurements. A small range of diameter measurements, which was limited to about 6 cm, was their disadvantage Radar devices had much bigger range (even up to 75 cm). However, devices that are available on market do not offer the possibility of diameter measurements. In the world, some methods for diameter determining on the basis of radargram measurements obtained from the radar tests are being developed. Radar devices with the option of diameter measuring are likely to appear soon on the market.

3. Tested Specimens

3.1. Autoclaved Areated Concrete (AAC) Precast Lintels

To verify the accuracy of NDT methods, some tests were performed on widely used lintels that were made of AAC with the width of b = 180 mm, height h = 240 mm and the total length of L = 2000 mm. The strength of AAC elements was f = 4 N/mm^2. Detailed results from the material tests on lintels can be found in the papers [29–32]. The lintel reinforcement was made of steel with a yield strength of 500 N/mm^2 (B class according to EN 1992-1-1:2008 [33]). Longitudinal rebars had a diameter of 8 mm (three rebars down and two rebars up). Longitudinal reinforcement in the form of open stirrups that were made of rebars having a diameter of 4.5 mm. Stirrups were placed along the whole element at a constant spacing of 150 mm (Figure 4). The longitudinal reinforcement and stirrups were welded and then covered with corrosion-resistant protective coating made of resin.

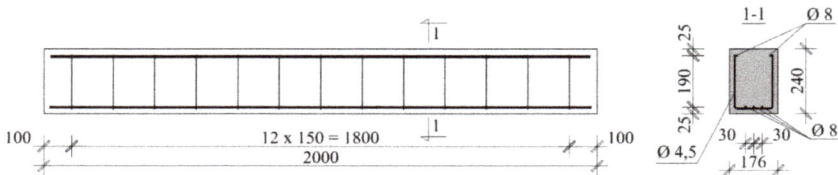

Figure 4. Dimensions and reinforcement of the tested lintel.

Two beams were selected for tests. To ensure the precise location of the reinforcement, both ends of tested beams were broken down and the reinforcement was measured, as in Figure 5. Tiny openings (diameter of 2 mm) were made in 11 places on lateral surfaces and at eight bottom points (between stirrups) to measure the real reinforcement cover—c_{obs} with the accuracy of $\Delta c_{obs} = \pm 0,1$ mm. As the location of both beams was the same, tests were only conducted on one of them.

Figure 5. Tested elements with the reinforcement broken down at its ends.

3.2. Lightweight Concrete Specimens

The impact of bar diameter and spacing on measurement accuracy was tested on nine models (Figure 6) that were prepared from lightweight concrete with density of 0.9 kN/m^3 and strength of 10.2 N/mm^2 after seven days, and of 18.1 N/mm^2 after 28 days. Each model had three rebars that were arranged to ensure the distance between rebars equal to 20, 30, and 40 mm (dimension "a" in Figure 6). The diameter of used rebars was ø = 10; 16; and, 20 mm. The dimensions of models in a plain view were 240 × 440 mm and thickness of 40, 60, and 80 mm. Different thickness was to ensure the correct (two-sided) cover of rebars with concrete assuming that rebar distance from one of the surface was 20 mm. It is the minimum thickness that is accepted by the standard EN 1996-1-1 [33] for the exposure class XC1 (indoor) and the structural class S3. The wooden elements were precisely cut to provide the cover thickness of 20 mm. Reinforcement was laid using wooden spacers of relevant thicknesses (Figure 7a), and then stabilized in washers with screws (Figure 7b). Concrete was laid, mechanically compacted, and properly cured. Figure 8 shows models before and after concreting. Specimens symbols contained a letter S, diameter *d*, and space between rebars *a*. For example, S-15-30 identified a specimen that was reinforced with three rebars having a diameter of 16 mm and space between rebars that is equal to 30 mm.

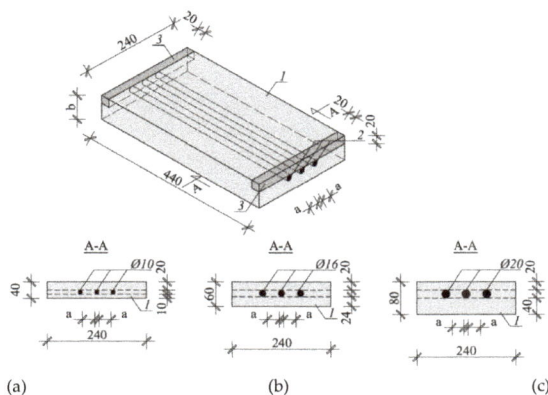

(a) (b) (c)

Figure 6. Lightweight concrete specimens: (**a**) specimen reinforced with rebars of 10 mm diameter, (**b**) specimen reinforced with rebars of 16 mm diameter, (**c**) specimen reinforced with rebars of 20 mm diameter, 1—lightweight concrete, 2—rebars, 3—wooden washer.

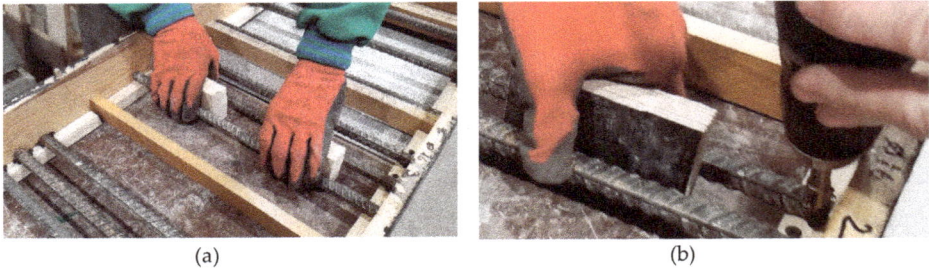

Figure 7. Stabilization of reinforcement in formwork: (**a**) distance resulting from wooden spacers, (**b**) rebar fastening to wooden washers with screws.

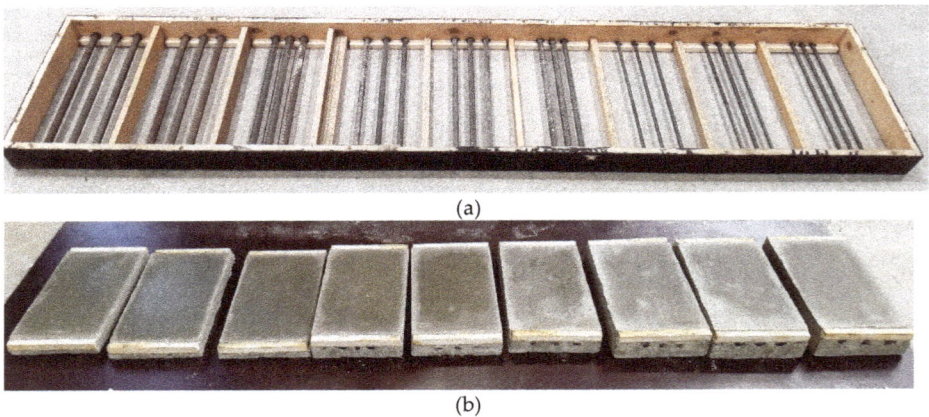

Figure 8. Specimen reinforcement in formwork (**a**) and specimens after concreting (**b**).

4. Applied Test Equipment

The tests were conducted using two electromagnetic scanners: PS 200 (manufacturer Hilti Corp., Schaan, Lichtenstein), Profometer 630 AI (manufacturer Proceq AG, Schwerzenbach, Switzerland) and one GPR device –GPR Live (manufacturer Proceq AG, Schwerzenbach, Switzerland). Figure 9 shows the testing apparatus. A scanning transducer of the electromagnetic device 1 was equipped with one circumferential transmitting coil and seven pairs of receiving coils. The receiving coils induce current with microammeters, and the received signal is then processed and analysed. The electromagnetic device no. 2 is similar in principle. The radar equipment 3 was equipped with antennas to perform tests, while using a signal with a variable frequency within a range of 0.2–4.0 GHz. Frequency was changed progressively in an automatic way during tests, and the max. acquisition time was 20 ns.

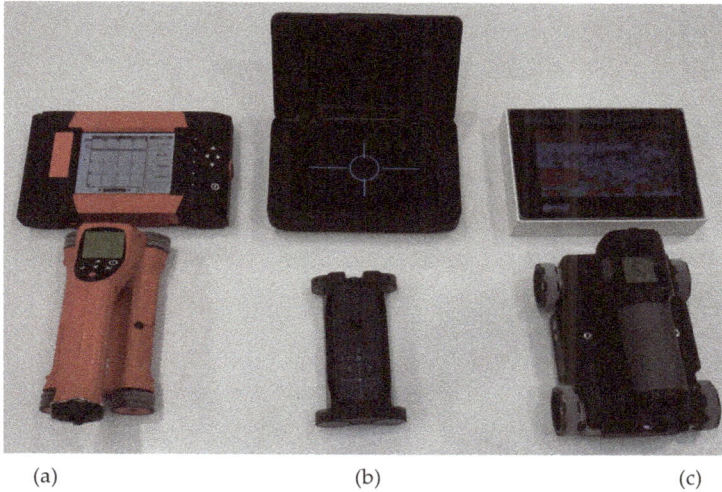

(a)   (b)   (c)

Figure 9. Devices used in tests, (**a**) electromagnetic device (1), (**b**) electromagnetic device (2), and (**c**) radar device (3).

The measuring accuracy of each device was the same, $\Delta c_{obs} = \pm 1.0$ mm (according to information provided by device manufacturers).

5. Testing Locations

5.1. AAC Precast Lintels

The tests were performed with each device by scanning the same points on the lateral and bottom surface of beams. Stirrup rebars were detected on the lateral surface, whereas the main reinforcement was tested from the bottom. During tests, line and area scans were conducted. Thus, the test consisted in moving the transducer along lines or areas that were routed on the surface of the tested element (Figure 10). Figure 11a illustrates scanned areas for stirrup testing and Figure 11b presents scanned areas for testing the main reinforcement. Figure 11c shows the location of a line scan and the point of measuring stirrups, whereas Figure 11d shows the location of a line scan for the main reinforcement.

Figure 10. Transducer on the surface of the tested element.

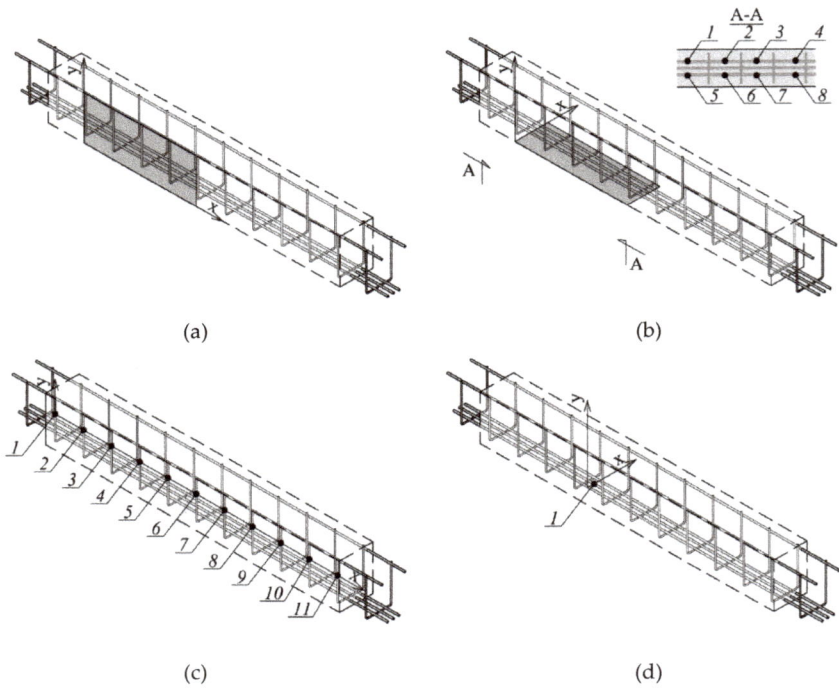

(a)   (b)

(c)   (d)

Figure 11. Places of lintel tests: (**a**) Stirrup tests—area scan (measurement in the grey-coloured field), (**b**) main reinforcement test—area scan (measurement in the grey-coloured field, measurement at points 1–8), (**c**) Stirrup tests—line scan (measurement at points 1–11), and (**d**) main reinforcement test—line scan (measurement at point 1).

5.2. Lightweight Concrete Specimens

Tests on lightweight concrete specimens, just as tests on lintels, consisted in scanning along lines or on surfaces that were routed on a given element using the transducer. The lightweight concrete specimens were tested using the electromagnetic Equipment (1) over the area of 150 × 300 mm illustrated in Figure 12a. The electromagnetic Equipment (2) was used to perform line scans at the mid-length of the specimen (Figure 12b), while area scans were conducted with the radar equipment over the area of 200 × 300 mm, as shown in Figure 12c. At the first stage, scans were conducted on the element surface where the concrete thickness was 20 mm, and then the element was turned to conduct further scans. Concrete cover was 10, 24, and 40 mm for rebars having a diameter of 10, 16, and 20 mm. Figure 13 shows the types of used equipment.

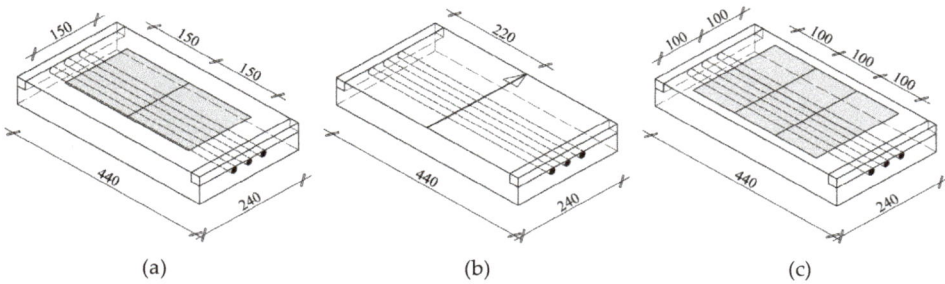

Figure 12. Location of tests performed with: (**a**) electromagnetic device (1), (**b**) electromagnetic device (2), (**c**) radar device (3).

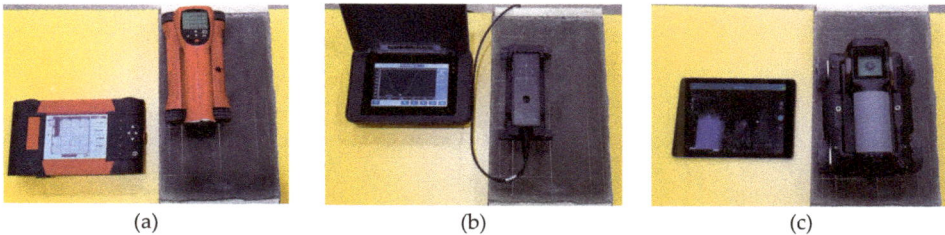

Figure 13. Testing specimens using: (**a**) electromagnetic device (1), (**b**) electromagnetic device (2), and (**c**) radar device (3).

6. Test Results

6.1. AAC Precast Lintels (AAC)

6.1.1. Stirrups Tests

The stirrups were tested at the beam side as line and area scans. Line scans could be only used to determine the reinforcement location (spacing and concrete cover). Figure 14 illustrates the comparison of exemplary results from line scans that were taken with electromagnetic devices and the radar device (radargram). A great conformity was found between the test results and real measurements of rebar location. Figure 15 presents examples of area scans taken with the electromagnetic device 1 and the radar device 3. The area scans conducted with two methods very clearly showed both stirrup reinforcement and longitudinal reinforcement.

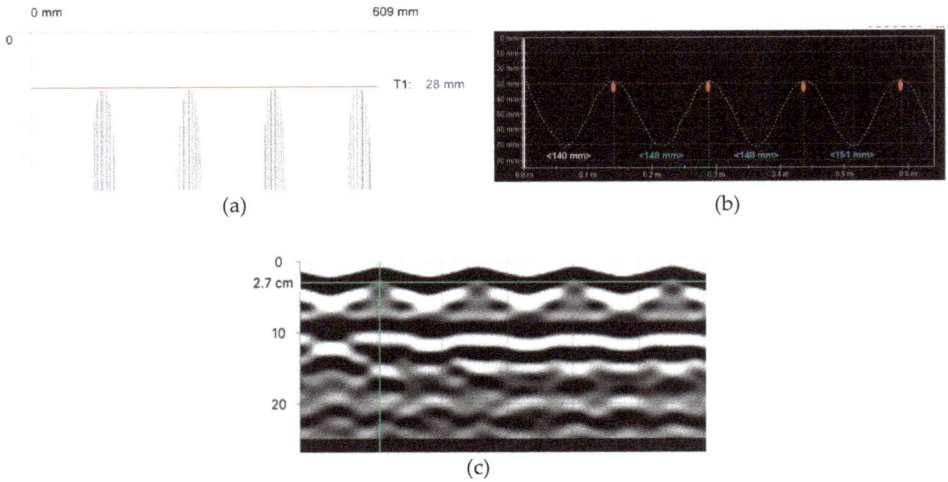

(a)

(b)

(c)

Figure 14. Line scans (measurement at points shown in Figure 10c) with devices: (**a**) electromagnetic 1, average cover of visible results—27.8 mm, (**b**) electromagnetic 2, average cover—28.6 mm, (**c**) radar 3, average cover 27 mm.

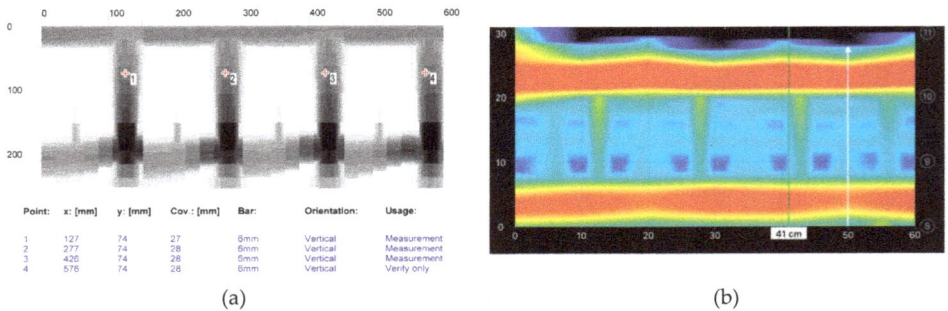

Point:	x: [mm]	y: [mm]	Cov : [mm]	Bar:	Orientation:	Usage:
1	127	74	27	6mm	Vertical	Measurement
2	277	74	28	6mm	Vertical	Measurement
3	426	74	28	6mm	Vertical	Measurement
4	576	74	28	6mm	Vertical	Verify only

(a)

(b)

Figure 15. Area scans conducted with electromagnetic devices (measurement on the surface shown in Figure 10a), (**a**) electromagnetic 1, average cover of visible results– 27.8 mm, (**b**) radar 3,– 27 mm.

6.1.2. Tests on Longitudinal Rebars

The aim of testing the main reinforcement was to determine the cover size, the number, and diameters of rebars in the main reinforcement (diameters were determined only in case of electromagnetic scans). Figure 16 shows the comparison of results from electromagnetic tests. The area scan (Figure 16a) was taken with the electromagnetic device 1 and the line scan (Figure 10b) was taken with the electromagnetic device 2. Both of the devices did not detect the correct number of rebars. Device 1 detected two rebars instead of three, but it measured the diameter quite correctly. However, device 2 detected only one rebar and the measured diameter was equal to 18 mm. Figure 16 shows the results from tests performed with the GPR device. Area and line scans were taken. The line scan is presented as a radargram (Figure 17a) and the area scan as a map (Figure 17b). Only one longitudinal rebar was detected during both tests while using the radar method.

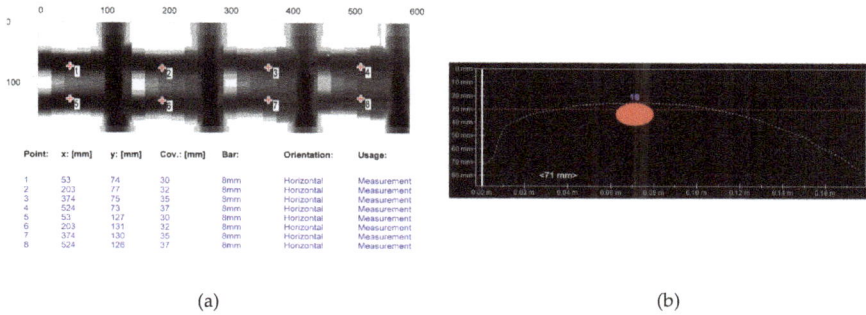

Point:	x: [mm]	y: [mm]	Cov.: [mm]	Bar:	Orientation:	Usage:
1	53	74	30	8mm	Horizontal	Measurement
2	203	77	32	8mm	Horizontal	Measurement
3	374	75	35	8mm	Horizontal	Measurement
4	524	73	37	8mm	Horizontal	Measurement
5	53	127	30	8mm	Horizontal	Measurement
6	203	131	32	8mm	Horizontal	Measurement
7	374	130	35	8mm	Horizontal	Measurement
8	524	128	37	8mm	Horizontal	Measurement

(a)   (b)

Figure 16. Scans taken with electromagnetic devices, (**a**) device 1, two rebars detected, average cover of visible results—33.6 mm, diameter 6-8 mm (measurement on the surface shown in Figure 10b), (**b**) device 2, one rebar detected, average cover of visible results—28 mm, diameter 18 mm (measurement at the point shown in Figure 11d).

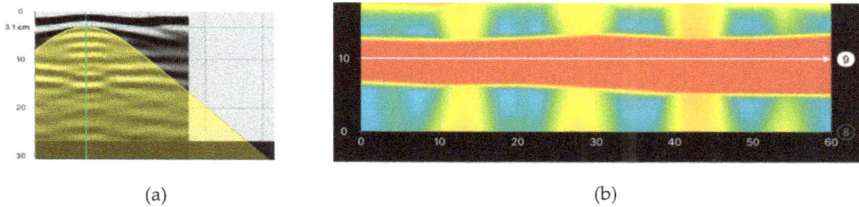

(a)   (b)

Figure 17. Scans taken with the radar (3), one rebar detected, (**a**) line scan (radargram—measurement at the point shown in Figure 10d), (**b**) area scan (measurement on the surface shown in Figure 10b).

To sum it up, the test results for the longitudinal reinforcement, in which the rebars were closer to each other (an axial spacing of rebars was 30 mm, as shown Figure 4) were poorer than in the case of tests that were performed on stirrup reinforcement as described in point 6.1.1. None of devices used during various testing methods, correctly detected the number of longitudinal rebars. The most satisfactory result was observed for tests that were performed with the electromagnetic instrument 1.

6.2. Lightweight Concrete Specimens (Tests with Concret Cover 20 mm)

6.2.1. Tests Using a 20 mm Cover

Tests on lintel beams indicated significant errors in measuring the quantity of reinforcement and the number of rebars in the main reinforcement. Thus, additional tests were performed on lightweight models that were prepared for this purpose. The aim of those tests was to capture the relationship between rebar diameters and spacing and the accuracy of measurements. Figure 18 shows the results from tests conducted with the electromagnetic equipment (1). They contain the measured number and diameter $\varnothing_{obs}$ of rebars and average cover c_{obs}.

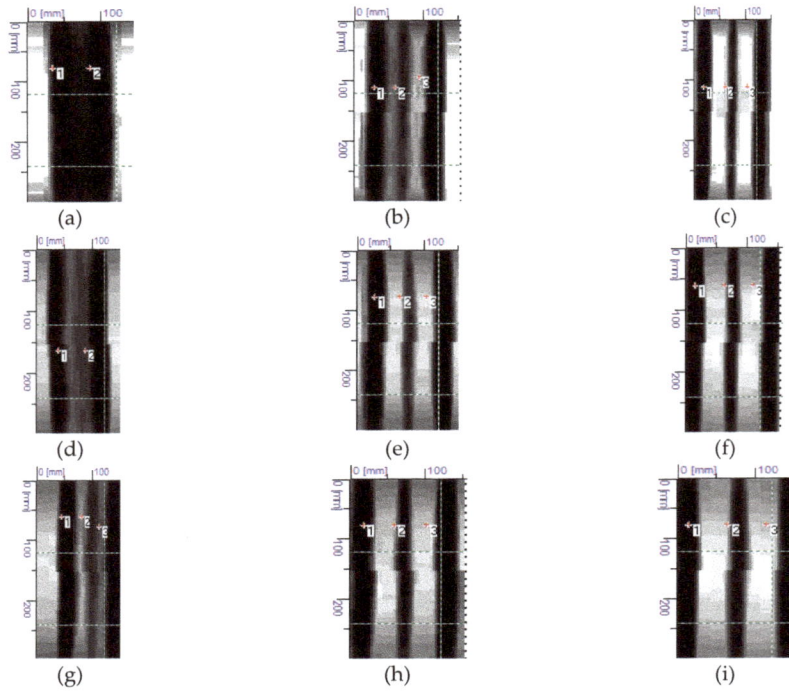

Figure 18. Results from tests conducted with the electromagnetic device (1): (**a**) S-10-20 (result: two rebars, $\varnothing_{obs}$ = 8 mm, c_{obs} = 25 mm), (**b**) S-10-30 (three rebars, $\varnothing_{obs}$ = 8 mm, c_{obs} = 20 mm), (**c**) S-10-40 (three rebars, $\varnothing_{obs}$ = 10 mm, c_{obs} = 21 mm), (**d**) S-16-20 (two rebars, $\varnothing_{obs}$ = 30 mm, c_{obs} = 25 mm), (**e**) S-16-30 (three rebars, $\varnothing_{obs}$ = 16 mm, c_{obs} = 18 mm), (**f**) S-16-40 (three rebars, $\varnothing_{obs}$ = 20 mm, c_{obs} = 20 mm), (**g**) S-20-20 (three rebars, $\varnothing_{obs}$ = 20 mm, c_{obs} = 25 mm), (**h**) S-20-30 (three rebars, $\varnothing_{obs}$ = 20 mm, c_{obs} = 18 mm), and (**i**) S-20-40 (three rebars, $\varnothing_{obs}$ = 20 mm, c_{obs} = 22 mm).

As in case of lintel beams, the tests failed to detect the correct number of rebars spaced at a_{nom} = 20 mm for specimens that were reinforced with rebars having a diameter $\varnothing_{nom}$ = 10 mm. In the specimen with rebars at the similar spacing, but reinforced with rebars $\varnothing_{nom}$ = 16 mm, three rebars could be noticed, but only two of them were detected with the equipment, which also overestimated the diameter of 30 mm. When rebars with a 20 mm diameter were used, the device identified as many as three rebars, even at the minimum spacing; however, the indicated diameter was twice smaller (10 mm) than the real value. At a larger spacing a_{nom} = 30 and 40 mm, the electromagnetic equipment (1) detected the correct diameter of rebars, but some inaccuracies were found while measuring the covers.

Figure 19 shows the results from tests conducted with the electromagnetic equipment (2). As above, the results include the number of rebars, their measured diameters d, and the average size of tested concrete cover c. For the smallest diameter and spacing, the equipment detected only one rebar, just as in case of tests on precast beam. The equipment found three rebars in specimens that were reinforced with rebars of 16 and 20 mm diameter, spaced at more than 20 mm. Some deviations were observed in measured diameters and covers. Contrary to the electromagnetic equipment (1), the equipment (2) provided more accurate diameters with less precisely measured thickness of concrete cover. Differences in cover measurements were even in the order of up to 5 mm.

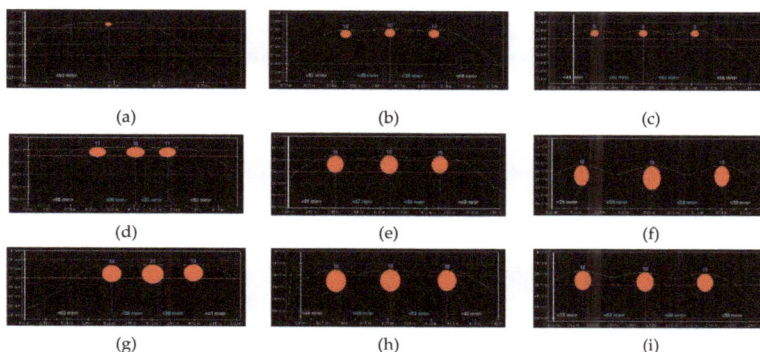

Figure 19. Results from tests performed with the electromagnetic device (2): (**a**) S-10-20 (result: one bar, $\varnothing_{obs}$ = 8 mm, c_{obs} = 20 mm), (**b**) S-10-30 (three rebars, $\varnothing_{obs}$ = 10 mm, c_{obs} = 19.3, mm), (**c**) S-10-40 (three rebars, $\varnothing_{obs}$ = 9 mm, c_{obs} = 19.3 mm), (**d**) S-16-20 (three rebars, $\varnothing_{obs}$ = 17 mm, c_{obs} = 15.5, mm), (**e**) S-16-30 (three rebars, $\varnothing_{obs}$ = 15 mm, c_{obs} = 16.4, mm), (**f**) S-16-40 (three rebars, $\varnothing_{obs}$ = 14 mm, c_{obs} = 16.5, mm), (**g**) S-20-20 (three rebars, $\varnothing_{obs}$ = 19 mm, c_{obs} = 15 mm), (**h**) S-20-30 (three rebars, $\varnothing_{obs}$ = 19 mm, c_{obs} = 15.4 mm), and (**i**) S-20-40 (three rebars, $\varnothing_{obs}$ = 18 mm, c_{obs} = 18.3 mm).

Figure 20 shows results from tests that were conducted with the electromagnetic equipment (3). The results include the number and average size of tested concrete cover c_{obs}. The correct number of rebars was detected in all of the tested specimens. The measured value of covers also showed high compliance with the true value (20 mm) The radar equipment could not directly measure diameters, which was its drawback. Therefore, there are no rebar diameters in Figure 20.

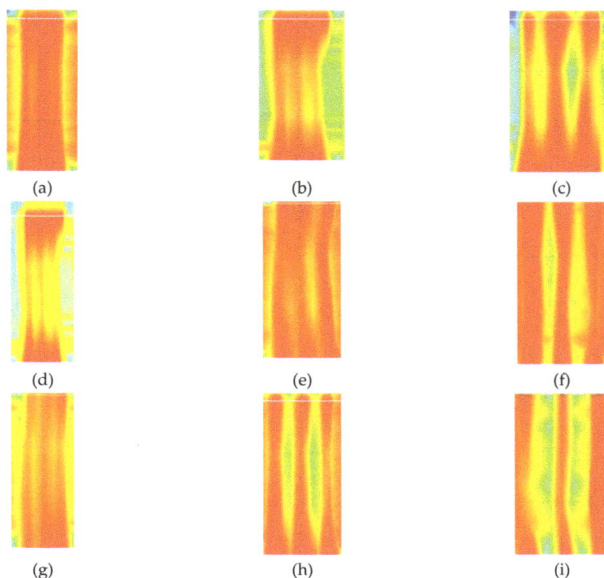

Figure 20. Results from tests performer with the radar equipment (3): (**a**) S-10-20 (result: three rebars, c_{obs} = 22 mm), (**b**) S-10-30 (three rebars, c_{obs} = 22 mm), (**c**) S-10-40 (three rebars, c_{obs} = 21 mm), (**d**) S-16-20 (three rebars, c_{obs} = 22 mm), (**e**) S-16-30 (three rebars, c_{obs} = 21 mm), (**f**) S-16-40 (three rebars, c_{obs} = 20 mm), (**g**) S-20-20 (three rebars, c_{obs} = 22 mm), (**h**) S-20-30 (three rebars, c_{obs} = 20 mm), (**i**) S-20-40 (three rebars, c_{obs} = 20 mm).

6.2.2. Tests Using Covers of Variable Thickness

Figure 21 shows the results from tests that were conducted with the electromagnetic equipment (1). The obtained results were comparable to tests using the cover with constant thickness (Figure 18). However, some differences were noticed. All of the rebars were detected in specimens with rebars of 10 mm diameter and concrete thickness reduced by 50%. However, at a spacing of 20 mm (Figure 21a) and 30 mm (Figure 21b), the rebar diameter was identified with a serious error of the order of 40% and 20%, respectively. Correct results were obtained for rebars at a clear spacing of 40 mm, just as for twice as big cover.

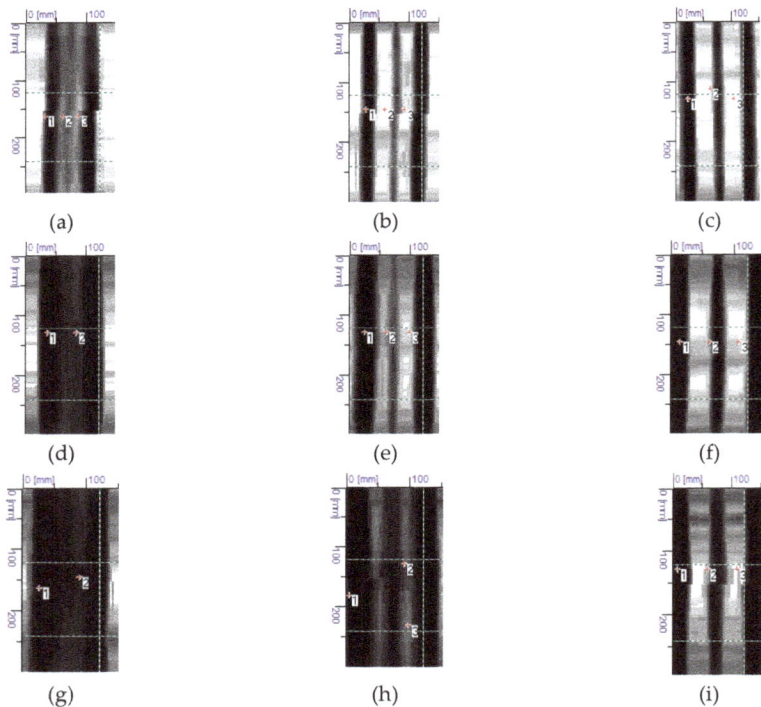

Figure 21. Results from tests conducted with the electromagnetic device (1): (**a**) S-10-20 (result: three rebars, $\varnothing_{obs}$ = 6 mm, c_{obs} = 11 mm), (**b**) S-10-30 (three rebars, $\varnothing_{obs}$ = 12 mm, c_{obs} = 12 mm), (**c**) S-10-40 (three rebars, $\varnothing_{obs}$ = 10 mm, c_{obs} = 11 mm), (**d**) S-16-20 (two rebars, $\varnothing_{obs}$ = 30 mm, c_{obs} = 32 mm), (**e**) S-16-30 (three rebars, $\varnothing_{obs}$ = 16 mm, c_{obs} = 27 mm), (**f**) S-16-40 (three rebars, $\varnothing_{obs}$ = 16 mm, c_{obs} = 24 mm), (**g**) S-20-20 (two rebars, $\varnothing_{obs}$ = 20 mm, c_{obs} = 48 mm), (**h**) S-20-30 (three rebars, $\varnothing_{obs}$ = 20 mm, c_{obs} = 47 mm), (**i**) S-20-40 (three rebars, $\varnothing_{obs}$ = 20 mm, c_{obs} = 35 mm).

Tests on specimens with rebar of 16 mm diameter and the cover of 24 mm produced results that were comparable to those for the same specimens, but on the other side (at cover thickness of 20 mm). For rebars at a closer spacing, the device wrongly identified two rebars having a diameter of 30 mm (Figure 21d). It overmeasured the thickness of concrete cover by more than 30%. For rebars at a moderate spacing, the measurement errors were comparable to those from testing the concrete cover of 20 mm thickness. In the specimens with rebars at the largest spacing, the measured diameters of rebars were the same.

A doubled increase in cover thickness in specimens with rebars of 20 mm caused a more serious errors event at the closest spacing of rebars (Figure 21g) because their number was incorrectly identified.

Higher measurement deviations were found in two other specimens when compared to tests using the cover with twice as small thickness.

Figure 22 shows the results from tests that were conducted with the electromagnetic equipment (2). Similarly as for the electromagnetic device (1), some differences were noticed in test results when compared to tests using the concrete cover of 20 mm thickness. For rebars with a diameter of 10 mm at the closest spacing (Figure 22a), their correct number was not identified. At a greater spacing, the equipment identified the correct number but overmeasured the rebar diameters. The results for specimens with rebars having a diameter of 16 mm were similar to those from testing the other side of the element at the concrete cover of 20 mm. Tests on specimens with rebars having a diameter of 20 mm indicated more significant inaccuracies in measurements in comparison to tests on rebars having a diameter of 10 mm. The device failed to identify the correct number of rebars at spacing of 20 mm (Figure 22g), whereas their correct number was detected when twice the concrete thickness was used. The test results were more satisfactory at larger spacing of rebars. The diameters were slightly undermeasured and the thickness of concrete cover was correct.

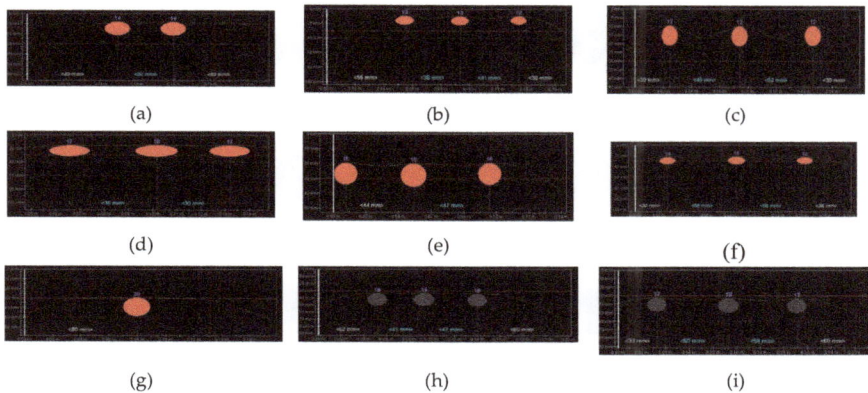

Figure 22. Results from tests performed with the electromagnetic device (2): (**a**) S-10-20 (result: two rebars, $\varnothing_{obs}$ = 14 mm, c_{obs} = 7.8 mm), (**b**) S-10-30 (three rebars, $\varnothing_{obs}$ = 13 mm, c_{obs} = 9.7 mm), (**c**) S-10-40 (three rebars, $\varnothing_{obs}$ = 12 mm, c_{obs} = 9.7 mm), (**d**) S-16-20 (three rebars, $\varnothing_{obs}$ = 17 mm, c_{obs} = 19.5, mm), (**e**) S-16-30 (three rebars, $\varnothing_{obs}$ = 16 mm, c_{obs} = 19.8 mm), (**f**) S-16-40 (three rebars, $\varnothing_{obs}$ = 14 mm, c_{obs} = 21.9 mm), (**g**) S-20-20 (one rebar, $\varnothing_{obs}$ = 20 mm, c_{obs} = 29 mm), (**h**) S-20-30 (three rebars, $\varnothing_{ob}$ = 18 mm, c_{obs} = 33.5 mm), (**i**) S-20-40 (three rebars, $\varnothing_{obs}$ = 17 mm, c_{obs} = 33.1 mm).

Figure 23 shows results from tests that were conducted with the radar equipment (3). The number of rebars was wrongly identified only in the case of specimens with rebars having a diameter of 10 mm and 20 mm, and at the closest spacing. However, measurements of concrete cover were quite accurate.

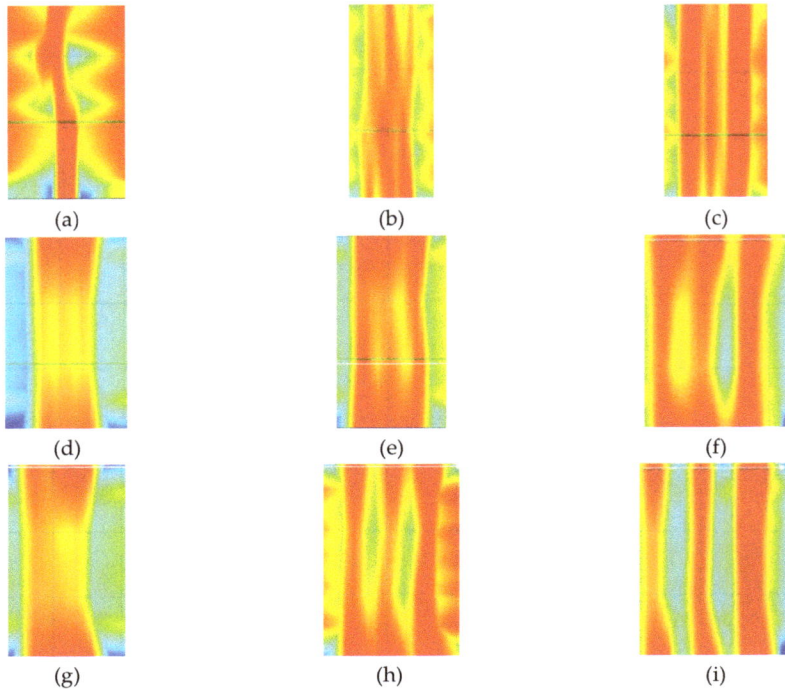

Figure 23. Results from tests performed with the radar equipment (3): (**a**) S-10-20 (result: one rebar, c_{obs} = 11 mm), (**b**) S-10-30 (three rebars, c_{obs} = 12 mm), (**c**) S-10-40 (three rebars, c_{obs} = 11 mm), (**d**) S-16-20 (three rebars, c_{obs} = 23 mm), (**e**) S-16-30 (three rebars, c_{obs} = 24 mm), (**f**) S-16-40 (three rebars, c_{obs} = 23 mm), (**g**) S-20-20 (two rebars, c_{obs} = 38 mm), (**h**) S-20-30 (three rebars, c_{obs} = 40 mm), (**i**) S-20-40 (three rebars, c_{obs} = 41 mm).

7. Discussion of Results

7.1. AAC Precast Lintels

7.1.1. Stirrups Tests

Direct measurements were taken to analyse the accuracy of conducted tests. The thickness of reinforcement cover in holes made with a 2 mm drill was subjected to direct measurements at the tested locations (Figure 7). Results from these measurements are presented in column 2 of Table 1.

Table 1. Measurement results for stirrup reinforcement using each method.

Measuring Point no.	Cover Measured with the Method, mm			
	Direct Method, c_{obs}	Electromagnetic Device (1)	Electromagnetic Device (2)	Radar Device (3)
1	30.1	27	28.9	27
2	28.1	28	28.6	27
3	28.3	28	29.1	27
4	28.2	28	28.1	27
5	29.1	27	28.6	28
6	27.3	27	28.2	27
7	27.9	27	28.7	27
8	26.9	28	28.3	27
9	28.3	27	28.4	27
10	28.2	27	28.6	27
11	29.0	28	28.7	28
Average cover, c_{obs}:	28.3	27	28.6	27
Nominal cover, c_{nom}:	$15 \text{ mm} \leq c_{nom} \leq 25 \text{ mm}$			
Calibration uncertainties Δc_{obs}	0.1	1.0	0.1	1.0
Standard deviation, s:	0.87	0.52	0.30	0.40
Standard deviation of the mean value $\bar{s} = \frac{s}{\sqrt{n}}$	0.3	0.2	0.1	0.1
Standard deviation of the calibration $S_d = \frac{\Delta c_{obs}}{\sqrt{3}}$	0.1	0.6	0.1	0.6
Standard uncertainty $S_c = \sqrt{S_d^2 + \bar{s}^2}$	0.3	1	0.1	1
Result with standard uncertainty $c_{mv} \pm S_c$	28.3 ± 0.3	27 ± 1	28.6 ± 0.1	27 ± 1
Maximum uncertainty $\Delta c = \Delta c_{obs} + 3\bar{s}$	0.9	1	0.4	1
Result with maximum uncertainty $c_{mv} \pm \Delta c$	28.3 ± 0.9	27 ± 1	28.6 ± 0.4	27 ± 1
Minimum cover c_{min}	27.4	26	28.2	26

Results from non-destructive testing of 11 inner stirrups of the beam are presented in Table 1 (three last columns of the Table), along with estimators of uncertainty according to JCGM 100:2008 [34]. The uncertainty of calibration was assumed to be equal to the accuracy of particular devices, and the uncertainty of an experimenter and the random uncertainty were neglected. The real average cover of stirrups was c_{obs} = 28.3 mm, with a standard deviation of 0.87 mm and variation coefficient of 3%. The estimated average reinforcement cover with the standard uncertainty was 3 ± 0.3 mm (the confidence level 68.3%), and, while taking into account the maximum uncertainty at the confidence level of 99.7%, the cover was equal to 28.3 ± 0.9 mm. The reinforcement nominal cover was c_{nom} = 25 mm, and the execution deviation allowed by the standard EN 13670 [35] was Δ_{minus} = 10 mm. The result obtained from direct measurements was within the determined limits. When considering other methods, the obtained average cover was 27 mm, 28.6 mm, and 27 mm, while the standard and maximum deviations did not exceed ±0.3 mm and ±1.0 mm. The obtained minimum covers measured with every NDT methods were bigger than the determined minimum cover while taking into account the dimension deviation. The reinforcement diameter was only determined with the electromagnetic scanner 1. The obtained result was equal to 6 mm, with the real diameter of 4.5 mm. However, the device manufacturer declared the detection of rebars with a diameter from 6 mm.

The diameters of stirrups were calculated on the basis of the obtained radargram and while using the Equations (1)–(3). Moving time t_0 was determined with the device, and other quantities that are required for equations were read from the radargram. For Equations (1) and (2), the calculated stirrup diameter was 5.6 mm, and in case of the Equation (3), the calculated diameter was 5.9 mm. Thus, the determined diameter was greater by 33% and 41%, respectively, than the real diameter.

7.1.2. Tests on Longitudinal Rebars

Similarly, as in case of testing stirrups, rebar cover was directly measured and column 2 of Table 2 shows the results. Measurement results are presented in Table 2 (three last columns of the Table) along with estimators of uncertainty according to JCGM 100:2008 [34]. The same assumptions were made as for the uncertainty analysis of stirrup location. There were some problems with the comparison of measurement results with the real results, because each device detected fewer rebars than the real number. Finally, it was decided to compare the test results of the electromagnetic device 1 (Figure 10a) to real measurements made on extreme rebars in holes that were drilled in places where electromagnetic measurements were taken (eight measuring points). In case of devices 2 and 3 devices that detected only one rebar, values from holes were compared to the value measured for one rebar. Thus, the same values are presented for measuring points 1 and 5, 2 and 6, 3 and 7, 4, and 8 (Figure 10a) in Table 2. The real average cover of longitudinal rebars was c_{obs} = 29.2 mm, with the standard deviation of 0.7 mm and the variation coefficient of 2.4 mm. The estimated average reinforcement cover with the standard uncertainty was 29.2 $\pm$ 0.3 mm (the confidence level 68.3%), and taking into account the maximum uncertainty at the confidence level of 99.7%, the cover was equal to 29.2 $\pm$ 0.8 mm.

The nominal cover of reinforcement was c_{nom} = 25 mm, and the execution deviation allowed by the standard EN 13670:2011 [35] was Δ_{minus} = 10 mm. Thus, the obtained result from the direct tests was within the determined limits. For other methods, medium-sized covers equal to 34 mm, 27.5 mm, and 31 mm were obtained. Standard deviations mostly did not exceed $\pm$ 1 mm and at a maximum $\pm$ 4 mm. The obtained minimum covers were bigger than minimum values while taking into account executive deviations.

In case of main reinforcement testing (beams at the bottom), the obtained deviations were considerably greater than in the case of stirrup tests (beams on the lateral surface). It should be emphasized that every device that was used in the tests did not detect the real number of rebars in the main reinforcement. Consequently, the result was burdened with a serious error resulting from the wrong reading of rebars number by research devices.

Table 2. Measurement results for the main reinforcement using each method.

Measuring Point no.	Cover Measured with the Method, mm			
	Direct Method, c_{obs}	Electromagnetic Device (1)	Electromagnetic Device	Radar Device (3)
1	29.1	30	26.2	31
2	28.8	32	27.3	30
3	28.7	35	28.1	31
4	29.2	37	28.4	32
5	29.1	30	26.2	31
6	28.9	32	27.3	30
7	29.2	35	28.1	31
8	30.9	37	28.4	32
Average cover, c_{obs}:	**29.2**	**34**	**27.5**	**31**
Nominal cover, c_{nom}:	15 mm $\leq c_{nom} \leq$ 25 mm			
Calibration uncertainties Δc_{obs}	0.1	1.0	0.1	1.0
Standard deviation, s:	0.70	2.9	0.9	0.8
Standard deviation of the mean value, $\bar{s} = \frac{s}{\sqrt{n}}$	0.2	1.0	0.3	0.3
Standard deviation of the calibration $S_d = \frac{\Delta c_{obs}}{\sqrt{3}}$	0.1	0.6	0.1	0.6
Standard uncertainty $S_c = \sqrt{S_d^2 + \bar{s}^2}$	0.3	1	0.3	0.6
Result with standard uncertainty $c_{mv} \pm S_c$	**29.2 $\pm$ 0.3**	**34 $\pm$ 1**	**27.5 $\pm$ 0.3**	**31 $\pm$ 0.6**
Maximum uncertainty $\Delta c = \Delta c_{obs} + 3\bar{s}$	0.8	4	1.1	2
Result with maximum uncertainty $c_{mv} \pm \Delta c$	**29.2 $\pm$ 0.8**	**34 $\pm$ 4**	**27.5 $\pm$ 1.1**	**31 $\pm$ 2**
Minimum cover c_{min}	28.4	30	26.4	29

7.2. Lightweight Concrete Specimens

Table 3 presents the results for lightweight concrete models that were tested on the side where the cover thickness was constant and equal to 20 mm, whereas Table 4 shows the results for the variable thickness of the cover. A shaded field shows measured values that were the same as the nominal values in models. The analysis of values in the table indicates that the number of rebars having a diameter of 10, 16, and 20 mm could be identified at a spacing greater than 20 mm and cover thickness greater than 10 mm while using the electromagnetic equipment (1) and (2). Measured values closest to nominal values of diameters were observed at a rebar spacing of 30 mm. The equipment showed similar diameters, but the false number of rebars at a closer spacing. The average ratio $\varnothing_{obs}/\varnothing_{nom}$ in accordance with electromagnetic methods (1) and (2), was 1.1 and 0.93, respectively, for the cover of 20 mm, and 1.08 and 1.02 for the cover of variable thickness. If the accuracy of indications of the electromagnetic equipment of the order of ca. ±1.0 mm was taken into account during the analysis of measurements, then the results from measuring diameters with these methods could be considered as reliable at a spacing larger than 20 mm and for diameters that are greater than 10 mm. Rebar diameters could not be read in the case of using the radar equipment (3). Therefore, the diameters were calculated in a similar way to precast lintel beam, while using Equations (1)–(3). More satisfactory compliance with nominal values was achieved using Equations (1) and (2). Table 3 shows those values. The measured diameters were greater by ca. 15% than nominal values.

Table 3. Test results for lightweight concrete specimens (at constant cover thickness of 20 mm).

Geometry of Models				Results from Tests Conducted with the Equipment (1), (2) or (3)								
Reinforcement Diameter (mm)	Number of Rebars n_{nom}	Rebar Spacing a_{nom}, mm	Reinforcement Cover c_{nom}, mm	$\varnothing_{obs}$, mm			n_{obs}, mm			c_{obs}, mm		
				(1)	(2)	(3) *	(1)	(2)	(3)	(1)	(2)	(3)
		20		8	8	11.9	2	1	3	25	20	22
10	3	30	20	8	10	11.8	3	3	3	20	19.3	22
		40		10	9	11.5	3	3	3	21	19.3	21
		20		30	17	18.8	2	3	3	25	15.5	22
16	3	30	20	16	15	18.2	3	3	3	18	16.4	21
		40		20	14	17.8	3	3	3	20	16.5	20
		20		20	19	23.3	3	3	3	25	15	22
20	3	30	20	20	19	23.1	3	3	3	18	15.4	20
		40		20	18	22.8	3	3	3	22	18.3	20

(1)– electromagnetic device (Figure 13a), (2) – electromagnetic device (Figure 13b), (3) – radar device (Figure 13c), * – values obtained from Equations (1) and (2).

Table 4. Test results for lightweight concrete specimens (at variable cover).

Geometry of Models				Results from Tests Conducted with the Equipment (1), (2) or (3)								
Reinforcement Diameter (mm)	Number of Rebars n_{nom}	Rebar Spacing a_{nom}, mm	Reinforcement Cover c_{nom}, mm	$\varnothing_{obs}$, mm			n_{obs}, mm			c_{obs}, mm		
				(1)	(2)	(3) *	(1)	(2)	(3)	(1)	(2)	(3)
		20		6	14	12.0	3	2	1	11	7.8	11
10	3	30	10	12	13	11.9	3	3	3	12	9.7	12
		40		10	12	11.6	3	3	3	11	9.7	11
		20		30	17	18.9	2	3	3	32	19.5	23
16	3	30	24	16	16	18.8	3	3	3	27	19.8	24
		40		16	14	18.2	3	3	3	24	21.9	23
		20		20	20	23.4	2	1	2	48	29	38
20	3	30	40	20	18	23.2	3	3	3	47	33.5	40
		40		20	17	22.7	3	3	3	35	33.1	41

(1) – electromagnetic device (Figure 13a), (2) – electromagnetic device (Figure 13b), (3) – radar device (Figure 13c), * – values obtained from Equations (1) and (2).

The rebar covers could be measured while using all of those methods. At the nominal cover with thickness of 10 mm, average values of 11.3, 9.07, and 11.3 could be read from the equipment (1), (2), and (3). For specimens with a 20 mm cover, the average measured values of cover thickness were 21.6, 17.3, and 21.1 mm, depending on the used device. for slightly greater cover (24 mm), the identified average values were 27.7, 20.4, and 23.3 mm; whereas, a double increase in concrete thickness to 40 mm resulted in average values that were equal to 43.3, 31.9, and 39.7 mm. Thus, the radar equipment was found to measure concrete thickness with the greatest precision.

In all measurements of covers, the difference from the nominal value was ±6,5 mm, that is, within the limits of allowable execution deviations Δ_{minus} = 10 mm specified by the standard EN 13670:2011 [35].

Tests that were conducted on lightweight concrete specimens confirmed the results obtained from tests on precast lintel beam. The greatest inaccuracies of measurements were observed for rebars with lower diameters ($\leq$10 mm) at close spacing ($\leq$20 mm), and with great diameters (20 mm) at close spacing (20 mm) and great depth (40 mm). For greater diameters ($\geq$16 mm), even close spacing did not affect the detection of the reinforcement, including the number of rebars, their diameter, and the thickness of concrete cover. The radar equipment provided a more precise size of concrete cover when compared to the electrical equipment. Rebar diameters that were calculated from the radargram were overcalculated by ca. 15 in comparison to nominal values. However, considerably better results in measuring diameters were noticed in the case of electromagnetic diameters.

8. Conclusions

The conducted tests demonstrated the occurrence of some limitations of two popular methods for detecting reinforcement location in the structure. The devices were able to correctly detect rebars, even those of small diameters, providing that they were located at an adequate spacing. Devices for scanning reinforcement had some problems with defining the correct quantity and diameter of rebars placed close to each other, at a distance shorter than 2–3 diameters.

The results obtained for covers measured with electromagnetic and radar devices did not significantly differ in terms of an average value. For the real cover, the differences did not exceed the acceptable level of few percent (<5%). There was no clear tendency to state that any of the methods artificially distorted indications, for instance, due to the simplification for measurement method validation. However, it should be emphasized that the minimal covers determined by all non-destructive techniques were bigger than the minimal cover specified with regard to size deviations (accepted by standard EN 13670:2011 [35]) and obtained by direct measurements. Thus, the tests on existing building can generate an error that is greater than the acceptable execution deviation. This can lead to wrong conclusions regarding the accuracy of a building. Measurements that were taken in the existing structure should be analysed with caution because the incorrect interpretation of measurement results for reinforcement location in the cross-section can cause a reduction in safety coefficient for steel $\gamma_{S,red1}$ from 1.15 to 1.1 (acc. to EC-2 [33]) and an unintentional reduction of safety level in particularly significant support zones of reinforced concrete structures and prestressed structures.

The tests indicated that a measurement error was significant, particularly for measuring structures that were reinforced with small diameter bars at a small spacing. This could lead to improper interpretation of test results and, consequently, to wrong calculation analyses for structures. A similar error could be observed in the tests on reinforcement of greater diameter, with rebars at a closer spacing when the concrete thickness was twice the rebar diameter or spacing.

To sum it up, the following recommendations for performing tests on reinforcement location can be considered as the practical application of the obtained results:

a) while detecting the cover with thickness < 20 mm, errors in determining the number and diameter of rebars can be expected regardless of the used test equipment,

b) while detecting the reinforcement at a depth > 20 mm, the employed equipment can correctly determine the number of rebars if the spacing is greater than 20 mm.

c) NDT equipment is not suitable for detecting the reinforcement in support zones of beams and slabs, and in joints where reinforcement is the densest, and

d) this equipment is perfect for detecting the longitudinal reinforcement in slabs, and stirrups in beams.

Studies aiming at connecting both methods, consisting in connecting scans that were taken with devices operating according to the electromagnetic and radar method seem to be reasonable. Hybrid devices are likely to generate accurate results, especially in terms of measuring the reinforcement diameters.

Author Contributions: Conceptualization, Ł.D. and R.J.; methodology, Ł.D., R.J. and W.M.; formal analysis, Ł.D. and R.J. and W.M.; investigation, Ł.D., W.M. and R.J.; writing—original draft preparation, Ł.D.; writing—review and editing, Ł.D. and R.J.; visualization, W.M. and Ł.D.; supervision, R.J. and Ł.D.

Acknowledgments: The authors would like to express particular thanks to Solbet Sp. z o.o. company, for supply of materials used during the research works. We would also like to thank Hilti, Proceq and Viateco for providing research equipment.

Conflicts of Interest: The authors declare no conflicts of interest.

References

1. Malhorta, V.M.; Carino, N.J. *Handbook on Nondestructive Testing of Concrete*, 2nd ed.; CRC Press LCC, ASTM International: Boca Raton, FL, USA; London, UK; New York, NY, USA; Washington, DC, USA, 2004.
2. Bungey, J.H.; Millard, S.G.; Grantham, M.G. *Testing of Concrete in Structures*, 4th ed.; Taylor & Francis: London, UK; New York, NY, USA, 2006.

3.   Drobiec, Ł.; Jasiński, R.; Piekarczyk, A. *Diagnostic Testing of Reinforced Concrete Structures*; Methodology, Field Tests, Laboratory Tests of Concrete and Steel; Wydawnictwo Naukowe PWN: Warszawa, Poland, 2013. (In Polish)

4.   Hoła, J.; Schabowicz, K. State-of-the-art non-destructive methods for diagnostic testing of building structures–anticipated development trends. *Arch. Civ. Mech. Eng.* **2010**, *10*, 5–18. [CrossRef]

5.   Drobiec, Ł. *Diagnosis of Industrial Structures*; Materiały Budowlane No 2015/2; Sigma-Not: Warsaw, Poland, 2015; pp. 32–34.

6.   Zima, B.; Rucka, M. Detection of debonding in steel bars embedded in concrete using guided wave propagation. *Diagnostyka* **2016**, *3*, 27–34.

7.   Hoła, J.; Bień, J.; Schabowicz, K. Non-destructive and semi-destructive diagnostics of concrete structures in assessment of their durability. *Bull. Pol. Acad. Sci.* **2015**, *63*, 87–96. [CrossRef]

8.   Ma, X.; Liu, H.; Wang, M.L.; Birken, R. Automatic detection of steel rebar in bridge decks from ground penetrating radar data. *J. Appl. Geophys.* **2018**, *158*, 93–102. [CrossRef]

9.   ISO/IEC. *Guide 99:2007 International Vocabulary of Metrology—Basic and General Concepts and Associated Terms*; ISO: Geneva, Switzerland, 2007.

10.   Chady, T.; Frankowski, P.K. Electromagnetic evaluation of reinforced concrete structure. In Proceedings of the AIP Conference, Denver, CO, USA, 15–20 July 2013; Volume 32, pp. 1355–1362.

11.   Drobiec, Ł.; Górski, M.; Krzywoń, R.; Kowalczyk, R. Comparison of non-destructive electromagnetic methods of reinforcement detection in RC structures. In Proceedings of the Challenges for Civil Construction (CCC 2008), Porto, Portugal, 16–18 April 2008.

12.   Ombres, L.; Verre, S. Shear performance of FRCM strengthened RC beams. *ACI Spec. Publ.* **2018**, *324*, 7.1–7.22.

13.   Micelli, F.; Cascardi, A.; Marsano, M. Seismic strengthening of a theatre masonry building by using active FRP wires. Brick and Block Masonry—Trends, Innovations and Challenges. In Proceedings of the 16th International Brick and Block Masonry Conference, Padova, Italy, 26–30 June 2016; pp. 753–761.

14.   Sivasubramanian, K.; Jaya, K.P.; Neelemegam, M. Covermeter for identifying cover depth and rebar diameter in high strength concrete. *Int. J. Civ. Struct. Eng.* **2013**, *3*, 557–563.

15.   Lachowicz, J.; Rucka, M. Application of GPR method in diagnostics of reinforced concrete structures. *Diagnostyka* **2015**, *16*.

16.   Lachowicz, J.; Rucka, M. 3-D finite-difference time-domain modelling of ground penetrating radar for identification of rebars in complex reinforced concrete structures. *Arch. Civ. Mech. Eng.* **2018**, *18*, 1228–1240. [CrossRef]

17.   Agred, K.; Klysz, G.; Balayssac, J.P. Location of reinforcement and moisture assessment in reinforced concrete with a double receiver GPR antenna. *Constr. Build. Mater.* **2018**, *188*, 1119–1127. [CrossRef]

18.   Shaw, M.R.; Millard, S.G.; Molyneaux, T.C.K.; Taylor, M.J.; Bungey, J.H. Location of steel reinforcement in concrete using ground penetrating radar and neural networks. *Ndt E Int.* **2005**, *38*, 203–212. [CrossRef]

19.   Chang, C.W.; Lin, C.H.; Lien, H.S. Measurement radius of reinforcing steel bar in concrete using digital image GPR. *Constr. Build. Mater.* **2009**, *23*, 1057–1063. [CrossRef]

20.   Shihab, S.; Al-Nuaimy, W. Radius estimation for cylindrical objects detected by ground penetrating radar. *Subsurf. Sens. Technol. Appl.* **2005**, *6*, 151–166. [CrossRef]

21.   Ristic, A.V.; Petrocacki, D.; Govedarica, M. A New Method to Simultaneously Estimate the Radius of a Cylindrical Object and the Wave Propagation Velocity from GPR Data. *Comput. Geosci.* **2009**, *35*, 1620–1630. [CrossRef]

22.   Idi, B.Y.; Kamarudin, M.N. Utility Mapping with Ground Penetrating Radar: An Innovative Approach. *J. Am. Sci.* **2011**, *7*, 644–649.

23.   Zanzi, L.; Arosio, D. Sensitivity and accuracy in rebar diameter measurements from dual-polarized GPR data. *Constr. Build. Mater.* **2013**, *48*, 1293–1301. [CrossRef]

24.   Mechbal, Z.; Khamlichi, A. Determination of concrete rebars characteristics by enhanced postprocessing of GPR scan raw data. *Ndt E Int.* **2017**, *89*, 30–39. [CrossRef]

25.   Wei, J.S.; Hashim, M.; Marghany, M. New approach for extraction of subsurface cylindrical pipe diameter and material type from ground penetrating radar image. In Proceedings of the 1st Asian Conference on Remote Sensing (ACRS 2010), Hanoi, Vietnam, 1–5 November 2010; Volume 2, pp. 1187–1193.

26.   Alhsanat, M.B.; Wan Hussin, W.M.A. A New Algorithm to Estimate the Size of an Underground Utility via Specific Antenna. In Proceedings of the PIERS, Marrakesh, Maroko, 20–23 March 2011; pp. 1868–1870.

27. Wiwatrojanagul, P.; Sahamitmongkol, R.; Tangtermsirikul, S.; Khamsemanan, N. A new method to determine locations of rebars and estimate cover thickness of RC structures using GPR data. *Constr. Build. Mater.* **2017**, *140*, 257–273. [CrossRef]

28. Wiwatrojanagul, P.; Sahamitmongkol, R.; Tangtermsirikul, S. A method to detect lap splice in reinforced concrete using a combination of covermeter and GPR. *Constr. Build. Mater.* **2018**, *173*, 481–494. [CrossRef]

29. Drobiec, Ł.; Jasiński, R.; Mazur, W. Precast lintels made of autoclaved aerated concrete—Test and theoretical analyses. *Cem. Wapno Beton* **2017**, *5*, 339–413.

30. Mazur, W.; Drobiec, Ł.; Jasiński, R. Research of Light Concrete Precast Lintels. *Procedia Eng.* **2016**, *161*, 611–617. [CrossRef]

31. Mazur, W.; Drobiec, Ł.; Jasiński, R. Research and numerical investigation of masonry—AAC precast lintels interaction. *Procedia Eng.* **2017**, *193*, 385–392. [CrossRef]

32. Mazur, W.; Jasiński, R.; Drobiec, Ł. Shear Capacity of the Zone of Supporting of Precast Lintels Made of AAC. *IOP Conf. Ser. Mater. Sci. Eng.* **2019**, *471*, 052070. [CrossRef]

33. EN 1992-1-1. *Eurocode 2: Design of Concrete Structures—Part 1-1: General Rules and Rules for Buildings*; European Committee for Standardization (CEN): Brussels, Belgium, 2004.

34. JCGM 100:2008. *Evaluation of Measurement Data. Guide to the Expression of Uncertainty in Measurement*; Joint Committee for Guides in Metrology: Sevres, France, 2008.

35. EN 13670:2011. *Execution of Concrete Structures*; European Committee for Standardization (CEN): Brussels, Belgium, 2011.

Article

Application of Nanoindentation and 2D and 3D Imaging to Characterise Selected Features of the Internal Microstructure of Spun Concrete

Jarosław Michałek *, Michał Pachnicz and Maciej Sobótka

Faculty of Civil Engineering, Wrocław University of Science and Technology, 50-370 Wrocław, Poland;
michal.pachnicz@pwr.edu.pl (M.P.); maciej.sobotka@pwr.edu.pl (M.S.)
* Correspondence: jaroslaw.michalek@pwr.edu.pl; Tel.: +71-320-22-64

Received: 13 February 2019; Accepted: 25 March 2019; Published: 27 March 2019

Abstract: The spinning of concrete is a process in which concrete mixture is moulded and compacted under the action of the centrifugal force arising during the fast rotational motion of the mould around its longitudinal axis. As a result of the spinning of the liquid concrete mixture, an element annular in cross section, characterised by an inhomogeneous layered wall structure, is produced. The heavier constituents tend towards the cross-section wall's outer side, while the lighter components tend towards its inner side. The way in which the particular constituents are distributed in the element's cross section is of key importance for the macro properties of the manufactured product. This paper presents procedures for investigating spun concrete and interpreting the results of such investigations, which make it possible to characterise the microstructure of the concrete. Three investigative methods were used to assess the distribution of the constituents in the cross section of the element: micro-computed tomography (μCT), 2D imaging (using an optical scanner) and nanoindentation. A procedure for interpreting and analysing the results is proposed. The procedure enables one to quantitatively characterise the following features of the microstructure of spun concrete: the mechanical parameters of the mortar, the aggregate content, the pore content, the cement paste content, the aggregate grading and the size (dimensions) of the pores. Special attention is devoted to the determination of the variation of the analysed quantities in the cross section of the element. The result of the application of the investigative procedures is presented for an exemplary spun concrete element. The proposed procedures constitute a valuable tool for evaluating the process of manufacturing spun concrete elements.

Keywords: spun concrete; micro-computed tomography; nanoindentation; deconvolution; mathematical morphology

---

1. Introduction

The main aim of this study was to assess the capabilities of three different investigative methods, namely nanoindentation, micro-computed tomography (μCT) and 2D optical scanning, as applied to evaluate selected properties of the microstructure of spun concrete. In particular, it was attempted to use these methods to describe the internal microstructure of concrete, e.g. local strength parameters, pore space morphology, spatial distribution of aggregate and cement paste. The performance of utilised methods was demonstrated on the example of spun concrete samples. This type of concrete has an internal structure different from the commonly used cast in place concrete. Due to the production process, the spun concrete is characterised by a layered structure across the wall of the annular cross section. Therefore, the parameters describing the internal microstructure vary within the cross-section. Proposed methodology for determining aforementioned parameters can be a practical tool helpful, for example, in the development of the optimal technology for the production of spun concrete elements.

The spinning of concrete is a process in which a concrete mixture is shaped and compacted under the action of a normal (radial) force arising during the fast (500–700 rpm) rotational motion of the mould around its longitudinal axis. Because of its peculiarities, this method can be used to produce exclusively hollow elements. As a result of spinning, the concrete mixture introduced into the mould consolidates and retains the acquired shape as the process continues. The moulding pressure produced by spinning is not uniform across the layer being consolidated. The moulding pressure on the inner layer of the product wall is close to zero, while, on its outer side, it reaches a maximum value. A graph of the pressure across the product wall can be presented in the form of a triangle.

The generated normal pressure is initially taken by the water present between the particles of the concrete mixture. As a result, the hydrostatic pressure increases and under this pressure the water begins to displace (through the forming filtration channels directed radially towards the mould axis) from one layer to another. At the same time, the concrete mixture consolidates. The displacing water takes the finest concrete mixture constituents (mainly cement) with it. In the course of this process, the cohesion of the mixture increases and the mechanical bond between the mixture constituents acquires some strength and in certain conditions a state of equilibrium is reached, whereby the filtration stops. Since at this moment the moulding pressure is fully transferred to the solid particles of the already considerably compacted concrete mixture, the ultimate effect of the compaction will depend on the composition of the concrete mixture (to a large extent on the aggregate grading).

Under such a distribution of pressure during spinning, water is not carried away from the cross-section wall thickness equally (the element's wall layers being characterised by different w/c values and thus by different porosity). Consequently, the mixture layers closer to the mould axis begin to consolidate only after the limit density of the outer layer is reached while the moulding pressure increases. This means that, unlike in the case of vibrated concrete, the structure of spun concrete across the element wall is inhomogeneous and layered. The structure is characterised by the fact that the heavier constituents (large particles) tend towards the cross section's outer side, while the lighter components (cement paste) tend towards its inner side. As a result, the outer layer can become highly compact and after the concrete sets it can acquire high strength and resistance to chemical and mechanical impacts. The inner layer will consist of highly consolidated cohesive cement paste and after setting it can become highly impervious and resistant to the impact of flowing water. According to Ref. [1], the 2–4 mm thick inner layer can contain >25% more cement than the intermediate layers.

The deformation and strength characteristics of the layered structure of spun concrete have been the subject of only a few studies. Mainly macroscopic investigations of spun concrete are carried out. Marquardt was one of the first researchers [1] who described the structure of spun concrete used for the production of pipes. He experimentally found that the larger is the difference between the specific weight of the concrete mixture constituents, the faster and more completely does the mixture fractionate. For the concrete mixture, he recommended well sorted mixed aggregate particles with similar petrographic properties and a maximum diameter of 15 mm. He also described the variation in cement content across the wall of a pipe made of spun concrete. He found that cement across 83% of the wall thickness is quite uniformly distributed and an increase in cement content takes place in only the 2–5 mm thick inner layer of the cross section. To reduce cross-sectional lamination, he proposed to use layered spinning at lower mould rotational speeds. After spinning each of the layers and before feeding the next concrete mixture batch, one should pour out the spun out water from the mould and smooth the inner layer.

Achverdov, investigating the cross section of a spun concrete pipe, found [2] that the 2–3 mm thick inner layer can contain >25% more cement than the intermediate layers. The cement content in the inner layer of the cross section depends on the kind and amount of cement and on the initial water content in the concrete mixture. He also stated that the principles used for selecting a filler particle size distribution for vibrocompacted regular concretes cannot be applied to spun concretes. The total fine aggregate content should be considerable and in the case of pipes, the amount of 0/2 mm sand should constitute 40–50% by weight of the total filler content. Achverdov (similarly to Marquardt [1]) noted that, to obtain higher integrity and a higher qualitatively structure of spun concrete, it is necessary to use a multilayer moulding system. In the case of multilayer compaction, two rotation phases (the initial one and the proper one) are used for each of the layers and, when the moulding of a particular layer is finished, the centrifuge is stopped and the water drained from the mixture is removed.

Dilger, Ghali and Krishna Mohan Rao, when investigating spun concrete poles [3,4], found that, because the cross section's inner layer has a high water and cement content, the shrinkage inside the pole's cross section is greater than on its outside, which results in shrinkage cracks developing vertically and extending deep into the pole wall. They also found that segregation can be reduced or even eliminated through a proper concrete mixture design precluding an excess of fine-grain fractions in the concrete mixture. This means that only mortar necessary to fill the gaps between coarse aggregate particles should be supplied. Because of the high compaction energy generated during spinning, the amount of mortar is considerably smaller than in the case of normal concrete mixtures. Based on this research, it was determined that the ratio of the mass of the sand in the mixture to the mass of the coarse aggregate should not exceed 25–30%. However, because of such a low fine-grained parts content in the mixture, the latter is difficult to mix and compact in the centrifuge.

Adesiyun, Kamiński, Kubiak and odo [5,6] studied the influence of such parameters as sand equivalent (25–50%), cement content (410–530 kg/m^3), plasticiser amount (0–2%), rotational speed (400–700 rpm) and spinning time (5–11 min) on the structure of spun concrete. They conducted tests on 230 mm high annular spun concrete samples with the inside diameter of 45–60 mm. The test results confirm the conclusions drawn in Ref. [3,4], concerning the difference in shrinkage between the inner and outer surface of the wall of the tested cross section. They used computer image analysis, which makes it possible to extract the information contained in the image of the cross section of the element wall, to describe the structure of spun concrete. To precisely describe the structure, the sample wall thickness was divided into twenty 2.25–3.0 mm wide strips (depending on the thickness of the cross section wall). They carried out tests for different combinations of the parameters mentioned above. Using computer image analysis, they graphically presented the distribution of aggregate, cement paste and air voids across the wall for the particular samples. They found that the aggregate content decreases from the spun concrete sample's outer zone to its outer zone, whereas the cement paste content changes in the opposite direction (in the inner zone, the cement paste content amounts to almost 100%). The air content in the wall cross section is higher for the inner layers than for the outer ones. Composition segregation occurs in all samples. The sample with the lowest sand equivalent value (25%), the lowest speed (400 rpm) and the shortest spinning time (5 min) has the least fractionated composition. The test result confirms the ones reported in [2–4].

The above investigations of spun concrete mainly focus on the macroscopic description and strength and deformation tests of the sample (usually the whole sample). It is only in [5,6] that modern methods based on computer image analysis are reported, whereby the peculiar structure of spun concrete across the cross section wall can be more precisely described. However, examples reported in the literature indicate that such methods can be used to determine spatial distributions of air void [7–9] aggregate [7,10] or fibres [11–13]. Furthermore, size and shape of the mentioned constituents of cementitious composites can be characterised [14]. Thus, desiring to probe even deeper into the structure of spun concrete, the present authors used innovative (as applied to this field) investigative methods, such as nanoindentation, micro-computed tomography (μCT) and 2D optical scanning. Thanks to the use of these methods, the following features of the inner concrete microstructure were successfully quantitatively determined: the distribution of concrete components, the variation of the mortar mechanical parameters across the cross section wall and the spatial distribution of pores.

2. Preparation of Samples for Tests

A sample of the concrete mixture used for the manufacture of spun concrete power poles was prepared for these investigations. The sample was made in a little mould with an inside diameter of 150 mm and a height of 300 mm (Figure 1), attached to the steel mould used for manufacturing spun concrete power poles in one of the precast concrete plant in Poland [15]. The steel mould together with the attached little mould was placed in a centrifuge (Figure 2) and subjected to spinning for 8 min at the maximum speed of 600 rpm. After concrete mixture spinning, the excess of evaporable water was removed from the little mould and the latter, together with the moulded sample, was transferred to a steam box. After about 4 h, concrete spinning steam was fed gradually (the temperature rising at a rate not higher than 10 °C/h) into the steam box. The temperature in the steam box did not exceed 60–70 °C and the concrete was cured for 8 h. Then, the supply of steam to the steam box was turned off to allow the moulds to naturally cool down to a temperature below 40 °C. Subsequently, the sample was extracted from the mould and was left in laboratory conditions for two weeks. An about 10 mm thick slice (Figure 3), in which the characteristic structure of spun concrete is visible, was cut out from the sample to be used in the tests.

Figure 1. Little mould attached to steel mould used for manufacturing spun concrete power poles.

Figure 2. Steel mould with attached little mould is being placed in centrifuge.

Figure 3. About 10 mm slice cut out of principal sample. Division of slice into smaller samples for tests proper is shown.

Prior to the tests proper, smaller samples to be tested in the particular devices were marked off on the slice (Figure 3). Samples A1–D2 (Figure 3) were tested in the nanoindenter, while the remaining part of the cross section was examined using a computer microtomograph and an optical scanner. The smaller samples were cut out using a high-speed diamond saw made by Struers Labotom-5 (Struers, Shanghai, China) (Figure 4). A series of preliminary tests was carried out on the prepared samples to determine: the indentation parameters (the force, the spacing of test points, and the necessary number of tests), the parameters of the scanning in the microtomograph (the radiation intensity, the exposure time and the filters used) and the optical scanning parameters (the way of preparing the surface and the method of imaging the tested surface). Based on the preliminary test results, a test plan was adopted. The aim of the tests was to determine the following three parameters of the tested cross section:

- the distribution of aggregate across the wall of the cross section;
- the variation of the mortar's mechanical parameters across the wall of the cross section; and
- the distribution of pores across the wall of the cross section.

Figure 4. The slice mounted in the chamber of precision saw.

The identified quantities were related to conventional coordinate R in the radial direction (perpendicular to the axis of rotation of the spun concrete element), where coordinate $R = 0$ applies to the peripheral edge of the element.

3. Analysis of Aggregate Distribution Across Wall

3.1. Preparation of Samples and Scanning

The investigations presented in this section consisted in analysing the images obtained from the optical scanning of the cross section of the spun concrete element. The investigative procedure included the following steps:

1. Cutting out samples
2. Preparing their surface for scanning
3. Scanning
4. Segmenting aggregate from obtained images
5. A morphometric analysis of the aggregate based on its binary image

First, samples were cut out of the tested element and their surfaces were levelled so that they could be scanned. The most effective method of preparing the surface of the samples was sought to obtain the best result of aggregate segmentation in the next step. Different ways of grinding, etching and dyeing (using various inks and dyes (including fluorescent ones)) the matrix were tried. Ultimately, the most effective of the ways was found to be grinding, etching with 10% hydrochloric acid solution and repeated dyeing with acrylic ink and grinding again until a flat surface was obtained owing to the filling of the etched cement matrix volume with the ink. Grinding took place in a Struers Labo-Pol-5 grinding (Struers, Shanghai, China) and polishing machine using an MD Piano disc (Struers, Shanghai, China).

Scanning with a resolution of 600 dpi (which in pixel size terms amounts to 42.33 mm/pix) was carried out in a Brother MFC L5750DW device (Brother, Bridgewater, NJ, USA). The scanning result for an exemplary sample is shown in Figure 5.

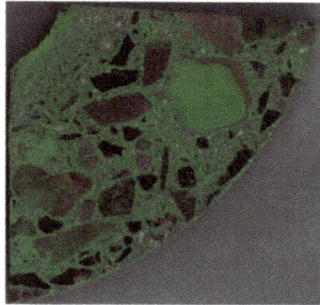

Figure 5. Scan of spun concrete element cross section after matrix dyeing (scale 1:1).

3.2. Segmentation of Aggregate

The next step consisted in aggregate segmentation. Without going into details (presented below), the aim of segmentation was to obtain a binary image in which white pixels indicate the area occupied by aggregate, while pixels in the remaining area (constituting the background) are black. The aggregate segmentation procedure proposed below is based on image processing methods. GIMP 2.10.4, ImageJ 1.52e (Fiji distribution) and CTAn 1.17.1.7 + were used for this purpose. Similar to in the case of the preparation of samples for scanning, the segmentation procedure was developed through many trials. The procedure deemed the best consisted of the following steps:

1. Gaussian blurring (smoothing) with a radius of 1 pixel (3 times)
2. The manual "retouching" of large aggregate fragments that were found to be susceptible to etching (Figure 6)
3. The segmentation of the "dark" aggregate
4. The segmentation of the "light" aggregate
5. The product of the images from Step 3 and Step 4
6. The removal of "pores"

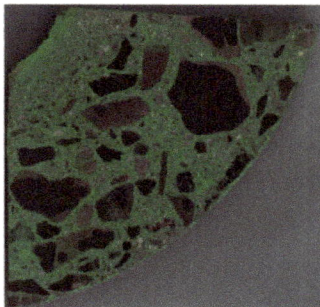

Figure 6. Image of sample after segmentation procedure Step 2.

To segment the "dark" (i.e., darker than the dye used in the selected RGB channel) aggregate (Step 3), first a mask was created from the green channel (G) of the image and by superimposing it on the white background a new image was created, this time in shades of grey (Figure 7). Then, thresholding was applied, assigning the value of 1 to the pixels with a brightness of 0–87/255 and the value of 0 to the other pixels (Figure 8).

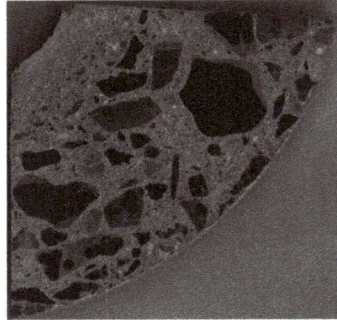

Figure 7. Mask created from green channel (G).

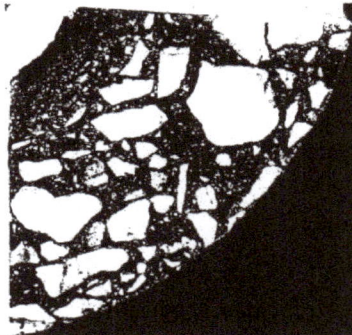

Figure 8. Image of sample after segmentation procedure Step 3.

The "light" aggregate (Step 4) was segmented similarly as in Step 2, this time superimposing red channel (R) and blue channel (B) masks (Figure 9). Then, thresholding according to the 20–255/255 range was applied. The result of this operation is shown in Figure 10.

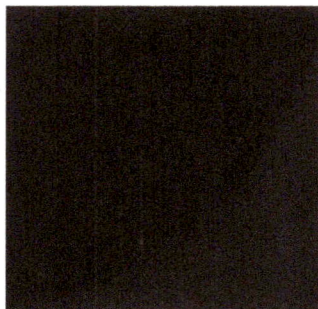

Figure 9. Superimposition of masks created from red (R) and blue (B) channels.

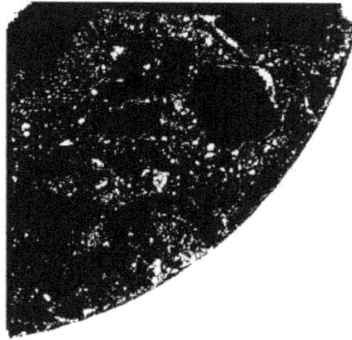

Figure 10. Image of sample after segmentation procedure Step 4.

The product of the images from Step 3 and Step 4 (Step 5) is shown in Figure 11. The removal of pores (i.e., black areas completely surrounded by the white area) (Step 6) and the use of a mask confining the area to the surface of the scanned sample are shown in Figure 12.

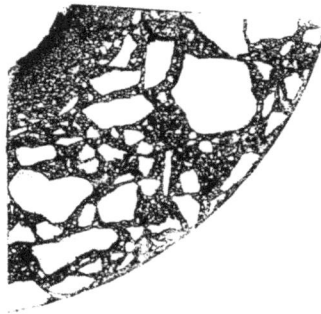

Figure 11. Step 5—product of images from Steps 3 and 4.

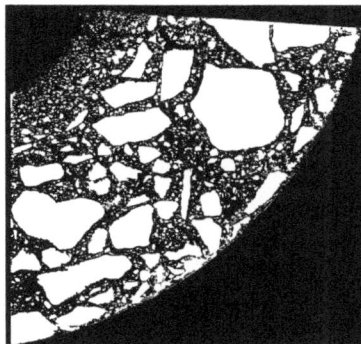

Figure 12. Step 6—the removal of "pores".

3.3. Morphometric Analysis

3.3.1. Basic Terms and Measures

Generally, the so-called morphologic microstructure measures are used to characterise the spatial distribution of the constituents of composite materials [16]. As part of this research, the aggregate content was determined and a procedure for calculating this quantity as a function of coordinate R was proposed. Moreover, the so-called local structure thickness was used to describe the fractionation of aggregate across the wall. The methodology presented below is based on the fact that a binary image can be understood as a discrete approximation of the indicator function (as defined in [17–19]) on a uniform grid of rows and columns of pixels. Then, the fractional content (of aggregate in this case) in a certain considered image area Ω can be defined as the sum of pixels in the constituent, divided by the sum of pixels in area Ω.

$$\phi = \frac{\sum\limits_{\Omega(i,j)} I^{(k)}(i,j)}{\sum\limits_{\Omega(i,j)} 1} \tag{1}$$

where $I^{(k)}$ is the binary indicator function of constituent k, i is the row number, j is the column number, and

$$\Omega(i,j) = \{(i,j) : \mathbf{x}(i,j) \in \Omega\} \tag{2}$$

stands for the number of pixels in area Ω.

To describe the variation in aggregate content as a function of element wall thickness R, a series of subareas, constituting a narrow circumferential band characterised by a fixed value of coordinate R in a range from $R - 1/2 \cdot \Delta R$ to $R + 1/2 \cdot \Delta R$, was selected (Figure 13).

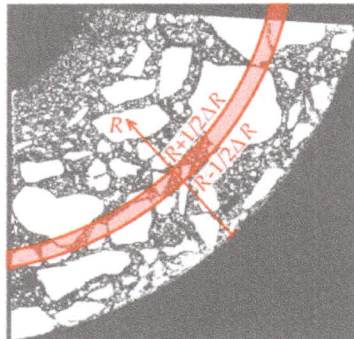

Figure 13. Way of selecting circumferential band.

Then, the value of fractional content $\phi(R)$ in the band whose centre line is described by coordinate R can be defined as:

$$\phi(R) = \frac{\sum\limits_{\Omega_R(i,j)} I^{(k)}(i,j)}{\sum\limits_{\Omega_R(i,j)} 1} \tag{3}$$

where

$$\Omega_R(i,j) = \left\{ (i,j) : R(i,j) \in \left[R - \frac{1}{2}\Delta R, R + \frac{1}{2}\Delta R \right) \right\} \tag{4}$$

runs through the set of pixels contained in the considered band. The linear dimension of a single pixel, i.e., $\Delta R = 42.3$ µm, was assumed as the basic value of ΔR in the analysis.

Local structure thickness is a scalar field defined in the area occupied by the considered constituent. In general, the procedure for determining local structure thickness consists in filling the area occupied by the considered constituent with circles with the possibly largest diameter. Then, the local structure thickness in given point x is defined as the largest diameter of the circle that is fully contained in the considered constituent and at the same time contains point x [20,21]. This is shown schematically in Figure 14.

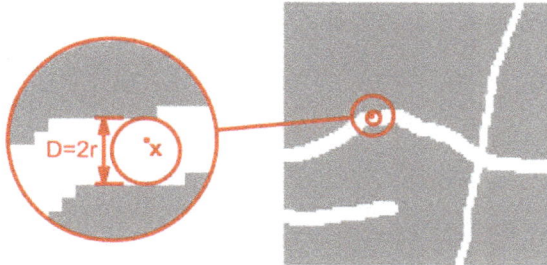

Figure 14. Element width in point x, as maximum diameter of inscribed circle [21].

Calculations are first performed for the whole image, whereby a map of the local structure thickness of considered constituent, $D^{(k)}(i,j)$, is obtained. A similar procedure as for the fractional content was used to characterise the variation in aggregate size across the concrete element wall (i.e., relative to coordinate R). This time the average size of the aggregate in circumferential band is defined as:

$$D(R) = \frac{\sum\limits_{\Omega_R(i,j)} D^{(k)}(i,j)}{\sum\limits_{\Omega_R(i,j)} I^{(k)}(i,j)} \tag{5}$$

i.e., as the weighted average of all the $D^{(k)}(i,j)$ values of the pixels found simultaneously in band and in the aggregate. It should be noted here that the decided advantage of this way of determining aggregate size (as local structure thickness) over other methods consisting, e.g., in counting pixels, is the fact that the result in this case does not depend on an accidental "merger" of a few aggregate particles into one geometric object in the analysed image.

3.3.2. Results

The authors' own procedure written in the Wolfram Language in Mathematica was used for the calculations. Selected morphological transformations were performed in the programs GIMP, ImageJ (Fiji distribution) and Bruker CTAn. A graph of the variation in aggregate content in the considered sample is shown in Figure 15.

A graph of mean aggregate size is shown in Figure 16, while a map of local structure thickness is shown in Figure 17b. Moreover, Figure 17c shows the sum of the images presenting the map of aggregate thickness, the applied ROI mask and the outermost circumferential bands taken into account in the analysis.

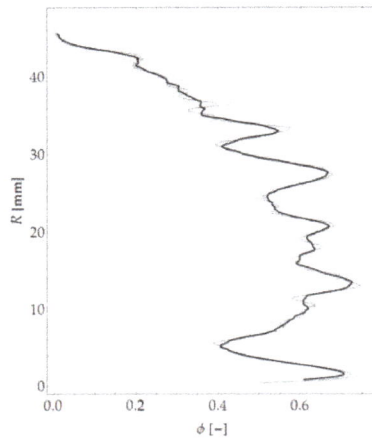

Figure 15. Aggregate content across element thickness.

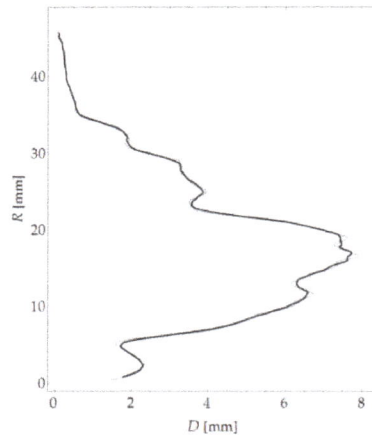

Figure 16. Mean aggregate size versus coordinate *R*.

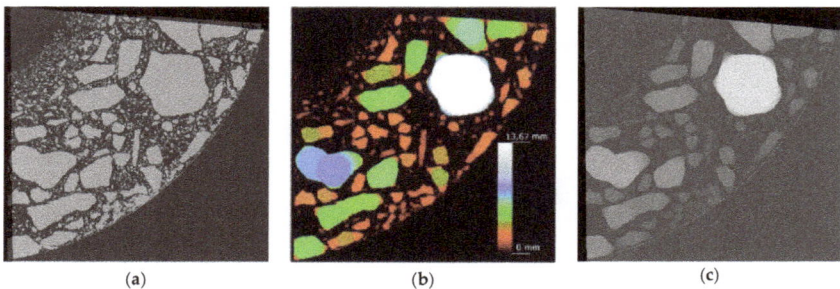

Figure 17. Results of applied morphological transformations: (**a**) image of sum of analysed binary image, applied mask defining ROI and outermost circumferential bands taken into account in analysis; (**b**) map of local aggregate size; and (**c**) image of sum of local aggregate size, applied map defining ROI and outermost circumferential bands taken into account in analysis.

4. Analysis of Variation of Mortar Mechanical Parameters Across Wall

The nanoindentation method [22] was used to determine the mechanical parameters of the mortar, understood within the particular analysis as cement paste together with fine aggregate, i.e. finer than sand particles. Hardness (HIT), indentation modulus (MIT) and surface aggregate content (ϕ) in the mortar were determined on the basis of the test results.

The concrete samples were specially prepared for tests in the nanoindenter. The preparation included: embedding concrete pieces in epoxy resin (Figure 18a) and levelling and polishing the surface tested (Figure 18b). The grinding and polishing procedure was individually fitted to each sample on the basis of the authors' experience in the testing of concrete samples.

(a)

(b)

Figure 18. Samples for testing in nanoindenter: (**a**) samples embedded in epoxy resin; and (**b**) samples with surface polished for testing.

The proper preparation of the samples was verified by evaluating the images obtained from the optical microscope (Figure 19) and carrying out a series of trial indentations. The absence of artefacts (cracks or streaks) in the images indicated that the surface had been properly prepared for testing [23,24].

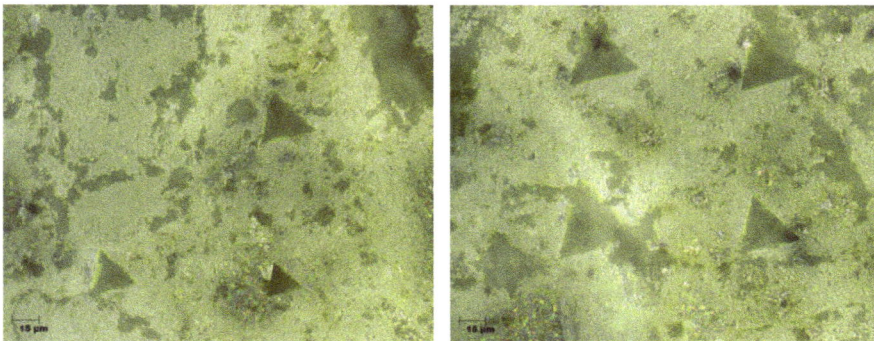

Figure 19. Exemplary photos of surface with trial indentations.

The indentation test was carried out in accordance with the standard procedure [25]. An indenter with known geometry and known mechanical parameters was pressed into the tested material and simultaneously the characteristic values of the applied load and the depth of penetration of the indenter tip were measured. In the course of the test, force F continuously grew, whereby penetration depth h continuously increased. Hence, the *F-h* dependence (Figure 20a) could be determined, based on which the maximum penetration depth (h_{max}) of the indenter tip and the range of elastic W_e and plastic W_p deformations of the material could also be determined (Figure 20b) [22,26,27].

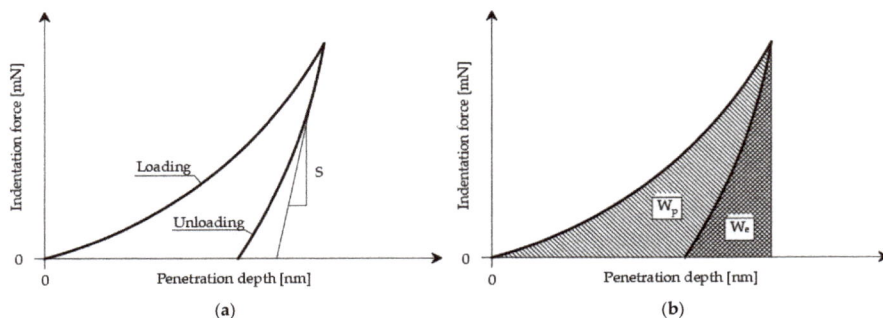

Figure 20. Graphical representation of standard indentation test parameters: (**a**) exemplary *F-h* dependence obtained during nanoindentation test; and (**b**) way of determining plastic behaviour (W_p) and elastic behaviour (W_e).

The basic parameters determined by the test are hardness (H_{IT}) and indentation modulus (M_{IT}). Hardness is defined as follows:

$$H_{IT} = \frac{F_{max}}{A} \tag{6}$$

where A is a projection of the indenter contact surface onto the surface of the sample. This quantity is usually determined as a function of maximum penetration depth h_{max} [28,29].

Indentation modulus M_{IT} is calculated using the Sneddon solution [30] describing the pressing of an axially symmetric rigid cone into an elastic half-space. Then, M_{IT} is defined as follows:

$$M_{IT} = \frac{1}{2} \frac{dF}{dh} \frac{\sqrt{\pi}}{\sqrt{A}} \tag{7}$$

Based on trials, a loading procedure comprising two forces (10 mN and 250 mN) was adopted. The pattern of indenter loading during each test is shown in Figure 21.

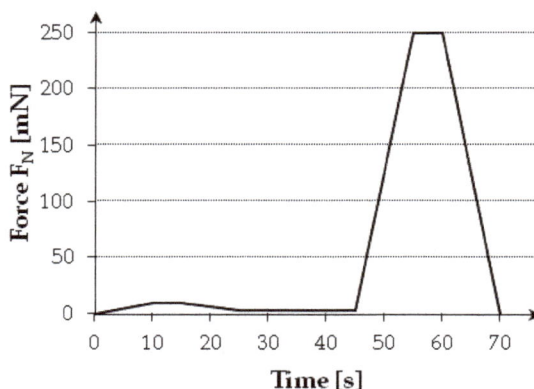

Figure 21. Loading function in each of the tests.

The grid indentation technique (GIT) was used to determine the parameters of the sample surface [31]. According to the assumptions of GIT, penetration depth h, interindentation spacing l

(mesh dimension) and number of carried out tests N should satisfy the following conditions (see, e.g., [32]):

$$3h < D < l(N)^{1/2} \tag{8}$$

$$d < \Omega < h \tag{9}$$

$$R_q < 3h \tag{10}$$

where D represents the characteristic dimension of nonuniformity on a given scale, d is the maximum dimension of inclusions in the tested constituent, Ω is the characteristic dimension of the so-called representative elementary volume [33,34] and R_q is the mean square deviation of surface roughness.

Using the CSM TTX-NHT nanoindenter with the Berkovich tip [35], a series of mortar hardness measurements on seven measurement grids (Figure 22) was carried out. One hundred indentations were made at every 50 μm on each of the grids.

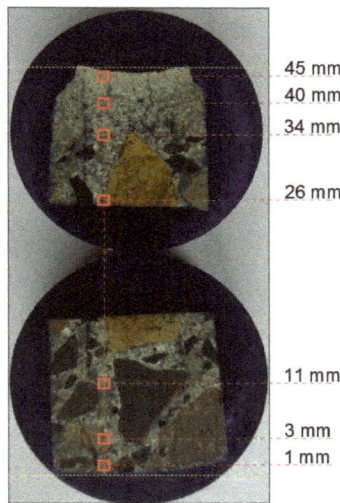

Figure 22. Arrangement of particular test grids—schematic representation.

The results of the nanoindentation tests were processed using a Mathematica script written by the authors. The mean value and the standard deviation of hardness HIT and indentation modulus MIT for the indentation force of 250 mN were determined for each of the grids. Because of the high inhomogeneity of the tested material, it was necessary to considerably increase the number of measurements to determine the parameters for the load level of 10 mN. The determined values are presented in Table 1 and in the diagrams of the variation of the mechanical parameters along radius R (Figure 23). The solid line marks the calculated mean value, while the broken lines correspond to the (upper and lower) boundaries of the values.

Table 1. Mean values and standard deviations of measured hardness H_{IT} and indentation modulus M_{IT}.

R	μM_{IT}	σM_{IT}	μH_{IT}	σH_{IT}
(mm)	(GPa)	(GPa)	(GPa)	(GPa)
1	46.26	23.86	4.49	5.00
3	45.52	25.83	4.29	5.07
11	66.72	30.00	7.87	5.67
26	50.76	27.12	5.75	5.28
34	52.09	23.75	5.01	4.84
40	40.04	21.53	3.26	4.26
45	23.66	12.31	1.00	1.53

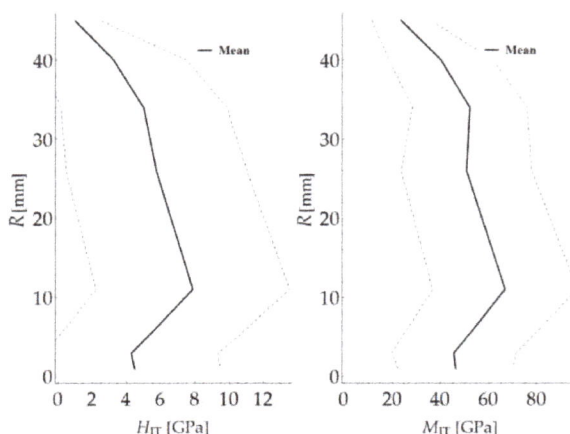

Figure 23. Variation in hardness HIT and indentation modulus MIT across wall of cross section.

In addition, for the above measurements, the segmentation of the mortar constituents was carried out assuming two material phases (cement paste and aggregate). The deconvolution technique was used for this purpose [31,36]. According to this method, each individual indentation is considered to be an independent random event, and its results (M_{IT} and H_{IT}) are considered to be random variables. The values of the cumulative distribution function for the measured M_{IT} and H_{IT} values (respectively, F_M and F_H) can be calculated as follows:

$$\left. \begin{array}{l} F_M\left(M_{IT(i)}\right) = \frac{i}{N} - \frac{1}{2N} \\ F_H\left(H_{IT(i)}\right) = \frac{i}{N} - \frac{1}{2N} \end{array} \right\} \text{ for } i \in [1, N] \tag{11}$$

Assuming that the distribution of the mechanical parameters of the particular constituents can be described using Gaussian distributions, expressed by mean values $\mu_j^{M_{IT}}$ and $\mu_j^{H_{IT}}$ and standard deviations $\sigma_j^{M_{IT}}$ and $\sigma_j^{H_{IT}}$, for, respectively, indentation modulus M_{IT} and hardness H_{IT}, the cumulative distribution function for each segmented constituent has the form:

$$F\left(X_{IT(i)}; \mu_j^{X_{IT}}, \sigma_j^{X_{IT}}\right) = \frac{1}{\sigma_j^{X_{IT}}} \frac{1}{\sqrt{2\pi}} \int\limits_{-\infty}^{X_{IT(i)}} \exp\left(\frac{-\left(u - \mu_j^{X_{IT}}\right)^2}{2\left(\sigma_j^{X_{IT}}\right)^2}\right) du; \ z \ X_{IT} = (M_{IT}, H_{IT}) \tag{12}$$

The sought values $\left\{ f_j, \mu_j^{M_{IT}}, \sigma_j^{M_{IT}}, \mu_j^{H_{IT}}, \sigma_j^{H_{IT}} \right\}, j = 1, n$ are determined by minimising the difference between the cumulative distribution function for the experimental results and the one assumed in the form of a Gaussian distribution, i.e.,

$$
\min \left[\sum_{i=1}^{N} \left(\sum_{j=1}^{n} f_j F\left(M_{IT(i)}; \mu_j^{M_{IT}}, \sigma_j^{M_{IT}} \right) - F_M\left(M_{IT(i)} \right) \right)^2 \right.
$$
$$
\left. + \sum_{i=1}^{N} \left(\sum_{j=1}^{n} f_j F\left(H_{IT(i)}; \mu_j^{H_{IT}}, \sigma_j^{H_{IT}} \right) - F_H\left(H_{IT(i)} \right) \right)^2 \right]
$$

(13)

Thanks to this approach, one can estimate the aggregate content in the mortar. An exemplary result of segmentation for one measurement grid is shown in Figures 24 and 25.

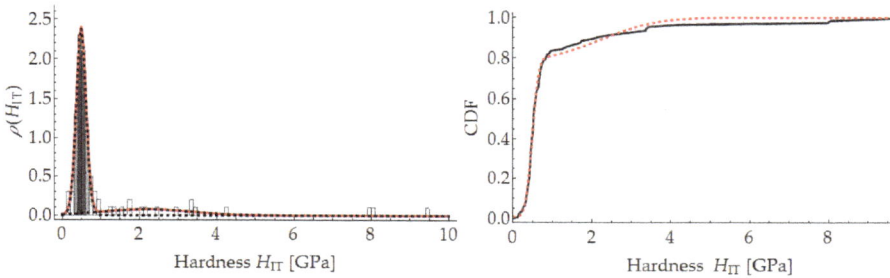

Figure 24. Segmentation of mortar components through deconvolution based on hardness histogram.

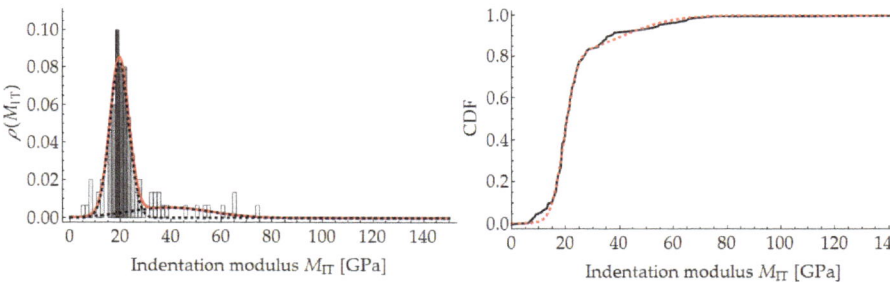

Figure 25. Segmentation of mortar constituents through deconvolution based on indentation modulus histogram.

The segmentation results for the whole sample are presented in Table 2 and in the diagrams of the variation of the mechanical parameters along radius R (Figure 26).

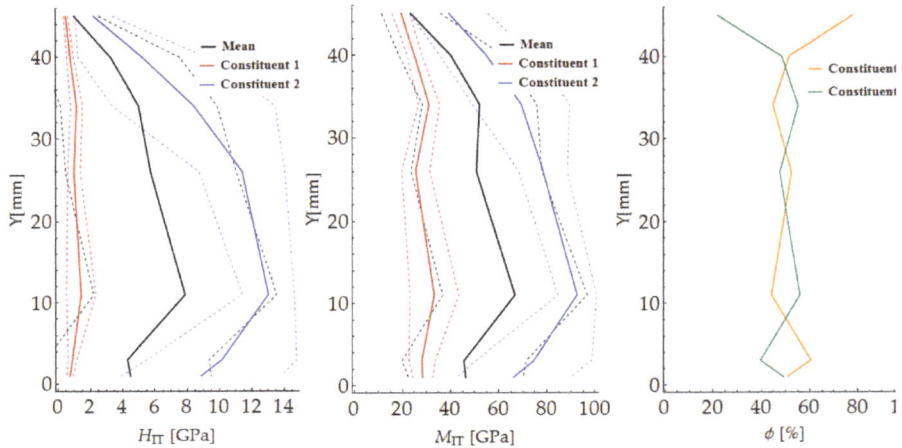

Figure 26. Variation in hardness H_{IT}, indentation modulus E_{IT} and constituent content across wall.

Table 2. Hardness H_{IT}, indentation modulus E_{IT} and fractional content ϕ of tested material constituents.

R	Constituent	μM_{IT}	σM_{IT}	μH_{IT}	σH_{IT}	ϕ
(mm)		(GPa)	(GPa)	(GPa)	(GPa)	(%)
1	1	28.42	4.16	0.74	0.21	50.90
	2	66.13	23.69	8.87	4.98	49.10
3	1	28.05	5.53	0.89	0.26	60.40
	2	74.51	24.12	10.16	4.64	39.60
11	1	33.23	10.17	1.43	0.91	44.10
	2	92.63	7.99	13.02	1.62	55.90
26	1	25.54	5.88	0.99	0.39	52.40
	2	78.44	10.14	11.41	2.66	47.60
34	1	31.12	4.38	1.16	0.36	44.80
	2	69.36	20.21	8.42	5.04	55.20
40	1	25.06	3.04	0.77	0.22	51.50
	2	54.80	19.31	5.23	3.97	48.50
45	1	19.52	3.76	0.49	0.13	78.00
	2	39.42	16.14	2.21	1.15	22.00

5. Analysis of Distribution of Pores Across Wall

The non-destructive technique of micro-computed tomography (μCT) was used to characterise pore space variation. The adopted approach was similar to those presented in [37–40]. The sample was trimmed to a rectangular prism whose dimensions ensured the analytical resolution of 10 μm/pix. Then, the rectangular prism was mounted on a base and placed in the microtomograph chamber (Figure 27).

Figure 27. Sample mounted on base and placed in microtomograph chamber.

Scanning was carried out using the Bruker Skyscan 1172 scanner. The scanning consists in subjecting the prepared sample to a series of exposures and reconstructing the material structure from the obtained projections (Figure 28).

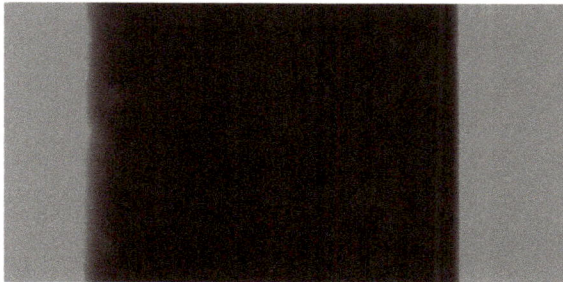

Figure 28. Exemplary projection of material structure.

The scanning parameter values shown in Table 3 were selected on the basis of trial scans and the earlier tests of the samples.

Table 3. Selected major scanning parameters.

Parameter	Value
Source Voltage (kV)	100
Source Current (μA)	100
Image Pixel Size (μm)	9.88
Filter	Al + Cu
Exposure (ms)	1100
Rotation Step (deg)	0.24
Frame Averaging	ON (8)
Random Movement	ON (10)
Use 360 Rotation	YES
Geometrical Correction	ON

The NRecon program based on the Feldkamp algorithm [41,42] was used for image reconstruction. The set of reconstruction parameters is shown in Table 4.

Table 4. Selected major reconstruction parameters.

Parameter	Value
Post-alignment	−1.00
Pixel Size (µm)	9.87947
Object Larger than FOV	OFF
Ring Artefact Correction	18
Beam Hardening Correction (%)	10
Threshold for defect pixel mask (%)	50
CS Static Rotation (deg)	5.00
Minimum for CS to Image Conversion	0.000
Maximum for CS to Image Conversion	0.040

The reconstructed structure of the tested sample is shown in Figure 29. To quantitatively and qualitatively evaluate the material structure, one should additionally determine the "volume" of interest (VOI) in the reconstructed model. A rectangular region of interest, marked red in the exemplary cross sections of the sample (Figure 30), was adopted.

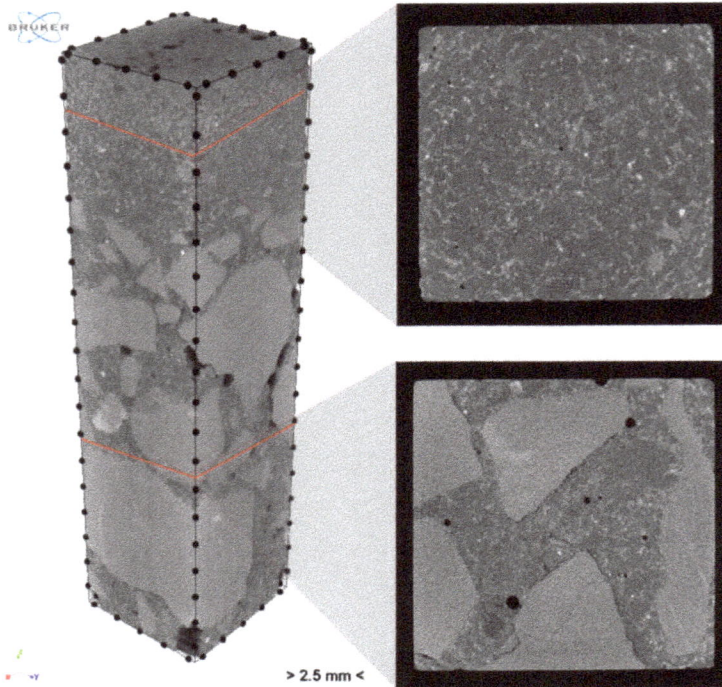

Figure 29. Reconstructed structure of sample with exemplary cross sections.

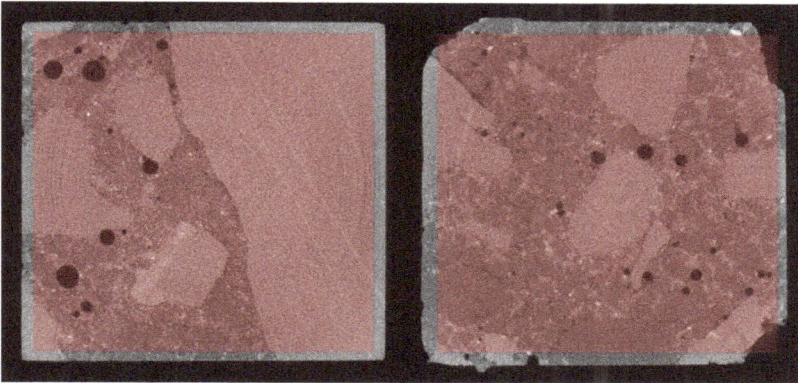

Figure 30. Region of interest (ROI) for selected cross sections.

The obtained results were analysed using the Bruker software (CTAn and CTVox) and a Mathematica script written by the authors.

To estimate the porosity of the tested material, it was necessary to segregate the pore space from the images. For this purpose, the images were binarised (using thresholding) for the particular cross sections. An exemplary result of the segmentation for selected cross sections is shown in Figure 31.

Selected VOIs

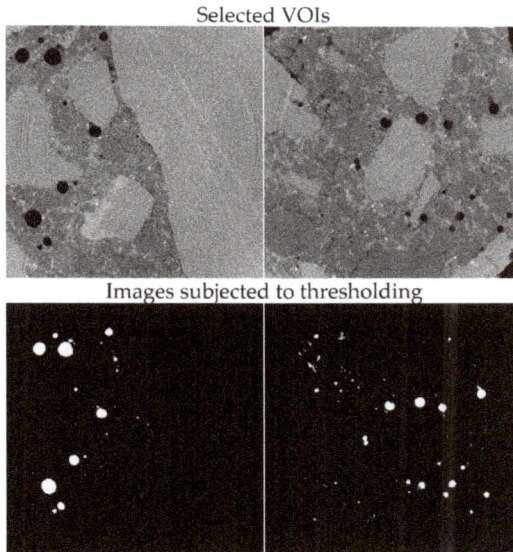

Images subjected to thresholding

Figure 31. Segmentation of pores for selected cross sections.

The isolated pore space could be presented in a 3D model and quantitatively analysed (Figure 32). The graph in Figure 32 represents porosity versus sample height. Figure 33 shows the spatial distribution and a graph of pore local thickness versus sample height. It should be noted that pore thickness is understood here similarly as described in Section 3.3. However, this time, the calculations were performed as three-dimensional and local thickness was defined as the maximum diameter of a sphere contained in the pore volume.

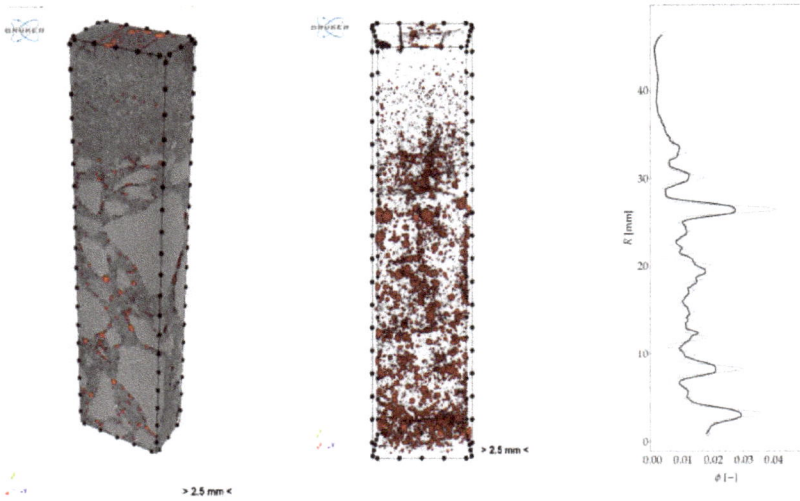

Figure 32. Cross section through 3D model of concrete sample with detailed pores, and pore space within VOI.

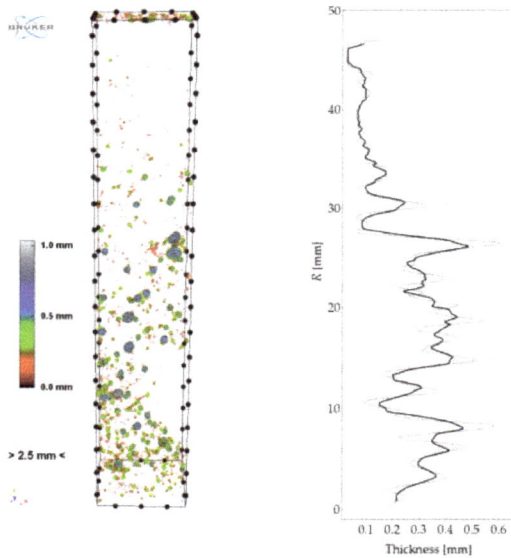

Figure 33. Spatial distribution of pore "thickness" and graph of mean thickness along sample height.

Considering the results obtained from 2D and 3D scanning and knowing the aggregate and pore content, the cement paste content was calculated. A graph of this quantity is shown in Figure 34. In this case, it was assumed that the analysed material consisted of three constituents, i.e., aggregate, mortar and pores.

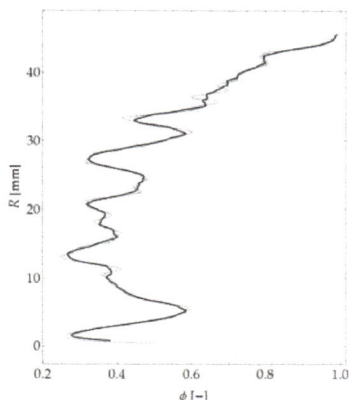

Figure 34. Graph of mortar content along sample height.

6. Discussion

A test methodology comprising three different testing methods, namely nanoindentation, micro-computed tomography (μCT) and two-dimensional optical scanning, is presented. The methodology was used to describe selected features of the inner microstructure of spun concrete, such as the distribution of its constituents, the variation of the mechanical parameters of the mortar across the cross-section wall, and the spatial distribution of pores. Tests were carried out on a sample moulded from the mixture used for the manufacture of spun concrete power poles in one of the precast concrete plants in Poland.

The variation of the aggregate content and the mean aggregate size across the wall of the cross section of a spun concrete sample was characterised using an aggregate segmentation procedure based on image processing methods, and a morphometric analysis employing mathematical morphology. The aggregate segmentation method is similar to the computer image analysis method used in [5,6]. The obtained results are satisfactory, providing not only a qualitative, but also quantitative description of the distribution of aggregate across the wall of a spun concrete cross section.

Thanks to the use of the nanoindentation method, it was possible to determine the mechanical parameters of the mortar (cement paste and fine aggregate) across the wall of the cross section of a spun concrete sample. The hardness (HIT), the indentation modulus (MIT) and the surface concrete content (ϕ) in the tested mortar were determined. Because of the size of the nanoindenter tip and the indentation forces used, testing was limited to places where there was no coarse aggregate in the sample concrete. This means that acquired information about the mechanical characteristics of spun concrete across the wall of the cross section is not full. However, in addition to the measurements of the mechanical parameters, the authors succeeded in segmenting the mortar constituents (cement paste and aggregate) using the deconvolution technique. Based on the obtained results, it was possible to trace the distribution of cement paste (and thus of cement) across the wall of the cross section. The findings reported by Marquardt [1] that cement is quite uniformly distributed across the wall and that only in the inner layer of the cross section its content increases were confirmed.

The variation in pore space was characterised using micro-computed tomography (μCT). Thanks to this method, it was possible to spatially (in 3D) describe the distribution of pores in the sample as well as their distribution across the wall of the cross section. Mainly closed air voids were found to occur. The results of the analyses are surprising and they do not corroborate the literature results (e.g., [5,6]), where it is suggested that the air content in the cross section of the wall is rather higher in the inner layers than in the outer layers.

The investigative methodology comprising nanoindentation, micro-computer tomography (μCT) and two-dimensional optical scanning seems to be useful for evaluating selected features of the internal microstructure of spun concrete. The methodology allows one to gain an insight into the interior of concrete and it yields qualitatively and quantitatively valuable results. The developed procedures for investigating the internal structure of spun concrete can be successfully used, e.g., to evaluate the process of manufacturing spun concrete elements, by verifying the porosity of the produced material, the fractionation of the aggregate, and the cement content, taking into account the variation of the parameters across the wall.

Taken together, the main contribution of the work to the field is the development of a procedure for the quantitative description of selected microstructure parameters. It should be emphasised that special attention was paid to the evaluation of the variability of these parameters in the cross section of the element. In the future perspective, the developed procedure will serve as a base for upcoming studies planned by the authors. For instance, the procedure allows for a comparative assessment of elements manufactured under different technological regimes. Thus, the methodology proposed in the article may prove to be a useful tool in the optimisation of manufacturing process concerning, i.e. selection of time and speed of mould spinning. Moreover, the procedure can be utilised in the analysis of the spun concrete microstructure changes along the traction pole, for which the diameter decreases towards its top.

Author Contributions: Conceptualisation, J.M.; methodology, J.M., M.P., and M.S.; software, M.P. and M.S.; validation, J.M., M.P., and M.S.; formal analysis, J.M.; investigation, J.M.; resources, J.M., M.P., and M.S.; data curation, M.P. and M.S.; writing—original draft preparation, J.M., M.P., and M.S.; writing—review and editing, J.M., M.P., and M.S.; visualisation, M.P. and M.S.; supervision, J.M.; project administration, J.M.; and funding acquisition, J.M. and M.S.

Funding: This research received no external funding.

Conflicts of Interest: The authors declare no conflict of interest.

References

1. Marquardt, E. Geschleuderte Beton-und Eisenbetonrohre. *Die Bautechnik* **1930**, *40*, 587–602.
2. Achverdov, I.N. Novoe v technologii železobetonnych centrifugirovannych rastrubnych trub. *Beton i železobeton* **1961**, *5*, 195–200.
3. Dilger, W.H.; Ghali, A.; Krishna Mohan Rao, S.V. Improving the Durability and Performance of Spun-Cast Concrete Poles. *PCI J.* **1996**, *41*, 68–90. [CrossRef]
4. Dilger, W.H.; Krishna Mohan Rao, S.V. High Performance Concrete Mixtures for Spun-Cast Concrete Poles. *PCI J.* **1997**, *42*, 82–96. [CrossRef]
5. Adesiyun, A.; Kamiński, M.; Kubiak, J.; odo, A. Laboratory test on the properties of spun concrete. In Proceedings of the Third Interuniversity Research Conference, Szklarska Poręba, Poland, 22–26 November 1994; pp. 3–8.
6. Adesiyun, A.; Kamiński, M.; Kubiak, J.; odo, A. *Investigation of Spun-Cast Concrete Structure, School of Young Research Methodology of Concrete Structures*; Dolnośląskie Wydawnictwo Edukacyjne: Wrocław, Poland, 1996; pp. 13–19.
7. Wong, R.C.K.; Chau, K.T. Estimation of air void and aggregate spatial distributions in concrete under uniaxial compression using computer tomography scanning. *Cem. Concr. Res.* **2005**, *35*, 1566–1576. [CrossRef]
8. Skarżyński, .; Tejchman, J. Experimental investigations of fracture process in concrete by means of X-ray micro-computed tomography. *Strain* **2016**, *52*, 26–45. [CrossRef]
9. Du Plessis, A.; Olawuyi, B.J.; Boshoff, W.P.; Le Roux, S.G. Simple and fast porosity analysis of concrete using X-ray computed tomography. *Mater. Struct.* **2016**, *49*, 553–562. [CrossRef]
10. Ostrowski, K.; Sadowski, .; Stefaniuk, D.; Wałach, D.; Gawenda, T.; Oleksik, K.; Usydus, I. The Effect of the Morphology of Coarse Aggregate on the Properties of Self-Compacting High-Performance Fibre-Reinforced Concrete. *Materials* **2018**, *11*, 1372. [CrossRef]
11. Schabowicz, K.; Jóźwiak-Niedźwiedzka, D.; Ranachowski, Z.; Kudela, S.; Dvorak, T. Microstructural characterization of cellulose fibres in reinforced cement boards. *Arch. Civ. Mech. Eng.* **2018**, *18*, 1068–1078. [CrossRef]

12. Ponikiewski, T.; Gołaszewski, J.; Rudzki, M.; Bugdol, M. Determination of steel fibres distribution in self-compacting concrete beams using X-ray computed tomography. *Arch. Civ. Mech. Eng.* **2015**, *15*, 558–568. [CrossRef]

13. Ponikiewski, T.; Katzer, J.; Bugdol, M.; Rudzki, M. Steel fibre spacing in self-compacting concrete precast walls by X-ray computed tomography. *Mater. Struct.* **2015**, *48*, 3863–3874. [CrossRef]

14. Garboczi, E.J. Three-dimensional mathematical analysis of particle shape using X-ray tomography and spherical harmonics: Application to aggregates used in concrete. *Cem. Concr. Res.* **2002**, *32*, 1621–1638. [CrossRef]

15. Kubiak, J.; odo, A.; Michałek, J. Produkcja wirowanych żerdzi elektroenergetycznych w formach nieotwieranych podłużnie. *Mater. Bud.* **2015**, *1*, 40–41. [CrossRef]

16. ydżba, D.; Rajczakowska, M.; Stefaniuk, D.; Kmita, A. Identification of the carbonation zone in concrete using X-ray microtomography. *Stud. Geotech. Mech.* **2014**, *36*, 47–54. [CrossRef]

17. Torquato, S. *Random Heterogeneous Materials: Microstructure and Macroscopic Properties*; Springer Science & Business Media: Berlin, Germany, 2013.

18. ydżba, D. Zastosowania metody asymptotycznej homogenizacji w mechanice gruntów i skał. *Prace Naukowe Instytutu Geotechniki i Hydrotechniki Politechniki Wrocławskiej. Monografie* **2002**, *74*, 274.

19. Różański, A. Sur la représentativité, la taille minimale du VER et les propriétés effectives de transport des matériaux composites aléatoires. Ph.D. Thesis, Lille 1 University, Villeneuve-d' Ascq, France.

20. Hildebrand, T.; Rüegsegger, P. A new method for the model-independent assessment of thickness in three-dimensional images. *J. Microsc.* **1997**, *185*, 67–75. [CrossRef]

21. Cała, M.; Cyran, K.; Kawa, M.; Kolano, M.; ydżba, D.; Pachnicz, M.; Rajczakowska, M.; Różański, A.; Sobótka, M.; Stefaniuk, D.; et al. Identification of Microstructural Properties of Shale by combined Use of X-ray Micro-CT and Nanoindentation Tests. *Proced. Eng.* **2017**, *191*, 735–743. [CrossRef]

22. Oliver, W.C.; Pharr, G.M. An Improved Technique for Determining Hardness and Elastic Modulus Using Load and Displacement Sensing Indentation Experiments. *J. Mater. Res.* **1992**, *7*, 1564–1583. [CrossRef]

23. Miller, M.; Bobko, C.; Vandamme, M.; Ulm, F.-J. Surface Roughness Criteria for Cement Paste Nanoindentation. *Cem. Concr. Res.* **2008**, *38*, 467–476. [CrossRef]

24. Struers-Ensuring Certainty. Available online: https://www.struers.com/en/Knowledge/Grinding-and-polishing# (accessed on 2 November 2019).

25. *ISO 14577-1:2015, Metallic Materials-Instrumented Indentation Test for Hardness and Materials Parameters-Part 1: Test Method ISO Central Secretariat*; rue de Varembé, 1211: Geneva, Switzerland, 2015.

26. Doerner, M.F.; Nix, W.D. A Method for Interpreting the Data from Depth-Sensing Indentation Instruments. *J. Mater. Res.* **1986**, *1*, 601–609. [CrossRef]

27. Fischer-Cripps, A.C. Illustrative Analysis of Load-Displacement Curves in Nanoindentation. *J. Mater. Res.* **2007**, *22*, 3075–3086. [CrossRef]

28. Fischer-Cripps, A.; Nicholson, D. Nanoindentation. Mechanical Engineering Series. *Appl. Mech. Rev.* **2004**, *57*, B12. [CrossRef]

29. Oliver, W.C.; Pharr, G.M. Measurement of Hardness and Elastic Modulus by Instrumented Indentation: Advances in Understanding and Refinements to Methodology. *J. Mater. Res.* **2004**, *19*, 3–20. [CrossRef]

30. Sneddon, I.N. Boussinesq's problem for a rigid cone. *Math. Proc. Camb. Philos. Soc.* **1948**, *44*, 492–507. [CrossRef]

31. Constantinides, G.; Ravi Chandran, K.S.; Ulm, F.-J.; Van Vliet, K.J. Grid Indentation Analysis of Composite Microstructure and Mechanics: Principles and Validation. *Mater. Sci. Eng. A* **2006**, *430*, 189–202. [CrossRef]

32. Bobko, C.; Ulm, F.-J. The Nano-Mechanical Morphology of Shale. *Mech. Mater.* **2008**, *40*, 318–337. [CrossRef]

33. Kanit, T.; Forest, S.; Galliet, I.; Mounoury, V.; Jeulin, D. Determination of the Size of the Representative Volume Element for Random Composites: Statistical and Numerical Approach. *Int. J. Solids Struct.* **2003**, *40*, 3647–3679. [CrossRef]

34. ydżba, D.; Różański, A. Microstructure Measures and the Minimum Size of a Representative Volume Element: 2D Numerical Study. *Acta Geophys.* **2014**, *62*, 1060–1086. [CrossRef]

35. Berkovich, E.S. Three Faceted Diamond Pyramid for Micro-Hardness Testing. *Ind. Diam. Rev.* **1951**, *11*, 129.

36. Sorelli, L.; Constantinides, G.; Ulm, F.-J.; Toutlemonde, F. The Nano-Mechanical Signature of Ultra High Performance Concrete by Statistical Nanoindentation Techniques. *Cem. Concr. Res.* **2008**, *38*, 1447–1456. [CrossRef]

37. Rajczakowska, M.; Stefaniuk, D.; ydżba, D. Microstructure Characterization by Means of X-Ray Micro-CT and Nanoindentation Measurements. *Stud. Geotech. Mech.* **2015**, *37*, 75–84. [CrossRef]
38. Al-Raoush, R.; Papadopoulos, A. Representative Elementary Volume Analysis of Porous Media Using X-Ray Computed Tomography. *Powder Technol.* **2010**, *200*, 69–77. [CrossRef]
39. Cnudde, V.; Cwirzen, A.; Masschaele, B.; Jacobs, P.J.S. Porosity and Microstructure Characterization of Building Stones and Concretes. *Eng. Geol.* **2009**, *103*, 76–83. [CrossRef]
40. Peyton, R.L.; Haeffner, B.A.; Anderson, S.H.; Gantzer, C.J. Applying X-ray CT to Measure Macropore Diameters in Undisturbed Soil Cores. *Geoderma* **1992**, *53*, 329–340. [CrossRef]
41. Feldkamp, L.A.; Davis, L.C.; Kress, J.W. Practical Cone-Beam Algorithm. *J. Opt. Soc. Am. A JOSAA* **1984**, *1*, 612–619. [CrossRef]
42. Rodet, T.; Noo, F.; Defrise, M. The Cone-Beam Algorithm of Feldkamp, Davis, and Kress Preserves Oblique Line Integrals. *Med. Phys.* **2004**, *31*, 1972–1975. [CrossRef]

materials

MDPI

Article

Assessment of the Condition of Wharf Timber Sheet Wall Material by Means of Selected Non-Destructive Methods

Tomasz Nowak[iD]**, Anna Karolak *[iD], Maciej Sobótka**[iD] **and Marek Wyjadłowski**[iD]

Wroclaw University of Science and Technology, Wybrzeze Wyspianskiego 27, 50-370 Wroclaw, Poland;
tomasz.nowak@pwr.edu.pl (T.N.); maciej.sobotka@pwr.edu.pl (M.S.); marek.wyjadlowski@pwr.edu.pl (M.W.)
* Correspondence: anna.karolak@pwr.edu.pl; Tel.: +48-71-320-38-79

Received: 26 March 2019; Accepted: 7 May 2019; Published: 10 May 2019

check for
updates

Abstract: This paper presents an assessment of the condition of wood coming from a wharf timber sheet wall after 70 years of service in a (sea) water environment. Samples taken from the structure's different zones, i.e., the zone impacted by waves and characterised by variable water-air conditions, the zone immersed in water and the zone embedded in the ground, were subjected to non-destructive or semi-destructive tests. Also, the basic parameters of the material, such as its density and moisture content, were determined. Moreover, the ultrasonic, stress wave and drilling resistance methods were used. Then, an X-ray microtomographic analysis was carried out. The results provided information about the structure of the material on the micro and macroscale, and the condition of the material was assessed on their basis. Also, correlations between the particular parameters were determined. Moreover, the methods themselves were evaluated with regard to their usefulness for the in situ testing of timber and to estimate, on this basis, the mechanical parameters needed for the static load analysis of the whole structure.

Keywords: timber structures; non-destructive methods; ultrasonic wave; stress wave; drilling resistance; X-ray micro-computed tomography

1. Introduction

Timber as a universal building material has been used in many kinds of structures in, e.g., geotechnical engineering. In this field, one of the principal uses of structural timber is as timber pile foundations in low bearing capacity building lands situated in river deltas and beds and on peat bogs, which often were important locations because of their strategic position. As a foundation method, timber piles have been used almost everywhere in Europe in, i.a., the Scandinavian countries, the countries of Eastern and Middle Europe, such as Poland, the Baltic countries and Russia, as well as Germany, Great Britain and the Netherlands, where the foundation of historical buildings is most fully documented. Also, in the south of Europe, for instance in Venice, deep foundation timber piles may be found in nearly all the historical buildings erected since the 12th century [1].

Wood is a heterogeneous, hygroscopic, cellular and anisotropic material. Its mechanical properties depend on many factors, such as density and water content, which means that the creation of a constitutive model of wood poses a great challenge [2].

Wood biodegrades over time. Under the impact of external factors, wooden members undergo chemical and physical changes. Wood can be regarded as a durable material when it is completely immersed in water, and so protected against decomposition caused by aerobic fungi.

In nature, five basic chemical processes occur which reconvert a wooden material into carbon dioxide and water: oxidation, hydrolysis, dehydration, reduction and free radical reactions [3].

Table 1 shows major degradation pathways and the chemistries involved in the pathways [3].

Table 1. Major degradation pathways and chemistries.

Biological Degradation	Fungi, bacteria, insects, termites, enzymatic reactions, oxidation, hydrolysis, reduction, free radical reactions
Chemical Reactions	Oxidation, hydrolysis, reduction, free radical reactions
Mechanical Degradation	Chewing dust, wind, hail, snow, sand, stress, crack, fracture, abrasion, compression
Thermal Degradation	Lightning, fire, sun
Water Degradation	Rain, sea, ice, acid rain, dew
Water Interactions	Swelling, shrinking, freezing, cracking, erosion
Weather Degradation	Ultraviolet radiation, water, heat

All the above factors can be significant when testing wood samples taken from a retaining structure since the latter is exposed to the air, water and soil environment. Therefore, wood analyses should be based on more than one research technique to gain a deeper insight into the change of wood parameters over time in different environmental conditions [3].

In the considered case, since the structure was in service in the water environment and in saturated soils, the hazards can be divided into the following three main groups:

- a low or variable water level,
- excessive loading,
- decomposition of wood in the water environment.

If timber members are above the groundwater table, access to oxygen makes the activation of wood decomposing fungi possible. The decomposition rate is determined by the time during which the timber members are above groundwaters and the member's length situated above the groundwater table [4]. It is estimated that the maximum rate of decomposition caused by fungi attacking water-saturated wood amounts to approximately 10 mm/year [1]. However, the long-lasting service of timber structures below the groundwater table does not prevent decomposition [5], and examinations have shown some of the pile foundations in Venice to be in extremely bad condition.

This paper presents research methods which enable one to determine the condition and some parameters of a material which has been in service in the water-soil environment. For the best results, wood should be tested on different levels of detail, i.e., macro, submacro and microlevels. On the macrolevel, acoustic methods, drilling resistance method or laboratory tests of basic material parameters, such as density and moisture content, are used. On the microlevel the cell wall of the wood is tested and different elements, such as hardwood, sapwood and annual rings, are identified [6], using, e.g., X-ray micro-computed tomography.

The considered timber sheet wall was made of tongue-and-groove jointed timber piles. The history of this structure is not well known because of its previous military use. The timber sheet wall had been in service for about 70 years. The wharf is in the Swina straight connecting the Szczecin Lagoon with the Baltic Sea. In the Swina strait, fresh water (fully or partially) mixes with seawater due to the stratification. The salinity of the Swina strait ranges from 1‰ to 8‰. The average salinity of the Baltic Sea amounts to about 7‰, generally ranging from 2 to 12‰. It can be assumed that the water environmental conditions correspond to low salinity seawater.

After the timber sheet wall had been dug out and dismantled, its members were closely examined with regard to their original and current cross-sectional dimensions and to the quality of the wood. In the photograph (Figure 1a,b) of the dismantled members of the wall, one can see pile surfaces which were in service in diverse environmental conditions: completely embedded in the ground, stayed in water and stayed in the variable water-air environment. One can notice that the timber embedded in the ground, under the groundwater table, has preserved its constant volume. Bacteria destroy the cellulose very slowly, while the lignin remains constant, and water replaces the large cellulose molecules. The original waterfront layout has been reconstructed, see Figure 1c.

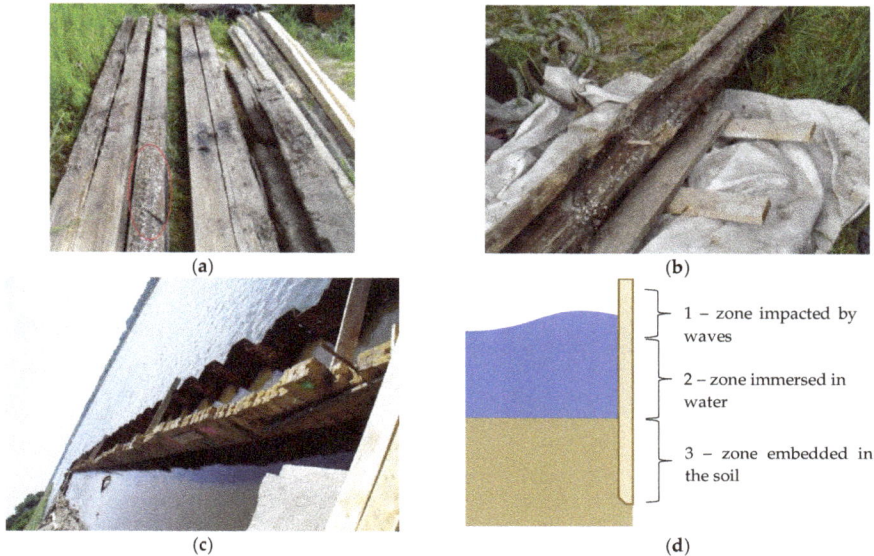

Figure 1. (**a,b**) View of wharf timber sheet pile wall from which samples were taken for testing (the zone impacted by waves is marked with red ellipse); (**c**) view of rebuilt wharf timber sheet pile wall; (**d**) scheme of the global structure with the location of the specimens.

The main objective of the work is to develop a methodology for testing of wooden structural members using non-destructive techniques. This is aimed at obtaining information which is necessary to assess the technical condition of the material in wooden members and to conduct a global structure analysis. In particular, to carry out such analysis, it is required to estimate the values of mechanical parameters and to assess possible zones of destruction. The detailed aim is to compare the quality of wood subjected to various environmental conditions (Figure 1d).

2. Selected Non-destructive Methods for Wood Assessment

2.1. Brief Survey of Methods

Material tests for wood can be divided into three groups: destructive tests (DT), semi-destructive tests (SDT) [7] and non-destructive/quasi-non-destructive tests (NDT) [8]. Unlike destructive tests, the tests belonging to the latter two groups do not affect or only slightly affect the properties of the tested sample, whereby the parameters of a wooden member can be determined with no detriment to its value. Their undeniable advantage is also the mobility of the testing equipment, whereby tests can be carried out in situ when it is not possible to take samples for laboratory tests (as in the case of heritage assets). Among the non-destructive and semi-destructive methods one can distinguish global testing methods (e.g., ultrasonic and stress wave techniques) and local testing methods (e.g., the drilling resistance method).

In order to acquire detailed data on the values of the physical and mechanical parameters of wood both non-destructive and destructive methods should be used. If the results yielded by the two testing methods are found to correlate, the data acquired in this way are fully sufficient for further static load analyses of the structural members or the whole building structure. Nevertheless, even using only non-destructive methods (as in the case of, e.g., heritage assets), one can obtain some information about the properties of the tested member's material, assess the technical condition of the structure or acquire some data helpful in evaluating this condition or in the design of possible repairs or upgrades. Thanks to the use of non-destructive methods one can also detect internal damage or flaws in the wood [9].

Among the non-destructive and quasi-non-destructive testing methods used to assess and diagnose timber structures, the most common are the ones presented in Table 2, and also described in detail in, i.a., [7,8,10,11]. The non-destructive and quasi-non-destructive methods can be divided into two groups: global testing methods (e.g., visual evaluation and ultrasonic and stress wave techniques) and local testing methods (e.g., the drill resistance method, the core drilling method and the hardness test method).

Table 2. Selected methods available for assessing timber in building structure.

Organoleptic Methods	Acoustic Methods	Quasi-Non-destructive (Semi-Destructive) Methods	Radiographic Methods	Other Methods
Visual evaluation	Stress waves	Drilling resistance	X-rays	Computed tomography
Acoustic evaluation	Ultrasonic technique	Core drilling	Gamma rays	Ground penetrating radar
Fragrance evaluation	Acoustic emission	Screw withdrawal		Near infrared
		Hardness tests		spectrometry
		Needle penetration		
		Pin pushing		
		Tension microspecimens		

2.2. Acoustic Methods

2.2.1. Idea of Acoustic Test

Using acoustic testing methods, such as the ultrasonic and stress wave techniques, one can evaluate the properties of wood by analysing the velocity of wave propagation in the tested material. The methods can be used to estimate selected mechanical properties (e.g., the modulus of elasticity) of a material and to detect its internal structural discontinuities.

The basic parameter used in the acoustic methods is sound wave propagation velocity (V), defined as follows:

$$V = L/T, \qquad (1)$$

where L is the distance (between two measuring points) covered by the excited sound wave, and T is the time needed to cover this distance.

Knowing the velocity of wave propagation and the wood density (ρ), one can determine the dynamic modulus of elasticity (MOE_{dyn}), which can be interrelated with the static modulus of elasticity (MOE_{stat}) [10]. The dynamic modulus of elasticity is calculated from the formula:

$$MOE_{dyn} = V^2 \times \rho, \qquad (2)$$

where V is the velocity of sound wave propagation, and ρ is the density of the tested element.

The velocity of sound wave propagation largely depends on the structure of the material. In the case of wood, it depends on the grain direction being several times higher (usually 3–5 times higher) along than across the grain [10,12].

According to [12], for wood with no significant structural flaws the velocity of sound wave propagation amounts to 3500–5000 m/s along the grain and to 1000–1500 m/s across the grain. Other values than the above ones may indicate internal discontinuities in the material structure. The lower values of the velocity across the grain are due to the internal structure of this material (on its way the wave encounters more cell walls, whereby the time in which it covers the distance increases, whereas in the longitudinal direction there are fewer barriers or they do not occur, whereby the velocity is higher).

2.2.2. Description of Testing Methods and Devices

Several kinds of devices are used for testing by means of ultrasonic or stress wave methods. In this case study, two of them were used and the test results are presented in Section 3.

The Fakopp Microsecond Timer (Figure 2a) uses the stress wave technique. The test consists of exciting a stress wave with a single strike of a special hammer. The device probes are driven directly into the tested sample. There is no need to drill holes as in the case of other devices (e.g., Sylvatest Trio). The device measures the time of wave propagation between the two probes (the receiving probe and the transmitting probe).

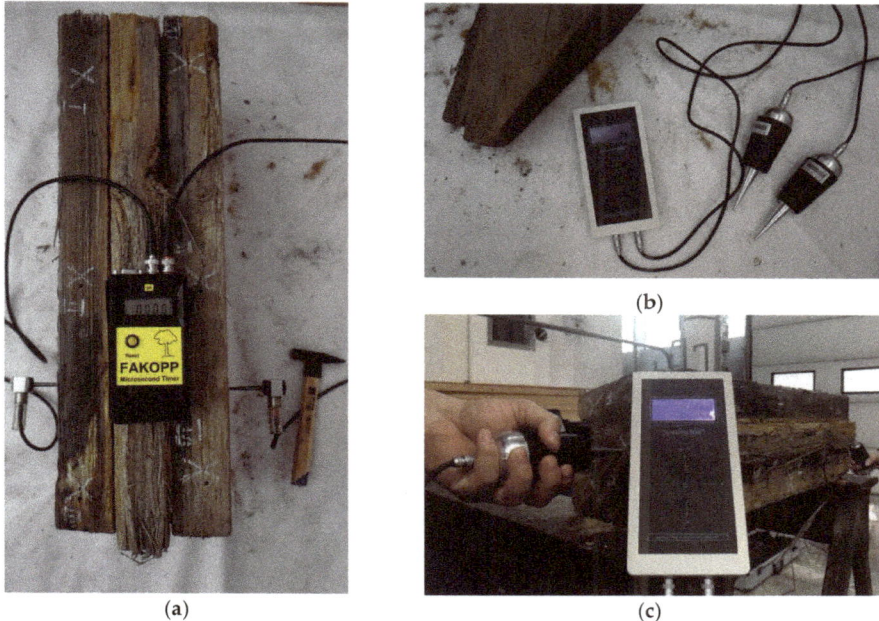

Figure 2. Devices for testing by acoustic method: (**a**) Fakopp Microsecond Timer using stress wave; (**b**) Sylvatest Trio device using ultrasonic wave; (**c**) Sylvatest Trio device during test.

In the case of the ultrasonic technique, the measurement can be performed in two ways: directly and indirectly. The first way consists of transmitting a signal from the transmitting probe to the receiving probe, with the probes placed on the opposite sides of the tested sample. As regard the second way, there is no need to place probes on the opposite sides of the sample because the signal is registered as reflected (the echo method). Owing to this, the range of the applicability of this test widens since only a unilateral access is required (which is useful when testing, e.g., historical monuments in situ).

Sylvatest Trio (Figure 2b), manufactured by the firm CBS-CBT, is another device which one can use to non-destructively evaluate the properties of wood [13]. The device measures the time needed for an ultrasonic wave to pass between transmitting-receiving probes placed against the tested sample, and the energy of this wave. In order to carry out the test the tips of the probes should be inserted into previously drilled holes each 5 mm in diameter and 10 mm deep. One should bear in mind that because of the high sensitivity of the device other mechanical waves excited near the test site can affect the test results. Also, the material's moisture content and internal stresses can significantly influence the results.

In order to obtain exhaustive results, it is recommended, for both methods, to perform a large number of measurements in different points and directions.

2.2.3. Correlation between Physical and Mechanical Properties of Wood and Results Yielded by Acoustic Methods

In many studies (e.g., [14–20]) based on acoustic methods attempts were made to assess the effectiveness of the methods and to find a correlation between the physical and mechanical properties of wood and the parameter values obtained from measurements.

According to the above studies, there is a strong correlation between the dynamic modulus of elasticity (MOE_{dyn}) and the static modulus of elasticity (MOE_{stat}). According to [18], for sound wood free of flaws the determination coefficient for the static and dynamic modulus of elasticity amounts to 0.96.

Also, comparative analyses of the effectiveness of the Fakopp Microsecond Timer (Fakopp Enterprise Bt., Agfalva, Hungary) and the Sylvatest Trio device (-CBS-CBT, Choisy-le-Roi, France) were carried out. They showed the two devices to be highly effective [18,19] and confirmed the correlation between the value of MOE_{dyn} and that of MOE_{stat} [20].

As part of other investigations, the decrease in the value of the velocity of the ultrasonic wave and the stress wave was analysed. According to [12], a reduction in the velocity by about 30% can correspond to a 50% fall in the load bearing capacity, while a reduction in the level of velocity by more than 50% can indicate considerable damage and the loss of load bearing capacity by the tested element. According to the results of the above research the relative decrease in the value of the velocity of wave propagation between two measuring points (ΔV_{rel}) describes the degree of damage to the material. The value of ΔV_{rel} is defined by Equation (3):

$$\Delta V_{rel} = [(V_{ref} - V_{mes})/V_{ref}] \cdot 100\%, \tag{3}$$

where ΔV_{rel} is the relative decrease in velocity, V_{ref} is the reference velocity (value of the velocity for a sound wood, taken from tests or literature), and V_{mes} is the measured velocity.

The relation between the relative decrease in velocity and the degree of damage is shown in Table 3 [21].

Table 3. Relation between relative velocity decrease and degree of damage.

Relative Velocity Decrease [%]	Degree of Damage [%]
0–10	no destruction
10–20	10
20–30	20
30–40	30
40–50	40
≥50	≥50

2.3. Drilling Resistance Method

2.3.1. Description of Test

One of the semi-destructive (SDT) methods is the drilling resistance method.

After the test a small borehole, below 3.0 mm in diameter, (not larger than the exit hole of most of the woodworm) remains in the sample material, but with no detriment to the properties of the element, whereby the test can be regarded as semi-destructive [7].

The test consists of measuring the energy needed to drill the resistance drilling device's metal needle into the material. The test makes it possible to detect structural discontinuities, damage, knots and other flaws and also to estimate the density and strength of the material [22]. The device measures the drilling resistance of a drill with a diameter of 1.5–3.0 mm and a length of 300–500 mm, rotating at a constant speed of about 1500 rpm (Figure 3).

Figure 3. IML RESI PD-400S device used in tests.

Drilling resistance is closely connected with the difference in density between the zones of early and late wood [22], the structure of the annual rings [23,24], changes in wood density caused by, i.a., biological decomposition, and the drilling angle [25]. The device registers the measurement results at every 0.1 mm, in the form of drilling resistance-depth graphs. The peaks in the graph correspond to the high resistance and high density of the material while the declines represent its low resistance and low density. The flatline in the diagram indicates places where the material does not show any drilling resistance, which means that the material is completely decomposed. During drilling the measurement in the entry and exit zones is disturbed because of the time needed for the drill to assume the proper position and rotational speed. Consequently, the graph in these zones usually has the form of a smoothly rising or declining curve.

Using the device one can detect structural flaws and discontinuities in timber elements without adversely affecting their useful properties (see, e.g., [14,26,27]).

2.3.2. Correlation between Physical and Mechanical Properties of Wood and Drilling Resistance Results

Attempts have been made to correlate drilling resistance results with strength test results in order to estimate the mechanical parameters of wood in the structure (e.g., [27–33]). Diagrams of relative resistance (RA) versus drilling depth (H) make it possible to evaluate the parameters of wood through the correlation between the average value of the resistance measure (RM) parameter and the density, strength and the modulus of elasticity of the wood. The value of RM can be calculated from formula 4 [25]:

$$RM = \frac{\int_0^H RA \cdot dh}{H},$$
(4)

where $\int_0^H RA \cdot dh$ is the area under the drilling resistance graph, and H is the drilling depth.

Attempts have also been made to correlate the resistance measure with different material parameters (density, longitudinal modulus of elasticity, transverse modulus of elasticity, longitudinal compressive strength, transverse compressive strength and bending strength) for different wood species, new wood and old wood. The results of some of the endeavours presented high determination coefficients amounting to 0.78 for the transverse compressive strength and to 0.67 for the modulus of elasticity [16] as well as to 0.64 for the modulus of elasticity and the longitudinal compressive strength [25]. As regards density, the determination coefficients of 0.71 [25], 0.75 for Pine, 0.74 for Spruce, 0.65 for Fir [28], 0.70 [29], 0.80 [30] or even 0.88 [31] were obtained. However, some researchers [32,34,35] did not obtain such a good correlation.

In general, tests performed on non-decayed, defect free, small sized laboratory specimens provide high values of correlation coefficients. On the other hand, results of the onsite tests of full-sized elements must be analysed with greater caution due to the possible presence of defects. In this context the drilling resistance method should be perceived to be a qualitative method.

The RM parameter value is influenced by many factors, such as the tree species, the condition of the wood and its moisture content and the drilling direction [32]. The results should be treated as not a quantitative, but qualitative assessment and the resistance drilling method test can be a complement to other tests or the starting point for a preliminary inspection of timber members or the location of damage inside the cross section.

2.4. X-Ray Micro-Computed Tomography

X-ray micro-computed tomography (Skyscan 1172, Bruker, Kontich, Belgium) is a state-of-the-art non-destructive technique for visualizing the inner structure of the tested object [36,37]. In essence, the tests consists of mathematically reconstructing the three-dimensional microstructure of the tested material on the basis of a series of high-resolution X-ray pictures. The scanning consists of recording a series of projections taken during the slow rotation of the sample placed on the scanner's rotary fixture [38]. The Bruker SkyScan 1172 microtomograph used in the tests and a view of a sample placed in the scanning chamber are shown in Figure 4 below.

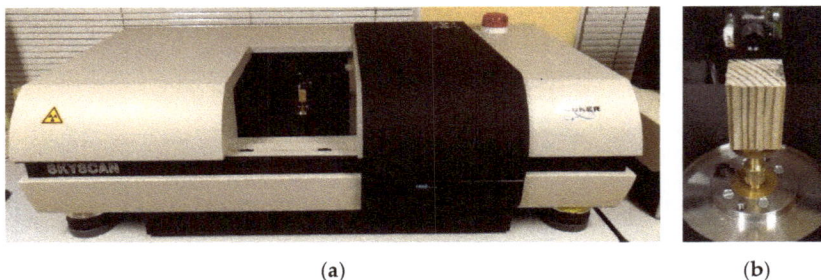

(a)   **(b)**

Figure 4. (a) Bruker SkyScan 1172 device; (b) sample of timber mounted on stage inside scanning chamber.

A single projection taken at a set sample rotation angle shows (in greyscale) the distribution (registered by the detector) of the intensity of the X-ray radiation emitted by the source and attenuated by passing through the sample. After a series of projections is recorded, the mathematical reconstruction of the tested material is carried out. The fact that according to the Lambert-Beer law, radiation absorption depends on the material's attenuation coefficient and on the thickness of the layer which the radiation must penetrate, is exploited for this purpose. The most commonly used algorithms are based on back projection, e.g., the Feldkamp algorithm [39] used in the present study. The result of the reconstruction is a series of images representing the cross sections of the examined object. The images show (in greyscale) the distributions of the attenuation coefficient, i.e., a characteristic of the material. The cross sections, arranged one above the other in space, make up a three-dimensional image of the internal microstructure of the examined object. Using image analysis techniques, quantitative and qualitative analyses of the material's microstructure can be carried out on the basis of the reconstructed images [40,41].

Since the absorption coefficient depends mainly on density, the greyscale of the resulting image can be treated as a monotonic density function. Consequently, the correlation between the level of brightness of the pixels (voxels) and the density of the tested material can be determined. For example, in [41,42] the linear correlation between density and the brightness level of the resultant tomographic image was used to determine the density of wood.

When testing heterogenous materials (composites), a morphometric analysis is carried out [38,43]. Its aim is to quantitatively characterise the morphology of a given component of the composite, particularly by determining the shape parameters and form of the geometrical objects constituting the area (in space) occupied by the considered component. Such an analysis is carried out on binary images obtained from segmentation.

3. Materials and Methods

In this case study, non-destructive and semi-destructive testing methods, i.e. the ultrasonic and stress wave techniques and the drilling resistance technique, were used to estimate the parameters of samples taken from the structure. Also c.a. 20 mm × 20 mm × 30 mm samples were cut out from the same structural members and subjected to scanning in the microtomograph in order to augment the non-destructive test results. The direct result of the scanning is a 3D image of the microstructure of the tested material. The level of brightness in the images is approximately proportional to the local density of the tested material. Owing to this a semi-quantitative comparison of wood density for the different zones of the timber sheet wall could be made.

Using the testing methods and devices described in Section 2, a series of tests were carried out on samples taken from the wharf timber sheet wall. The samples dimensions were about 18 cm × 20 cm cross-section and 60 cm length (Figure 5). The samples come from different locations in the wharf structure, which means that in the course of their service they were submerged to different levels and exposed to the variable impact of water. The material of the samples is pinewood (*Pinus sylvestris L*). The tests were carried out in a laboratory at Wroclaw University of Science and Technology.

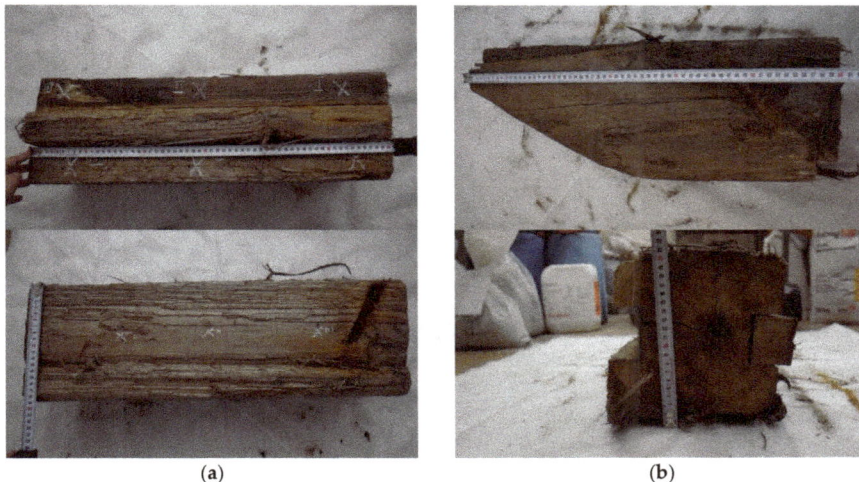

(a) (b)

Figure 5. Samples and their dimensions: (a) sample 1 cross section 18 cm × 20 cm and length 60 cm, (b) sample 3 cross-section 18 cm × 20 cm and length 60 cm with cut.

The acoustic tests and the stress wave tests were carried out using respectively the Sylvatest Trio device and the Fakopp Microsecond Timer.

A new generation device IML RESI PD400 with a drill length of about 400 mm was used for drilling resistance testing. The device can register both drilling resistance and the feed force at every 0.1 mm. Five-millimetre deep entry and exit zones were assumed when calculating mean drilling resistance RM from formula (5). The zones were not taken into account in the calculations. The places where drilling resistance (RA) amounted to less than 5% and where the under-five-per-cent values in the diagram extended for minimum 5 mm were regarded as zones with flaws.

Moreover, the moisture content in the samples was determined using an FMW moisture meter (of the resistance type with a hammer probe).

Density was determined using 20 mm × 20 mm × 400 mm flawless samples prepared from the tested timber samples (16, 12 and 12 samples from each of the member, altogether 40 samples) with a moisture content of 18%. In accordance with the standard procedure [44], density was calculated from formula (5):

$$\rho = \rho(u) \cdot [1 - 0.005 \, (u - u_{ref})], \tag{5}$$

where ρ is density, u is the sample's moisture content during testing, and u_{ref} is the reference moisture content = 12%.

Also, the correction due to the size of the sample is needed; the value should be divided by 1.05 [44].

Density determinations and the drilling resistance tests were carried out for the wood moisture content of about 18%. Acoustic tests were carried out three times for chosen different sample moisture content levels: about 30% (direct after taking samples from the structure), 24–28% and 17–18% to examine the effect of moisture content on the measurement results.

Also, a series of scans of the small samples were performed using the Bruker SkyScan 1172 microtomograph.

The total number of three samples were scanned using X-ray micro-computed tomography (see Figure 6 below):

- "1" a portion of timber from the zone impacted by waves—a sample with c.a. 20 mm × 20 mm × 30 mm dimensions,
- "2" a portion of timber from the zone submerged in water)—a sample with c.a., 20 mm × 20 mm × 30 mm dimensions,
- "3" a portion of timber from the zone sunk in the ground—a sample with c.a., 20 mm × 20 mm × 30 mm dimensions.

Figure 6. Samples prepared for scanning: "1", "2" and "3" (from left to right).

The samples were scanned in the Bruker SkyScan 1172 device (Figure 4a). The same set of scanning parameters was used for each of the samples in order to ensure identical scanning conditions. The selected major scanning parameters are summarised in Table 4 below.

Table 4. Scanning parameters.

Parameter	Value
Source Voltage	59 kV
Source Current	167 μA
Projection image size	2000 × 1333 pix
Image Pixel Size	13,56 μm
Filter	Al foil
Exposure	750 ms
Rotation Step	0.24°
Frame Averaging	ON (6)
Random Movement	ON (10)
Use 360 Rotation	YES

4. Results and Discussion

4.1. Density and Moisture Content

Density was determined for the 40 flawless small samples and the results are presented in Table 5.

Table 5. Determined densities.

Sample	Number of Measurements	Density ρ				
		Mean Value from Tests [kg/m³]	Mean value Calculated according to [44] [kg/m³]	Range [kg/m³]	Standard Deviation [kg/m³]	Coefficient of Variation [%]
1	16	511.5	472.5	449.6–547.0	27.2	5.3
2	12	514.4	475.2	484.4–549.1	25.8	5.0
3	12	623.0	575.5	598.8–671.1	20.5	3.3
summary	40	545.8	504.2	449.6–671.1	58.5	10.8

On the basis of the measurements performed by means of a resistance-type moisture meter the moisture content in the samples was determined to amount to 18 ± 1%. In addition, testing by acoustic methods was carried out for two more different moisture content levels, i.e., about 30% and 24–28%.

4.2. Drilling Resistance

Drilling resistance tests were carried out on 3 samples with a moisture content of 18 ± 1%. Twenty measurements were performed on each of the samples. An exemplary drilling resistance curve and a feed force curve are shown in Figure 7. The depth to which the biological corrosion of the wood extends (3 mm in this case) can be easily read off the diagram. The drilling resistance test results for the particular samples are presented in Tables 6 and 7.

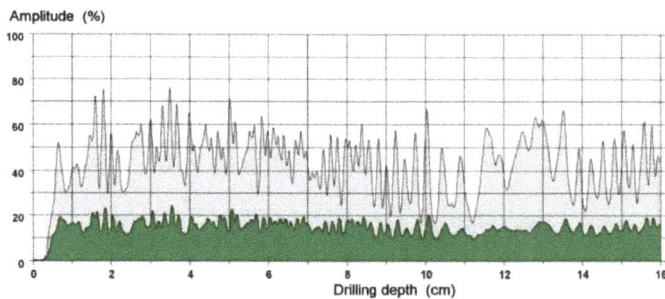

Figure 7. Exemplary drilling resistance curve (green) and feed force curve (grey).

Table 6. Mean drilling resistance tests results.

Sample	Number of Measurements	Drilling Resistance *RM* [%]			
		Mean	Range	Standard Deviation	Coefficient of Variation
1	20	16.5	14.9–18.3	1.2	7.4
2	20	16.3	13.5–19.2	2.3	14.1
3	20	16.9	15.2–20.1	1.6	9.5
summary	60	16.6	13.5–20.1	1.8	10.6

Table 7. Mean feed force test results.

Sample	Number of Measurements	Feed Force *FM* [%]			
		Mean	Range	Standard Deviation	Coefficient of Variation
1	20	47.4	42.3–53.4	3.5	7.3
2	20	47.7	35.9–61.7	9.8	20.6
3	20	55.3	44.3–70.6	7.7	13.9
summary	60	50.1	35.9–70.6	8.2	16.4

Only 60 measurements were carried out as part of the laboratory tests, but the number of samples in in situ tests is usually not larger because of the not fully non-destructive character of the testing method. Since the test has a pointwise character and wood is a heterogenous material, it is necessary to perform numerous measurements to assess its condition and density. Therefore, one cannot responsibly evaluate wood on the basis of single measurements.

Many factors have a bearing on drilling resistance, e.g., moisture content, drill sharpness, drilling angle and direction and battery charge status [45]. Moreover, wood flaws, such as knots (resulting in very high drilling resistance) and damaged zones (zero or close to zero drilling resistance) affect the RM value, which was taken into account in the analysis. The places with knots were neglected in drilling resistance and feed force calculations.

Despite the quite good correlation (Figure 8) between drilling resistance and feed force ($R^2 = 0.8114$), no correlation between these quantities and density was found (Figures 9 and 10), which casts doubt on the correlativity between them [32,46]. The results obtained using the drilling resistance method can be used to estimate the depth of wood damage in static load analyses to reduce the cross sections of the members.

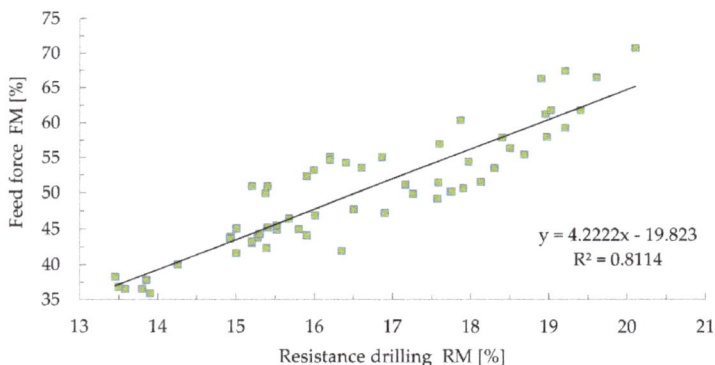

Figure 8. Correlation between drilling resistance and feed force.

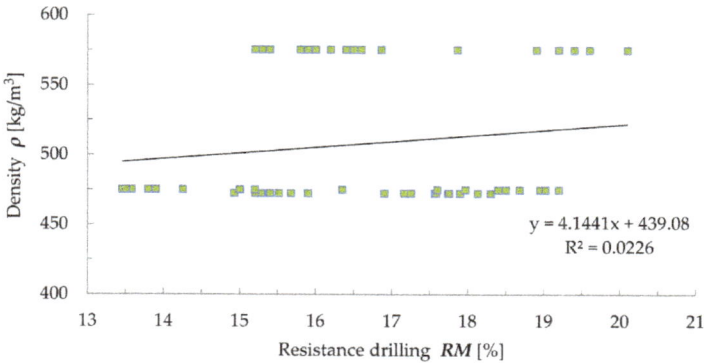

Figure 9. Correlation between drilling resistance and density.

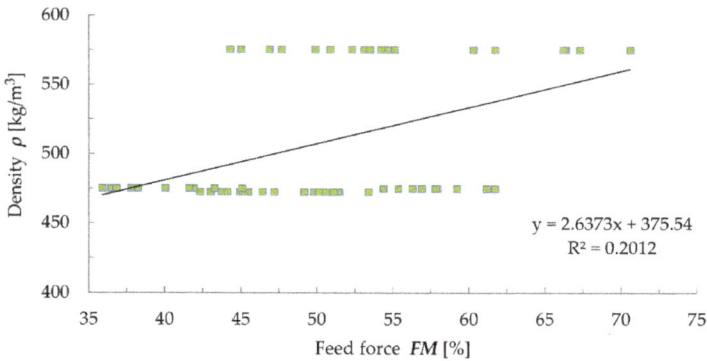

Figure 10. Correlation between feed force and density.

4.3. Stress and Ultrasonic Waves

The propagation times of the stress wave and the ultrasonic wave and the length of the distance covered by the waves in both the directions (along and across) relative to the grain were registered and used to calculate the velocity of wave propagation in the material. Then the dynamic moduli of elasticity were calculated using the densities measured for the particular samples (sample 1—472.5 kg/m^3, sample 2—475.2 kg/m^3, sample 3—575.5 kg/m^3). The results are shown in Tables 8 and 9. Also the dynamic elasticity modulus values parallel and perpendicular to the grain, yielded by the two methods were correlated for selected samples. The results are presented in Figure 11.

Table 8. Fakopp Microsecond Timer test results: velocity of stress wave propagation and dynamic modulus of elasticity depending on moisture content and direction relative to grain.

Sample	Direction Relative to Grain	V [m/s]			MOE$_{dyn}$ [GPa]		
		Moisture Content			Moisture Content		
		~30%	24–28%	~18%	~30%	24–28%	~18%
1	parallel	4872.8	5376.6	5644.8	11.22	13.66	15.06
	perpendicular	1153.7	1439.9	1443.8	0.63	0.98	0.98
2	parallel	-	-	-	-	-	-
	perpendicular	1218.5	1372.8	1270.6	0.71	0.90	0.77
3	parallel	-	-	-	-	-	-
	perpendicular	1406.1	1447.7	1665.2	1.14	1.21	1.60

Table 9. Sylvatest Trio test results: velocity of ultrasonic wave propagation and dynamic modulus of elasticity depending on moisture content and direction relative to grain.

Sample	Direction Relative to Grain	V [m/s]			MOE$_{dyn}$ [GPa]		
		Moisture Content			Moisture Content		
		~30%	24–28%	~18%	~30%	24–28%	~18%
1	parallel	5128.9	5855.4	6035.2	12.43	16.20	17.21
	perpendicular	1118.4	1311.5	1356.9	0.59	0.81	1.04
2	parallel	-	-	-	-	-	-
	perpendicular	1030.2	1311.4	1481.1	0.50	0.82	1.04
3	parallel	-	-	-	-	-	-
	perpendicular	1199.8	1251.1	1422.3	0.68	0.90	1.04

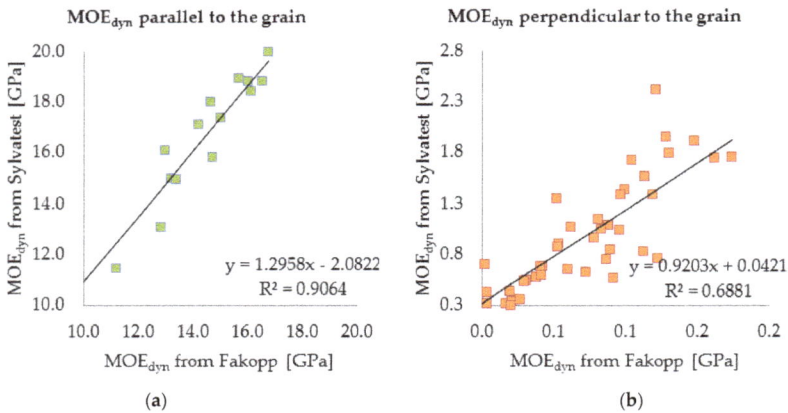

Figure 11. Correlation between dynamic modulus of elasticity determined using ultrasonic wave method (Sylvatest) and stress wave method (Fakopp) for all the samples for all moisture content levels for both directions relative to grain: (**a**) parallel; (**b**) perpendicular.

As one can see above, the results yielded by the two measuring methods using the Sylvatest Trio device and the Fakopp Microsecond Timer are similar. For the selected samples the correlation coefficient along and across the grain amounts to respectively $R^2 = 0.9064$ and $R^2 = 0.6881$. This result can be regarded as satisfactory and it indicates the two methods can be used complementarily to estimate the material parameters of wood.

4.4. X-Ray Micro-Computed Tomography

3D images of the samples were reconstructed on the basis of a series of X-ray projections, using the Feldkamp algorithm in the NRecon software (version 1.7.1.0) by Bruker. Selected major reconstruction parameters are summarised in Table 10 below.

Table 10. Reconstruction parameters.

Parameter	Value
Pixel Size	13.53217 μm
Smoothing	2 pix
Ring Artefact Correction	19
Beam Hardening Correction	41%
Minimum for CS to Image Conversion	0.000
Maximum for CS to Image Conversion	0.030

Exemplary projections for all of the tested samples are shown in Figure 12.

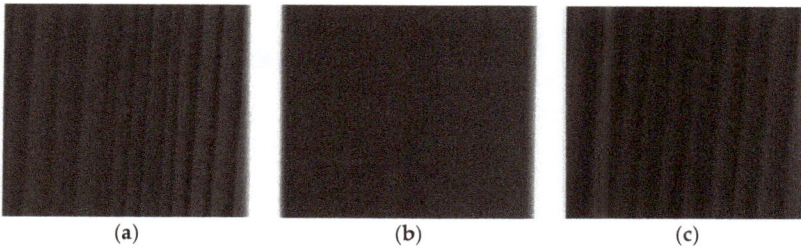

Figure 12. Exemplary projections for samples: (a) "1"; (b) "2"; (c) "3" (scale 200%).

Exemplary cross sections of the samples, obtained by reconstructing the 3D sample model are shown in Figures 13 and 14, at a scale of respectively 200% and 800% (an enlarged fragment of the image).

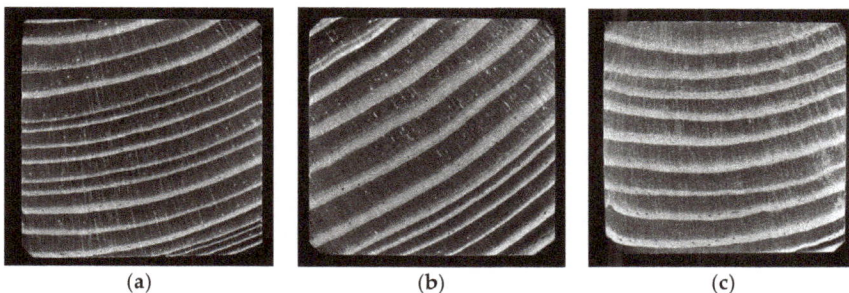

Figure 13. Exemplary cross sections of samples: (a) "1"; (b) "2"; (c) "3" (scale 200%).

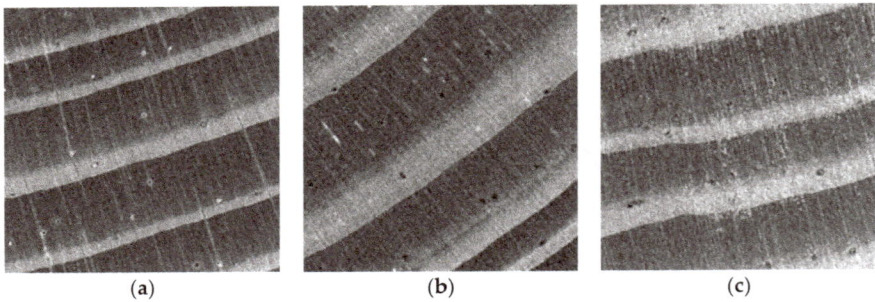

Figure 14. Exemplary cross section (scale 800%): (**a**) sample "1"; (**b**) sample "3"; (**c**) sample "2".

Figure 15 shows the rendering of the 3D model of the samples.

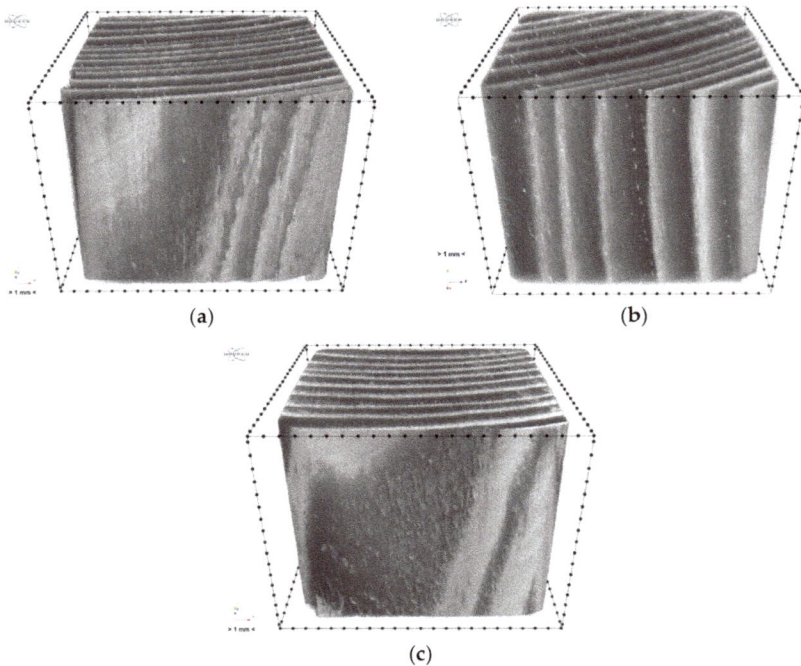

Figure 15. Reconstruction—3D view: (**a**) sample "1", (**b**) sample "2", (**c**) sample "3".

The cell structure is practically invisible due to the adopted scanning resolution. Only the early and late wood with local flaws (small microcrack in sample "1") and higher-density inclusions (samples "1" and "2") can be distinguished. In the images obtained from scanning, sample "3" is generally brighter than the other two, which unambiguously indicates its higher density. This is particularly visible in the cross sections (Figures 13 and 14). It also appears from the reconstruction that the late growth rings, which are denser (brighter in the imaging results), are thicker and occupy more material volume in samples "2" and "3", whereas in sample "1" the volume fraction of late growth rings is clearly smaller. Moreover, small highly dense (probably mineral) inclusions are noticeable in samples "1" and "2".

A cubic area of $(1200 \text{ vox})^3$ completely contained within the volume of the tested material, i.e., the so-called volume of interest (VOI), was selected in order to quantitatively characterise the above observations. The selection of VOI is shown in Figure 16 below.

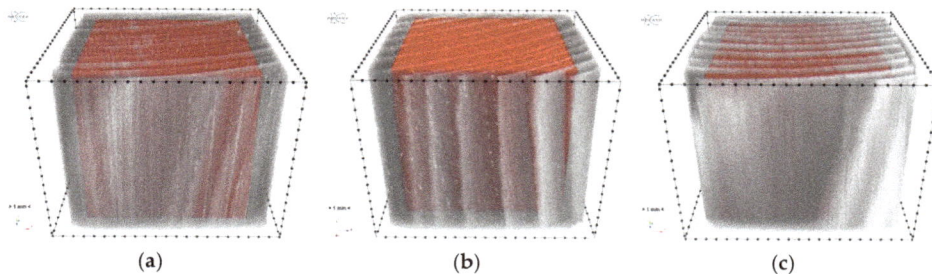

Figure 16. Selection of VOI: (**a**) sample "1"; (**b**) sample "2"; (c) sample "3".

As mentioned, there is a correlation between greyscale and density. Thanks to the use of such a correlation in the linear form as in [41,42], the spatial distributions of local density in the analysed samples (Figure 17) and the statistical distributions (histograms) of density in the VOI of the particular samples (Figure 18) were determined. The correlation coefficients were determined by comparing the mean density of a given sample with the average grey level in VOI, and the density of the air with the average grey level of area outside the sample (visible in the images obtained from X-ray micro-computed tomography). The coefficient of proportionality of this correlation, determined independently for each of the three samples, amounted to respectively 5.884, 5.808 and 5.906. The distributions presented below were obtained using the mean value of this coefficient, i.e., 5.866.

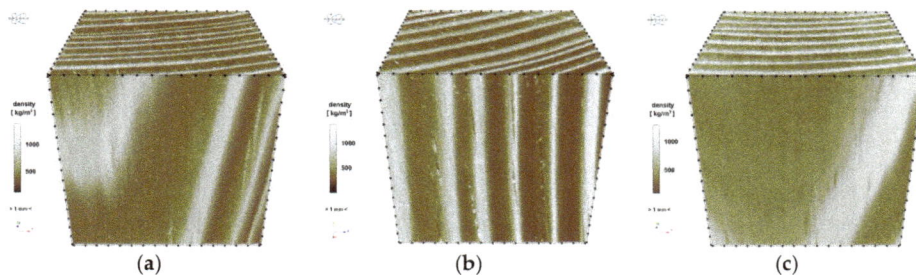

Figure 17. Spatial distribution of density: (**a**) "1"; (**b**) "2" and (**c**) "3".

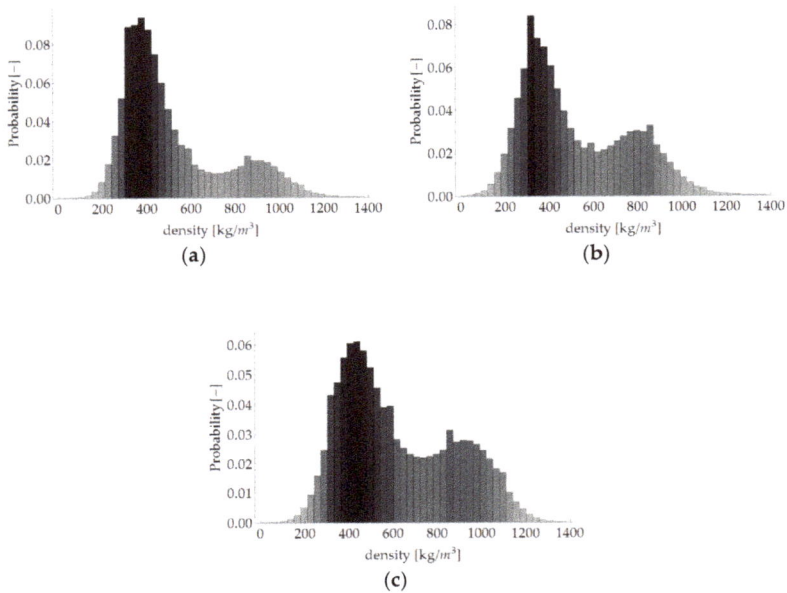

Figure 18. Histogram of local density in VOI: (a) sample "1"; (b) "2" and (c) "3".

As one can see, the above histograms represent bimodal probability distributions. This means that there occur two principal components. In the case of this analysis, these are the early and late growth rings. Segmentation, i.e. partitioning the image into segments occupied by the particular components, was performed using the thresholding preceded by a smoothing filter. The threshold value of brightness corresponds to the grey level value at which the image histogram reaches a minimum (sample "1": 135, "2": 122, "3": 149). A morphometric analysis was carried out for the two image components. In particular, the volume fraction of the component and the spatial distribution of its local thickness [47] were determined and then its mean thickness was calculated. The calculated values are shown in Table 11 and the thickness distributions are presented in Figures 19 and 20. In order to avoid the error connected with determining thickness at the boundary of VOI, the area of the latter was limited to the volume of $(1000 \text{ vox})^3$ by applying dilation with a radius of 100 voxels.

Table 11. Summary of analysis of micro-computed tomography images.

Parameter	Value in Sample:		
	"1"	"2"	"3"
Mean density [kg/m^3]	510	519	619
Mean density of early wood [kg/m^3]	410	367	463
Mean density of late wood [kg/m^3]	912	814	934
Volume fraction of early wood [%]	80.1	66.2	66.9
Volume fraction of late wood [%]	19.8	33.6	33.1
Mean thickness of early growth [mm]	1.29	1.34	1.26
Mean thickness of late growth [mm]	0.35	0.70	0.56

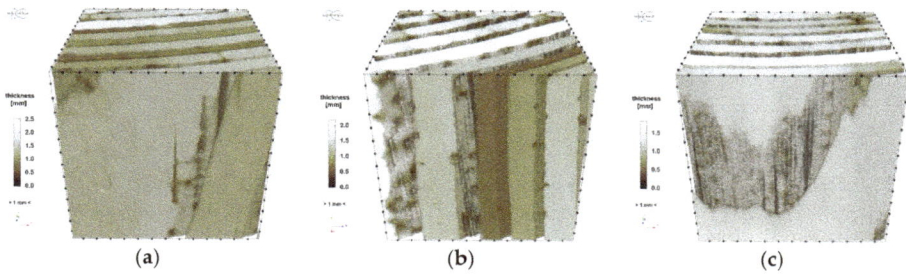

Figure 19. Distribution of local thickness of early growth rings: (**a**) sample "1"; (**b**) sample "2"; (**c**) sample "3".

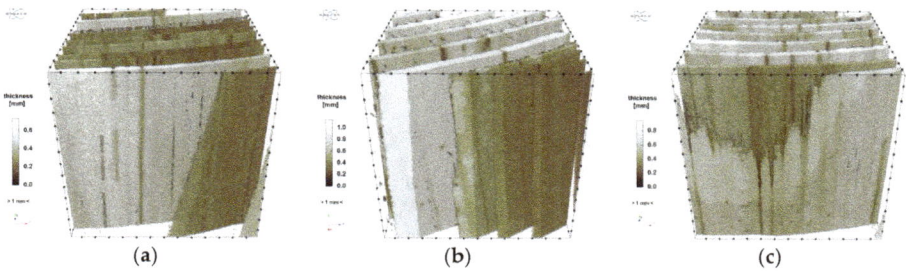

Figure 20. Distribution of local thickness of later growth rings: (**a**) sample "1"; (**b**) sample "2"; (**c**) sample "3".

4.5. Comparative Analysis for Samples from Different Zones

Results obtained for samples from different zones: the zone impacted by waves (sample 1), the zone immersed in water (sample 2) and the zone embedded in the ground (sample 3) are summarised in the Table 12.

Table 12. Selected results for different samples.

Testing Method	Parameter	Value in Sample:		
		"1"	"2"	"3"
Laboratory test	Mean density (at moisture content 18%) [kg/m³]	511.5	514.4	623.0
Acoustic method (Fakopp)	Mean MOE$_{dyn}$ perpendicular to the grain [GPa]	0.98	0.77	1.6
Resistance drilling method	Mean RM [%]	16.5	16.3	16.9
	Mean FM [%]	47.4	47.7	55.3

The obtained results confirm the engineering "intuition" about the impact of the environment on the degradation level of the material. The best condition of wood was observed for sample embedded in the soil (sample 3). The highest values of density were obtained for this sample. The same applies to the modulus of elasticity and the FM parameter, which are positively correlated with density. At the same time differences between parameters for samples from the zone impacted by waves (sample 1) and the zone immersed in water (sample 2) are not significant. Moreover, in the whole member there were no changes observed in the material within the surface layer exposed to direct environmental impact in the contact zone. Eventually, despite the noticeable differences between the different zones, it can be stated that in the tested timber members, after 70 years of operation, no significant destruction, reducing the safety of use, was found. In that sense, the condition of a wharf timber sheet wall's material may be described as fairly good.

5. Conclusions

For a reliable assessment of the technical condition of timber structures, the use of non-destructive examinations is recommended, in addition to visual evaluation. Still, there are no comprehensive studies in this area, which present correlations enabling estimation of the mechanical parameters of wood and the degree of destruction, although some attempts of predicting these parameters or instance by regression analysis were made (among others [48]).

None of the currently known non-destructive methods used to assess the condition of timber members does not allow for an unambiguous estimation of the strength characteristics of wood. This is not possible even when using the X-ray method [41], which enables relatively accurate measurement of wood density. It results from the material inhomogeneous internal structure, including different defects, f. ex. knots, which have a significant impact on the strength parameters of wood.

In the case of the resistance drilling tests, the obtained coefficients of determination between drilling resistance and density ($R^2 = 0.0226$) and between the feed force and density ($R^2 = 0.2012$) do not indicate any correlation between the results. Drilling resistance testing should be treated not as a quantitative, but qualitative assessment. The results obtained by means of this method can be used to estimate the depth of material damage in static load analyses to reduce the cross sections of the analysed members [14,49].

The acoustic testing (using the Sylvatest Trio device and the Fakopp Microsecond Timer) provided data on the value of the dynamic modulus of elasticity which can be correlated with the static modulus of elasticity. The latter is the basic mechanical parameter needed to carry out a global structural analysis. The obtained coefficients of determination between the values of MOE$_{dyn}$ yielded by the two measuring methods ($R^2 = 0.9064$ along the grain and $R^2 = 0.6881$ across the grain) for the selected samples are satisfactory, showing the results to be reliable. The velocity of acoustic wave propagation (and so the modulus of elasticity) clearly decreases as the moisture content in the wood increases. The acoustic testing methods can be regarded as useful for estimating material stiffness parameters (Young's modulus), but they require further research in order to develop correlations comprising wood moisture content. Currently, research is underway to correlate the dynamic modulus of elasticity with the static modulus of elasticity. On the basis of the determined value of modulus of elasticity, it is possible to estimate mechanical parameters of material, for example according to standard procedure [50].

One should bear in mind that the two methods supply information about the local state of the material. In order to determine the global parameters, one should perform the largest possible number of measurements, which is not always possible, especially in the case of in situ testing. Therefore, in order to obtain the most accurate data on the tested member or structure, it is recommended to combine several testing methods.

Despite their non-destructive character, in most cases the testing methods require samples to be taken to determine the density of the material.

Thanks to the use of X-ray micro-computed tomography, the internal microstructure of the wood could be imaged. The results of the laboratory density measurements were used as input data for determining the correlation between the greyscale of the tomography results and the local density of the wood. Consequently, it became possible, for example, to estimate the density of the early and late wood. It should be noted that the correlation was determined independently for each of the three tested samples and very good agreement was obtained. The values of the coefficient of proportionality do not differ by more than 2%. This means that such a correlation can be a tool for the precise evaluation of the local density of the tested material on the microscale. As a result of the morphomorphic analysis based on the scanning results the volume fractions and morphology of the particular wood components, i.e. the early and late growth rings, could be determined. The data acquired from the analysis in the microtomograph can be useful in micromechanical modelling aimed at estimating the effective parameters of the material on the basis of microscale information.

Summing up, resistance drilling tests enable to determine the depth of material decayed zones, acoustic methods provide estimation of mechanical parameters. The application of the X-ray microtomography allows detailed insight to be gained into the microstructure of the material in a different scale of observation. In particular, it makes it possible to determine the occurrence of microdefects and to determine the parameters (density) of wood constituents, i.e. early and late growths. It must be pointed out that the applied methods are not equivalent, but rather, they are complementary.

The paper presents the methodology for comprehensive wood testing in structural members using the described research methods. The set of results obtained from these methods makes it possible to assess the material, and consequently, to perform a global analysis of the structure. In particular, it is possible to estimate the value of mechanical parameters, whereas the qualitative evaluation makes it possible to determine the extent of possible material destruction and the location of any defects.

Author Contributions: Conceptualization, T.N., M.W.; methodology, A.K., T.N., M.S., M.W.; software, M.S.; validation, A.K., T.N., M.S., M.W.; formal analysis, A.K., T.N., M.S., M.W.; investigation, A.K., T.N., M.S.; resources, A.K., T.N., M.S., M.W.; data curation, A.K., T.N., M.S., M.W.; writing—original draft preparation, A.K., T.N., M.S., M.W.; writing—review and editing, A.K.; visualization, M.S.; supervision, T.N., M.W.; project administration, A.K., M.W.; funding acquisition, T.N., M.S.

Funding: This research received no external funding.

Acknowledgments: We thank Filip Patalas and Krzysztof Wujczyk for their help in preparing the samples.

Conflicts of Interest: The authors declare no conflict of interest.

References

1. Francesca, C.; Paolo Simonini, P.; Lionello, A. Long-term mechanical behavior of wooden pile foundation in Venice. In Proceedings of the 2nd International Symposium on Geotechnical Engineering for the Preservation of Monuments and Historic Sites, Napoli, Italy, 30–31 May 2013. [CrossRef]
2. Klaassen, K.W.M.; Creemers, J.G.M. Wooden foundation piles and its underestimated relevance for cultural heritage. *J. Cult. Herit.* **2012**, *135*, 123–128. [CrossRef]
3. Nilson, T.; Rowell, R. Historical wood—Structure and properties. *J. Cult. Herit.* **2012**, *135*, 55–59. [CrossRef]
4. Klaassen, K.W.M. Life Expectation of Wooden Foundations—A Non-Destructive Approach. In Proceedings of the International Symposium Non-Destructive Testing in Civil Engineering (NDT-CE), Berlin, Germany, 15–17 September 2015; pp. 775–779.
5. Klaassen, R.K.M. Bacterial decay in wooden foundation piles—Pattern and causes: A study of historical foundation piles the Netherlands. *Int. Biodeterior. Biodegrad.* **2008**, *61*, 45–60. [CrossRef]
6. Panshin, A.J.; de Zeeuw, C. *Textbook of Wood Technology. Structure, Identification, Properties, and Uses of the Commercial Woods in the United States and Canada*, 4th ed.; Mc Graw-Hill Book Company: New York, NY, USA, 1980.
7. Tannert, T.; Anthony, R.; Kasal, B.; Kloiber, M.; Piazza, M.; Riggio, M.; Rinn, F.; Widmann, R.; Yamaguchi, N. In situ assessment of structural timber using semi-destructive techniques. *Mater. Struct.* **2014**, *47*, 767–785. [CrossRef]
8. Riggio, M.; Anthony, R.; Augelli, F.; Kasal, B.; Lechner, T.; Muller, W.; Tannert, T. In situ assessment of structural timber using non-destructive techniques. *Mater. Struct.* **2014**, *47*, 749–766. [CrossRef]
9. Dolwin, J.A.; Lonsdale, D.; Barnet, J. Detection of decay in trees. *Arboric. J.* **1999**, *23*, 139–149. [CrossRef]
10. Kasal, B.; Lear, G.; Tannert, T. Stress waves. In *In Situ Assessment of Structural Timber. RILEM State-of-the-Art Reports*; Kasal, B., Tannert, T., Eds.; Springer: Dordrecht, The Netherlands, 2010; Volume 7, pp. 5–24, ISBN 978-94-007-0559-3.
11. Dackermann, U.; Crews, K.; Kasal, B.; Li, J.; Riggio, M.; Rinn, F.; Tannert, T. In situ assessment of structural timber using stress-wave measurements. *Mater. Struct.* **2014**, *47*, 787–803. [CrossRef]

12. Wang, X.; Divos, F.; Pilon, C.; Brashaw, B.K.; Ross, R.J.; Pellerin, R.F. *Assessment of Decay in Standing Timber Using Stress Wave Timing Nondestructive Evaluation Tools: A Guide for Use and Interpretation*; Gen. Tech. Rep. FPL-GTR-147; US Department of Agriculture, Forest Service, Forest Products Laboratory: Madison, WI, USA, 2004. [CrossRef]

13. Sandoz, J.L. Grading of construction timber by ultrasound. *Wood Sci. Technol.* **1989**, *23*, 95–108. [CrossRef]

14. Lechner, T.; Nowak, T.; Kliger, R. In situ assessment of the timber floor structure of the Skansen Lejonet fortification, Sweden. *Constr. Build. Mater.* **2014**, *58*, 85–93. [CrossRef]

15. García, M.C.; Seco, J.F.G.; Prieto, E.H. Improving the prediction of strength and rigidity of structural timber by combining ultrasound techniques with visual grading parameters. *Mater. Constr.* **2007**, *57*, 49–59. [CrossRef]

16. Lourenço, P.B.; Feio, A.O.; Machado, J.S. Chestnut wood in compression perpendicular to the grain: Non-destructive correlations for test results in new and old wood. *Constr. Build. Mater.* **2007**, *21*, 1617–1627. [CrossRef]

17. Ilharco, T.; Lechner, T.; Nowak, T. Assessment of timber floors by means of non-destructive testing methods. *Constr. Build. Mater.* **2015**, *101*, 1206–1214. [CrossRef]

18. Íñiguez, G.; Martínez, R.; Bobadilla, I.; Arriaga, F.; Esteban, M. Mechanical properties assessment of structural coniferous timber by means of parallel and perpendicular to the grain wave velocity. In Proceedings of the 16th International Symposium on Nondestructive Testing of Wood, Beijing, China, 11–13 May 2009.

19. Esteban, M.; Arriaga, F.; Íñiguez, G.; Bobadilla, I. Structural assessment and reinforcement of ancient timber trusses. In Proceedings of the International Conference on Structures & Architecture, Guimarães, Portugal, 21–23 July 2010.

20. Nowak, T.; Hamrol-Bielecka, K.; Jasieńko, J. Experimental testing of glued laminated timber members using ultrasonic and stress wave techniques. In Proceedings of the International Conference on Structural Health Assessment of Timber Structures, SHATIS '15, Wroclaw, Poland, 9–11 September 2015; pp. 523–533.

21. Fakopp Enterprise Microsecond Timer. Available online: http://www.fakopp.com/site/microsecond-timer (accessed on 20 November 2018).

22. Rinn, F. Practical application of micro-resistance drilling for timber inspection. *Holztechnologie* **2013**, *54*, 32–38.

23. Hiroshima, T. Applying age-based mortality analysis to a natural forest stand in Japan. *J. For. Res.* **2014**, *19*, 379–387. [CrossRef]

24. Wang, S.Y.; Chiu, C.M.; Lin, C.J. Application of the drilling resistance method for annual ring characteristics: Evaluation of Taiwania (Taiwania cryptomerioides) trees grown with different thinning and pruning treatments. *J. Wood Sci.* **2003**, *49*, 116–124. [CrossRef]

25. Feio, A.O.; Machado, J.S.; Lourenço, P.B. Compressive behavior and NDT correlations for chestnut wood (Castanea sativa Mill.). In Proceedings of the 4th International Seminar on Structural Analysis of Historical Constructions, Padova, Italy, 10–13 November 2004; pp. 369–375.

26. Jasieńko, J.; Nowak, T.; Bednarz, Ł. Baroque structural ceiling over the Leopoldinum Auditorium in Wrocław University: Tests, conservation, and a strengthening concept. *Int. J. Archit. Herit.* **2014**, *8*, 269–289. [CrossRef]

27. Branco, J.M.; Piazza, M.; Cruz, P.J. Structural analysis of two King-post timber trusses: Non-destructive evaluation and load-carrying tests. *Constr. Build. Mater.* **2010**, *24*, 371–383. [CrossRef]

28. Kloiber, M.; Tippner, J.; Hrivnák, J. Mechanical properties of wood examined by semi-destructive devices. *Mater. Struct.* **2014**, *47*, 199–212. [CrossRef]

29. Morales-Conde, M.J.; Rodríguez-Liñán, C.; Saporiti-Machado, J. Predicting the density of structural timber members in service. The combine use of wood cores and drill resistance data. *Mater. Constr.* **2014**, *64*, 1–11. [CrossRef]

30. Acuña, L.; Basterra, L.A.; Casado, M.M.; López, G.; Ramón-Cueto, G.; Relea, E.; Martínez, C.; González, A. Application of resistograph to obtain the density and to differentiate wood species. *Mater. Constr.* **2011**, *61*, 451–464. [CrossRef]

31. Tseng, Y.J.; Hsu, M.F. Evaluating the mechanical properties of wooden components using drill resistance method. In Proceedings of the 10th World Conference on Timber Engineering, Miyazaki, Japan, 2–5 June 2008; pp. 303–310.

32. Nowak, T.; Jasieńko, J.; Hamrol-Bielecka, K. In situ assessment of structural timber using the resistance drilling method–evaluation of usefulness. *Constr. Build. Mater.* **2016**, *102*, 403–415. [CrossRef]

33. Jasieńko, J.; Nowak, T.; Hamrol, K. Selected methods of diagnosis of historic timber structures–principles and possibilities of assessment. *Adv. Mater. Res.* **2013**, *778*, 225–232. [CrossRef]

34. Piazza, M.; Riggio, M. Visual strength-grading and NDT of timber in traditional structures. *J. Build. Apprais.* **2008**, *3*, 267–296. [CrossRef]

35. Sousa, H. Methodologies for Safety Assessment of Existing Timber Structures. Ph.D. Thesis, Department of Civil Engineering, University of Minho, Minho, Portugal, 2013.

36. Salvo, L.; Cloetens, P.; Maire, E.; Zabler, S.; Blandin, J.J.; Buffière, J.Y.; Ludwig, W.; Boller, E.; Bellet, D.; Josserond, C. X-ray micro-tomography an attractive characterisation technique in materials science. *Nucl. Instrum. Methods Phys. Res. Sect. B Beam Interact. Mater. At.* **2003**, *200*, 273–286. [CrossRef]

37. Schabowicz, K.; Jóźwiak-Niedźwiedzka, D.; Ranachowski, Z.; Kudela, S.; Dvorak, T. Microstructural characterization of cellulose fibres in reinforced cement boards. *Arch. Civ. Mech. Eng.* **2018**, *18*, 1068–1078. [CrossRef]

38. Cała, M.; Cyran, K.; Kawa, M.; Kolano, M.; Łydżba, D.; Pachnicz, M.; Rajczakowska, M.; Różański, A.; Sobótka, M.; Stefaniuk, D.; et al. Identification of Microstructural Properties of Shale by combined Use of X-Ray Micro-CT and Nanoindentation Tests. *Procedia Eng.* **2017**, *191*, 735–743. [CrossRef]

39. Feldkamp, L.A.; Davis, L.C.; Kress, J.W. Practical cone-beam algorithm. *J. Opt. Soc. Am. A* **1984**, *1*, 612–619. [CrossRef]

40. Elliott, J.C.; Dover, S.D. X-ray microtomography. *J. Microsc.* **1982**, *126*, 211–213. [CrossRef]

41. Lechner, T.; Sandin, Y.; Kliger, R. Assessment of density in timber using X-ray equipment. *Int. J. Archit. Herit.* **2013**, *7*, 416–433. [CrossRef]

42. Lazarescu, C.; Watanabe, K.; Avramidis, S. Density and moisture profile evolution during timber drying by CT scanning measurements. *Dry. Technol.* **2010**, *28*, 460–467. [CrossRef]

43. Rajczakowska, M.; Stefaniuk, D.; Łydżba, D. Microstructure Characterization by Means of X-Ray Micro-CT and Nanoindentation Measurements. *Studia Geotech. Mech.* **2015**, *37*, 75–84. [CrossRef]

44. *PN-EN 384:2016-10—Structural Timber. Determination of Characteristic Values of Mechanical Properties and Density*; Polish Committee for Standarization: Warsaw, Poland, 2016.

45. Kraft, U.; Pribbernow, D. *Handbuch der Holzprüfung. Anleitungen und Beispiele*; Verlag Bau+Technik GmbH: Düsseldorf, Germany, 2006.

46. Feio, A.O.; Lourenço, P.B.; Machado, J.S. Non-destructive evaluation of the mechanical behavior of chestnut wood in tension and compression parallel to grain. *Int. J. Archit. Herit.* **2007**, *1*, 272–292. [CrossRef]

47. Hildebrand, T.; Rüegsegger, P. A new method for the model-independent assessment of thickness in three-dimensional images. *J. Microsc.* **1997**, *185*, 67–75. [CrossRef]

48. Sousa, H.S.; Branco, J.M.; Machado, J.S.; Lourenço, P.B. Predicting mechanical properties of timber elements by regression analysis considering multicollinearity of non-destructive test results. In Proceedings of the International Conference on Structural Health Assessment of Timber Structures, SHATIS '17, Istanbul, Turkey, 20–22 September 2017; pp. 485–493.

49. Cuartero, J.; Cabaleiro, M.; Sousa, H.S.; Branco, J.M. Tridimensional parametric model for prediction of structural safety of existing timber roofs using laser scanner and drilling resistance tests. *Eng. Struct.* **2019**, *185*, 58–67. [CrossRef]

50. *PN-EN 338:2016-06—Structural Timber. Strength Classes*; Polish Committee for Standarization: Warsaw, Poland, 2016.

materials

MDPI

Article

Characteristic Curve and Its Use in Determining the Compressive Strength of Concrete by the Rebound Hammer Test

Dalibor Kocáb *[ORCID]**, Petr Misák and Petr Cikrle**

Faculty of Civil Engineering, Brno University of Technology, Veveří 331/95, 602 00 Brno, Czech Republic
* Correspondence: dalibor.kocab@vutbr.cz; Tel.: +420-54114-7811

Received: 28 June 2019; Accepted: 21 August 2019; Published: 23 August 2019

check for updates

Abstract: During the construction of concrete structures, it is often useful to know compressive strength at an early age. This is an amount of strength required for the safe removal of formwork, also known as stripping strength. It is certainly helpful to determine this strength non-destructively, i.e., without any invasive steps that would damage the structure. Second only to the ultrasonic pulse velocity test, the rebound hammer test is the most common NDT method currently used for this purpose. However, estimating compressive strength using general regression models can often yield inaccurate results. The experiment results show that the compressive strength of any concrete can be estimated using one's own newly created regression model. A traditionally constructed regression model can predict the strength value with 50% reliability, or when two-sided confidence bands are used, with 95% reliability. However, civil engineers usually work with the so-called characteristic value defined as a 5% quantile. Therefore, it appears suitable to adjust conventional methods in order to achieve a regression model with 95% one-sided reliability. This paper describes a simple construction of such a characteristic curve. The results show that the characteristic curve created for the concrete in question could be a useful tool even outside of practical applications.

Keywords: rebound hammer; SilverSchmidt; concrete; compressive strength; non-destructive testing

---

1. Introduction

Concrete structures have always been built for a long service life, safety, durability, load-bearing capacity, stability, and working and functional reliability. During construction or usage, however, there may arise a need to determine or verify the properties of the concrete. This is why accurate diagnostics of concrete elements or whole structures are essential. Non-destructive testing methods (NDT) are often used for this purpose [1]. Besides the ultrasonic pulse velocity test, the rebound hammer test is a popular method, because it is easy to use and is practically non-destructive [2–5]. There are two basic ways of using it to test concrete. The first is using a rebound hammer to diagnose older structures with the primary purpose of classing the concrete according to strength or uniformity [6,7]. Testing older structures often requires removing both plaster and the top layer of concrete, as well as combining the measurement with destructive compressive strength tests performed on cores [8–10]. Another use of the rebound test is assessing the quality of new concrete structures, especially those with a smooth surface.

The rebound hammer test is currently the most common method of testing the hardness of concrete and was first developed in Switzerland in the 1950s as the Schmidt rebound hammer (sometimes known as the Swiss hammer) [11]. In 1950 Ernst Schmidt created the first rebound hammer, which proved to be superior to indentation methods used until that time (the most common one involved pressing a steel ball into the surface) [12]. Testing hardness with the Schmidt hammer was gradually accepted as the best, and since then concrete hardness has been determined by measuring the rebound number instead of examining the indentation produced by the ball. Today, the Schmidt hammer is the most common concrete sclerometer worldwide [13]. One of the strong points of the Schmidt hammer is its ability to test materials other than concrete (see [14–19]).

Its benefits notwithstanding, the rebound hammer test has some weak points, which should be kept in mind during testing and evaluation. Its primary purpose has always been to determine the quality of new concrete elements or structures. The general relationships between hardness and compressive strength were adjusted for this purpose, having been created by the manufacturers and then carried over to technical standards; they apply mostly to concrete of 14 to 56 days of age. The measurement range is then designed for structural concrete used during the second half of the 20th century; i.e., compressive strength of approximately 10 to 70 N/mm^2 [20], or, more realistically, 15 to 60 N/mm^2 [21]. When testing older concretes, it is necessary to take into account the severity of carbonation or surface damage. Kim et al. [22] concerned themselves with the influence of concrete carbonation on the rebound value and compressive strength. They discovered the depth of carbonation affects compressive strength very differently than the rebound value, which implies that the same dependencies will not apply to differently carbonated concrete. This is why they used regression analysis to create an equation that includes a factor which corrects for a reduction in compressive strength due to age, and should only be a function of the rebound value and compressive strength. The influence of concrete age and carbonation was studied in greater detail by Szilágyi et al. [11], who say that even though the rebound hammer test has been available for over 60 years, available literature lacks models that understand surface hardness as an age-dependent property. Their paper [11] presents a SBZ model; a phenomenological constructive model, which operates with concrete hardness measured by the rebound hammer test in relation to time. The SBZ model is based on the development of the capillary pore system in hardened cement paste, which, for reasons of simplicity, is replaced by the w/c ratio. The model includes the relationship between w/c ratio and 28-day compressive strength, development of compressive strength over time, relationship between compressive strength and rebound number at the age of 28 days, and the progress of carbonation depth over time and its influence on the rebound number. As with carbonation, high temperatures also affect surface hardness and compressive strength. Panedpojaman and Tonnayopas [23] found that high temperatures are detrimental to compressive strength. If concrete is exposed to fire, up to about 420 °C the rebound number does not change in any major way, but compressive strength is lost. This is because calcium carbonate crystals form in the pore structure, causing hardness to decrease at a much slower rate and thus rendering the rebound hammer test unusable for determining compressive strength.

Correct statistical analysis of the measured values is critical for evaluating Schmidt hammer tests. Alwash et al. [24] analysed the influence of several factors on the reliability of rebound hammer test evaluation, such as within-test variability, variability of true compressive strength, number of test locations and cores used to determine the relationship between strength and the rebound number, way of choosing the test locations (random or conditional) and the model identification programme (regression or bi-objective). El Mir and Nehme [1] tested several hundred specimens focusing on the coefficient of variance in rebound values. They found that greater carbonation depth, higher w/c ratio, or higher porosity of the concrete cause the coefficient of variance to increase and reduce repeatability. However, concrete containing additions, such as metakaolin or silica fume, high-strength and high-performance self-compacting concrete, exhibit a lower coefficient of variance in rebound values. Szilágyi et al. [25] conducted an extensive statistical analysis of the variability of concrete hardness using a database of both laboratory and in situ measurements spanning over the past 60 years. The study covers several thousand tests (over eighty thousand individual rebound numbers) and shows that current sources and standards leave much to be desired in terms of the evaluation of concrete using a Schmidt impact hammer. The normality tests (the Shapiro–Wilk normality test) yielded rather inconsistent results: "the hypothesis of normality can only be accepted at very low levels of probability for individual test locations." Given the density function of the coefficient of variance of the rebound values, a strong positive skewness can be seen. Szilágyi et al. [25] speak against directly correlating the average rebound number with compressive strength as univariate functions, and argue for a "series of multivariate functions with independent variables of the degree of hydration, type and amount of cement and aggregate, environmental conditions, and testing conditions."

Besides assessing the quality of mature concrete or measuring its compressive strength, which, of course, carries certain associated problems, more modern types of impact hammers are capable of estimating the very early strength (also known as stripping strength) of concrete of only a few hours or days of age. This paper focuses primarily on creating a conversion formula to assess this early compressive strength, which is closely related to e.g., determining a time when it is safe to remove formwork. In fact, this is close to the original purpose of Schmidt hammers–testing concrete which is only several months old.

Many types of sclerometers have been developed over the past 70 years; however, the most common type used presently is the Schmidt type. There are several types available, differing from one another in the impact energy, shape and size of the plunger, or mechanical construction. In the past, and possibly even today, the most common Schmidt hammer is the Original Schmidt: a traditional rebound hammer, which became the basis for all the major rebound tests worldwide [26]. The basic type is the Schmidt N with impact energy of 2.207 Nm, but there is also the L type with energy of 0.735 Nm. A little over ten years ago a new model was introduced: the SilverSchmidt, which uses optical sensors to measure the impact and rebound velocity immediately before and after impact. It does not return the rebound number, but the Q-value. The different construction of the SilverSchmidt gives it the ability to measure concretes of lower as well as higher strength, and is supplied in the N and L configuration. A SilverSchmidt L can be fitted with a mushroom plunger accessory that enables measuring compressive strength as low as 5 N/mm². It is designed primarily for determining concrete strength uniformity and identifying inferior areas [27]. Upon introduction to the market, the manufacturer promised better quality of measurement, but this may not be entirely true: Viles et al. [17] consider the SilverSchmidt better than its predecessor, while Szilágyi et al. [25] rank it worse than the Original Schmidt. This may be why the newest model of the Schmidt hammers can only measure the rebound number and not the Q-value: it is the Original Schmidt Live supplied with a complex application (Apple iOS and Android) for measurement, reporting, and analysis [28].

The experiment described below was performed with an Original Schmidt N, SilverSchmidt N, SilverSchmidt L, and SilverSchmidt L with a mushroom plunger (MP) accessory.

2. Models of Relationship between Hardness Tests and Compressive Strength

2.1. Common Dependence Models

This section discusses some models of dependence commonly used in contemporary civil engineering for estimating compressive strength based on rebound hammer tests. They will be further compared with models created using the experimental data.

The standard [21] defines two models of dependence between test results obtained by an Original Schmidt N and compressive strength. These are two lines for which two different ranges apply.

Line A:

$$f_c = 1.75a - 29,$$ (1)

where $a = 25 - 40$ [-] is the rebound value and
Line B:

$$f_c = 1.786a - 30.44,$$ (2)

for rebound number range $a = 41 - 54$ [-].

The document [29] presents dependence models for the SilverSchmidt. The relationship for test results obtained by the SilverSchmidt N is discussed in two ways: First is the median relationship with 50% reliability:

$$f_c = 1.8943e^{0.064Q} \tag{3}$$

for $Q = 20 - 62$ [-]. This curve was created based on test results obtained by the BAM institute (Federal Institute for Materials Research and Testing in Berlin, Germany) with three different kinds of concrete, which differed in the w/c ratio and type of cement, covering a strength range of $f_c = 10 - 100 \text{ N/mm}^2$ [29].

Based on the results obtained by the BAM institute, The Shaanxi Province Construction Science Research Institute, China, and Hunan University, China, the document [29] defines a curve with 90% reliability

$$f_c = 2.77e^{0.048Q} \tag{4}$$

for a range of $Q = 22 - 75$ [-]. The curve was created on the basis of recommendations in EN 13791 [30], ASTM C805 [31] and ACI 228.1 [32]. They state that the dependence model should be created so that 90% of the experimental data would lie above the curve.

The [29] also describes a model for results obtained by the SilverSchmidt L for a range of $Q = 20 - 62$ [-] as

$$f_c = 1.9368e^{0.0637Q}. \tag{5}$$

Measurements by the SilverSchmidt L can also be made with the mushroom attachment, which should enable measurements of young concrete with low compressive strength. The relationship

$$f_c = 0.0108Q^2 + 0.2236Q \tag{6}$$

is defined by [27] for a range of $Q = 13 - 44$ [-] and $f_c = 5 - 30 \text{ N/mm}^2$.

2.2. Characteristic Curve

Drawing a general relationship between the rebound number and compressive strength can be a complex task. The aim of this part of the article is to show a new approach to the evaluation of NDT test results by designing the so-called characteristic curve, which ensures 95% reliability.

Several papers were published saying that formulating strength as a single parameter can be misleading [11]. Single-parameter formulation means that determining compressive strength requires knowing only the rebound number. Several publications [11,13] warn that the model of the relationship should also include e.g., the type of cement, aggregate, or w/c ratio. These and other parameters have an undeniable influence on the strength of concrete and therefore the rebound number measured by a Schmidt hammer as well.

The actual method of seeking the optimal model of relationship also requires attention. The most common method, known as regression analysis, or the least squares method, involves several assumptions, some of which may be violated during data evaluation. A determination of an optimal relationship between the rebound test and compressive strength often violates the assumption of homoscedasticity, i.e., homogeneity of variance [11]. A violation of the homogeneity of variance means that changes in the rebound number (x axis) change variation in compressive strength (y axis) as well. In other words, the higher the rebound number measured, the higher the variance in compressive strength. If the violation of homoscedasticity is ignored, the resulting model of relationship can underrepresent the value of compressive strength, especially in higher-strength concrete.

However, the experimental data presented in this paper shows that under certain circumstances the above-mentioned issues can be legitimately dismissed. These are cases where the goal is not to find a general model for every concrete, but a specific model for only one. Moreover, such a model is not intended for measuring compressive strength across the whole spectrum, but for estimating the stripping rebound number, which should correspond to stripping compressive strength (see below).

Most previously published models of the relationship between the rebound number and compressive strength are designed to estimate the median value of compressive strength. The model is therefore designed as a median curve plotted through the experimental data. In theory, a model thus designed is 50% reliable: the measured value of compressive strength is 50% likely to be higher or lower.

In the civil engineering practice, however, most cases do not use 50% reliability, but 95%. Such a value of a material property is called a characteristic value. Concerning compressive strength, there is the term characteristic strength [30]. It is essentially always a 5% quantile, meaning that 95% of test results should be higher than this value. It is, therefore, a sort of one-sided interval estimate of the parameter value. Even when assessing concrete strength based on NDT results, it would often be useful to have such a one-sided estimate in the form of a curve. Points on this curve, which could be called a characteristic curve, would determine 95% one-sided interval estimates of compressive strength; i.e., characteristic strength. This section presents one of the possible ways how to construct such a curve using experimental data.

This is done using the above-mentioned method of least squares in its simplest form of a linear regression model [33]:

$$y = b_1 + b_2 \cdot x, \tag{7}$$

where y is the conventional dependent variable (compressive strength), x is the independent variable (rebound number) and b_1 and b_2 are the regression coefficients being determined. As will be shown later, this simple model appears suitable for all the experimental data examined here. For a better understanding of how the characteristic curve is constructed, we shall demonstrate the principle of the least squares method.

The experimental data is essentially pairs (x_i, y_i), where x_i are the rebound number values and y_i are the corresponding values of compressive strength. Further, $i = 1, \ldots, n$, where n is the number of measurements, i.e., value pairs. An important role in the least squares method is played by the matrix

$$\mathbf{H} = \begin{bmatrix} n & \sum x_i \\ \sum x_i & \sum x_i^2 \end{bmatrix}, \tag{8}$$

where $\sum$ signifies $\sum_{i=1}^{n}$, and its determinant, which can be expressed as

$$\det \mathbf{H} = n \sum x_i^2 - \left(\sum x_i \right)^2. \tag{9}$$

The median values of regression coefficients can then be expressed using the following formulas:

$$b_2 = \frac{n \sum x_i y_i - \sum x_i \sum y_i}{\det \mathbf{H}}, \tag{10}$$

$$b_1 = \bar{y} - b_2 \bar{x}, \tag{11}$$

where $\bar{y}$ and $\bar{x}$ are the mean values of the properties being measured [34].

At this point the model has a curve passing through the experimental data. It is generally recommended to supplement this curve with so-called confidence bands. This means a confidence band for the median curve, which is normally determined by 95 % interval estimates of b_1 and b_2, and a prediction band, which determines the 95% interval estimate of compressive strength for the given rebound number. In order to construct such bands it is necessary to express the minimum value of the sum of square errors [33,34]

$$S_{min}^* = \sum_{i=1}^{n} (y_i - b_1 - b_2 x_i)^2 \tag{12}$$

and point estimator of variance

$$s^2 = \frac{S_{min}^*}{n - 2}. \tag{13}$$

Then, for any fixed x, it is necessary to denote the value of h^*:

$$h^* = \frac{1}{n} + \frac{n(x - \bar{x})^2}{\det \mathbf{H}}. \tag{14}$$

The confidence band for the median value is then determined using the formula

$$\left\langle (b_1 + b_2 x) - t_{(1-\alpha/2)} s \sqrt{h^*}; (b_1 + b_2 x) + t_{(1-\alpha/2)} s \sqrt{h^*} \right\rangle, \tag{15}$$

where $t_{(1-\alpha/2)}$ is $(1 - \alpha/2)$ quantile of Student's t-distribution with $n - 2$ degrees of freedom [33,34]. Next, the confidence band for the individual values (prediction band) is determined by the following formula

$$\left\langle (b_1 + b_2 x) - t_{(1-\alpha/2)} s \sqrt{1 + h^*}; (b_1 + b_2 x) + t_{(1-\alpha/2)} s \sqrt{1 + h^*} \right\rangle. \tag{16}$$

Both regression bands are shown in Figure 5 through Figure 11. It is useful to expand this traditional regression analysis method with regression coefficient testing and assessing the overall aptitude of the model with the multiple correlation coefficient, which in our case equals the correlation coefficient r. The aptitude of the model is most commonly denoted by the coefficient of determination r^2. The number $r^2 \times 100\%$ (conventionally) signifies the percentage of y_i, which is explained by the regression model.

It is important to remember that the confidence bands are constructed as two-sided interval estimates of y for every fixed x. This method of denotation works for most applications of regression analysis. However, seeing as civil engineering usually works with the characteristic value (5% quantile), this method is not quite ideal.

We construct the characteristic curve using the relationship for determining the prediction band. The breadth of the band is determined by variance s^2, regression coefficients b_1 and b_2, the h^* value for every fixed x, and Student's t-distribution. This quantile determines, among others, the aptitude of the confidence bands, i.e., the probability with which one could expect that the true y-value is indeed located within this band. Adjusting the $t_{(1-\alpha/2)}$ quantile to the $t_{(1-\alpha)}$ quantile with the same degree of freedom $n - 2$ at $\alpha = 0.05$ makes it possible to obtain the desired one-sided interval estimate. The characteristic curve $y_{0.05}$ can then be written as

$$y_{0.05} = (b_1 + b_2 x) - t_{(1-\alpha)} s \sqrt{1 + h^*}. \tag{17}$$

This curve essentially determines the value of compressive strength for every fixed x, i.e., for every fixed result of a rebound test. It is, therefore, a half-plane

$$\langle +\infty; y_{0.05} \rangle. \tag{18}$$

The following sections show the use of this characteristic curve for determining the characteristic stripping hardness with real-world data.

3. Experiment

The goal of the measurements and their evaluation was to create a new conversion relationship for determining the compressive strength of concrete by rebound hammer tests. These relationships concern mainly early-age strength: the so-called stripping strength, i.e., in the range of 5 to 10 N/mm^2. The experiment took place in three stages.

3.1. Determining the Conversion Relationship for Compressive Strength of Concrete Used in Bridge Construction

The goal of the first stage was two-fold; to determine a conversion relationship for the rebound number to compressive strength for the selected concrete and to ascertain whether the Original Schmidt N and SilverSchmidt rebound hammers are suitable for this purpose: whether the statistical evaluation would confirm that the more advanced SilverSchmidt suffers from greater measurement variability [25].

The experiment required making 18 cubic specimens of 150 mm in size. The specimens were cast on-site, during the construction of an arch bridge on Svitavská street in Brno, see Figure 1. The concrete in question was C 30/37 XF4; its composition is shown in Table 1. The properties of used cement are shown in Tables 2 and 3. Basic properties of all used aggregates (including aggregates used in Sections 3.2 and 3.3) are shown in Table 4. The admixtures are described in Table 1 by name and further information can be found in the manufacturer's technical data [35,36].

The specimens were made from concrete taken from three concrete mixer trucks. Six cube specimens were made from each concrete sample, see Figure 2a. The fresh-state properties, determined according to EN 12350 [37–39], are in Table 5, and the slump test is pictured in Figure 2b. After pouring, the specimens were covered with a PE sheet and left at the construction site for 24 h. At the age of 48 h they were transported to a laboratory at the Faculty of Civil Engineering, BUT, where they were removed from moulds and divided into 6 groups of 3 so that each group would contain a cube from truck 1, truck 2, and truck 3. The first group of 3 specimens was tested (compressive strength according to EN 12390-3 [40]) and the remaining 15 were placed under water at a temperature of (20 ± 2) °C. The other groups were tested at the age of 3, 7, 14, 28, and 90 days. The reason the concrete was tested at different ages was to obtain data that show how the concrete's properties develop over time. Because the concrete's composition is known, it is possible to use the single-parameter formulation of compressive strength using the rebound number.

Figure 1. Bridge construction during which concrete C 30/37 XF4 was sampled.

Table 1. Composition of concrete C 30/37 XF4.

Component	Amount (kg Per 1 m^3)
Cement CEM I 42.5 R (Mokrá cement plant)	400
Aggregate 0-4 mm—Ledce (see Table 4)	700
Aggregate 8-16 mm—Olbramovice (see Table 4)	669
Aggregate 11-22 mm—Lomnička (see Table 4)	284
Water	172
Air-entraining admixture—Sika$^{®}$ Aer 200	0.60
Plasticiser—Sika$^{®}$ ViscoCrete$^{®}$-5-800 Multimix (AT)	2.40

Table 2. Chemical properties of Cement CEM I 42.5 R (Mokrá cement plant) according to EN 196-2 [41].

Component/Property	Average Value (%)
CaO	63.7
SiO_2	19.6
Al_2O_3	4.8
Fe_2O_3	3.3
MgO	1.4
SO_3	3.1
Cl^-	0.040
K_2O	0.75
Na_2O	0.19
Na_2O Equivalent	0.69
Insoluble residue	0.7
Loss of ignition	3.4

Table 3. Basic physical and mechanical properties of Cement CEM I 42.5 R (Mokrá cement plant) according to EN 196-6 [42] and EN 196-8 [43].

Parameter	Average Value
Blain (m^2/kg)	408
Density (kg/m^3)	3110
Heat of hydration (7 days) (J/g)	310

Table 4. Basic properties of used aggregates.

Aggregate	Type	Rock	Specific Density (Mg/m^3)	Loose Bulk Density (Mg/m^3)
Ledce	Natural quarried	Gravel sand	2.553	1.408
Lípa	Natural quarried	Gravel sand	2.583	1.508
Bratčice	Natural quarried	Gravel sand	2.610	1.537
Olbramovice	Natural crushed	Granodiorite	2.640	1.450
Lomnička	Natural crushed	Gneiss	2.690	1.530
Litice	Natural crushed	Granodiorite	2.700	1.600

Table 5. Fresh-state properties of concrete C 30/37 XF4.

Mix Truck	Slump Test (mm)	Air Content (%)	Bulk Density (kg/m^3)
1	240	4.2	2320
2	180	3.8	2370
3	220	3.9	2360

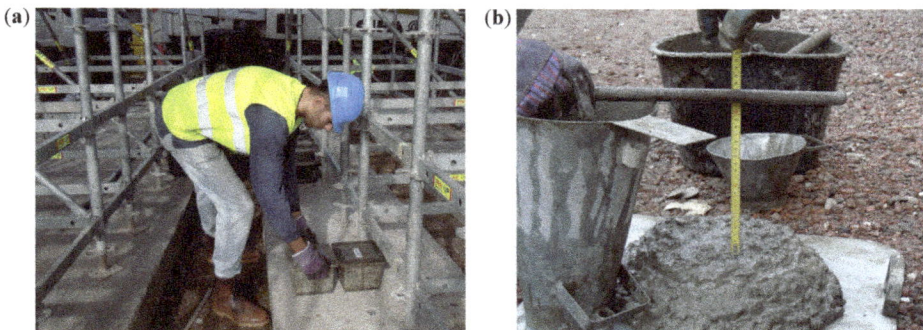

Figure 2. (**a**) cube specimens being made; (**b**) consistency test-slump.

Each cube was measured and weighed, and afterwards mounted in a testing press and compressed with a force of 50 kN. Two opposing sides were tested for hardness with a rebound hammer. A total of 5 readings of the rebound number R were taken using an Original Schmidt N and 5 readings of of the Q-value using a SilverSchmidt N, see Figure 3. Finally, compressive strength was determined according to [40].

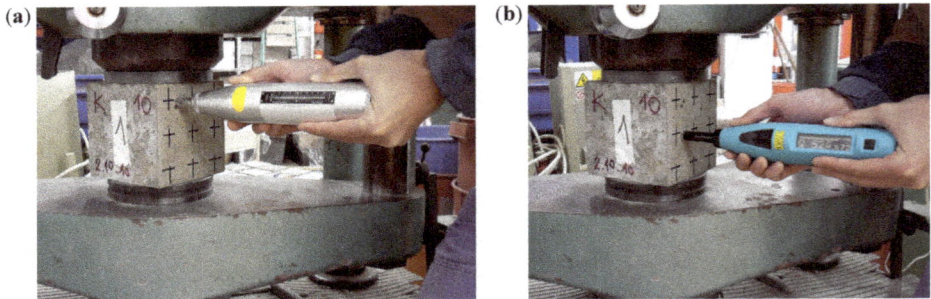

Figure 3. (**a**) testing a cube with an Original Schmidt N; (**b**) with a SilverSchmidt N.

3.2. Determining the Conversion Relationship for the Stripping Strength of Precast Concrete

The goal of the second part of the experiment was to determine the relationship between the rebound number and compressive strength for concrete C 50/60, which is used in precast prestressed girders. Effort was made to capture low (stripping) strength. The formula used for making the V03 girders is detailed in Table 6. V03 girders have an asymmetrical I cross-section, length of 25.346 m, height of 1.5 m and were used in the construction of a shopping centre.

Table 6. Composition of concrete C 50/60.

Component	Amount (kg Per 1 m^3)
Cement CEM I 42.5 R (Mokrá cement plant)	450
Aggregate 0-4 mm—Lípa (see Table 4)	690
Aggregate 4-8 mm—Litice (see Table 4)	215
Aggregate 8-16 mm—Litice (see Table 4)	845
Water	180
Plasticiser—Stachement 2180	4.50

While the girders were being made, concrete C 50/60 was sampled and made into 18 cube-shaped specimens with the size of 150 mm. While still freshly poured in steel moulds, they were covered with a PE sheet and left near the girders in the manufacturing hall, where the ambient temperature did not exceed 15 °C. Because the experiment aimed to test compressive strength at a very young age, the first three cubes were tested 27 h after cement was mixed with water. The other specimens were unmoulded at this age, placed under water and tested at the age of 42, 48, 68, 144, and 656 h. A total of 15 cubes were tested during the first 6 days and the last three at 28 days of age.

The testing was conducted similarly to the first stage of the experiment with the sole difference that besides the Original Schmidt N and SilverSchmidt N, the SilverSchmidt L both with and without the mushroom attachment was used. The SilverSchmidt L with the attachment is primarily used to test very young concrete, which is why it was used only during the first 2 days (i.e., first three measurement times), but not beyond. Figure 4 shows the manufacturing and testing of the cube specimens.

Figure 4. (**a**) making of the specimens; (**b**) manufacture of the V03 girders.

3.3. Determining the Conversion Relationship for the Stripping Strength of Concretes of Similar Composition

While the previous parts of the experiment were carried out with only one kind of concrete, the third involved making cube specimens from six different concrete mixtures. They only differed in the amount of cement, water, and admixtures; in essence, the w/c ratio. The actual components of the six concretes (marked I through VI) were identical. The amount of water was always balanced against the amount of admixtures so as to achieve the same workability in all the mixtures. Table 7 shows the composition of the concretes.

Table 7. Composition of the concretes.

Component	Amount (% Per 1 m^3)					
	I	II	III	IV	V	VI
Cement CEM I 42.5 R (Mokrá cement plant)	87.50	100.00	87.50	100.00	87.50	100.00
Aggregate 0/4 mm—Bratčice (see Table 4)	218.75	206.25	218.75	206.25	218.75	206.25
Aggregate 4/8 mm—Olbramovice (see Table 4)	46.25	46.25	46.25	46.25	46.25	46.25
Aggregate 8/16 mm—Olbramovice(see Table 4)	173.75	173.75	173.75	173.75	173.75	173.75
Water	44.75	44.75	43.50	43.50	41.00	41.00
Plasticiser—Sika® ViscoCrete® 4035	0.220	0.250	0.438	0.500	0.438	0.500
Air-entraining admixture—Sika® LPS A 94	0.000	0.000	0.000	0.000	0.188	0.188

The basic fresh-state properties determined in compliance with [37–39] are detailed in Table 8.

Table 8. Composition of the concretes.

Property	Concrete					
	I	II	III	IV	V	VI
Bulk Density (kg/m^3)	2300	2300	2270	2300	2190	2260
Slump test (mm)	50	60	70	50	60	50
Air content (%)	2.8	3.2	3.5	3.0	6.2	5.7

Each concrete was used to make 6 cube specimens sized 150 mm. This made a total of 36 test specimens. While still in polyurethane moulds, the specimens were covered with a PE sheet and stored at standard laboratory conditions (ambient temperature of (20 ± 2) °C and humidity of (60 ± 10)%), where they aged until testing. All the specimens were tested during the next day after being cast; the age ranged between 16 and 36 h. Immediately after unmoulding, the specimens were measured and weighed, and then mounted in a testing press and tested for the Q-value ten times using a SilverSchmidt PC L with the mushroom accessory attached; five measurements on two opposing sides. Finally, the compressive strength test was performed according to [40]. The goal was to determine whether it is possible to create one conversion relationship for the stripping compressive strength of several concretes of similar composition.

4. Results and Discussion

Test results presented in Section 3.1 (concrete bridge) and Section 3.2 (precast concrete) each time represent one concrete of known composition. The hardness and strength tests were designed so that the experimental data described the development of compressive strength over time. This justifies the single-parameter formulation of compressive strength based on the rebound number using linear regression model. The same type of model was used by the authors of the article [22], while most authors (see [3,7,10,19,44]) use rather polynomial or exponential function curves. In addition, authors of [11] refer to dozens of other publications, all of which use non-linear functions. The linear model was used in Section 3.3 as well, where there were several different kinds of concrete, but of very similar composition. These concretes varied in the amount of admixtures, aggregate, and w/c ratio, but the raw materials were the same.

The data evaluation was focused on the stripping rebound number, which should correspond to stripping compressive strength with 95% reliability; i.e., the value read from the characteristic curve (cf. Section 2.2). Stripping compressive strength, meaning strength at which formwork can be safely removed, is set here at 5 N/mm^2.

The evaluation of results of the bridge concrete (Section 3.1) focused on the aptitude of the linear regression model for tests performed with the Original Schmidt N and SilverSchmidt N impact hammers (Figures 5 and 6). The values of the coefficient of determination r^2 indicate that both these models show high aptitude, which, compared to Szilágyi's et al. conclusions [25] is slightly higher in the case of the SilverSchmidt N. The results also show that the linear regression model used herein is more suitable for the evaluation of measurements made with a SilverSchmidt N than the exponential model proposed in [29] (see Figure 6). In both cases, a characteristic curve was created and the stripping rebound numbers evaluated. Given the range of the data obtained, approx. 30 to 80 N/mm^2, these values are strongly extrapolated and thus are for illustration only. Table 7 summarises the results.

The results of tests described in Section 3.2 (precast concrete) are plotted in Figures 7–10 including linear regressions. In this case the linear regression model also showed aptitude, exhibiting high values of the coefficient of determination r^2 (see Table 9). The results of tests performed with an Original Schmidt N on this particular concrete show that relationships described in [21] may undervalue compressive strength (see Figure 7). The non-linear character of the dependence of rebound tests performed with a SilverSchmidt N, L, and L-MP [27,29] was not confirmed.

The greatest aptitude of the linear model was demonstrated by test results for the similar concretes described in Section 3.3 ($r^2 = 0.963$). These tests were performed with a SilverSchmidt L with the mushroom plunger accessory. As with the previous cases, these results indicated no need to use an exponential or other model.

Table 9. Summary.

Rebound Hammer	Concrete Type	$Q_{0.50}$ ($a_{0.50}$)	$Q_{0.05}$ ($a_{0.05}$)	r^2	Figure
Original Schmidt N	Bridge concrete	14.3	17.9	0.878	Figure 5
SilverSchmidt N	Bridge concrete	24.3	28.2	0.922	Figure 6
Original Schmidt N	Precast concrete	11.9	17.2	0.949	Figure 7
SilverSchmidt N	Precast concrete	17.6	26.7	0.922	Figure 8
SilverSchmidt L	Precast concrete	18.2	24.6	0.957	Figure 9
SilverSchmidt L–MP	Precast concrete	17.4	23.6	0.953	Figure 10
SilverSchmidt L–MP	Concretes of similar composition	13.5	15.9	0.963	Figure 11

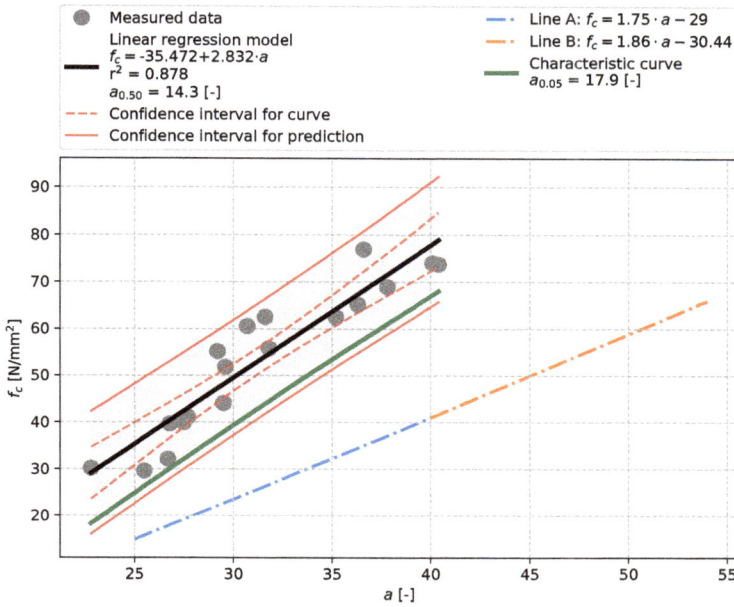

Figure 5. Bridge concrete described in Section 3.1—model of the dependence of the rebound number [-] vs. compressive strength f_c [N/mm^2]—Original Schmidt N.

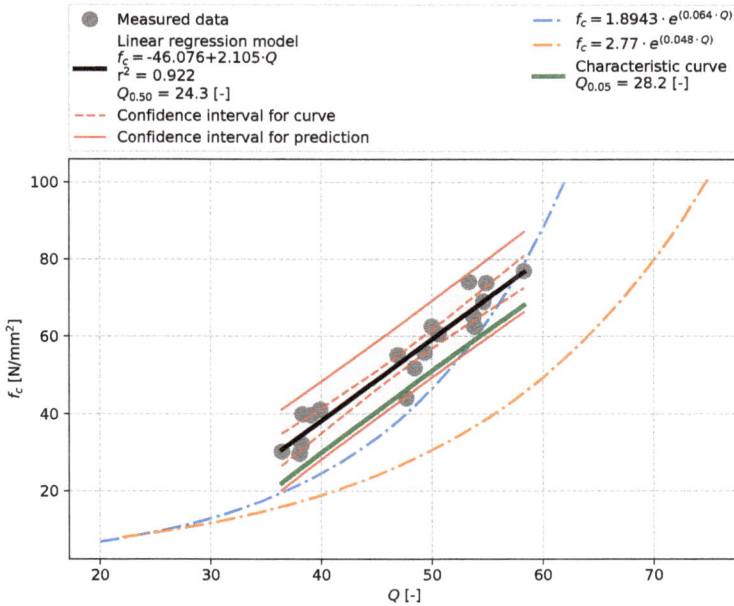

Figure 6. Bridge concrete described in Section 3.1—model of the dependence of the Q-value [-] vs. compressive strength f_c [N/mm^2]—SilverSchmidt N.

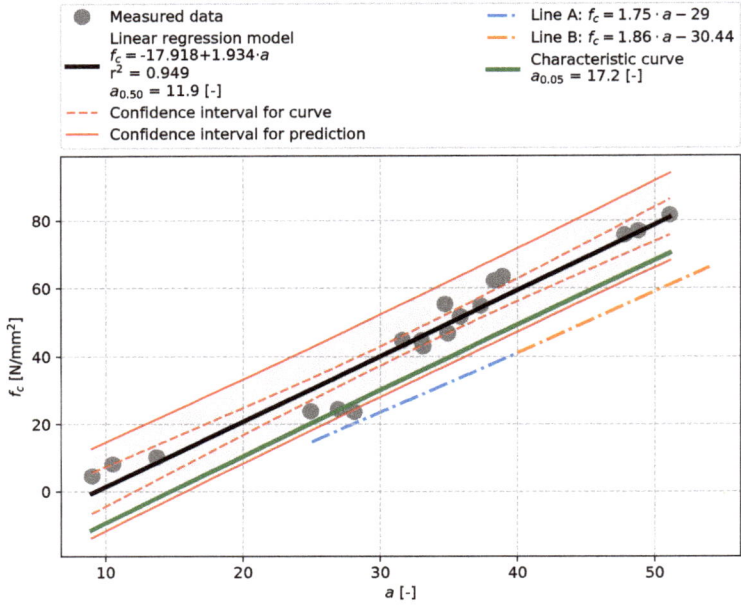

Figure 7. Precast concrete described in Section 3.2—model of the dependence of a [-] vs. compressive strength f_c [N/mm^2]—Original Schmidt N.

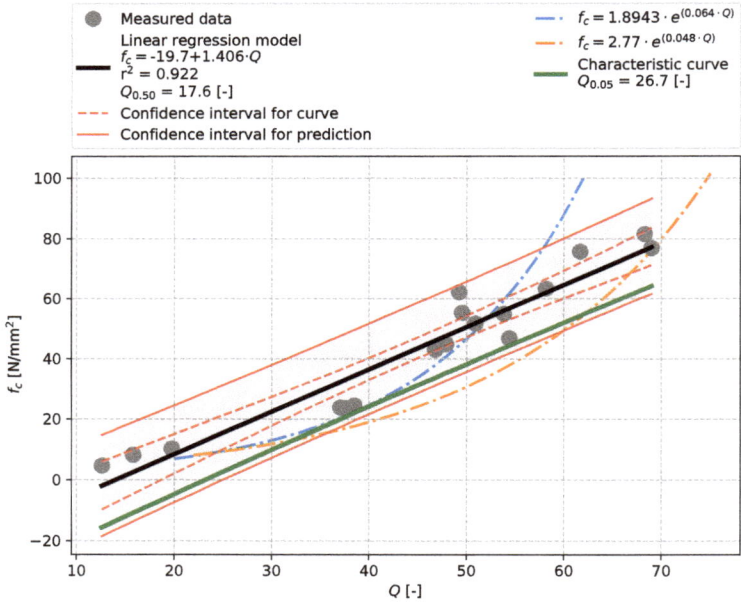

Figure 8. Precast concrete described in Section 3.2—model of the dependence of Q-value [-] vs. compressive strength f_c [N/mm^2]—SilverSchmidt N.

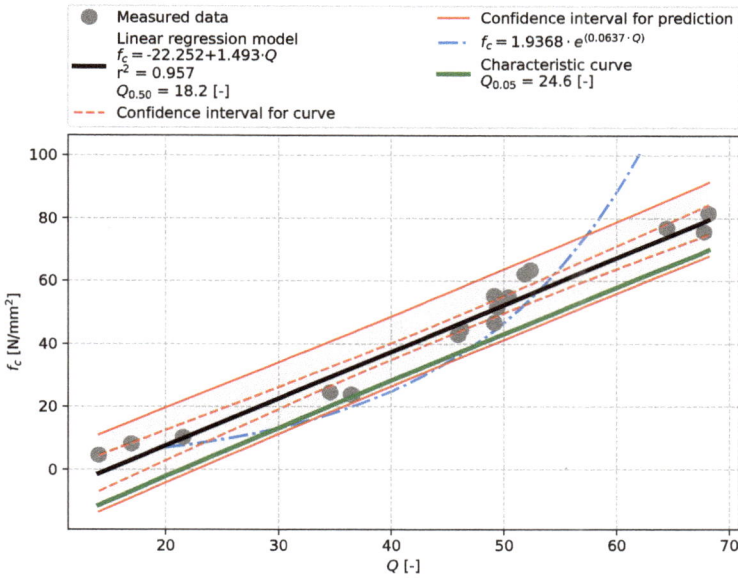

Figure 9. Precast concrete described in Section 3.2—model of the dependence of Q-value [-] vs. compressive strength f_c [N/mm^2]—SilverSchmidt L.

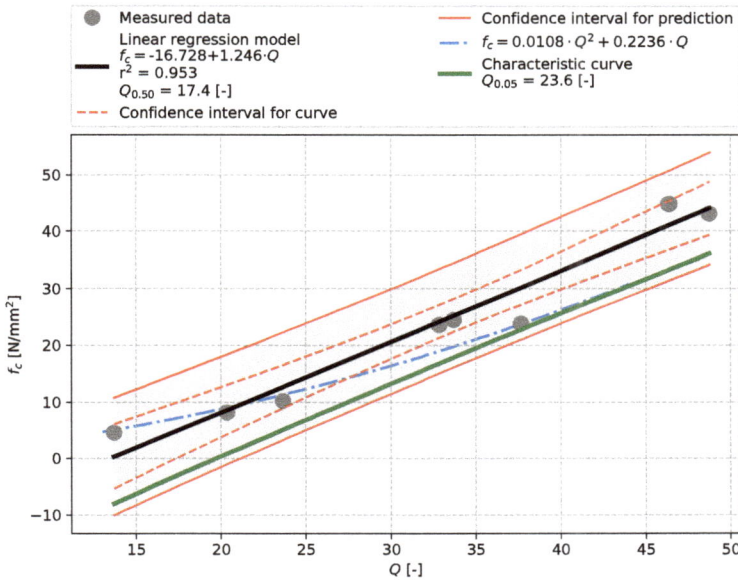

Figure 10. Precast concrete described in Section 3.2—model of the dependence of Q-value [-] vs. compressive strength f_c [N/mm^2]—SilverSchmidt L—MP.

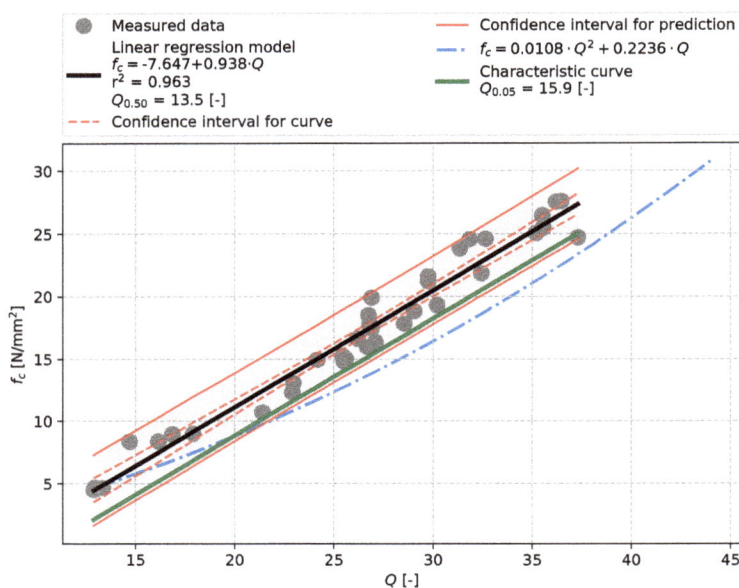

Figure 11. Precast concrete described in Section 3.3—model of the dependence of Q-value [-] vs. compressive strength f_c [N/mm^2]—SilverSchmidt L – MP.

5. Conclusions

The experiments indicate the following:

- The paper describes an innovative design of constructing a characteristic curve, which can be used to find the stripping rebound number obtained by different types of rebound hammers. Using the characteristic curve appears suitable for finding a specific rebound number which would indicate, with 95% reliability, that the concrete has the required stripping strength (in this paper it is 5 N/mm^2).
- The data shows that the concretes tested herein (including concretes with different formulas but the same raw materials) do not require a regression model more complex than a simple line. Moreover, this simplest shape of regression model enables a relatively simple formulation of the characteristic curve.
- We certainly do not question the validity of relationships described elsewhere (Section 2.1), or their ability to represent the relationship between the rebound number and compressive strength. It should also be remembered that these general models were created for concretes of virtually any composition, and a simple line would not be enough.
- Our regression models and rebound numbers representing stripping strength always apply to the one specific concrete they were designed for and do not work with others. The goal of the paper was to show how such regression models may be created for any concrete.

Author Contributions: Conceptualization, D.K. and P.M.; methodology, D.K., P.M. and P.C.; software, P.M.; validation, D.K. and P.M.; formal analysis, D.K. and P.M.; investigation, D.K. and P.C.; resources, D.K.; data curation, P.M.; writing–original draft preparation, D.K. and P.M.; writing–review and editing, D.K. and P.M.; visualization, P.M.; supervision, P.C.; project administration, D.K.; funding acquisition, D.K.

Funding: This paper has been written as a part of project No. 19-22708S, supported by the GAČR—Czech Science Foundation and project No. FAST-S-19-6002, supported by Brno University of Technology.

Acknowledgments: Authors would like to thank to Zahrada, Bílek and Žalud for support.

Conflicts of Interest: The authors declare no conflict of interest.

References

1. Mir, A.E.; Nehme, S.G. Repeatability of the rebound surface hardness of concrete with alteration of concrete parameters. *Constr. Build. Mater.* **2017**, *131*, 317–326. [CrossRef]
2. Bungey, J.H.; Millard, S.G.; Grantham, M. *Testing of Concrete in Structures*, 4th ed.; CRC Press: London, UK, 2006.
3. Ali-Benyahia, K.; Sbartaï, Z.M.; Breysse, D.; Kenai, S.; Ghrici, M. Analysis of the single and combined non-destructive test approaches for on-site concrete strength assessment. *Case Stud. Constr. Mater.* **2017**, *6*, 109–119. [CrossRef]
4. *Handbook on Nondestructive Testing of Concrete*, 2nd ed.; CRC Press: Boca Raton, FL, USA, 2004.
5. Xu, T.; Li, J. Assessing the spatial variability of the concrete by the rebound hammer test and compression test of drilled cores. *Constr. Build. Mater.* **2018**, *188*, 820–832. [CrossRef]
6. Review of the Rebound Hammer Method Estimating Concrete Compressive Strength on Site. In Proceedings of the International Conference on Architecture And Civil Engineering (ICAACE'14), Dubai, UAE, 25–26 December 2014; pp. 118–127. [CrossRef]
7. Breysse, D. Nondestructive evaluation of concrete strength. *Constr. Build. Mater.* **2012**, *33*, 139–163. [CrossRef]
8. Balayssac, J.P.; Garnier, V. *Non-Destructive Testing and Evaluation of Civil Engineering Structures*; Elsevier: Kidlington, UK, 2018.
9. Breysse, D.; Martínez-Fernández, J.L. Assessing concrete strength with rebound hammer. *Mater. Struct.* **2014**, *47*, 1589–1604. [CrossRef]
10. Jain, A.; Kathuria, A.; Kumar, A.; Verma, Y.; Murari, K. Combined Use of Non-Destructive Tests for Assessment of Strength of Concrete in Structure. *Procedia Eng.* **2013**, *54*, 241–251. [CrossRef]
11. Szilágyi, K.; Borosnyói, A.; Zsigovics, I. Rebound surface hardness of concrete. *Constr. Build. Mater.* **2011**, *25*, 2480–2487. [CrossRef]
12. Schmidt, E. *Rebound Hammer for Concrete Testing*; Schweiz Bauztg: Berlin, Germany, 1950.
13. Szilágyi, K.; Borosnyói, A.; Dobó, K. Static indentation hardness testing of concrete: A long established method revived. *Epa.-J. Silic. Based Compos. Mater.* **2011**, *63*, 2–8. [CrossRef]
14. Basu, A.; Aydin, A. A method for normalization of Schmidt hammer rebound values. *Int. J. Rock Mech. Min. Sci.* **2004**, *41*, 1211–1214. [CrossRef]
15. Bilgin, N.; Dincer, T.; Copur, H. The performance prediction of impact hammers from Schmidt hammer rebound values in Istanbul metro tunnel drivages. *Tunn. Undergr. Space Technol.* **2002**, *17*, 237–247. [CrossRef]
16. Liang, R.; Hota, G.; Lei, Y.; Li, Y.; Stanislawski, D.; Jiang, Y. Nondestructive Evaluation of Historic Hakka Rammed Earth Structures. *Sustainability* **2013**, *5*, 298–315. [CrossRef]
17. Viles, H.; Goudie, A.; Grab, S.; Lalley, J. The use of the Schmidt Hammer and Equotip for rock hardness assessment in geomorphology and heritage science. *Earth Surf. Process. Landforms* **2011**, *36*, 320–333. [CrossRef]
18. Bui, Q.B. Assessing the Rebound Hammer Test for Rammed Earth Material. *Sustainability* **2017**, *9*, 1904. [CrossRef]
19. Mohammed, B.S.; Azmi, N.J.; Abdullahi, M. Evaluation of rubbercrete based on ultrasonic pulse velocity and rebound hammer tests. *Constr. Build. Mater.* **2011**, *25*, 1388–1397. [CrossRef]
20. Original Schmidt Live Cue Cards. Available online: https://www.proceq.com/compare/schmidt-hammers/ (accessed on 25 June 2019).
21. ÚNMZ, Prague. *ČSN 73 1373: Non-Destructive Testing of Concrete—Determination of Compressive Strength by Hardness Testing Methods*, 1st ed.; ÚNMZ: Prague, Czech Republic, 2011.
22. Kim, J.K.; Kim, C.Y.; Yi, S.T.; Lee, Y. Effect of carbonation on the rebound number and compressive strength of concrete. *Cem. Concr. Compos.* **2009**, *31*, 139–144. [CrossRef]
23. Panedpojaman, P.; Tonnayopas, D. Rebound hammer test to estimate compressive strength of heat exposed concrete. *Constr. Build. Mater.* **2018**, *172*, 387–395. [CrossRef]
24. Alwash, M.; Breysse, D.; Sbartaï, Z.M.; Szilágyi, K.; Borosnyói, A. Factors affecting the reliability of assessing the concrete strength by rebound hammer and cores. *Constr. Build. Mater.* **2017**, *140*, 354–363. [CrossRef]

25. Szilágyi, K.; Borosnyói, A.; Zsigovics, I. Extensive statistical analysis of the variability of concrete rebound hardness based on a large database of 60years experience. *Constr. Build. Mater.* **2014**, *53*, 333–347. [CrossRef]
26. Schmidt Rebound Hammers. Available online: https://www.proceq.com/compare/schmidt-hammers/ (accessed on 21 June 2019).
27. SilverSchmidt. Available online: https://www.proceq.com/uploads/tx_proceqproductcms/import_data/files/SilverSchmidt_Operating%20Instructions_English_high.pdf (accessed on 21 June 2019).
28. Original Schmidt Live. Available online: https://www.proceq.com/product/original-schmidt-live/ (accessed on 21 June 2019).
29. The SilverSchmidt Reference Curve. Available online: https://www.pcte.com.au/silver-schmidt-rebound-hammer (accessed on 21 June 2019).
30. CEN. *EN 13791: Assessment of In-Situ Compressive Strength In Structures and Precast Concrete Components;* CEN: Brussels, Belgium, 2006.
31. ASTM International. *ASTM C805/C805M-18: Standard Test Method for Rebound Number of Hardened Concrete;* ASTM International: West Conshohocken, PA, USA, 2018.
32. American Concrete Institute. *ACI 228.1R-03: In-Place Methods to Estimate Concrete Strength;* American Concrete Institute: Farmington Hills, MI, USA, 2003.
33. Chatterjee, S.; Simonoff, J.S. *Handbook of Regression Analysis;* Wiley: Hoboken, NJ, USA, 2013.
34. Pedhazur, E.J.; Kerlinger, F.N. *Multiple Regression in Behavioral Research*, 2nd ed.; Holt, Rinehart and Winston: New York, NY, USA, 1982.
35. Sika Company. Available online: https://cze.sika.com/ (accessed on 21 June 2019).
36. Stachema. Available online: https://prisadydobetonu.stachema.cz/ (accessed on 21 June 2019).
37. CEN. *EN 12350-2: Testing Fresh Concrete—Part 2: Slump-Test*, 1st ed.; CEN: Brussels, Belgium, 2009.
38. CEN. *EN 12350-6: Testing Fresh Concrete—Part 6: Density;* CEN: Brussels, Belgium, 2009.
39. CEN. *EN 12350-7: Testing Fresh Concrete—Part 7: Air Content—Pressure Methods;* CEN: Brussels, Belgium, 2009.
40. CEN. *EN 12390-3: Testing Hardened Concrete—Part 3: Compressive Strength of Test Specimens;* CEN: Brussels, Belgium, 2009.
41. CEN. *EN 196-2: Method of Testing Cement—Part 2: Chemical Analysis of Cement;* CEN: Brussels, Belgium, 2013.
42. CEN. *EN 196-6: Methods of Testing Cement—Part 6: Determination Of Fineness;* CEN: Brussels, Belgium, 2018.
43. CEN. *EN 196-8: Methods of Testing Cement—Part 8: Heat of Hydration—Solution Method;* CEN: Brussels, Belgium, 2010.
44. Pucinotti, R. Reinforced concrete structure. *Constr. Build. Mater.* **2015**, *75*, 331–341. [CrossRef]

materials

MDPI

Article

Effect of Freeze–Thaw Cycling on the Failure of Fibre-Cement Boards, Assessed Using Acoustic Emission Method and Artificial Neural Network

Tomasz Gorzelańczyk * and **Krzysztof Schabowicz**

Faculty of Civil Engineering, Wrocław University of Science and Technology, Wybrzeże Wyspiańskiego 27, 50-370 Wrocław, Poland
* Correspondence: tomasz.gorzelanczyk@pwr.edu.pl

Received: 9 June 2019; Accepted: 4 July 2019; Published: 7 July 2019

check for
updates

Abstract: This paper presents the results of investigations into the effect of freeze–thaw cycling on the failure of fibre-cement boards and on the changes taking place in their structure. Fibre-cement board specimens were subjected to one and ten freeze–thaw cycles and then investigated under three-point bending by means of the acoustic emission method. An artificial neural network was employed to analyse the results yielded by the acoustic emission method. The investigations conclusively proved that freeze–thaw cycling had an effect on the failure of fibre-cement boards, as indicated mainly by the fall in the number of acoustic emission (AE) events recognized as accompanying the breaking of fibres during the three-point bending of the specimens. SEM examinations were carried out to gain better insight into the changes taking place in the structure of the tested boards. Interesting results with significance for building practice were obtained.

Keywords: fibre-cement boards; non-destructive testing; acoustic emission; artificial neural networks; SEM

---

1. Introduction

Fibre-cement boards have been used in construction since the beginning of the last century. Their inventor was the Czech engineer Ludwik Hatschek, who developed and patented the technology of producing fibre cement, then called "Eternit". The material was strong, durable, lightweight, moisture-resistant, freeze–thaw resistant and non-combustible [1]. Fibre cement became one of the most popular roofing materials in the world in the 20th century. This was so until one of its components (i.e., asbestos) was found to be carcinogenic. In the 1990s, asbestos was replaced with environment-friendly fibres, mainly cellulose fibres. The fibre-cement boards produced today are made up of cement, cellulose fibres, synthetic fibres and various additives and admixtures. They are a completely different building product than the original one [2], and still require investigation and improvement. The additional components and fillers of fibre-cement boards are lime powder, mica, perlite, kaolin, microspheres and recycled materials [3,4], whereby fibre-cement boards can be regarded as an innovative product which fits into the sustainable development strategy. At present, such boards are used in construction mainly as ventilated façade cladding [5], as illustrated in Figure 1. In the course of their service life, fibre-cement boards are exposed to various factors, such as chemical (acid rains) and physical aggressiveness (ultraviolet radiation), but mainly to variable environmental impacts, including sub-zero temperatures in the winter season.

Figure 1. Exemplary uses of fibre-cement boards as ventilated façade cladding.

After a few winter seasons, the effect of sub-zero temperatures—especially of temperature (freeze–thaw) cyclicity—needs to be determined in order to establish whether the fibre-cement boards can remain in service as ventilated façade cladding. The knowledge of this effect is essential not only from the scientific point of view, but also for building practice. It is worth noting that research on fibre-cement boards has so far been mostly limited to determining—solely through the bending strength (modulus of rupture, *MOR*) test—their standard physicomechanical parameters and the effect of in-service factors (e.g., soaking–drying cycles, heating–raining cycles and high temperatures) and the various fibres and production processes [6]. Only a few cases of testing fibre-cement boards by non-destructive methods, limited to imperfections arising during the production process, can be found in the literature [7–10]. Besides the effects of high temperature and fire described in [11,12], the impact of sub-zero temperatures is one of the most destructive in-service factors to many building products, particularly composite products containing reinforcement in the form of various fibres, especially fibres of organic origin (to which cellulose fibres belong). In the authors' opinion, freeze–thaw cycles can very adversely affect the durability of such composites. Experiments were carried out in order to prove this thesis. The experiments consisted of subjecting fibre-cement board specimens to 1 and 10 freeze–thaw cycles and then investigating them under three-point bending by means of the acoustic emission (AE) method. Artificial intelligence in the form of an artificial neural network [13] was employed to analyse the experimental results. Previous studies by the authors [11,12,14,15] presented the assessment of the effect of freeze–thaw cycling based solely on bending strength to be inadequate. Whereas in this study, using the acoustic emission technique and analysing the degradation of the specimens, the authors were able to describe the degrading changes in the structure of the tested boards on the basis of not only the mechanical parameters, but also the acoustic phenomena. The registered AE signals provided the basis for developing reference acoustic spectrum characteristics accompanying cement matrix cracking and fibre breaking during bending. Then, an artificial neural network was used to recognize the characteristics in the AE records. In the course of freeze–thaw cycling, the fibres in the boards gradually degraded, which manifested as a fall in the number of events recognized as accompanying the breaking of fibres. This is described in more detail later in this paper. In order to verify the results and gain better insight into the changes taking place in the structure of the fibre-cement boards, they were examined under a scanning electron microscope (SEM).

2. Literature Survey

To-date, research on fibre-cement boards has focused on the effect of in-service factors [16–18] and the effect of high temperatures, determined by testing the physicomechanical parameters of the boards—mainly their bending strength (*MOR*). Only a few cases of testing fibre-cement boards by non-destructive methods, including the acoustic emission method, have been reported in the literature. Ardanuy et al. [6] presented the results of investigations into the effect of high temperatures on fibre-cement boards, but were limited to the bending strength test. Li et al. [19] examined the effect of high temperatures on composites produced using the extrusion method, but solely on the basis of the mechanical properties of the composite. Schabowicz et al. [11,12] used non-destructive methods to assess the effects of high temperature and fire on the degree of degradation of fibre-cement boards on the basis of the physicomechanical parameters. Other reported investigations of fibre-cement boards were devoted to the detection of imperfections arising during the production process. Papers by Drelich et al. [8] and Schabowicz et al. [20] presented the possibility of exploiting Lamb waves in a non-contact ultrasound scanner to detect defects in fibre-cement boards at the production stage. A method of detecting delaminations in composite elements by means of an ultrasonic probe was presented in a study by Stark, Vistap et al. [21]. Ultrasonic devices and a method used to detect delaminations in fibre-cement boards were described by Dębowski et al. [7]. Berkowski et al. [22], Hoła and Schabowicz [23] and Davis et al. [24] proposed the use of the impact-echo method jointly with the impulse response method to recognize delaminations in concrete elements. However, it is not recommended to test fibre-cement boards in this way since the two methods are intended for testing elements which are thicker than 100 mm. The special hammer used in the impulse response method can damage the fibre-cement boards being tested, while in the impact-echo method multiple wave reflections cause disturbances which make it difficult to interpret the obtained image [22]. Therefore, it is inadvisable to use the two methods to test fibre-cement boards, which are about 8 mm thick. There is scant information in the literature on the use of other non-destructive methods to test fibre-cement boards. The preliminary research described in [9,25] showed the terahertz (T-Ray) method to be suitable for testing fibre-cement boards. The character of terahertz signals is very similar to that of ultrasonic signals, but their interpretation is more complicated. Schabowicz et al. [15] and Ranachowski et al. [26] used X-ray microtomography to identify delaminations and low-density regions in fibre-cement boards. This technique was found to precisely reveal differences in the microstructure of such boards. It can be a useful tool for testing the structure of fibre-cement boards in which defects can arise as a result of production errors, but it is applicable only to small boards. As already mentioned, only a few cases of testing fibre-cement boards by means of acoustic emission have been reported so far. Ranachowski et al. [26] carried out pilot tests on fibre-cement boards produced by extrusion, including boards exposed to the temperature of 230 °C for 2 h, using the acoustic emission method to determine the effect of cellulose fibres on the strength of the fibre-cement boards and tried to distinguish the AE events emitted by the fibres from the ones emitted by the cement matrix. The investigations showed this method to be suitable for testing fibre-cement boards. Schabowicz et al. [11,12] and Gorzelańczyk et al. [11] proposed the use of the acoustic emission method to study the effects of fire and high temperatures on fibre-cement boards. The effect of high temperatures on concrete has been studied using the acoustic method (e.g., by Ranachowski [27] and Ranachowski et al. [28–30]), and is described widely in the literature. A large quantity of data are recorded during acoustic emission measurements, and they need to be properly analysed and interpreted. For this purpose, it can be useful to combine the acoustic emission method and artificial intelligence, including artificial neural networks (ANNs). ANNs are used to analyse and recognize signals acquired during the failure of various materials [31]. In [32–34] ANNs were used to analyse the results of testing concrete by means of non-destructive methods. Łazarska et al. [35] and Woźniak et al. [36] in their investigations of steel successfully used the acoustic emission method and artificial neural networks to analyse the obtained results. Rucka and Wilde [37,38], Zielińska and Rucka [39] and Wojtczak and Rucka [40] successfully used the ultrasonic method to investigate damage to concrete structures and masonry pillars. ANNs were also successfully used by Schabowicz et al. [11,12] to analyse the results of tests consisting of exposing fibre-cement boards to fire and high temperature.

Considering the above information, the authors came to the conclusion that the acoustic emission method combined with artificial neural networks would be suitable for assessing the changes taking place in the structure of fibre-cement boards exposed to freeze–thaw cycling.

3. Strength Tests

Two series of fibre-cement boards, denoted respectively A and B, were tested to determine the effect of freeze–thaw cycling. Altogether 60 specimens were tested. The basic specifications of the boards in the two series are given in Table 1.

Table 1. Basic specifications of the tested fibre-cement boards of series A and B.

Series	Board Thickness e (mm)	Board Colour	Application	Board Bulk Density ρ (g/cm^3)	Flexural Strength *MOR* (MPa)
A	8.0	natural	exterior	1.65	24
B	8.0	full body coloured	exterior	1.58	38

The freeze–thaw cycling of the specimens was conducted as follows. First the specimens were cooled (frozen) at a temperature of -20 ± 4 °C in a freezer for 1–2 h and kept at this temperature for the next hour. The specimens were heated (thawed) in a water bath at a temperature of 20 ± 4 °C for 1–2 h and kept at this temperature for the next hour. One should note that the above temperatures apply to the conducting medium (i.e., air or water). Each freeze–thaw cycle lasted on average about 5–6 h. Air-dry reference specimens (not subjected to freeze–thaw cycling) were denoted as A_R and B_R. The denotations of exemplary series of boards are presented in Table 2. Figure 2 shows exemplary views of the tested (20×100 mm and 8 mm thick) specimens.

Table 2. Series of fibre-cement boards and test cases and their denotations.

Series Name/Test Case	Series A	Series B
Air-dry condition (reference board)	A_R	B_R
1 freeze–thaw cycles	A_1	B_1
10 freeze–thaw cycles	A_{10}	B_{10}

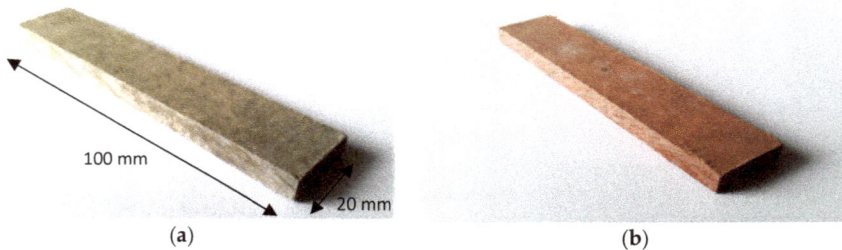

100 mm

20 mm

(a)   (b)

Figure 2. Tested fibre-cement board specimens: (**a**) board A, (**b**) board B.

In order to determine the effect of freeze–thaw cycling on the fibre-cement boards, the latter were subjected to three-point bending and investigated using the acoustic emission method. Breaking force F, strain ε and AE signals were registered in the course of the three-point bending. Figure 3 shows the three-point bending test stand and the acoustic emission measuring equipment.

Figure 3. (**left**) Three-point bending test stand and acoustic emission (AE) measuring equipment and (**right**) close-up of fibre-cement board specimen during test.

The curve of flexural stress σ_m, the flexural strength (*MOR*), the limit of proportionality (*LOP*) and strain ε were taken into account in the analysis of the experimental results. The *MOR* was calculated from the standard formula [41]:

$$MOR = \frac{3Fl_s}{2b\,e^2},\tag{1}$$

where:

F is the loading force (N);
l_s is the length of the support span (mm);
b is the specimen width (mm); and
e is the specimen thickness (mm).

Figure 4 shows σ–ε graphs under bending for the specimens of all the tested fibre-cement boards.

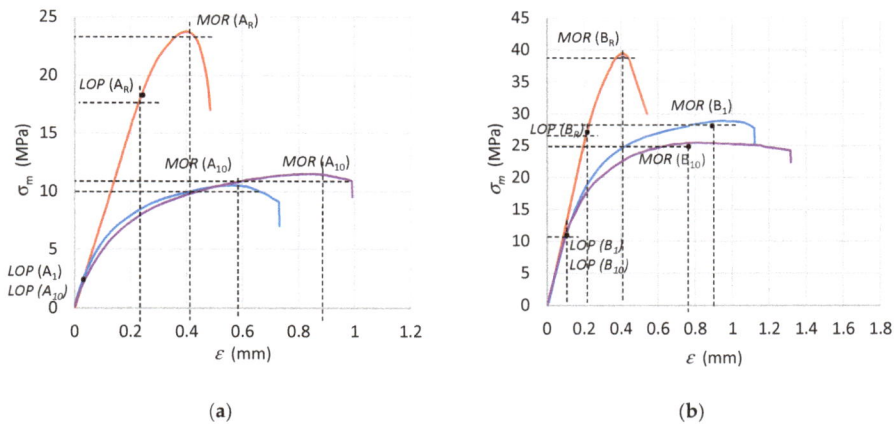

(**a**)   (**b**)

Figure 4. Bending σ–ε relation for specimens of fibre-cement boards: (**a**) series A, (**b**) series B. *LOP*: limit of proportionality; *MOR*: modulus of rupture.

Figure 4 shows that as a result of freeze–thaw cycling, flexural strength (*MOR*) decreased by 35%–50% in comparison with the flexural strength (*MOR*) of the reference fibre-cement boards. No significant difference in the change of bending strength (*MOR*) between the specimens subjected to one or ten freeze–thaw cycles was noticed. Note that the reference specimens were tested in air-dry condition at a mass moisture of 6%–8%. The effect of moisture on the value of flexural strength should be mainly ascribed to the weakening of the bonds between the crystals of the cement matrix structural lattice. The weakening is due to the fact that the bonds partially dissolve at a higher moisture content in the material, whereby the flexural strength (*MOR*) slightly decreases. As regards the path of the $\sigma-\varepsilon$, curve, one can see (Figure 4) that it clearly changed with the number of freeze–thaw cycles for both tested series of boards. Therefore it can be concluded that for the tested series of fibre-cement boards it was possible to determine the effect of the number of freeze–thaw cycles, as reflected in not only a reduction in flexural strength (*MOR*), but also in changes in the path of the $\sigma-\varepsilon$ curve. It was found that when the number of cycles was increased from 1 to 10, the stiffness of the fibre-cement board and its brittleness decreased. In the case of the tested series, as the number of cycles increased, so did the range of the nonlinear increase in flexural stress while the limit of proportionality (*LOP*) considerably decreased. Thus, one can conclude that destructive changes took place in the structure of the fibre-cement boards. However, in the authors' opinion, knowledge of the mechanical parameters is not enough to determine the damaging effect of freeze–thaw cycles on fibre-cement boards. Therefore, in order to better identify the changes taking place in the structure of the tested fibre-cement boards, the acoustic method and an artificial neural network were used in this research.

4. Investigations Conducted Using Acoustic Emission Method and Artificial Neural Network

The next step in this research on degrading changes in the structure of fibre-cement boards exposed to freeze–thaw cycling was an analysis of the AE signals registered in the course of the three-point bending test. The analysis was based on AE descriptors such as: events rate N_{ev}, events sum $\sum N_{ev}$ and events energy E_{ev}, and on the signal frequency distribution. Figure 5 shows exemplary values of events sum $\sum N_{ev}$ registered for the boards of series A_R–A_{10} and B_R–B_{10}.

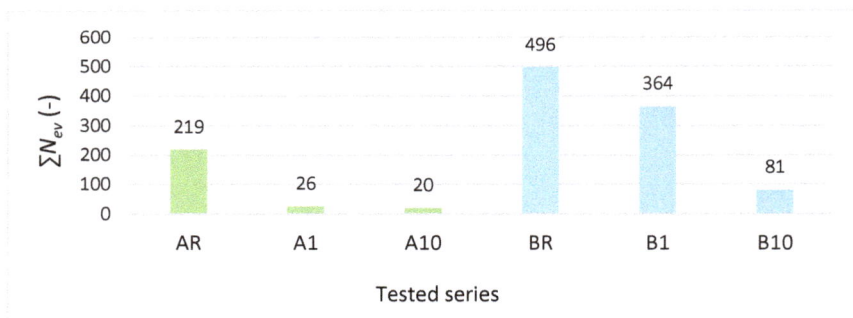

Figure 5. Exemplary events sum $\sum N_{ev}$ values registered for air-dry specimens and specimens subjected to freeze–thaw cycling.

For a more precise analysis of the failure of the boards under bending and the effect of freeze–thaw cycling, events sum $\sum N_{ev}$ and flexural stress σ_m versus time for selected cases are shown in Figure 6.

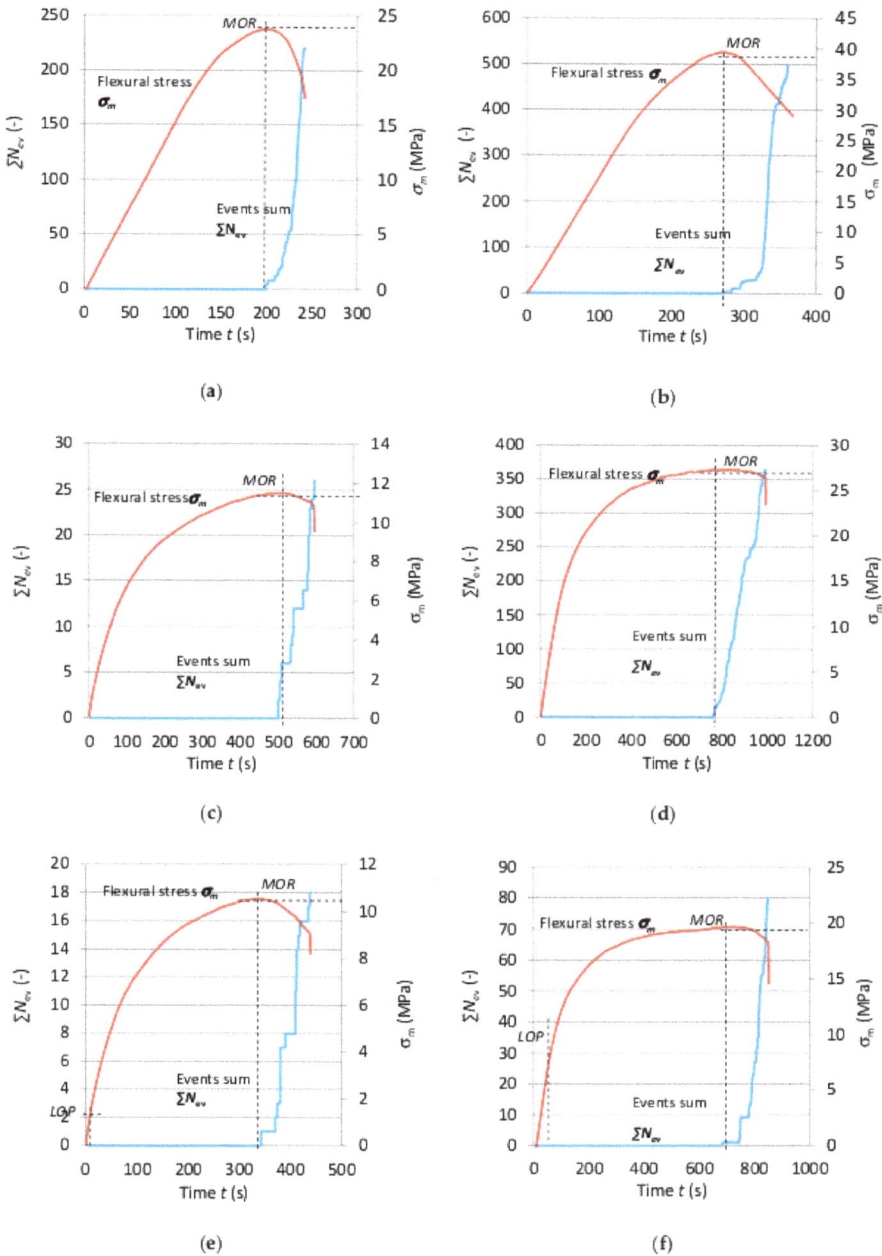

Figure 6. Flexural stress σ_m and events sum $\sum N_{ev}$ as function of time t for boards of: (**a**) series A_R, (**b**) series B_R, (**c**) series A_1, (**d**) series B_1, (**e**) series A_{10}, (**f**) series B_{10}.

A clear fall in registered events and a change in the path of events sum $\sum N_{ev}$ can be seen in Figure 6. The events were registered after the flexural strength (*MOR*) was exceeded. Note that a reduction in the number of events and a decrease in their energy were observed after subjecting the boards to freeze–thaw cycling. This does not mean that no AE events occurred, but one can suppose that the discrimination threshold for the registered events in the case of exposure to freeze–thaw cycling was too high. This can be connected with the measuring capability of the equipment used. Whereas the registered events with much lower energy could indicate a different process of destruction of the fibres—the latter can be pulled out of the wet cement matrix, whereby the event energy declines. These suppositions could be confirmed by SEM image analysis, which is presented further in this paper.

A spectral analysis of the AE event characteristics was carried out to more precisely identify the origins of the registered AE events. Reference acoustic spectra for cement matrix cracking, fibre breaking and the acoustic background were selected on this basis, which is described in detail in [12]. The reference acoustic spectra for cement matrix cracking were selected by analysing the record of acoustic activity in the time-frequency system during the bending of fibre boards previously fired in a laboratory furnace at a temperature of 230 °C for 3 h. It should be mentioned here that the investigations presented in this paper are part of a larger project devoted to the testing of fibre-cement boards. For example, in [11] it was observed that when fibre-cement boards were exposed to a temperature of 230 °C for 3 h, it resulted in the pyrolysis of their cellulose fibres, whereby the obtained fibre-cement board structure was completely devoid of fibres. Owing to this, reference acoustic spectra could be obtained for cement matrix cracking alone. The reference acoustic spectrum characteristic for fibre breaking was selected from the spectra obtained for air-dry reference boards of series A_R to B_R, characterized by a repetitively similar characteristic in the frequency range of 10–24 kHz, clearly distinct from the cement matrix characteristic. The characteristic of the background acoustic spectrum, originating from the testing machine, was determined by averaging the characteristics for all the tested boards of series A and B in the initial phase of bending.

One should note that the selected spectral fibre breaking characteristics are understood as the signal accompanying the cracking of the cement matrix together with the fibres, whereas the spectral matrix characteristic is understood as the signal accompanying the cracking of the cement matrix alone. The selected acoustic spectrum characteristic reference standards were recorded every 0.5 kHz in 80 intervals. Figure 7 shows a record of the reference acoustic spectrum characteristics of the signal accompanying respectively cement matrix cracking and fibre breaking, and of the background.

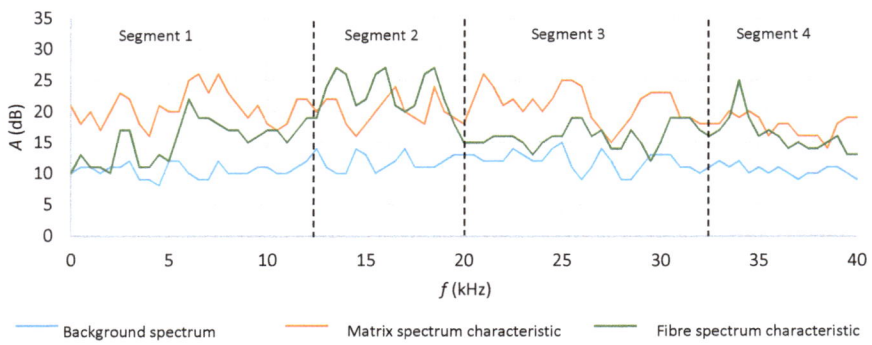

Figure 7. Background, fibre and cement matrix acoustic spectrum characteristics as function of frequency [12].

Figure 7 shows that the background acoustic activity was at the level of 10–15 dB. The cement matrix acoustic spectrum characteristic reached the acoustic activity of 25 dB within frequency ranges 5–10 kHz (segment 1) and 20–32 kHz (segment 3). The acoustic activity of over 25 dB within frequency ranges 12–18 dB (segment 2) and 32–38 kHz (segment 4) was read off for the fibres.

The cement matrix, fibre and background reference standards were implemented in an artificial neural network, and the training and testing of the latter began. A unidirectional multilayer backpropagation structure with momentum was adopted for the ANN. A model of the artificial neuron is shown in Figure 8. The model includes N inputs, one output, a summation block and an activation block.

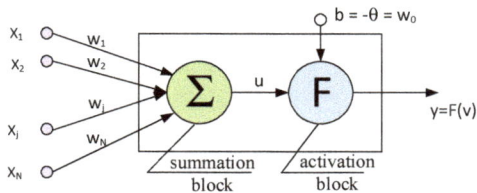

Figure 8. Model of an artificial neuron [13,42].

The following variables and parameters were used to describe the model shown in the Figure 8:

$$x_i = (x_1, x_2, \dots, x_N) \qquad \text{an input vector,} \tag{2}$$

$$w_i = (w_1, w_2, \dots, w_N) \qquad \text{a weight vector,} \tag{3}$$

$$b = -\theta = w_0 \qquad \text{a bias,} \tag{4}$$

$$v = u + b = \sum_{j=1}^{N} w_j x_j - \theta = \sum_{j=0}^{N} w_j x_j \quad \text{a network potential,} \tag{5}$$

$$F(v) \qquad \text{an activation function.} \tag{6}$$

A model of the artificial neural network with inputs, information processing neurons and output neurons is shown in Figure 9.

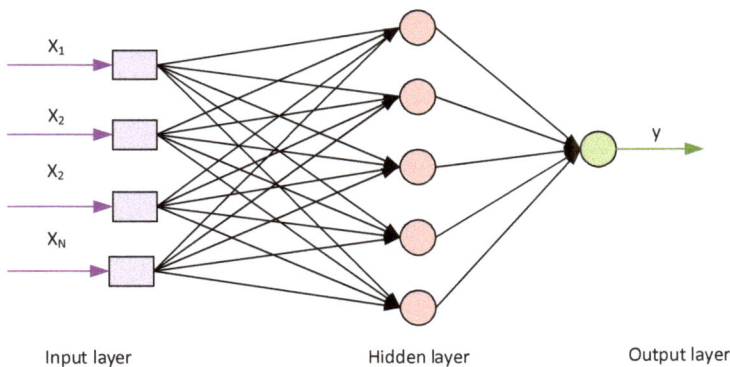

Figure 9. Model of the artificial neural network [13,43].

Eight appropriate learning sequences were adopted iteratively to achieve optimal compatibility of the learned ANN with the training pattern, as presented in Table 3.

Table 3. ANN learning sequences.

No.	Learning Sequences	No.	Learning Sequences
1	16,000 × A, 8000 × B, 4000 × C, 2000 × D	5	16,000 × A, 8000 × B, 4000 × C
2	16,000 × A, 8000 × B, 4000 × C, 2000 × D	6	16,000 × A, 8000 × B, 4000 × C
3	16,000 × A, 8000 × B, 4000 × C, 2000 × D	7	16,000 × A, 8000 × B, 1000 × C
4	16,000 × A, 8000 × B, 4000 × C	8	16,000 × A, 8000 × B

The spectral characteristics of fibre breaking were assigned to input A, the characteristics of cement matrix cracking were assigned to input B and the spectral characteristics of the background were assigned to input C. The spectral characteristics of fibre breaking were reproduced at input D.

After the ANN was trained on the input data, its mapping correctness was verified using the training and testing data. For this purpose, two pairs of input data were fed, that is, the data used for training the ANN, to check its ability to reproduce the reference spectra, and the one used for testing the ANN, to check its ability to identify the reference spectral characteristics originating from the fibres and the cement matrix during the bending test. For the eight performed training sequences, the ANN compliance with the training standard amounted to 0.995. Then, records of the ANN output in the form of recognised acoustic spectra for, respectively, fibre breaking, matrix cracking and the background were obtained. The learning coefficient (accelerating learning) was adopted at the level of 0.01 and momentum (increasing the stability of the obtained network configuration) at the level of 0.1. The following sigmoidal activation function was used: $1/(1 + \exp(-x_i))$.

Figure 10 shows the results of the recognition of the reference acoustic spectrum standards. They are superimposed on the record of events rate N_{ev} and bending stress σ_m versus time. The diagrams are for the reference specimens of series A_R and for the series subjected to 1 and 10 freeze–thaw cycles (respectively A_1 and A_{10}). In order to better illustrate the recognized acoustic spectra, the matrix reference standards are marked green while the fibre reference standards are marked light brown.

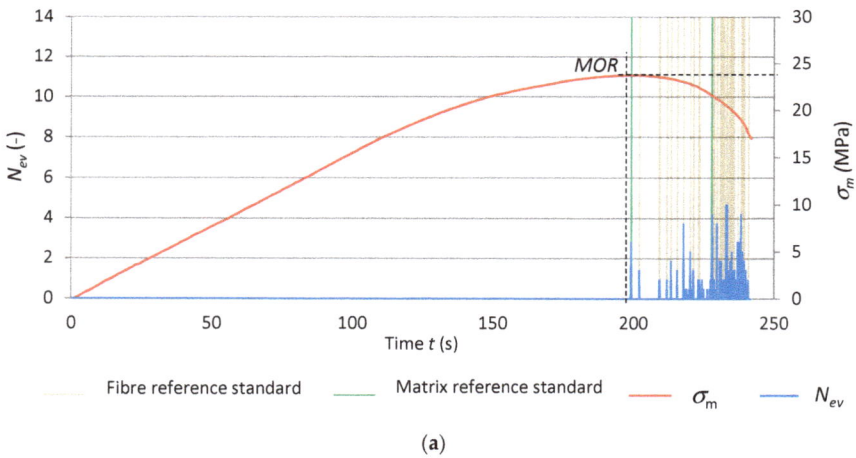

(a)

Figure 10. *Cont.*

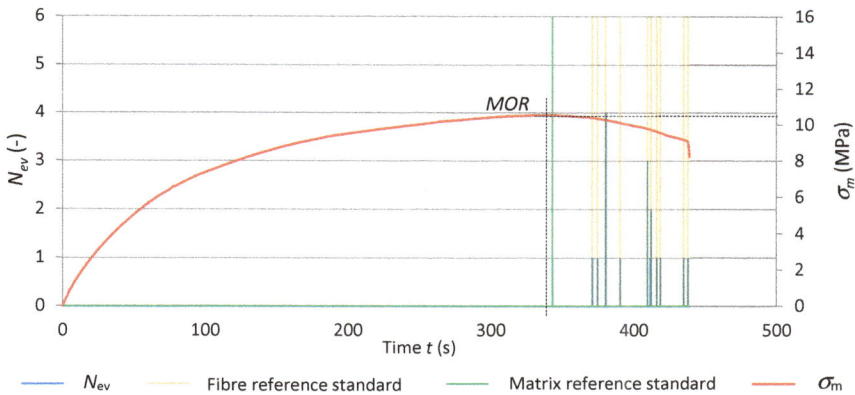

(b)

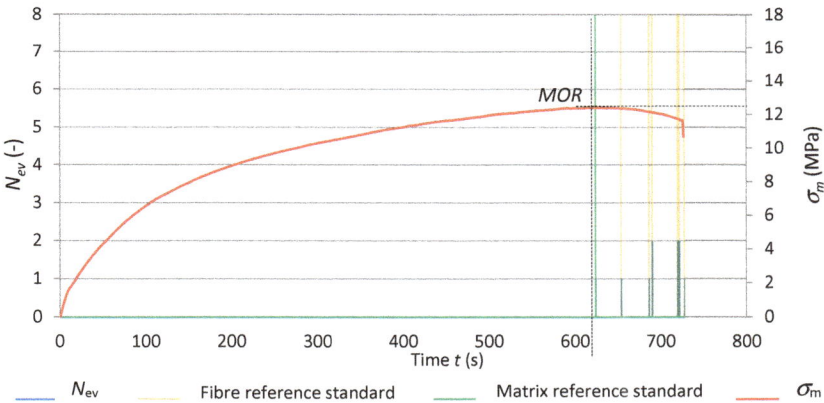

(c)

Figure 10. Freeze–thaw cycling diagrams of events rate N_{ev} and bending stress σ_m versus time, with superimposed identified reference spectral characteristics: (**a**) series A_R, (**b**) series A_1, (**c**) series A_{10}.

Figure 10 clearly shows that the number of registered events for the fibre-cement boards subjected to 1 and 10 freeze–thaw cycles was lower in comparison with the reference boards. Slightly fewer events were registered after ten freeze–thaw cycles. These events had very low energy E_{ev} ranging from 10 to 100 nJ. An event originating from a cement matrix fracture initiates next events originating from the breaking of fibres or from their pulling out of the matrix. Under the influence of moisture and freeze–thaw cycles, the cement-fibre board became more plastic, which manifested in the disappearance of the interval in which strains were proportional to stresses. The matrix cracked once the flexural strength (*MOR*) was exceeded. The ten freeze–thaw cycles limited the registered events to solely cement matrix cracking after the exceedance of the flexural strength (*MOR*).

Table 4 shows events recognized as respectively accompanying fibre breaking and cement matrix cracking for the tested fibre-cement boards of series A_R–A_{10}.

Table 4. Events recognized as respectively accompanying fibre breaking and cement matrix cracking for fibre-cement boards of series A_R–A_{10}.

Series	Events Sum $\sum N_{ev}$	Sum of Recognized Events $\sum N_{ev,r}$	Sum of Events Assigned to Fibre Breaking $\sum N_{ev,f}$	Sum of Events Assigned to Matrix Cracking $\sum N_{ev,m}$
A_R	219	201	193	8
A_1	26	24	20	4
A_{10}	20	19	16	3

Graphs of events rate N_{ev} and flexural stress σ_m versus time under freeze–thaw cycling, with superimposed identified reference spectral characteristics for the specimens of series B_R, B_1 and B_{10} are shown in Figure 11.

Figure 11 shows that the course of the AE signals registered for the fibre-cement boards of series B_1 was similar to that for series B_R, whereas the flexural stress σ_m curves differed. The limit of proportionality clearly fell and the interval in which the increment in stress relative to strain was nonlinear was wider. These changes can be ascribed to cement matrix yielding. The numerous events with the high energy of over 1000 nJ, originating from fibre breaking, indicate fibre destruction similar as in the reference boards. Therefore, it can be concluded that the fibres used in the tested boards did not change their properties under the influence of moisture and freeze–thaw cycling. This was confirmed by the high flexural strength (*MOR*) of about 27 MPa. In the case of the fibre-cement boards of series B_{10} subjected to ten freeze–thaw cycles, a slight decrease (about 20%) in flexural strength (*MOR*) and a drop in fibre breaking events in comparison with series B_1 were observed. This means that the long-lasting dampness in conjunction with the freeze–thaw cycling had a degrading effect on the boards of series B_{10}, reducing their strength and increasing their deformability. Moreover, the decrease in the sum of registered events in comparison with the reference boards could also be due to the lower acoustic activity of the events accompanying the breaking of the fibres, emitted below the discrimination threshold.

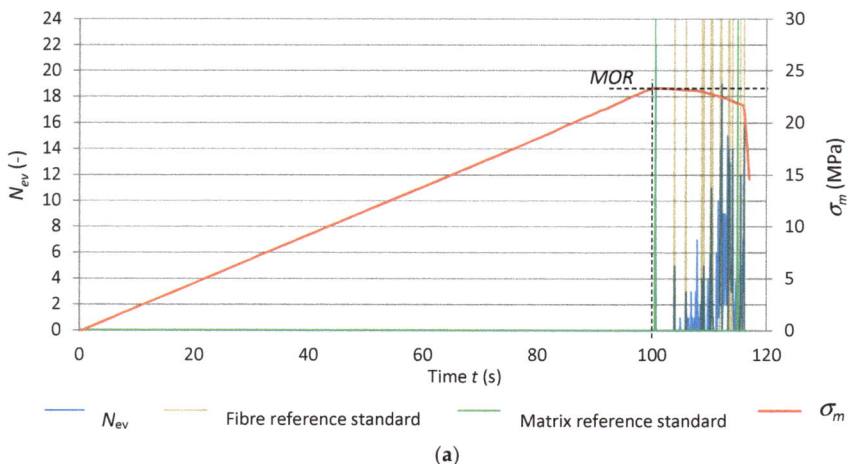

(**a**)

Figure 11. *Cont.*

(b)

(c)

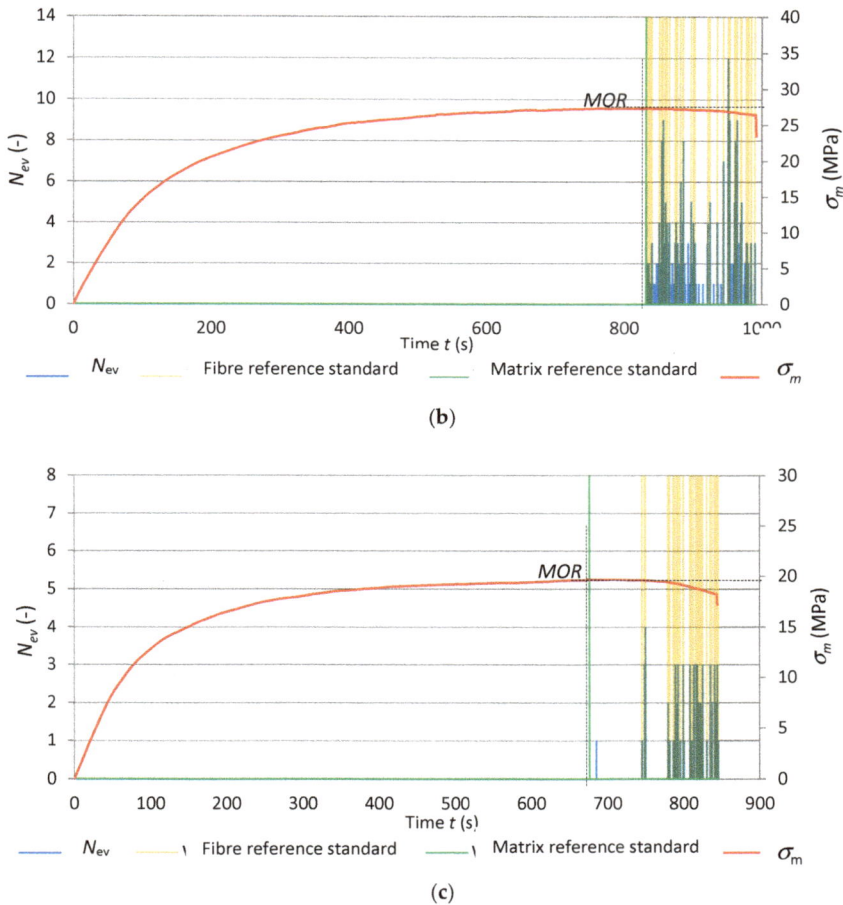

Figure 11. Events rate N_{ev} and flexural stress σ_m versus time, under freeze–thaw cycling, with superimposed identified reference spectral characteristics for: (**a**) series B_R, (**b**) series B_1, (**c**) series B_{10}.

Summing up the results of the investigations, one can conclude that the effect of freeze–thaw cycling manifested itself mainly in a decrease in flexural strength (*MOR*) for all the tested series. The graphs presented in Figures 10 and 11 show that when the flexural strength (*MOR*) was reached, the cement matrix cracked. The cracking of the cement matrix initiated events consisting of the breaking of the fibres. In all the series, the linear increment in stress relative to strain was found to clearly increase. The freeze–thaw cycling reduced the acoustic activity of the events taking place during the bending test. One can suppose that the different ways in which the fibres are destroyed as a result of freeze–thaw cycling can contribute to the decrease in the energy of fibre breaking events. Fibres can be damp and swollen and when undergoing destruction emit events characterized by very low energy. Whereas dampness-resistant PVA fibres (polyvinyl alcohol) instead of breaking can be pulled out of the damp cement matrix, which also will not generate high energy events. One can suppose that by optically examining the image of the fractured surface one can gain information about the mode of failure of the fibres.

Table 5 shows the events recognized as accompanying fibre breaking and cement matrix cracking for the boards of series B_R–B_{10}.

Table 5. Events recognized as accompanying fibre breaking and cement-matrix cracking for boards of series B_R–B_{10}.

Series	Events Sum	Sum of Recognized Events	Sum of Events Assigned to Fibre Breaking	Sum of Events Assigned to Matrix Cracking
	$\sum N_{ev}$	$\sum N_{ev,r}$	$\sum N_{ev,f}$	$\sum N_{ev,m}$
B_R	496	487	439	48
B_1	364	350	332	18
B_{10}	81	76	68	8

5. Investigations Conducted Using Scanning Electron Microscope (SEM)

A high-resolution environmental scanning electron microscope (SEM) Quanta 250 FEG, FEI with an EDS analyser was used for the investigations. Figures 12 and 13 show exemplary images obtained by means of the SEM for respectively series A_R–A_{10} and B_R–B_{10}. The examined fibre-cement boards had previously been subjected to the three-point bending test.

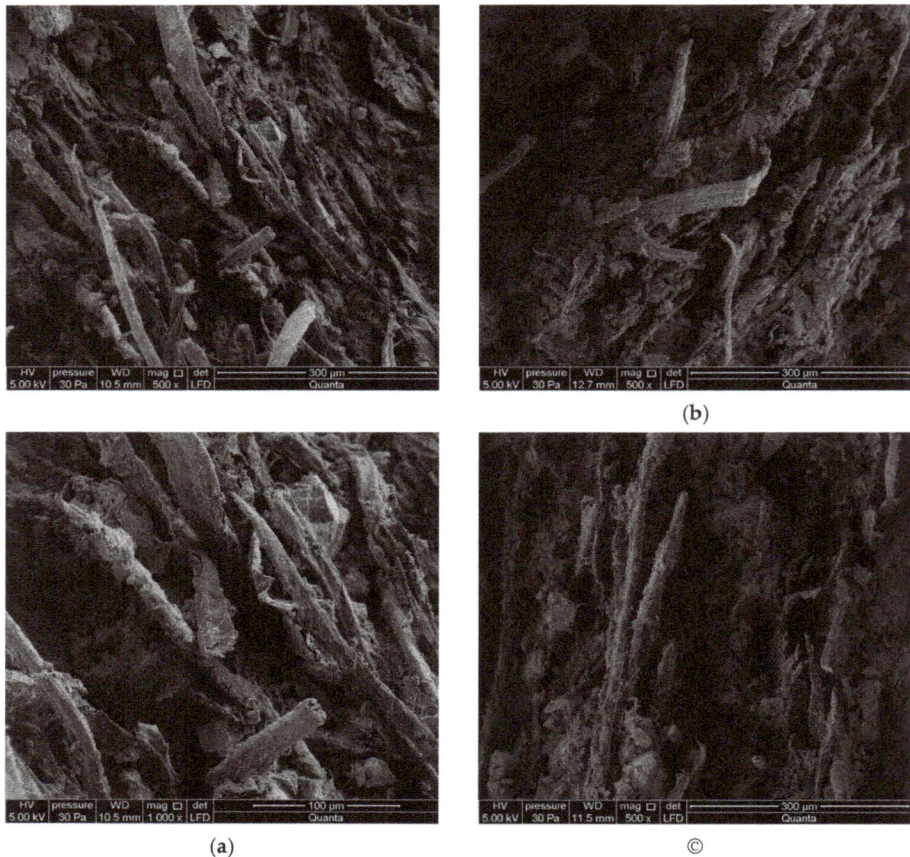

Figure 12. SEM images of boards: (**a**) series A_R (magnification 1000× at top, 500× at bottom), (**b**) series A_1, (**c**) series A_{10}.

Figure 13. SEM images of boards: (**a**) series B_R (magnification 1000× at top, 500× at bottom), (**b**) series B_1, (**c**) series B_{10}.

On the basis of an analysis of the SEM and EDS images, the macrostructure of the fibre-cement boards of series A_R and B_R can be described as compact. The microscopic examinations revealed the structure to be fine-pore, with pore size of up to 50 μm. Cavities up to 500 μm wide and grooves were visible in the places in the fractured surface where fibres had been pulled out. In the images one can see cellulose and PVA fibres. Various forms of hydrated calcium silicates of the C-S-H type occurred, with an "amorphous" phase and a phase comprised of strongly adhering particles predominating. An analysis of the composition of the fibres showed elements native to them and elements native to cement. The surface of the fibres was coated with a thin layer made up of cement matrix and hydration products. The fact that there were few places with a space between the fibres and the matrix indicates that they were strongly bonded. An examination of the fibre-cement boards subjected to one or ten freeze–thaw cycles reveals that most of the fibres were pulled out of the cement matrix. Numerous cavities, most left by the pulled-out cellulose fibres, and grooves were visible. The matrix structure was more granular and included numerous delaminations. Strongly compact structures having an irregular shape were found to be present.

Summing up the results of the investigations, the authors conclude that the freeze–thaw cycling of fibre-cement boards affected their structure. Fibres wholly pulled out of the matrix and the voids left by them in the cement matrix were visible in the fibre-cement boards subjected to freeze–thaw cycling.

This was especially noticeable in the fibre-cement boards subjected to 10 freeze–thaw cycles, mainly in the boards of series A. In the case of the reference boards, much more fibres (whose ends were firmly "anchored" in the cement matrix) were ruptured. This conclusively proves that freeze–thaw cycling significantly weakened the structure of fibre-cement boards.

6. Conclusions

The impact of low temperatures, in the form of freeze–thaw cycling, is by nature destructive to most building products. The degree of resistance to this impact is measured by the number of freeze–thaw cycles after which such products retain the properties specified by the standards. The investigations of the fibre-cement boards of series A and B carried out as part of this study showed that the boards differed in their degree of resistance to the impact of freeze–thaw cycling. Thanks to the use of the acoustic emission method during the three-point bending of the boards, the course of their failure for one or ten freeze–thaw cycles could be observed and compared with that of the reference boards. An artificial neural network was employed to analyse the results yielded by the acoustic emission method. The investigations conclusively proved that freeze–thaw cycling affected the way fibre-cement boards failed, as reflected mainly in the decrease in the number of AE events recognized as accompanying the breaking of fibres during the three-point bending of the specimens. SEM examinations were carried out in order to obtain better insight into the changes taking place in the structure of the tested boards. They confirmed that significant changes took place in the structure of the boards, especially after 10 freeze–thaw cycles. The structure became less compact (more granular). Most of the fibres were not ruptured during the bending test, but pulled out of the cement matrix, as confirmed by the low energy of the registered AE events and their small number. This was particularly evident in the case of the tested fibre-cement boards of series A. In the authors' opinion, the above findings are important for building practice since they indicate that it is inadvisable to use fibre-cement boards whose flexural strength (*MOR*) considerably decreases under the influence of freeze–thaw cycling for the cladding of ventilated façades, especially in high buildings located in zones of high wind load.

Author Contributions: T.G. prepared the specimens, completed the experiments, analysed the test results and performed paper editing; K.S. conceived and designed the experimental work.

Funding: This research received no external funding.

Conflicts of Interest: The authors declare no conflict of interest.

References

1. Schabowicz, K.; Gorzelańczyk, T. Fabrication of fibre cement boards. In *The Fabrication, Testing and Application of Fibre Cement Boards*, 1st ed.; Ranachowski, Z., Schabowicz, K., Eds.; Cambridge Scholars Publishing: Newcastle upon Tyne, UK, 2018; pp. 7–39. ISBN 978-1-5276-6.
2. Bentchikou, M.; Guidoum, A.; Scrivener, K.; Silhadi, K.; Hanini, S. Effect of recycled cellulose fibres on the properties of lightweight cement composite matrix. *Constr. Build. Mater.* **2012**, *34*, 451–456. [CrossRef]
3. Savastano, H.; Warden, P.G.; Coutts, R.S.P. Microstructure and mechanical properties of waste fibre–cement composites. *Cem. Concr. Compos.* **2005**, *27*, 583–592. [CrossRef]
4. Coutts, R.S.P. A Review of Australian Research into Natural Fibre Cement Composites. *Cem. Concr. Compos.* **2005**, *27*, 518–526. [CrossRef]
5. Schabowicz, K.; Szymków, M. Ventilated facades made of fibre-cement boards (in Polish). *Mater. Bud.* **2016**, *4*, 112–114. [CrossRef]
6. Ardanuy, M.; Claramunt, J.; Toledo Filho, R.D. Cellulosic Fibre Reinforced Cement-Based Composites: A Review of Recent Research. *Constr. Build. Mater.* **2015**, *79*, 115–128. [CrossRef]
7. Dębowski, T.; Lewandowski, M.; Mackiewicz, S.; Ranachowski, Z.; Schabowicz, K. Ultrasonic tests of fibre-cement boards (in Polish). *Przegląd Spaw.* **2016**, *10*, 69–71. [CrossRef]
8. Drelich, R.; Gorzelanczyk, T.; Pakuła, M.; Schabowicz, K. Automated control of cellulose fibre cement boards with a non-contact ultrasound scanner. *Autom. Constr.* **2015**, *57*, 55–63. [CrossRef]

9. Chady, T.; Schabowicz, K.; Szymków, M. Automated multisource electromagnetic inspection of fibre-cement boards. *Autom. Constr.* **2018**, *94*, 383–394. [CrossRef]

10. Schabowicz, K.; Jóźwiak-Niedźwiedzka, D.; Ranachowski, Z.; Kudela, S.; Dvorak, T. Microstructural characterization of cellulose fibres in reinforced cement boards. *Arch. Civ. Mech. Eng.* **2018**, *4*, 1068–1078. [CrossRef]

11. Schabowicz, K.; Gorzelańczyk, T.; Szymków, M. Identification of the degree of fibre-cement boards degradation under the influence of high temperature. *Autom. Constr.* **2019**, *101*, 190–198. [CrossRef]

12. Schabowicz, K.; Gorzelańczyk, T.; Szymków, M. Identification of the degree of degradation of fibre-cement boards exposed to fire by means of the acoustic emission method and artificial neural networks. *Materials* **2019**, *12*, 656. [CrossRef] [PubMed]

13. Leflik, M. Some aspects of application of artificial neural network for numerical modelling in civil engineering. *Bull. Pol. Acad. Sci. Tech. Sci.* **2013**, *61*, 39–50. [CrossRef]

14. Gorzelańczyk, T.; Schabowicz, K.; Szymków, M. Non-destructive testing of fibre-cement boards, using acoustic emission (in Polish). *Przegląd Spaw.* **2016**, *88*, 35–38. [CrossRef]

15. Adamczak-Bugno, A.; Gorzelańczyk, T.; Krampikowska, A.; Szymków, M. Non-destructive testing of the structure of fibre-cement materials by means of a scanning electron microscope (in Polish). *Bad. Nieniszcz. I Diagn.* **2017**, *3*, 20–23. [CrossRef]

16. Claramunt, J.; Ardanuy, M.; García-Hortal, J.A. Effect of drying and rewetting cycles on the structure and physicochemical characteristics of softwood fibres for reinforcement of cementitious composites. *Carbohydr. Polym.* **2010**, *79*, 200–205. [CrossRef]

17. Mohr, B.J.; Nanko, H.; Kurtis, K.E. Durability of kraft pulp fibre-cement composites to wet/dry cycling. *Cem. Concr. Compos.* **2005**, *27*, 435–448. [CrossRef]

18. Pizzol, V.D.; Mendes, L.M.; Savastano, H.; Frías, M.; Davila, F.J.; Cincotto, M.A.; John, V.M.; Tonoli, G.H.D. Mineralogical and microstructural changes promoted by accelerated carbonation and ageing cycles of hybrid fibre–cement composites. *Constr. Build. Mater.* **2014**, *68*, 750–756. [CrossRef]

19. Li, Z.; Zhou, X.; Bin, S. Fibre-Cement extrudates with perlite subjected to high temperatures. *J. Mater. Civ. Eng.* **2004**, *3*, 221–229. [CrossRef]

20. Schabowicz, K.; Gorzelańczyk, T. A non-destructive methodology for the testing of fibre cement boards by means of a non-contact ultrasound scanner. *Constr. Build. Mater.* **2016**, *102*, 200–207. [CrossRef]

21. Stark, W. Non-destructive evaluation (NDE) of composites: Using ultrasound to monitor the curing of composites. In *Non-Destructive Evaluation (NDE) of Polymer Matrix Composites. Techniques and Applications*, 1st ed.; Karbhari, V.M., Ed.; Woodhead Publishing: Limited, UK, 2013; pp. 136–181. ISBN 978-0-85709-344-8.

22. Berkowski, P.; Dmochowski, G.; Grosel, J.; Schabowicz, K.; Wójcicki, Z. Analysis of failure conditions for a dynamically loaded composite floor system of an industrial building. *J. Civ. Eng. Manag.* **2013**, *19*, 529–541. [CrossRef]

23. Hoła, J.; Schabowicz, K. State-of-the-art non-destructive methods for diagnostic testing of building structures–anticipated development trends. *Arch. Civ. Mech. Eng.* **2010**, *10*, 5–18. [CrossRef]

24. Davis, A.; Hertlein, B.; Lim, K.; Michols, K. Impact-echo and impulse response stress wave methods: Advantages and limitations for the evaluation of highway pavement concrete overlays. In Proceedings of the Conference on Nondestructive Evaluation of Bridges and Highways, Scottsdale, AZ, USA, 4 December 1996; pp. 88–96.

25. Chady, T.; Schabowicz, K. Non-destructive testing of fibre-cement boards, using terahertz spectroscopy in time domain (in Polish). *Bad. Nieniszcz. I Diagn.* **2016**, *1*, 62–66.

26. Schabowicz, K.; Ranachowski, Z.; Jóźwiak-Niedźwiedzka, D.; Radzik, Ł.; Kudela, S.; Dvorak, T. Application of X-ray microtomography to quality assessment of fibre cement boards. *Constr. Build. Mater.* **2016**, *110*, 182–188. [CrossRef]

27. Ranachowski, Z.; Ranachowski, P.; Dębowski, T.; Gorzelańczyk, T.; Schabowicz, K. Investigation of structural degradation of fibre cement boards due to thermal impact. *Materials* **2019**, *12*, 944. [CrossRef] [PubMed]

28. Ranachowski, Z.; Schabowicz, K. The contribution of fibre reinforcement system to the overall toughness of cellulose fibre concrete panels. *Constr. Build. Mater.* **2017**, *156*, 1028–1034. [CrossRef]

29. Ranachowski, Z. The application of neural networks to classify the acoustic emission waveforms emitted by the concrete under thermal stress. *Arch. Acoust.* **1996**, *21*, 89–98.

30. Ranachowski, Z.; Jóźwiak-Niedźwiedzka, D.; Brandt, A.M.; Dębowski, T. Application of acoustic emission method to determine critical stress in fibre reinforced mortar beams. *Arch. Acoust.* **2012**, *37*, 261–268. [CrossRef]
31. Yuki, H.; Homma, K. Estimation of acoustic emission source waveform of fracture using a neural network. *NDT E Int.* **1996**, *29*, 21–25. [CrossRef]
32. Schabowicz, K. Neural networks in the NDT identification of the strength of concrete. *Arch. Civ. Eng.* **2005**, *51*, 371–382.
33. Asteris, P.G.; Kolovos, K.G. Self-compacting concrete strength prediction using surrogate models. *Neural Comput. Appl.* **2019**, *31*, 409–424. [CrossRef]
34. Lee, S.C. Prediction of concrete strength using artificial neural networks. *Eng. Struct.* **2003**, *25*, 849–857. [CrossRef]
35. Łazarska, M.; Woźniak, T.; Ranachowski, Z.; Trafarski, A.; Domek, G. Analysis of acoustic emission signals at austempering of steels using neural networks. *Met. Mater. Int.* **2017**, *23*, 426–433. [CrossRef]
36. Woźniak, T.Z.; Ranachowski, Z.; Ranachowski, P.; Ozgowicz, W.; Trafarski, A. The application of neural networks for studying phase transformation by the method of acoustic emission in bearing steel. *Arch. Metall. Mater.* **2014**, *59*, 1705–1712. [CrossRef]
37. Rucka, M.; Wilde, K. Experimental study on ultrasonic monitoring of splitting failure in reinforced concrete. *J. Nondestruct. Eval.* **2013**, *32*, 372–383. [CrossRef]
38. Rucka, M.; Wilde, K. Ultrasound monitoring for evaluation of damage in reinforced concrete. *Bull. Pol. Acad. Sci. Tech. Sci.* **2015**, *63*, 65–75. [CrossRef]
39. Zielińska, M.; Rucka, M. Non-Destructive Assessment of Masonry Pillars using Ultrasonic Tomography. *Materials* **2018**, *11*, 2543. [CrossRef]
40. Wojtczak, E.; Rucka, M. Wave Frequency Effects on Damage Imaging in Adhesive Joints Using Lamb Waves and RMS. *Materials* **2019**, *12*, 1842. [CrossRef]
41. EN 12467–Cellulose Fibre Cement Flat Sheets. Product Specification and Test Methods. 2018. Available online: https://standards.globalspec.com/std/10401496/din-en-12467 (accessed on 9 June 2019).
42. Osowski, S. *Neural Networks for Information Processing (in Polish)*; OWPW: Warsaw, Poland, 2000.
43. Estêvão, J.M.C. Feasibility of using neural networks to obtain simplified capacity curves for seismic assessment. *Buildings* **2018**, *8*, 151. [CrossRef]

materials

MDPI

Article

Enhanced Singular Value Truncation Method for Non-Destructive Evaluation of Structural Damage Using Natural Frequencies

Qiuwei Yang [1], Chaojun Wang [1], Na Li [1], Wei Wang [1,*] and Yong Liu [2]

[1] School of Civil Engineering, Shaoxing University, Shaoxing, Zhejiang 312000, China;
    yangqiuwei79@gmail.com (Q.Y.); qtwcj123@163.com (C.W.); lina@usx.edu.cn (N.L.)
[2] State Key Lab Water Resources & Hydropower Engineering School, Wuhan University, Wuhan 430072,
    China; liuy203@whu.edu.cn
* Correspondence: wellswang@usx.edu.cn

Received: 21 February 2019; Accepted: 25 March 2019; Published: 28 March 2019

check for updates

Abstract: As natural frequencies can be easily and accurately measured, structural damage evaluation by frequency changes is very common in engineering practice. However, this type of method is often limited by data, such as when the available natural frequencies are very few or contaminated. Although much progress has been made in frequency-based methods, there is still much room for improvement in calculation accuracy and efficiency. To this end, an enhanced singular value truncation method is proposed in this paper to evaluate structural damage more effectively by using a few lower order natural frequencies. The main innovations of the enhanced singular value truncation method lie in two aspects: The first is the normalization of linear systems of equations; the second is the multiple computations based on feedback evaluation. The proposed method is very concise in theory and simple to implement. Two numerical examples and an experimental example are employed to verify the proposed method. In the numerical examples, it was found that the proposed method can successively obtain more accurate damage evaluation results compared with the traditional singular value truncation method. In the experimental example, it was shown that the proposed method possesses more precise and fewer calculations compared with the existing optimization algorithms.

Keywords: non-destructive evaluation; structural damage; natural frequency; singular value truncation; multiple feedbacks; data noise

---

1. Introduction

Structural damage often leads to changes in the dynamic response parameters of a structure. By testing the vibration parameters and observing their changes, structural damages can be monitored in a timely manner to avoid disastrous consequences. In recent decades, structural damage evaluation has become a key issue in the field of civil engineering, mechanical engineering, aerospace engineering and so on. The method based on natural frequency changes [1–11] is one of the mainstream methods for structural damage evaluation, since the natural frequencies are most easily and accurately measured in comparison with other dynamic characteristics of a structure. Messina et al. [3] proposed a damage detection method termed the multiple damage location assurance criterion by using the natural frequency sensitivity analysis. Yu et al. [4] made use of natural frequency perturbation theory and artificial neural network to detect small structural damage. Yang and Liu [5] proposed a frequency-based method with added masses to identify damages of the symmetrical structures. Khiem and Toan [6] proposed a method to calculate the natural frequencies of a multiple-cracked beam and detect an unknown number of multiple cracks from the measured natural frequencies. Ding et al. [7] presented an improved artificial bee colony algorithm for crack identification in beam structure. Krishnanunni et al. [8] defined an objective function using the frequency sensitivity equation and minimized it using a cuckoo search algorithm to evaluate structural damage. Choi and Han [9] studied frequency-based damage detection in cantilever beam by using a vision-based monitoring system with a motion magnification technique. Pan et al. [10] proposed a novel concept of noise response rate (NRR) to evaluate the sensitivity of each mode of the frequency shift to noise. It was shown that selecting vibration modes with low NRR values improves the prediction accuracy of frequency-based damage detection. Ercolani et al. [11] studied the inverse method of damage detection from the measurement of the first three natural frequencies of vibration on two experimental beams.

Although much progress has been made in frequency-based methods, there is still much room for improvement in the calculation accuracy and efficiency since the available natural frequencies are very few and contaminated. For the damage identification problem, the damaged elements in the structure are often only a small minority because the actual damage usually occurs only in a few local areas. This particularity of damage identification has not been fully utilized in the previous frequency-based methods. In this paper, an enhanced singular value truncation (ESVT) method is proposed for structural damage evaluation by using only a few natural frequencies. Central to the proposed method is the normalization of linear systems of equations and the multiple computations based on feedback evaluation. The above particularity of damage detection is fully utilized in the proposed procedure by removing many undamaged elements in each computation according to the feedback evaluation. This operation can significantly reduce the computational complexity and obtain more accurate damage evaluation results. The presentation of this work is organized as follows. In Section 2, the natural frequency sensitivity theory is brief reviewed and then an enhanced singular value truncation method is proposed for structural damage evaluation. Two numerical examples and an experimental example are used to demonstrate the feasibility and superiority of the developed method in Sections 3 and 4, respectively. From the numerical results, it was found that the proposed method can successively obtain more accurate damage evaluation results compared with the traditional singular value truncation method. From the experimental results, it was shown that the proposed method possesses more precise and fewer calculations compared with the existing optimization algorithms. The conclusions of this work are summarized in Section 5.

2. Theoretical Development

2.1. Natural Frequency Sensitivity for Damage Detection

As is well known, the low-order natural frequencies of structural vibration can be easily and accurately measured in engineering practice. Thus the natural frequency is the most commonly used parameter in structural model updating or damage detection. In this section, the basis for the natural frequency sensitivity technique [1–5] is briefly reviewed. According to the vibration theory, the modes of structural free vibration can be obtained theoretically by solving the following generalized eigenvalue problem:

$$K\phi_j = \lambda_j M\phi_j \tag{1}$$

where M and K are the mass and stiffness matrices of the structure, and λ_j and ϕ_j are the jth eigenvalue and eigenvector, respectively. Note that the eigenvalue λ_j can be obtained from the corresponding natural frequency f_j by

$$\lambda_j = (2\pi \cdot f_j)^2 \tag{2}$$

Generally, the mass matrix M is assumed constant in model updating or damage detection. Then the first-order sensitivity of the jth eigenvalue λ_j can be computed by

$$\frac{\partial \lambda_j}{\partial x_i} = \phi_j^T K_i \phi_j \tag{3}$$

where x_i and K_i are the ith elemental stiffness perturbed parameter (also called as damage parameter) and stiffness matrix, respectively. The goal of model updating or damage detection is to obtain the values of these stiffness perturbed parameters by the changes between the measured eigenvalues and the theoretical eigenvalues. Assuming λ_j^* is the jth measured eigenvalue, the eigenvalue change $\Delta\lambda_j$ can be calculated as

$$\Delta\lambda_j = \lambda_j^* - \lambda_j \tag{4}$$

On the other hand, the eigenvalue change $\Delta\lambda_j$ can be approximated using Taylor's series expansion and linear superposition principle as

$$\Delta\lambda_j = \sum_{i=1}^{N} x_i \frac{\partial \lambda_j}{\partial x_i} \tag{5}$$

where N is the number of total elements in structural finite element model (FEM). For m measured eigenvalues, the first-order sensitivity equation of natural frequencies can be obtained as

$$A \cdot x = b \tag{6}$$

$$A = \begin{bmatrix} \frac{\partial \lambda_1}{\partial x_1} & \cdots & \frac{\partial \lambda_1}{\partial x_N} \\ \vdots & \ddots & \vdots \\ \frac{\partial \lambda_m}{\partial x_1} & \cdots & \frac{\partial \lambda_m}{\partial x_N} \end{bmatrix} \tag{7}$$

$$x = \left\{ \begin{array}{c} x_1 \\ \vdots \\ x_N \end{array} \right\} \tag{8}$$

$$b = \left\{ \begin{array}{c} \Delta\lambda_1 \\ \vdots \\ \Delta\lambda_m \end{array} \right\} \tag{9}$$

By solving the linear Equation (6), the unknown stiffness perturbed parameters α_i can be obtained, which will be used for model updating or damage evaluation. For example, the generalized inverse [12–15] is used in many cases to compute x in Equation (6), that is

$$x = A^+ b \tag{10}$$

where the superscript "+" denotes the Moore–Penrose generalized inverse [16].

2.2. Enhanced Singular Value Truncation Method

In engineering practice, only a few lower order natural frequencies with noise can be obtained through structural vibration testing [17–20]. Thus the results obtained by Equation (10) are often very unstable and inaccurate. This leads to the failure of model updating and damage detection. Therefore it is very necessary to develop a new computational method to compute the stiffness perturbed parameters more reliably. Traditionally, the singular value truncation (SVT) method [21–28] can be used to replace the generalized inverse to solve Equation (6) more effectively. However, as will be shown in the next example, the results obtained by the common SVT are still undesirable for many cases. In view of this, an ESVT method is proposed in this section to obtain more accurate x for structural damage evaluation. The proposed ESVT method is very concise in theory and very easy in calculation. The innovations of the ESVT method lie in two aspects: (1) Normalization of linear systems of equations; (2) multiple computations based on feedback evaluation. The ESVT method is illustrated in detail as follows.

Using the similar idea of the total least squares method [29–33], the linear systems of Equation (6) can be normalized by the division operation as

$$A^* x = 1_v \tag{11}$$

$$1_v = \left\{ \begin{array}{c} 1 \\ \vdots \\ 1 \end{array} \right\} \tag{12}$$

$$A^* = \left[\begin{array}{ccc} \frac{a_{11}}{b_1} & \cdots & \frac{a_{1N}}{b_1} \\ \vdots & \ddots & \vdots \\ \frac{a_{m1}}{b_m} & \cdots & \frac{a_{mN}}{b_m} \end{array} \right] \tag{13}$$

where a_{ij} denotes the (i, j)th coefficient of A and b_i denotes the ith coefficient of b in Equation (6). The advantage of this normalization is that all errors, including measurement errors and model errors, are placed in the new coefficient matrix A^*. It will be found that this normalization process can improve the accuracy and robustness of the solution for the linear systems of equations. After the normalization process, Equation (11) can then be solved through the singular value truncation technique as follows. Performing the singular value decomposition on A^* in Equation (11), one has

$$U \Lambda V^T \cdot x = 1_v \tag{14}$$

$$U = [u_1, u_2, \cdots, u_n] \tag{15}$$

$$V = [v_1, v_2, \cdots, v_N] \tag{16}$$

$$\Lambda = \left[\begin{array}{cc} Z & 0 \\ 0 & 0 \end{array} \right], Z = diag(\sigma_1, \sigma_2, \cdots, \sigma_t) \tag{17}$$

where U and V are the orthogonal matrices, and $\sigma_1, \sigma_2, \cdots, \sigma_t$ are the nonzero singular values of A^* with $\sigma_1 \geq \sigma_2 \geq \cdots \geq \sigma_t$. By ignoring some smaller singular values, the singular value truncation solution of x for the first time can be obtained from Equation (14) as

$$x = \left(\sum_{y=1}^{s} \sigma_y^{-1} v_y u_y^T \right) \cdot 1_v \tag{18}$$

where s is the number of remained singular values, $s \leq t$. The suitable value of s is determined by the L-curve method [34–36]. The main steps of the L-curve method are as follows: (1) Compute all possible solutions of x by Equation (18) when s is taken from 1 to t. (2) For each solution of x, calculate the 2-norm of x and $Ax - b$ (or $A^*x - 1_v$). (3) Draw the scatter plot with $\|Ax - b\|_2$ as abscissa and $\|x\|_2$ as ordinate ($\| \cdot \|_2$ denotes the 2-norm). (4) Connect the resulting scatters with straight lines to form the L-curve. (5) Determine the suitable value of s according to the inflection point of the L-curve. The L-curve method will be further illustrated in Section 3.1.

For the damage evaluation problem, the perturbed elements in the FEM due to damages are often only a small minority. This particularity of damage detection problem has not been fully utilized in the published frequency-based algorithms. This particularity results in the existence of a large number of coefficients close to zero in the x obtained by Equation (18). Thus these coefficients close to zero in x should be seen as a product of data noise and set to zeros to simplify the equation (11) for the next recalculation. Generally, those values in x that satisfy $\frac{x_i}{\max(x)} \leq 0.05$ should be deemed to correspond to those undamaged elements in the structure. Then Equation (11) can be further simplified for the recalculation by removing some column vectors in A^* and coefficients in x corresponding to those undamaged elements. That is

$$A_2^* \cdot x' = 1_v \tag{19}$$

where A_2^* is the remained matrix of A^* after removing some column vectors related to those undamaged elements, x_2 is the remained vector of x after removing the corresponding coefficients. From Equation (19), the solution of x' can be obtained again using the similar singular value truncation progress between Equations (14) and (18) as

$$x' = (\sum_{y=1}^{s'} \sigma_y'^{-1} v_y' u_y'^T) \cdot 1_v \tag{20}$$

Note that the result obtained by Equation (20) is maybe still not the final solution. When x' is the same as the corresponding coefficients in x, the x' in Equation (20) is the final solution of the damage parameters. If not, the above recalculation process should be repeated and the new solution x'' of the stiffness perturbed parameters can be obtained. The above process should be repeated until the solutions of the two adjacent cases are exactly the same (for example, $x'' = x'$). At the last, structural damage evaluation can be carried out according to the final result. In the above process, it is important to note that the computational complexity of each computation in ESVT gradually decreases since the number of unknowns decreases gradually.

3. Numerical Examples

3.1. A Truss Structure

A cantilever truss structure as shown in Figure 1 was taken as the numerical example to demonstrate the effectiveness of the proposed method. The basic parameters of the structure were as follows: Young's modulus $E = 200$ GPa, density $\rho = 7.8 \times 10^3$ kg/m^3, and cross-sectional area $A = 3.14 \times 10^{-4}$ m^2. Two damage cases were studied in the example. The first one was a single damage case where element 10 has a 20% stiffness reduction. The second was a multiple damage case where elements 7 and 18 have 15% and 20% stiffness reductions, respectively. Only the first six natural frequencies (shown in Table 1) of the undamaged and damaged structures were used in the structural damage evaluation.

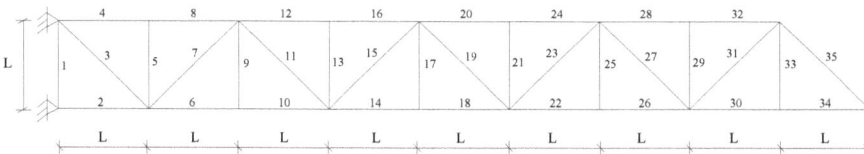

Figure 1. A cantilever truss structure (L = 0.5 m).

Table 1. The first six natural frequencies of the undamaged and damaged truss structures.

Natural Frequencies	Undamaged Structure	Damage Case 1	Damage Case 2
1	24.034	23.8588	23.9621
2	119.9908	119.3298	117.8529
3	195.9065	193.7641	194.9649
4	274.2121	273.7942	270.9942
5	436.9691	436.5461	436.3045
6	569.0272	568.812	565.0954

For each of damage cases, the evaluation results obtained by the SVT and ESVT are both given to illustrate the superiority of the ESVT method. For case 1, the SVT method was firstly employed to compute the damage parameters. As stated before, the suitable value of s in the computation process is determined by the L-curve as shown in Figure 2. Note that the scatters from right to left in Figure 2 correspond to s = 1, s = 2, etc. One can see from Figure 2 that the inflection point of the L-curve just corresponded to s = 5. Subsequently Figure 3 presents the damage evaluation result obtained by SVT method with s = 5. One can see from Figure 3 that the result was not satisfactory since element 10 cannot be uniquely determined as the damage element. Using the proposed ESVT method, Figures 4–8 give the damage evaluation results of the first to fifth calculations in ESVT. Apparently, the accuracy of damage evaluation result in Figures 4–8 was improving gradually and Figure 8 was the final result. It can be seen from Figure 8 that, after five operations, element 10 could be uniquely determined as the damage element. It was thus shown that the proposed ESVT method can achieve higher evaluation accuracy than the traditional SVT method.

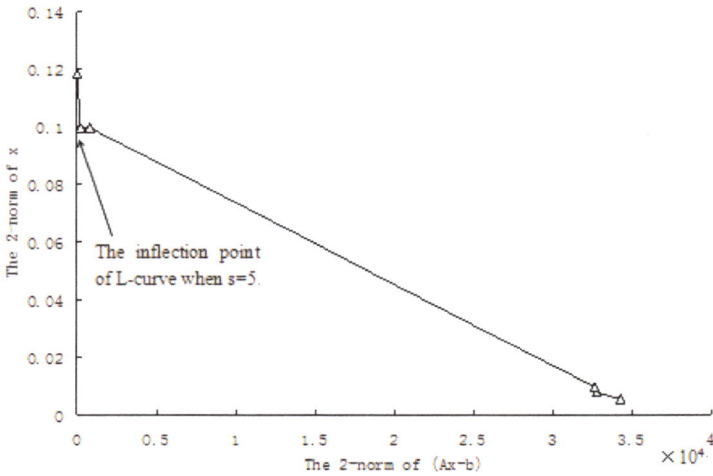

Figure 2. The L-curve used to determine the number of remained singular values in singular value truncation (SVT) for case 1.

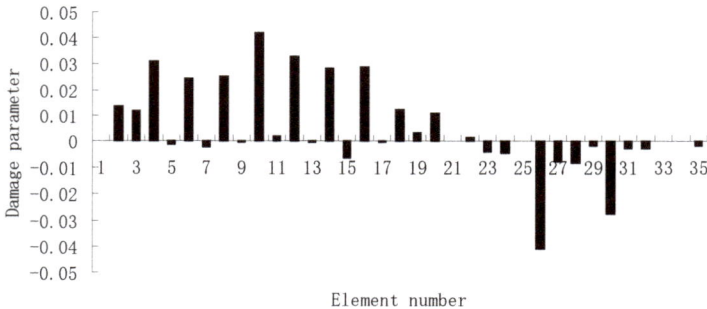

Figure 3. Damage evaluation result by the traditional SVT method for case 1 (element 10 had 20% stiffness reduction).

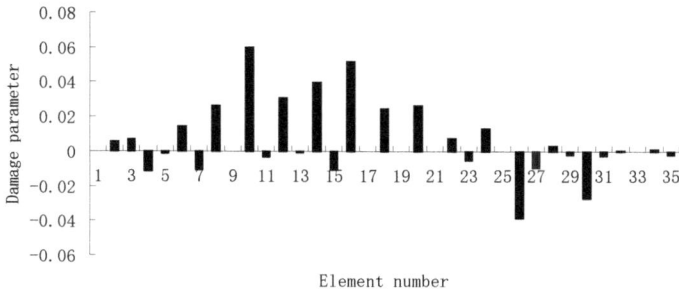

Figure 4. Damage evaluation result by the first computation of enhanced singular value truncation (ESVT) for case 1 (element 10 had 20% stiffness reduction).

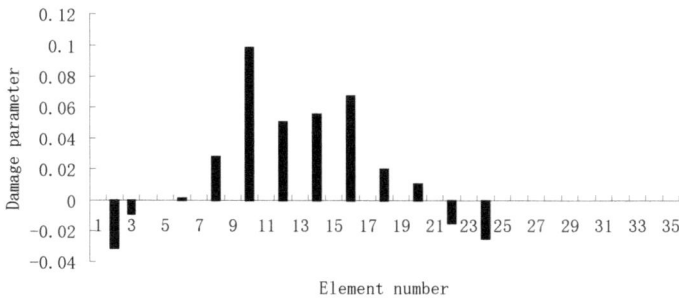

Figure 5. Damage evaluation result by the second computation of ESVT for case 1 (element 10 had 20% stiffness reduction).

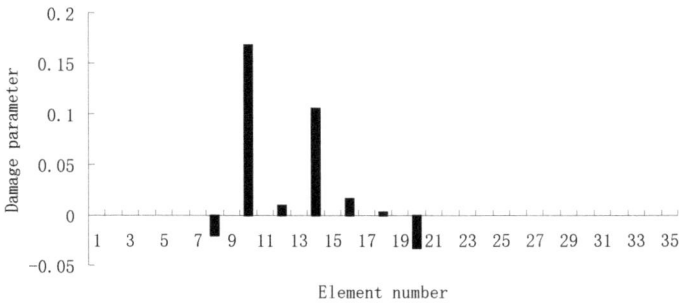

Figure 6. Damage evaluation result by the third computation of ESVT for case 1 (element 10 had 20% stiffness reduction).

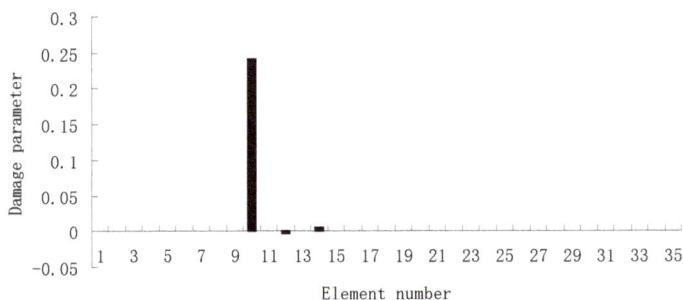

Figure 7. Damage evaluation result by the fourth computation of ESVT for case 1 (element 10 had 20% stiffness reduction).

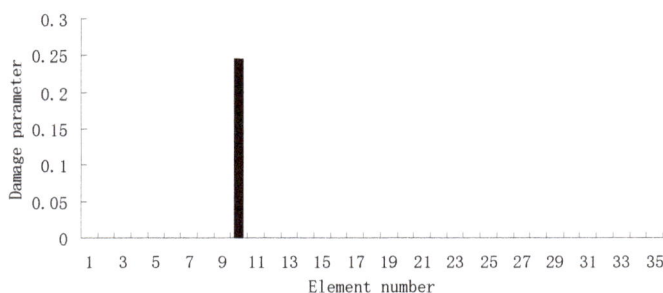

Figure 8. Damage evaluation result by the fifth computation of ESVT for case 1 (element 10 had 20% stiffness reduction).

For the second damage case, Figure 9 presents the damage evaluation result obtained by the traditional SVT method. From Figure 9, it was found that the result was not satisfactory since many elements besides 7 and 18 were determined as the damaged elements. Using the proposed ESVT method, Figures 10–14 provide the damage evaluation results of the first to fifth calculations in ESVT. It was clear that the accuracy of damage evaluation result in Figures 10–14 was improving gradually and Figure 14 was the final result for this case. The final result of Figure 14 clearly indicated that elements 7 and 18 were the true damaged elements. These results again show that the proposed ESVT method can achieve higher evaluation accuracy than the traditional SVT method.

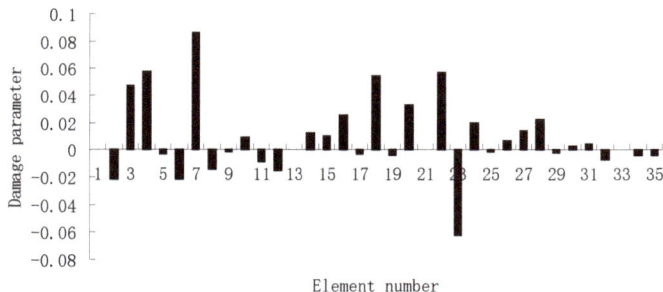

Figure 9. Damage evaluation result by the traditional SVT method for case 2 (elements 7 and 18 had 15% and 20% stiffness reductions).

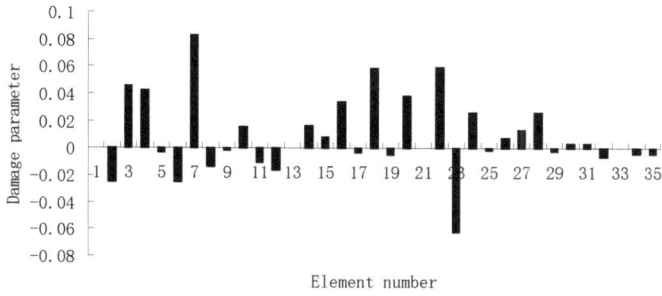

Figure 10. Damage evaluation result by the first computation of ESVT for case 2 (elements 7 and 18 had 15% and 20% stiffness reductions).

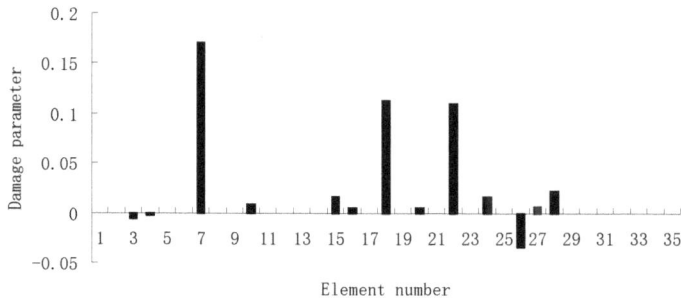

Figure 11. Damage evaluation result by the second computation of ESVT for case 2 (elements 7 and 18 had 15% and 20% stiffness reductions).

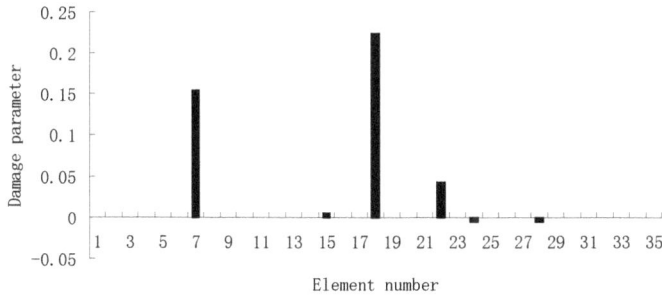

Figure 12. Damage evaluation result by the third computation of ESVT for case 2 (elements 7 and 18 had 15% and 20% stiffness reductions).

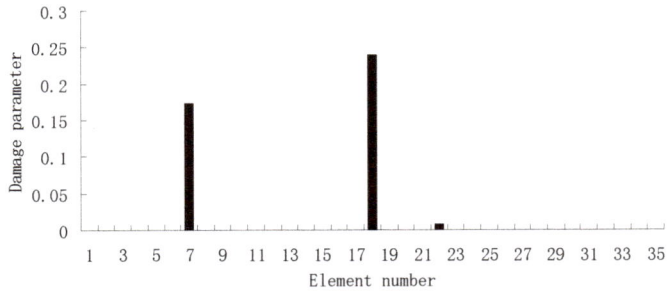

Figure 13. Damage evaluation result by the fourth computation of ESVT for case 2 (elements 7 and 18 had 15% and 20% stiffness reductions).

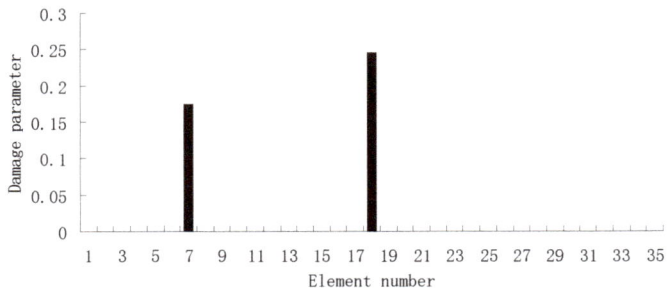

Figure 14. Damage evaluation result by the fifth computation of ESVT for case 2 (elements 7 and 18 had 15% and 20% stiffness reductions).

Next, Figures 15–17 present damage evaluation results using the first three, four and five frequencies to investigate the effect of the frequency number on the calculation results. From Figure 15, one can see that the result was not satisfactory when only three frequencies were used since element 10 cannot be uniquely determined as the damaged element in the final result of Figure 15e. From Figure 16, the result was also not satisfactory when only four frequencies were used since element 10 cannot be uniquely determined as the damage element in the final result of Figure 16f. When five frequencies were used, it can be seen from Figure 17 that the result was satisfactory since element 10 can be uniquely determined as the damaged element after six computations. It was thus shown that the results of damage evaluation become more accurate as the number of used frequencies increases. For this example, at least five frequencies were needed to obtain sufficient accurate damage evaluation results.

(1) Result of the first computation in ESVT

(2) Result of the second computation in ESVT

(a)

(b)

(3) Result of the third computation in ESVT

(4) Result of the fourth computation in ESVT

(c)

(d)

(5) Result of the fifth computation in ESVT

(e)

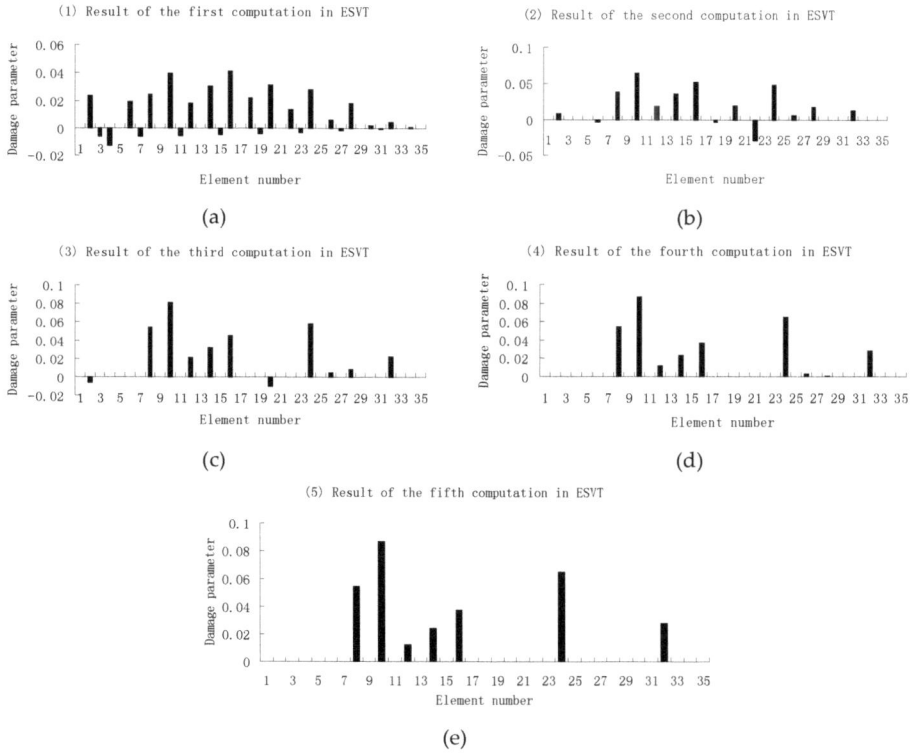

Figure 15. Damage evaluation results of the first to fifth computations in ESVT using the first three frequencies (element 10 had 20% stiffness reduction). (**a**) Result of the first computation in ESVT; (**b**) Result of the second computation in ESVT; (**c**) Result of the third computation in ESVT; (**d**) Result of the fourth computation in ESVT; (**e**) Result of the fifth computation in ESVT.

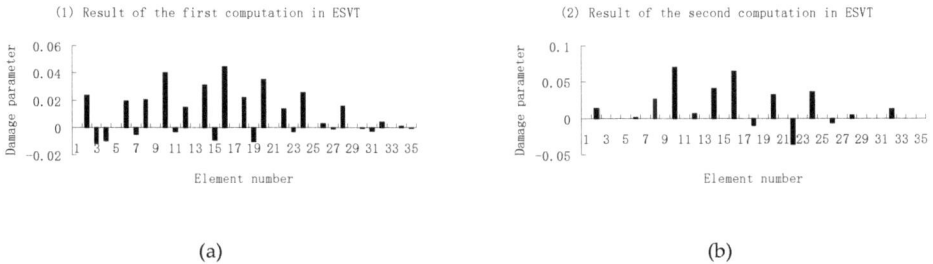

(1) Result of the first computation in ESVT

(2) Result of the second computation in ESVT

(a)

(b)

Figure 16. *Cont.*

(3) Result of the third computation in ESVT

(4) Result of the fourth computation in ESVT

(c)

(d)

(5) Result of the fifth computation in ESVT

(6) Result of the sixth computation in ESVT

(e)

(f)

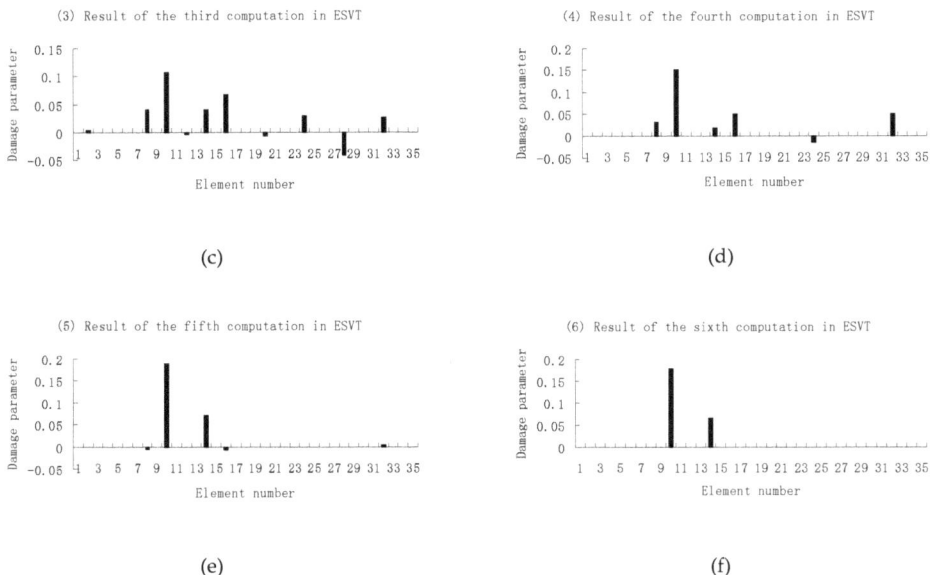

Figure 16. Damage evaluation results of the first to sixth computations in ESVT using the first four frequencies (element 10 had 20% stiffness reduction). (**a**) Result of the first computation in ESVT; (**b**) Result of the second computation in ESVT; (**c**) Result of the third computation in ESVT; (**d**) Result of the fourth computation in ESVT; (**e**) Result of the fifth computation in ESVT; (**f**) Result of the sixth computation in ESVT.

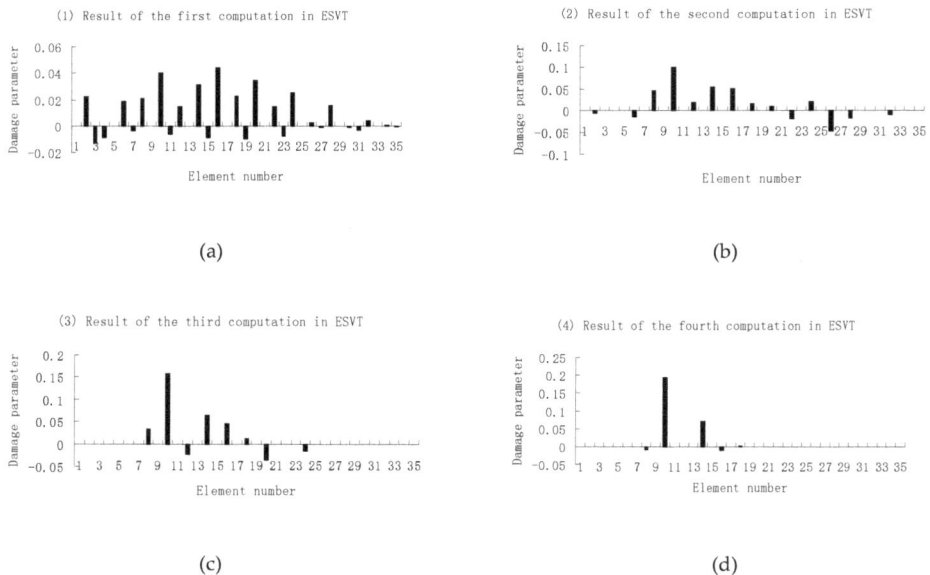

(1) Result of the first computation in ESVT

(2) Result of the second computation in ESVT

(a)

(b)

(3) Result of the third computation in ESVT

(4) Result of the fourth computation in ESVT

(c)

(d)

Figure 17. *Cont.*

(5) Result of the fifth computation in ESVT

(6) Result of the sixth computation in ESVT

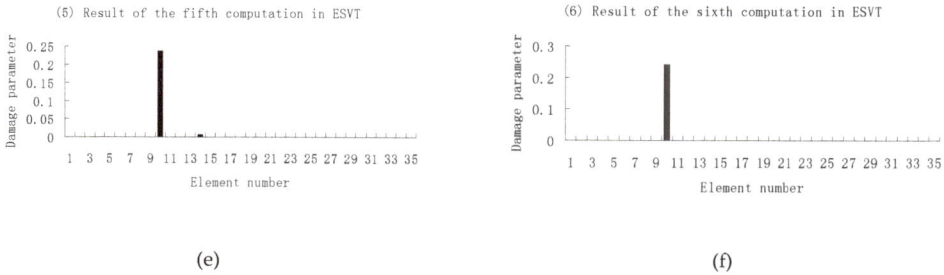

(e)

(f)

Figure 17. Damage evaluation results of the first to sixth computations in ESVT using the first five frequencies (element 10 had 20% stiffness reduction). (**a**) Result of the first computation in ESVT; (**b**) Result of the second computation in ESVT; (**c**) Result of the third computation in ESVT; (**d**) Result of the fourth computation in ESVT; (**e**) Result of the fifth computation in ESVT; (**f**) Result of the sixth computation in ESVT.

3.2. A Plate Structure

A plate structure as shown in Figure 18 was used as the second example to verify the proposed method. The main purpose of using this example was to verify the effectiveness of the proposed method for structures that require solid finite elements. The modulus of elasticity, mass density, and Poisson's ratio of this steel material were 200 GPa, 7800 kg/m^3, and 0.3, respectively. The plate was modeled using 50 identical solid elements as shown in Figure 18. In the following damage simulation, it was assumed that elements 12 and 20 had 20% and 15% stiffness reductions, respectively. Using the first eight frequencies, damage evaluation results obtained by the proposed ESVT method are shown in Figure 19. One can see from Figure 19 that the solution accuracy of the first to sixth computations was improving gradually and the sixth solution was the final result. The final result in Figure 19 clearly indicated that elements 12 and 20 were the true damaged elements. These results show that the proposed ESVT method can also be used successfully in structures that require solid finite elements.

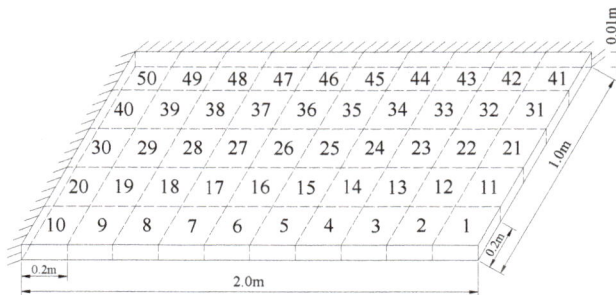

Figure 18. A plate structure.

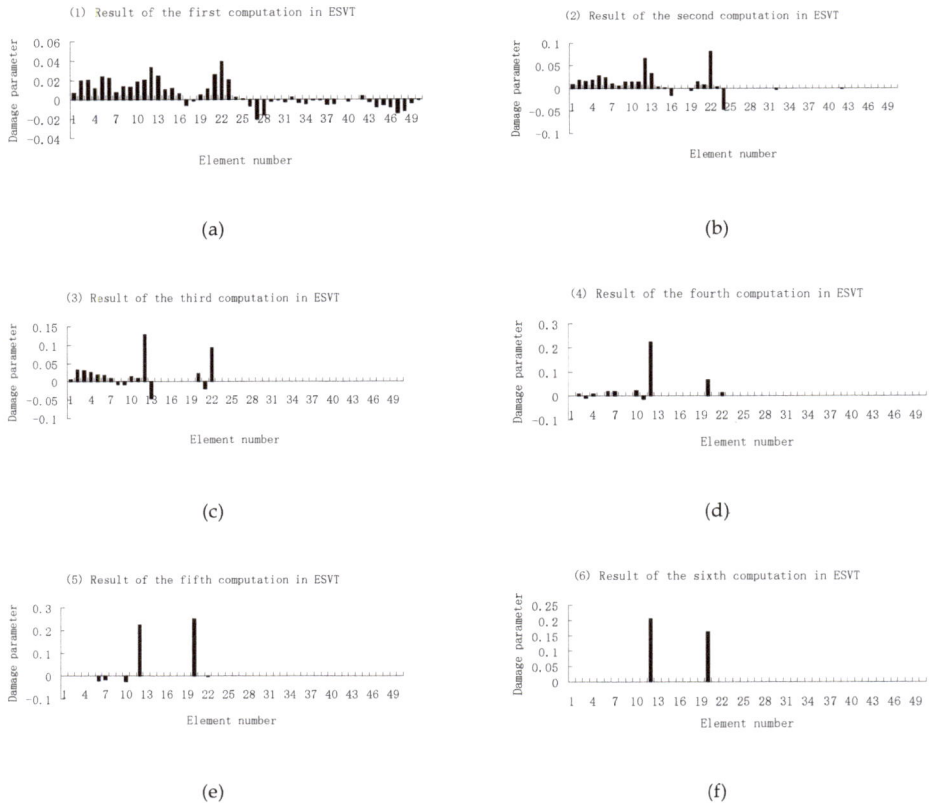

Figure 19. Damage evaluation results of the first to sixth computations in ESVT for the plate structure. (a) Result of the first computation in ESVT; (b) Result of the second computation in ESVT; (c) Result of the third computation in ESVT; (d) Result of the fourth computation in ESVT; (e) Result of the fifth computation in ESVT; (f) Result of the sixth computation in ESVT.

4. Experimental Validation

In this section, the experimental beam conducted by Yang et al. [37] was used as an example to verify the proposed method. As shown in Figure 20a, the length, width and height of the intact beam were 495.3 mm, 25.4 mm and 6.35 mm, respectively. The modulus of elasticity and mass density of this aluminium material were 71 GPa and 2210 kg/m³, respectively. The beam was modeled using 20 equal-length elements and the damage was induced in the ninth element by a saw cut as shown in Figure 20b. The analytical and experimental values of the first six natural frequencies for the undamaged and damaged structures are all shown in Table 2.

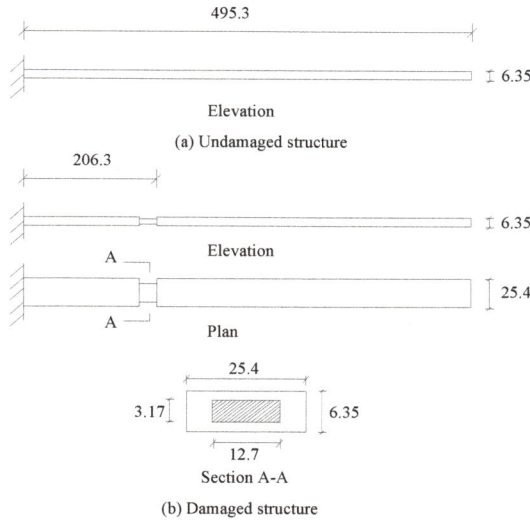

Figure 20. Configuration of the experimental beam [37]. (**a**) Undamaged structure; (**b**) Damage structure.

Table 2. The first six natural frequencies of the undamaged and damaged beams [37].

Natural Frequencies	Analytical Values (Hz)	Experimental Values (Undamaged)	Experimental Values (Damaged)
1	23.7	19.53	19.00
2	148.5	122.05	115.85
3	415.7	339.26	332.36
4	814.2	661.73	646.91
5	1345.3	1085.22	1037.46
6	2008.7	1594.59	1591.36

From columns 2 and 3 in Table 2, one can see that the differences between the analytical values obtained by FEM and the experimental values obtained by dynamic testing of the undamaged beam were very large. This means that the original FEM constructed by the software was not accurate enough to represent the undamaged beam. Thus the FEM of the undamaged beam was firstly corrected according to the natural frequency changes between the analytical values and the undamaged experimental values. Only the modified FEM could be used in the subsequent evaluation of structural damage. Note that the natural frequency sensitivity technique introduced in Section 2.1 can be used not only in damage evaluation but also in model updating. It should also be noted that the stiffness perturbed parameters of the modified FEM were computed only by one calculation process of the ESVT method in the model updating. This is the difference between the model updating problem and the damage identification problem. From the variations between column 2 and column 3 in Table 2, Figure 21 presents the stiffness perturbed parameters of the modified FEM and Table 3 gives the analytical values of the first six natural frequencies obtained by the modified FEM. From Table 3, one can see that the analytical values of the modified FEM were much closer to the undamaged experimental values than those of the original FEM. After model updating, structural damage evaluation can be subsequently carried out based on the modified FEM by using the gradual ESVT method. Figures 22–26 give the damage evaluation results of the first to fifth calculations in the ESVT. It was obvious that the accuracy of damage evaluation result in Figures 22–26 was improving gradually and Figure 26 was the final evaluation result. The final evaluation result of Figure 26 was very good since the true damage was correctly detected in element 9. For comparisons, the damage detection results reported by Krishnanunni et al. [8] and Hao et al. [38] are presented in Figure 27, obtained by Cuckoo Search algorithm (CSA) and Genetic algorithm (GA), respectively. Meanwhile, the result of Figure 26 is also shown in Figure 27 for easy comparison. From Figure 27, one can see that the damage evaluation result obtained by the proposed ESVT method had the highest accuracy among the three methods. Moreover, the computational complexity of the ESVT method was significantly lower compared to the other methods because both CSA and GA needed many iterations for good convergence. For example, the computation process using CSA reported by Krishnanunni et al. [8] was iterated 65,000 times for good convergence. Note that the proposed ESVT method only needed five calculations and the complexity of each calculation decreased gradually.

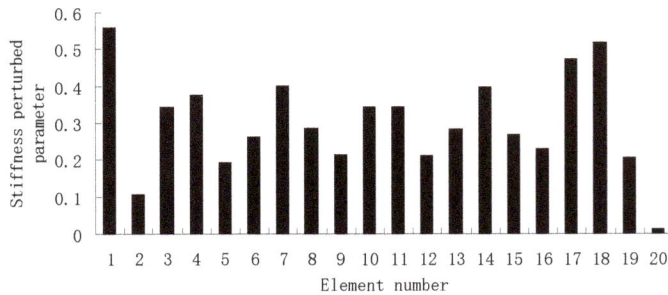

Figure 21. The stiffness perturbed parameters of the modified finite element model (FEM).

Table 3. Comparisons of natural frequencies obtained by the original FEM, the modified FEM and the experiment.

Natural Frequencies	Analytical Values of Original FEM (Hz)	Experimental Values (Undamaged Beam)	Analytical Values of Modified FEM (Hz)
1	23.7 (21.4%*)	19.53	19.0 (2.7%)
2	148.5 (21.7%)	122.05	119.8 (1.8%)
3	415.7 (22.5%)	339.26	333.7 (1.6%)
4	814.2 (23.0%)	661.73	651.2 (1.6%)
5	1345.3 (24.0%)	1085.22	1068.7 (1.5%)
6	2008.7 (26.0%)	1594.59	1582.6 (0.8%)

* The data in brackets denote the relative errors between analytical and experimental values.

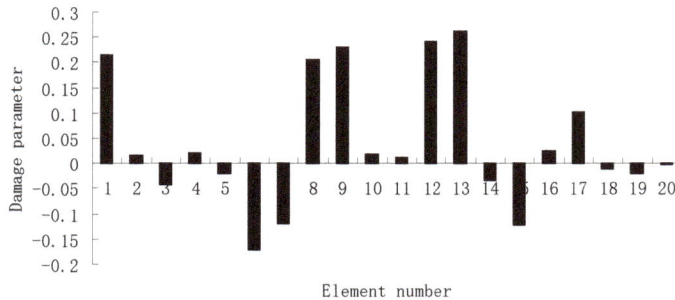

Figure 22. Damage evaluation result by the first computation of ESVT for the experimental beam.

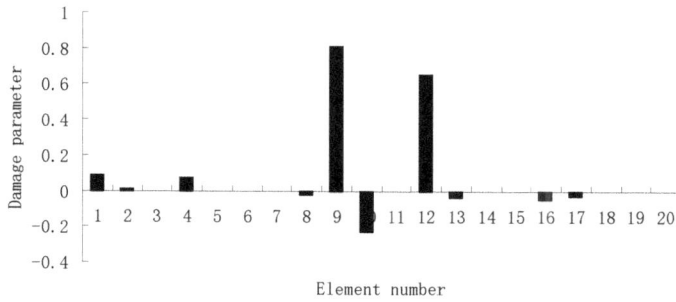

Figure 23. Damage evaluation result by the second computation of ESVT for the experimental beam.

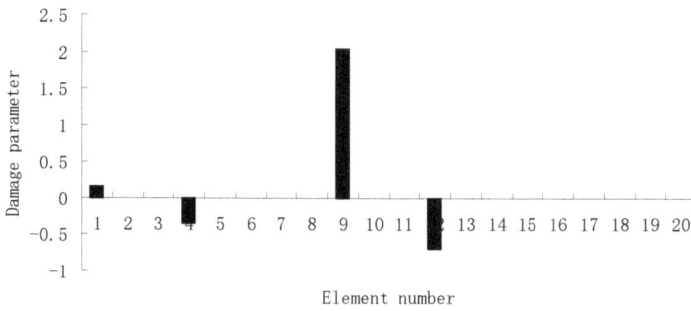

Figure 24. Damage evaluation result by the third computation of ESVT for the experimental beam.

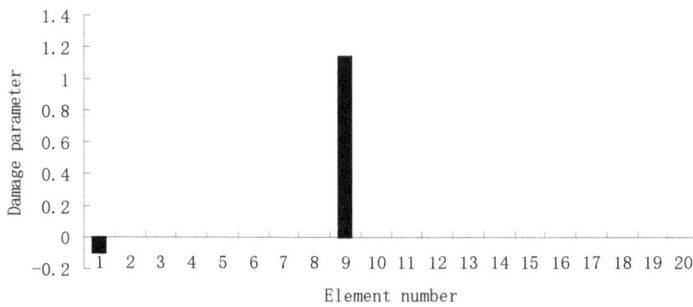

Figure 25. Damage evaluation result by the fourth computation of ESVT for the experimental beam.

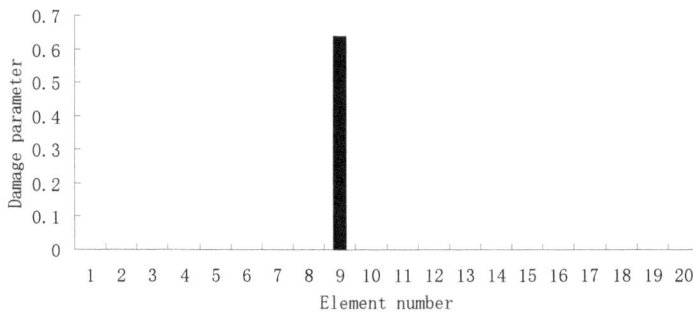

Figure 26. Damage evaluation result by the fifth computation of ESVT for the experimental beam.

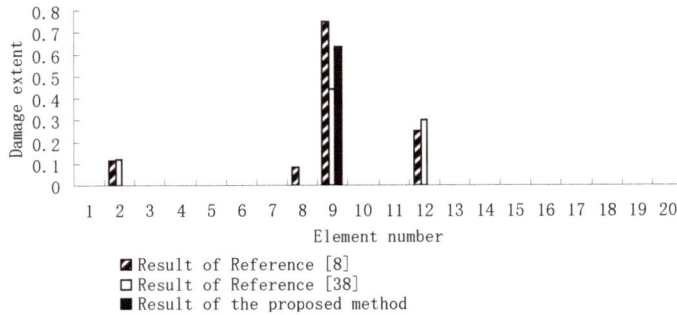

Figure 27. Comparison of damage evaluation results by the three methods.

5. Conclusions

For the damage evaluation problem, the damaged elements in the structure are often only a small minority because the actual damage usually occurs only in a few local areas. Using this particularity of damage evaluation, an ESVT method was proposed in this paper for structural damage detection using only a few lower order natural frequencies. Central to the ESVT method is the normalization of linear systems of equations and multiple computations based on feedback evaluation. In each computation of ESVT, many undamaged elements are removed according to the feedback evaluation to reduce the number of unknowns. This operation can significantly reduce the computational complexity and obtain more accurate damage evaluation results. The proposed method is very concise in theory and simple to implement. Two numerical examples and an experimental example were used to demonstrate the proposed method. From the numerical examples, it was found that the proposed method can successively obtain more accurate damage evaluation results compared with the traditional SVT method. From the experimental example, it was shown that the proposed method possesses more precise and fewer calculations compared with the existing optimization algorithms. It was shown that the proposed ESVT method may be a promising technique in non-destructive evaluation of structural damage. In practical applications, the proposed method can be applied to various types of structural damage such as reduction in elastic modulus and cracks, as long as these damages can cause observable frequency changes. Specific examples of crack detection by the proposed method will be further studied in the future.

Author Contributions: The authors confirm contribution to the paper as follow: Q.Y. and C.W. conducted the tests and analyzed the data; Q.Y. and W.W. proposed the idea and wrote the paper; N.L. revised the manuscript; Y.L. reviewed the results and approved the final version of the manuscript.

Funding: This research was funded by the National Natural Science Foundation of China (Grant numbers [11202138,41772311]) and the Zhejiang Provincial Natural Science Foundation of China (Grant number [LY17E080016]).

Conflicts of Interest: The authors declare no conflict of interest.

References

1. Salawu, O.S. Detection of structural damage through changes in frequency: a review. *Eng. Struct.* **1997**, *19*, 718–723. [CrossRef]
2. Yang, Q.W.; Liu, J.K. A coupled method for structural damage identification. *J. Sound Vib.* **2006**, *296*, 401–405. [CrossRef]
3. Messina, A.; Williams, J.E.; Contursi, T. Structural damage detection by a sensitivity and statistical-based method. *J. Sound Vib.* **1996**, *216*, 791–808. [CrossRef]
4. Yu, L.; Cheng, L.; Yam, L.H.; Yan, Y.J. Application of eigenvalue perturbation theory for detecting small structural damage using dynamic responses. *Compos. Struct.* **2007**, *78*, 402–409. [CrossRef]

5. Yang, Q.W.; Liu, J.K. Structural damage identification by adding given masses. *Eng. Mech.* **2009**, *26*, 159–163. (in Chinese).
6. Khiem, N.T.; Toan, L.K. A novel method for crack detection in beam-like structures by measurements of natural frequencies. *J. Sound Vib.* **2014**, *333*, 4084–4103. [CrossRef]
7. Ding, Z.; Lu, Z.; Huang, M.; Liu, J. Improved artificial bee colony algorithm for crack identification in beam using natural frequencies only. *Inverse Prob. Sci. Eng.* **2017**, *25*, 218–238. [CrossRef]
8. Krishnanunni, C.G.; Raj, R.S.; Nandan, D.; Midhun, C.K.; Sajith, A.S.; Ameen, M. Sensitivity-based damage detection algorithm for structures using vibration data. *J. Civ. Struct. Health Monit.* **2019**, *9*, 137–151. [CrossRef]
9. Choi, A.J.; Han, J.H. Frequency-based damage detection in cantilever beam using vision-based monitoring system with motion magnification technique. *J. Intell. Mater. Syst. Struct.* **2018**, *29*, 3923–3936. [CrossRef]
10. Pan, J.; Zhang, Z.; Wu, J.; Ramakrishnan, K.R.; Singh, H.K. A novel method of vibration modes selection for improving accuracy of frequency-based damage detection. *Composites Part B* **2019**, *159*, 437–446. [CrossRef]
11. Ercolani, G.D.; Felix, D.H.; Ortega, N.F. Crack detection in prestressed concrete structures by measuring their natural frequencies. *J. Civ. Struct. Health Monit.* **2018**, *8*, 661–671. [CrossRef]
12. Bicanic, N.; Chen, H.P. Damage identification in framed structures using natural frequencies. *Int. J. Numer. Methods Eng.* **1997**, *40*, 4451–4468. [CrossRef]
13. Xia, Y.; Hao, H.; Brownjohn, J.M.W.; Xia, P.Q. Damage identification of structures with uncertain frequency and mode shape data. *Earthquake Eng. Struct. Dyn.* **2002**, *31*, 1053–1066. [CrossRef]
14. Udwadia, F.E. Structural identification and damage detection from noisy modal data. *J. Aerosp. Eng.* **2005**, *18*, 179–187. [CrossRef]
15. Yang, J.; Li, P.; Yang, Y.; Xu, D. An improved EMD method for modal identification and a combined static-dynamic method for damage detection. *J. Sound Vib.* **2018**, *420*, 242–260. [CrossRef]
16. Rakha, M.A. On the Moore–Penrose generalized inverse matrix. *Appl. Math. Comput.* **2004**, *158*, 185–200. [CrossRef]
17. Farrar, C.R.; Doebling, S.W.; Nix, D.A. Vibration—Based structural damage identification. *Philos. Trans. R. Soc. Lond. Ser. A* **2001**, *359*, 131–149. [CrossRef]
18. Ruggiero, E.J.; Park, G.; Inman, D.J. Multi-input multi-output vibration testing of an inflatable torus. *Mech. Syst. Sig. Process.* **2004**, *18*, 1187–1201. [CrossRef]
19. Sodano, H.A.; Park, G.; Inman, D.J. An investigation into the performance of macro-fiber composites for sensing and structural vibration applications. *Mech. Syst. Sig. Process.* **2004**, *18*, 683–697. [CrossRef]
20. Foti, D.; Diaferio, M.; Giannoccaro, N.I.; Mongelli, M. Ambient vibration testing, dynamic identification and model updating of a historic tower. *NDT and E Int.* **2012**, *47*, 88–95. [CrossRef]
21. Neumaier, A. Solving ill-conditioned and singular linear systems: A tutorial on regularization. *SIAM Rev.* **1998**, *40*, 636–666. [CrossRef]
22. Basseville, M.; Mevel, L.; Goursat, M. Statistical model-based damage detection and localization: subspace-based residuals and damage-to-noise sensitivity ratios. *J. Sound Vib.* **2004**, *275*, 769–794. [CrossRef]
23. Chen, H.P. Application of regularization methods to damage detection in large scale plane frame structures using incomplete noisy modal data. *Eng. Struct.* **2008**, *30*, 3219–3227. [CrossRef]
24. Weber, B.; Paultre, P.; Proulx, J. Consistent regularization of nonlinear model updating for damage identification. *Mech. Syst. Sig. Process.* **2009**, *23*, 1965–1985. [CrossRef]
25. Li, X.Y.; Law, S.S. Adaptive Tikhonov regularization for damage detection based on nonlinear model updating. *Mech. Syst. Sig. Process.* **2010**, *24*, 1646–1664. [CrossRef]
26. Huang, Q.; Gardoni, P.; Hurlebaus, S.A. probabilistic damage detection approach using vibration-based nondestructive testing. *Struct. Saf.* **2012**, *38*, 11–21. [CrossRef]
27. Zhao, Z.; Lin, R.; Meng, Z.; He, G.; You, L.; Zhou, Y. A modified truncation singular value decomposition method for solving ill-posed problems. *J. Algorithms Comput. Technol.* **2018**, *1748301818813609*. [CrossRef]
28. Arcucci, R.; Mottet, L.; Pain, C.; Guo, Y.K. Optimal reduced space for Variational Data Assimilation. *J. Comput. Phys.* **2019**, *379*, 51–69. [CrossRef]
29. Golub, G.H.; Van Loan, C.F. An analysis of the total least squares problem. *SIAM J. Numer. Anal.* **1980**, *17*, 883–893. [CrossRef]
30. Markovsky, I.; Van Huffel, S. Overview of total least-squares methods. *Signal Process.* **2007**, *87*, 2283–2302. [CrossRef]

31. Weng, Y.; Xiao, W.; Xie, L. Total least squares method for robust source localization in sensor networks using TDOA measurements. *Int. J. Distrib. Sens. Netw.* **2011**, *7*, 172902. [CrossRef]

32. Tong, X.; Jin, Y.; Li, L. An improved weighted total least squares method with applications in linear fitting and coordinate transformation. *J. Surv. Eng.* **2011**, *137*, 120–128. [CrossRef]

33. Li, S.; Liu, L.; Liu, Z.; Wang, G. A robust total Kalman filter algorithm with numerical evaluation. *Surv. Rev.* **2019**, 1–8. [CrossRef]

34. Hansen, P.C. Analysis of discrete ill-posed problems by means of the L-curve. *SIAM Rev.* **1992**, *34*, 561–580. [CrossRef]

35. Hansen, P.C.; O'Leary, D.P. The use of the L-curve in the regularization of discrete ill-posed problems. *SIAM J. Sci. Comput.* **1993**, *14*, 1487–1503. [CrossRef]

36. Reichel, L.; Sadok, H. A new L-curve for ill-posed problems. *J. Comput. Appl. Math.* **2008**, *219*, 493–508. [CrossRef]

37. Yang, J.C.S.; Tsai, T.; PEVLIN, V. Structural damage detection by the system identification technique. *Shock Vibr. Bull.* **1985**, *55*, 57–65.

38. Hao, H.; Xia, Y. Vibration-based damage detection of structures by genetic algorithm. *J. Comput. Civ. Eng.* **2002**, *16*, 222–229. [CrossRef]

materials

MDPI

Article

Examining the Distribution of Strength across the Thickness of Reinforced Concrete Elements Subject to Sulphate Corrosion Using the Ultrasonic Method

Bohdan Stawiski [1] and Tomasz Kania [2,*]

[1] Faculty of Environmental Engineering and Geodesy, Wrocław University of Environmental and Life Sciences, pl. Grunwaldzki 24, 50-363 Wrocław, Poland
[2] Faculty of Civil Engineering, Wrocław University of Science and Technology, Wybrzeże Wyspiańskiego 27, 50-370 Wrocław, Poland
* Correspondence: tomasz.kania@pwr.edu.pl

Received: 23 June 2019; Accepted: 5 August 2019; Published: 7 August 2019

check for updates

Abstract: Sulphate corrosion of concrete is a complex chemical and physical process that leads to the destruction of construction elements. Degradation of concrete results from the transportation of sulphate compounds through the pores of exposed elements and their chemical reactions with cementitious material. Sulphate corrosion can develop in all kind of structures exposed to the corrosive environment. The mechanism of the chemical reactions of sulphate ions with concrete compounds is well known and described. Furthermore, the dependence of the compressive strength of standard cubic samples on the duration of their exposure in the sulphate corrosion environment has been described. However, strength tests on standard samples presented in the scientific literature do not provide an answer to the question regarding the measurement methodology and actual distribution of compressive strength in cross-section of reinforced concrete structures exposed to sulphate ions. Since it is difficult to find any description of this type of test in the literature, the authors undertook to conduct them. The ultrasonic method using exponential heads with spot surface of contact with the material was chosen for the measurements of concrete strength in close cross-sections parallel to the corroded surface. The test was performed on samples taken from compartments of a reinforced concrete tank after five years of operation in a corrosive environment. Test measurements showed heterogeneity of strength across the entire thickness of the tested elements. It was determined that the strength of the elements in internal cross-sections of the structure was up to 80% higher than the initial strength. A drop in the mechanical properties of concrete was observed only in the close zone near the exposed surface.

Keywords: concrete elements; concrete strength; reinforced concrete tanks; concrete corrosion; sulphate corrosion; ultrasound tests

---

1. Introduction

There are considerable quantities of effluents generated in chemical laboratories with varied pH, which is the measure of acidity or alkalinity of an aqueous solution. Neutral solutions have a pH of approximately 7.0. Neutralisation of effluents is performed by mixing acidic and alkaline compounds (if their compositions allow it) and by adding acidic or alkaline reagents. This takes place in various types of tanks. For neutralisation of laboratory effluents, reinforced concrete tanks are also used. The ratio of acidity and alkalinity is a critical factor in the chemistry of concrete [1]. The components of concrete are cement, aggregates, and water. Cement has a very alkaline pH, in order to bind all the components, it is important for it to remain near a pH of 12 [2]. In contact with effluents, concrete corrodes. Therefore, it should be characterized by the proper strength and tightness, and should be protected from the aggressive environment by the proper lining [3].

In the literature, one may encounter the opinion that after 1989 the quality of reinforced concrete structures in Poland improved radically [4], but problems with the materials, workmanship and design still exist, as can be seen in the latest research on Polish concrete structures [5,6]. The neutralisation tank presented later was made in 2012 and became corroded, which indicates that concrete insufficiently protected from corrosion will require repair, which should be preceded by a good evaluation of the condition of the damaged structure.

Concrete corrosion not only affects laboratory tanks but develops in all kind of structures. Concrete durability is the constant subject of challenges in the fields of science, design and workmanship [5,6]. As a consequence of concrete structures' exposure to corrosive environments, various substances are being transported into the concrete, causing its expansion, cracking, and strength degradation. Among the most destructive of the numerous corrosive substances are the sulphates [7].

Recent studies on sulphate corrosion of concrete are mainly focused on the mechanism of the chemical reactions of sulphate ions with the concrete compounds [8–16] and the distribution of strength over time of cubic samples stored in sulphate solutions. The corrosive reactions of sulphates in concrete have been well studied and evaluated [17–20]. The deterioration of concrete strength under sulphate corrosion is an essential basis for the prediction of concrete performance and durability. Existing studies indicate that sulphate ions in the environment chemically react with the internal composition of concrete by entering into the concrete through diffusion, convection, capillary adsorption, and other processes to generate expansive products such as ettringite, gypsum [17–20], and sodium sulphate crystals when concrete is corroded by sulphate solution in a dry-wet cycle. The expansive products continuously fill the internal pores of concrete, making the concrete more compact with improved concrete strength before deterioration. Ions of sulphuric acid react with the cement compounds e.g., according to Equations (1), (2) and (3) [7]:

$$H_2SO_4 + CaO \cdot SiO_2 \cdot 2H_2O \rightarrow CaSO_4 + Si(OH)_4 + H_2O \tag{1}$$

$$H_2SO_4 + CaCO_3 \rightarrow CaSO_4 + H_2CO_3 \tag{2}$$

$$H_2SO_4 + Ca(OH)_2 \rightarrow CaSO_4 + 2H_2O \tag{3}$$

The formation of gypsum leads to an increase in volume of approximately 124% in comparison with $Ca(OH)_2$, the main reactant of the process [17,18]. Gypsum stone, as the product of reactions (1), (2) and (3) reacts further with tricalcium aluminates (C3A) or hydrated calcium sulfoaluminate (monosulphate) to form the final chemical product Candlot's salt (ettringite), e.g., according to Reaction (4) [18]:

$$3CaO \cdot Al_2O_3 + 3(CaSO_4 \cdot 2H_2O) + 26H_2O \rightarrow 3CaO \cdot Al_2O_3 \cdot 3CaSO_4 \cdot 32H_2O \tag{4}$$

Formation of Candlot's salt is associated with volume expansion from 230% [9] to 820% [10]. According to [21] the following prerequisites must be reached for Candlot's salt crystallization leading to the concrete expansion:

- The volume of Candlot's salt must exceed some threshold value which depends on the capillary porosity of concrete,
- Only Candlot's salt formed after the hydration of cement leads to expansion,
- Candlot's salt must be formed at the boundaries of solid phases of concrete.

Candlot's salt crystallization pressure depends on the sulphate concentration and can reach the value of 35 N/mm^2 with sulphate concentration of 350 mol/m^3 [18].

The exemplary relationships of sulphate corrosion with strength of concrete samples immersed in sulphate solution have been established and described [21–23]. Zhou et al. in [21] stated that the compressive strength of cubic samples conditioned in dry-wet cycles in sulphate solution shows the rise period and decline area. The strength of concrete samples reached its peak at the 60th day of corrosion and increased by ~6.4% on the basis of its initial strength. With the increase in degradation period, the strength of concrete decreased continuously. The compressive strength decreased by ~4.4%, 18%, and 43.1% after 90, 120, and 150 days of corrosion. This research was done under laboratory conditions on standard cubic samples with a side length of 150 mm tested on a strength machine. Shi and Wang in [22] stated that strength of concrete samples conditioned in dry-wet cycles in 15% sodium sulphate solution reached its peak at the 15th day of corrosion and increased by 29% on the basis of its initial value. Du et al. in [23] tested the C25 concrete mixed with 20% fly ash placed in a sodium sulfate solution (20%) for the full-soaking corrosion test. Samples reached peak of strength at the 100th day of corrosion and increased by 10.6% on the basis of its initial value. Due to the methodology of these tests, the distribution of strength in individual cross-sections of samples which had been previously exposed to the corrosive environment was not determined experimentally.

Laboratory tests of sulphate attack on concrete materials that are based on submerging the specimens in sulphate solution and then measuring physical properties, such as strength, are effectively collecting all of these mechanisms into a single test. The result of such research is the characterization of a particular concrete sample's performance under specific, laboratory conditions. If the field conditions are variable, the performance of the concrete can also be different. Concrete compressive strength in structures is designed to withstand the designed forces. The durability of the structure depends on the fulfilment of the limit condition of concrete strength in the section that works in the state of compression. The question arises what is the compressive strength distribution in various cross sections parallel to the surface of a reinforced concrete structure under sulphate corrosion? Since tests of such type have not been performed so far, it is difficult to find appropriate literature references regarding possible methodology or results concerning the distribution of compressive strength across the thickness of reinforced concrete elements exposed to sulphate corrosion.

This paper presents research on the concrete samples taken from a concrete neutralization tank after five years of storing chemical effluents with sulphate compounds. The main purpose of the experimental tests was to determine the compressive strength of concrete in various cross sections parallel to the corroded surface. It has been stated that compressive strength of concrete subjected to the gaseous aggressive, sulphuric environment is variable across its thickness. Values have shown the increase in strength in the internal, exposed cross sections of the walls.

2. Materials and Methods

2.1. Materials

In the described case study, a neutralisation tank in a building with numerous chemical laboratories was examined. The described tank is being used for the neutralization of liquid wastes from a research program in the field of pharmaceutical production. The chemical composition of the effluents is variable over time and depends on the actual research program in the medical laboratory. A research object selected in this way allows assessment of the corrosion condition of concrete under unplanned conditions and evaluation of the change in the condition of samples of material taken from a real object in operation. The tank was designed as a reinforced concrete box, internally divided into three chambers. The tank was constructed of reinforced, water resistant concrete class C25/30, W8. The tank was made using monolithic technology and its walls were formed in the built-in (vertical) position.

The designer anticipated protecting the concrete using epoxy chemical-resistant lining. The reinforced concrete ceiling above the chambers was made on folded sheets (as composite stay-in-place formwork). Access to each of the chambers is via a manhole covered with a stainless steel lid. The ceiling slab across its thickness in the locations of the manholes was probably not protected from corrosion because during tests there were not any traces of any layer after five years of using the tank.

The described neutralisation tank operates in the batch mode. The chemical composition of the effluents that are being neutralised is variable over time and depends on the actual research and production program in the pharmaceutical laboratory. Effluents are pumped into the chamber where an electrode measures their pH. On that basis effluents are evaluated in terms of their compliance with the set point (neutral pH value). If the pH value is out of range, chemical pumps inject an acid or caustic reagent solution as required to bring the effluent to the correct level. The agitator keep the contents of the tank mixed, so the pH probe is always measuring a representative sample of the effluent and the added reagents are quickly distributed within the tank. For the caustic reagent solutions of caustic soda (NaOH, Chempur, Piekary Slaskie, Poland) are used. For the acid reagent a solution of sulphuric acid (H_2SO_4, Chempur, Piekary Slaskie, Poland) is used. A schematic method diagram of the described neutralization in the tank is shown on the Figure 1.

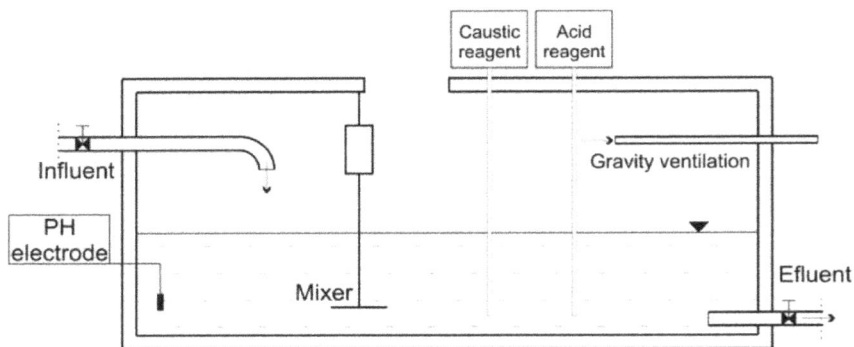

Figure 1. Method diagram of the neutralisation process in the concrete tank.

After neutralisation of effluents down to neutrality they are pumped out to the sewage system. The neutralisation process is accompanied by emission of gases which should be discharged outside, preferably using gravity ventilation or mechanical ventilation resistant to the aggressive environment. Gravity ventilation should have large cross-sections of ducts and small deviations from the vertical. Traditionally, they are openings made of brick 140 x 140 mm or round ϕ 150 mm. In the examined tank, the 'ventilation' was made of PVC pipes (Wiplast, Twardogora, Poland), diameter 50 mm, which were laid horizontally on the tank for a distance of approximately 2 m. These two parameters are sufficient to indicate its lack of effectiveness.

The tank became a natural experimental ground with respect to the effects of long-term exposure to liquid and gaseous aggressive environments on concrete, the internal surfaces of which were protected from liquid effluents using an asphalt rubber coating (probably made of Dysperbit) instead of the designed epoxy lining. No protective coating was applied for contact with the gaseous environment.

Tests were conducted on the operating tank. The lateral surfaces of the ceiling slab well visible in the manhole openings were highly corroded. The loss of concrete ranged from a dozen mm up to more than 20 mm. The folded sheet in the cutting location was also corroded. However, the corrosion rate was lower. Approximately 20 mm of sheet protrudes beyond the corroded concrete (Figure 2).

Figure 2. The folded sheet corroded more slowly than the concrete. Approximately 20 mm of sheet protrudes beyond the concrete. When built, concrete and steel were on one plane.

Such a high degree of concrete corrosion in a zone where it did not have any contact with effluents suggested that in the lower point where effluents were continuously contacting the concrete walls the situation would probably not be better, and might be even far worse. Although the tank walls had some traces of bituminous insulation, large areas lacked this coating because it had flaked off during use (Figure 3).

Figure 3. Walls below the ceiling slab were originally covered with insulation but in a considerable area of the walls this insulation had already flaked off.

At the time when the tests were performed the tank was in use and it was necessary to blind the holes after the completed tests. For this reason, in order to perform the measurements, it was decided to make one borehole with a diameter of 103 mm and other boreholes with diameters of 50 mm. The boreholes were made in the direction perpendicular to surfaces of the walls which had been formed in the built-in position (in the vertical direction). Since the boreholes were made at the same level (drilled perpendicularly to the element forming direction), in this case the variability of aggregate in the sample cross-section can be only random. The tests concerning strength distribution in concrete elements conducted by the authors and other researchers indicate the differentiation of strength with respect to the forming direction as a result of segregation of components in the gravitational field of the Earth and draining of water from concrete mix (bleeding) [24–26]. Such differentiation does not appear in the horizontal direction appropriate for the taken sample element.

2.2. Methods

The compressive strength of concrete samples in their various cross sections were determined with use of the ultrasound method on basis of longitudinal wave velocities [27]. The ultrasonic pulse velocity of a homogeneous solid can be related to its mechanical properties. Theoretical dependencies between ultrasonic wave velocity and elastic modulus and Poisson's ratio were investigated and described in the literature [28,29]. Based on the theory of elasticity applied to homogeneous and isotropic materials, for the method of testing used by the authors passing wave velocity C_L is directly proportional to the square root of the dynamic modulus of elasticity E_d, and inversely proportional to the square root of its density, ϱ, where v_d is the dynamic Poisson's ratio (5):

$$C_L = (E_d/\rho \cdot (1 - v_d)/((1 + v_d) \cdot (1 + 2v_d)))^{1/2} \text{ [km/s]} \tag{5}$$

Concrete is a heterogeneous material, so these assumptions are not strictly valid. High attenuation in concrete limits the ultrasonic pulse velocity method (UPV) to frequencies up to 100 kHz, which means that compressional waves do not interact with most concrete inhomogeneities [30]. Under this condition concrete can be regarded as a homogeneous material [31]. The tests conducted already in the seventies and eighties of the 20th century showed that there is a relationship between ultrasonic wave velocity and concrete strength. The possibility of using the correlation between these values was included both in scientific literature e.g., [28,29,32–38] as well as in norms e.g., [39–41]. In the study [38], Komlos and others compared eight basic methods of determining concrete strength based on measurements of ultrasound velocity. He concluded that the necessary requirement of such tests was to perform calibration of measurements with results of destructive tests, such are also research experiences of the authors [37,42]. The confirmation regarding the possibility of using the measurements of ultrasonic wave velocity to test compressive strength of concrete exposed to sulphate corrosion can be found in scientific literature from the beginning of this century and from the later years [43–45].

Measurements were performed using ultrasonic point probes of frequency equal to 40 kHz the testing results of which were presented in the study [37]. The structure of the probes is shown in the Figure 4. They were equipped with exponential, half wave concentrators with a length of 87 mm and base width of 42 mm. The diameter of the contact point of the concentrators was 1 mm.

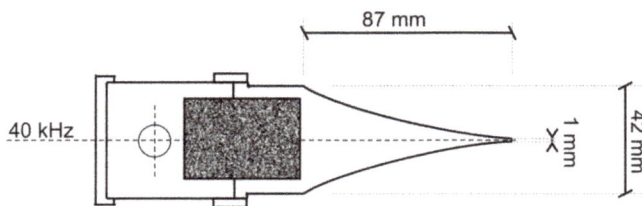

Figure 4. Ultrasonic spot head with the exponential concentrator.

The tests were performed with a UNIPAN 543 (Zaklady Aparatury Naukowej UNIPAN, Warsaw, Poland) ultrasonic pulse velocity (UPV) test instrument (Figure 5). Probe concentrators were applied from the two opposite sides of the examined concrete cylindrical samples, in planes parallel to the surface of the wall they were bored from. In that way the longitudinal wave velocities were measured.

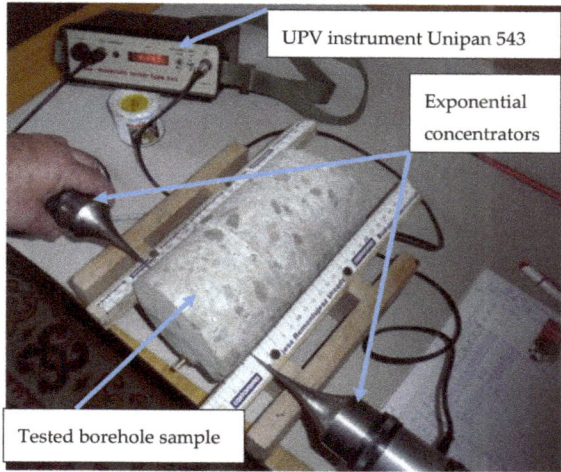

Figure 5. Borehole No. 1 during ultrasound tests.

The ultrasound rate was determined in two directions approximately perpendicular to each other, along the diameters, in planes located 10 mm from each other. Only the distance of the first measurement point from the external surface of the wall was 5 mm (Figure 6).

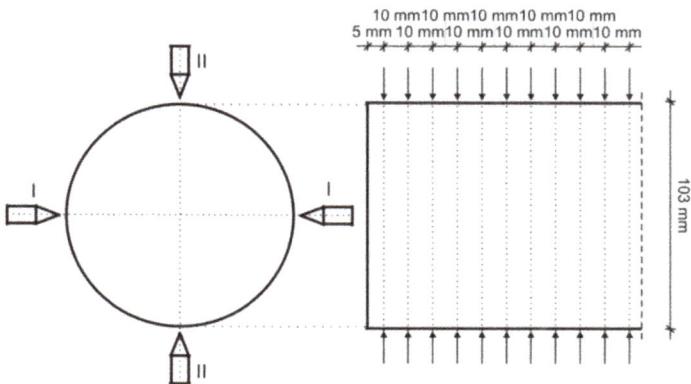

Figure 6. Layout of measuring points and sections on the tested boreholes.

The examined borehole materials were cut into samples of length equal to their diameter. Boreholes were cut thus obtaining samples with $\phi = h = 10.3$ cm and $\phi = h = 5.0$ cm. The strength of samples with $\phi = h = 10.3$ is equivalent to the strength tested on cubic samples, side 15 cm [29,41,42]. The coefficient for calculation of the strength of samples with $\phi = h = 05.0$ cm to the strength tested on cubic samples, side 15 cm is 1.08, what has been tested experimentally by Brunarski [29] and by the authors [42] in the expected strength range of the samples. To the ultrasound rate determined in the middle of the height of each sample, destructive strength was assigned as determined on the strength machine as a relation of destructive force P [N] to the surface area of cross-section A [mm^2] (6):

$$f_c = P/A \ [\text{N/mm}^2] \tag{6}$$

On that basis a hypothetical scaling curve was chosen and the pairs of results were obtained: compressive strength f_c [N/mm^2]-passing wave velocity C_L [km/s] according to the methodology described in the literature [29,39].

In order to confirm the salt formation in the concrete, a colorimetric semi-quantitative method has been used with the use of Merck test strips. Then, in order to determine the distribution of sulphate salts forming along the ultrasonic measurements, a gravimetric quantitative method has been used [46–49]. Tested samples of concrete were cut, dried and crushed. In the next step, samples were extracted in one molar hydrochloric acid. As the precipitating agent barium chloride was added to the pre-heated samples extract to precipitate barium sulphate, which was weighed after washing and calcination at 800 °C to constant weight. Determination of the sulphate content in the barium sulphate precipitate was determined on the basis of mass proportions according to the Equation (7):

$$x = a\ 96.064/233.400 = a\ 0.4116 \ [\text{g}] \tag{7}$$

where a is a mass of the barium sulphate [g], x is a mass of SO$_4{}^{2-}$ in the tested sample of material.

3. Results

3.1. Quantification of Sulphates in Tested Material

Concrete samples were taken from the walls and ceiling in order to determine salt content. It turned out that chlorides were available in the concrete in acceptable quantities (maximum in sample 1–0.085% of weight of the concrete), sulphates in high quantities above 1.2% by weight of the concrete. No nitrates or nitrites were found in the concrete. In the next step quantitative determination of sulfates using gravimetric method has been undertaken. Tests have been performed on the internal surface of the tank walls, and across its thickness, at distances of 20, 50 and 100 mm from the surface under sulphate attack. Quantity of the sulphates has been calculated with use of the Equation (7). The results of the analysis are presented in Table 1.

Table 1. Results of the gravimetric quantification of sulphates across the thickness of tested concrete elements.

Distance from Internal Wall Surface [mm]	Concrete Sample Mass [g]	Mass of BaSO$_4$ (a) [g]	Mass of SO$_4{}^{2-}$ (x) [g]	SO$_4{}^{2-}$ [% by Sample Weight]	SO$_4{}^{2-}$ Mean Value [%]
	10.2327	0.3796	0.1563	1.527	
0	10.1276	0.3873	0.1594	1.574	1.524
	10.3214	0.3691	0.1519	1.472	
	9.9885	0.3211	0.1321	1.323	
20	10.1445	0.3436	0.1414	1.394	1.345
	10.3417	0.3312	0.1363	1.318	
	10.1424	0.1163	0.0479	0.472	
50	9.9672	0.1480	0.0609	0.611	0.548
	10.1228	0.1382	0.0569	0.562	
	9.8276	0.0766	0.0315	0.321	
100	10.1412	0.1015	0.0418	0.412	0.315
	10.3429	0.0533	0.0219	0.212	

Performed quantitative analysis of sulphates on the thickness of the tested samples indicates their deep penetration and concentration in the entire cross-section of the walls. The highest concentration of SO$_4{}^{2-}$ has been examined in the zone near the inner wall surface (above 1.52%), at a depth of 20 mm

SO_4^{2-} percentage concentration was 1.35%, reaching 0.55% at a depth of 50 mm and 0.31% at a distance of 100 mm from the inner tank wall surface.

3.2. Calibration of Ultrasound Pulse Velocity-Compression Strength Curve Based on the Destructive Tests

Since the distribution of compressive strength across the thickness of reinforced concrete structures subject to sulphate corrosion from one side had not been tested so far, the authors performed such measurements using the ultrasonic method. In order to investigate the actual condition of concrete corrosion in the walls, three boreholes were made. The borehole location chosen was below the ceiling slab but in a manner ensuring that effluents did not overflow. The first and second borehole were made at the distance of 50 cm from the top wall edge, and the third borehole at the height of the top wall edge. Borehole No. 1 was of diameter 103 mm, and 2 & 3 diameter 50 mm. It was noted that on borehole No. 1 reinforcing meshes had moved towards the internal surface. For this reason, the thickness of lagging was reduced down to 10 mm and as a result of corrosion in the tested location 8 mm of concrete were missing, only a protective layer 2 mm thick (Figure 7) remained.

Figure 7. Thickness of non-corroded lagging in borehole No. 1 is only 2 mm.

After the measurements of ultrasonic pulse velocities of the tested samples they have been cut and tested in uniaxial loading on strength machine. On that base a hypothetical scaling curve with the following Equation (8) was chosen:

$$f_c = 53.6 \cdot C_L - 122.3 \, [\text{N/mm}^2] \tag{8}$$

where f_c is the compressive strength of concrete $[\text{N/mm}^2]$, and C_L is the ultrasound longitudinal wave velocity [km/s]. The results of destructive tests, measured pulse velocities and strength values calculated with use of Equation (8) are presented in Table 2.

Table 2. Results of uniaxial destructive tests, measured pulse velocities and strength values calculated with use of the chosen hypothetical scaling curve.

Sample No.	Core Size [cm × cm]	Ultrasound Longitudinal Wave Velocity C_L [km/s]	$f_{c,\varnothing}$-$f_{c,cube}$ Conversion Factor	Compression Strength [MPa]		
				Destructive Test		f_c from Equation (6)
				$f_{c,\varnothing}$	$f_{c,cube}$	
1	10.3 × 10.3	3.63	1.00	68.92	68.92	72.27
2	10.3 × 10.3	2.94		37.55	37.55	35.28
3	5.0 × 5.0	3.08		41.94	45.3	42.79
4	5.0 × 5.0	2.91	1.08	32.66	35.27	33.68
5	5.0 × 5.0	3.14		40.90	44.17	46.00
6	5.0 × 5.0	3.42		59.62	87.14	61.01
Mean value	-	3.19	-	46.20	48.47	48.51

In this way, scaling curves established hypothetically were used to convert the rate of ultrasound wave in the given cross-section at the borehole height into concrete compression strength in this cross-section.

3.3. Testing the Strength of Concrete across the Tank Wall Thickness

Passing times t [µs], calculated wave velocities C_L [km/s] and compressive strengths f_c [N/mm^2] in planes parallel to the surface of the boreholes are presented in Table 3. Results are presented starting with the ordinal number 1 (5 mm from the external side of the tested wall) in the direction of the internal side of the examined tank.

Table 3. Results of concrete compression strength test in borehole No. 1.

Ordinal Number	Ultrasound Netto Passing Time in Direction I-I $t_{n\,I\text{-}I}$ [µs]	Ultrasound Netto Passing Time in Direction II-II $t_{n\,II\text{-}II}$ [µs]	Mean Ultrasound Netto Passing Time t_n [µs]	Ultrasound Longitudinal Wave Velocity C_L [km/s]	Concrete Compression Strength f_c [N/mm^2]
1	33.00	34.30	33.65	3.09	43.36
2	33.40	34.20	33.80	3.08	42.62
3	34.20	35.70	34.95	2.98	37.20
4	35.30	35.40	35.35	2.94	35.39
5	33.50	32.60	33.05	3.15	46.37
6	33.00	31.20	32.10	3.24	51.36
7	30.60	33.50	32.05	3.25	51.63
8	32.20	33.80	33.00	3.15	46.62
9	30.30	32.00	31.15	3.34	56.65
10	29.60	33.80	31.70	3.28	53.55
11	28.40	29.10	28.75	3.62	71.59
12	28.50	29.00	28.75	3.62	71.59
13	28.70	29.00	28.85	3.61	70.92
14	28.70	28.60	28.65	3.63	72.27
15	28.70	28.60	28.65	3.63	72.27
16	28.70	30.30	29.50	3.53	66.66
17	32.20	32.30	32.25	3.23	50.55
18	31.40	31.00	31.20	3.33	56.37
Mean (1–18)	31.13	31.91	31.52	3.32	55.39
19	-	-	-	-	0.00

The dependencies of compressive strength as a function of depth for the borehole No. 1 are shown in Figure 8.

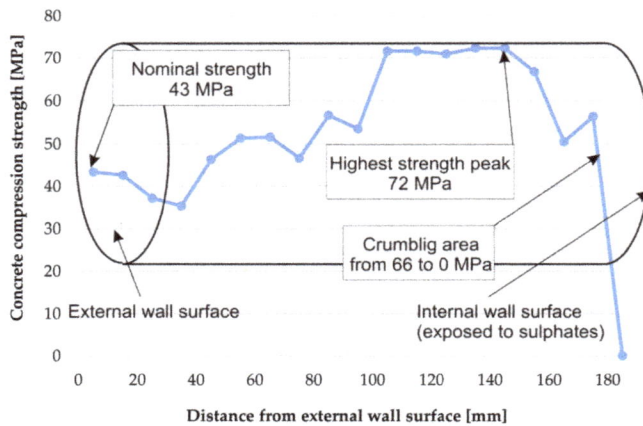

Figure 8. Change of concrete strength across the tank wall thickness in borehole No. 1.

Similar tests were performed on boreholes No. 2 & 3 which broke and reinforcement was not cut, hence their length is less than the thickness of the tank wall. The results of tests performed on borehole No. 2 are shown in Table 4.

Table 4. Results of concrete compression strength test in borehole No. 2.

Ordinal Number	Ultrasound Passing Time, Direction I-I $t_{n\ I-I}$ [μs]	Ultrasound Passing Time, Direction II-II $t_{n\ II-II}$ [μs]	Mean Ultrasound Passing Time t_n [μs]	Ultrasound Longitudinal Wave Velocity C_L [km/s]	Concrete Compression Strength f_c [N/mm²]
1	17.70	17.00	17.35	2.85	30.57
2	18.50	16.10	17.30	2.86	31.05
3	16.20	16.10	16.15	3.06	41.93
4	16.10	16.00	16.05	3.08	42.95
5	16.40	15.30	15.85	3.12	45.04
6	15.80	15.70	15.75	3.14	46.11
7	15.00	15.50	15.25	3.25	51.63
8	15.80	15.50	15.65	3.16	47.18
Mean (1–8)	16.44	15.90	16.17	3.07	42.06

Core no 2 was broken at the reinforcement mesh at a depth of 75 mm from the external surface of the wall. On the tested (not damaged) fragment of the core the growing dependency of strength as a function of depth was established. The results from Table 3 are depicted in Figure 9.

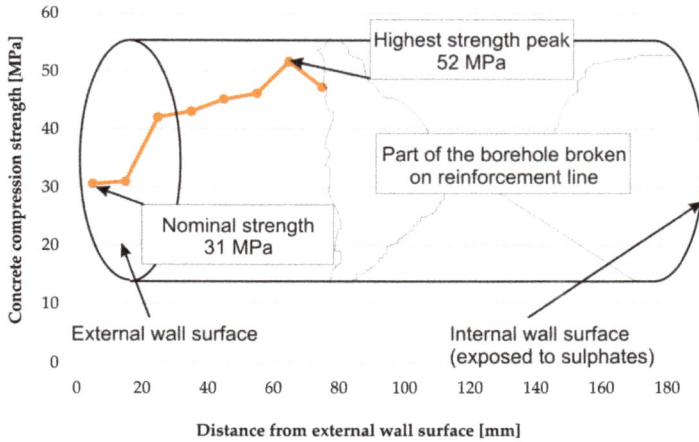

Figure 9. Change of concrete strength across the tank wall tested on borehole No. 2.

Results of tests performed on borehole No. 3 which was also broken in the middle of the tested section are shown in Table 5.

Table 5. Results of concrete compression strength test in borehole No. 3.

Ordinal Number	Ultrasound Passing Time, Direction I-I $t_{n\ I-I}$ [μs]	Ultrasound Passing Time, Direction II-II $t_{n\ II-II}$ [μs]	Mean Ultrasound Passing Time t_n [μs]	Ultrasound Longitudinal Wave Velocity C_L [km/s]	Concrete Compression Strength f_c [N/mm²]
1	16.00	17.00	16.50	3.00	38.5
2	16.30	16.80	16.55	2.99	38.0
3	16.60	17.40	17.00	2.91	33.7
4	16.00	16.60	16.30	3.04	40.4
5	16.90	17.60	17.25	2.87	31.5
6	16.90	16.90	16.90	2.93	34.7
7–9	-	-	-	-	
10	16.30	16.40	16.35	3.03	39.9
11	16.20	16.30	16.25	3.05	40.9
12	15.60	16.50	16.05	3.08	43.0
13	15.70	16.60	16.15	3.06	42.0
14	15.30	16.80	16.05	3.08	43.0
15	16.30	16.40	16.35	3.03	39.9
16	14.80	16.00	15.40	3.21	50.0
17	15.10	14.80	14.95	3.31	55.1
Mean (1–6,10–17)	16.00	16.58	16.29	3.04	40.8
18	-	-	-	-	0.00

Core No. 3 was cracked at the reinforcement at a distance of 60 to 80 mm from the external surface of the wall. For this reason the tests were not performed in this part of the core. On the tested (not damaged) fragment of the core the growing dependency of strength as a function of depth was established. The results from Table 2 are depicted in Figure 10.

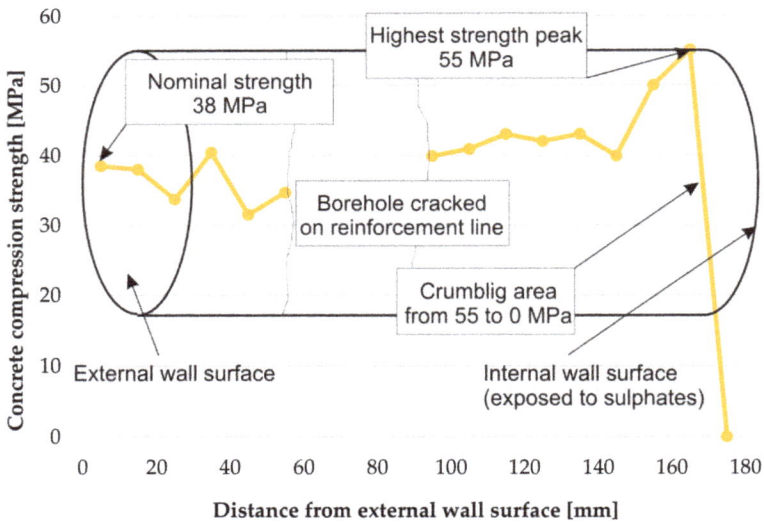

Figure 10. Change of concrete strength across the tank wall tested on borehole No. 3.

The method of measuring ultrasonic wave velocity using spot heads presented in this paper allowed determination of the distribution of strength in the cross-sections of reinforced concrete elements exposed to sulphate corrosion (with heterogeneous mechanical properties). The observations presented in this study show that the compressive strength of concrete subjected to a gaseous, aggressive, sulphuric environment is variable across its thickness. Concrete strength is variable across the wall thickness, however initially it was expected that by moving towards the tank interior, the strength would decrease, and from all boreholes the growing dependence was obtained, what has been shown in Figure 11.

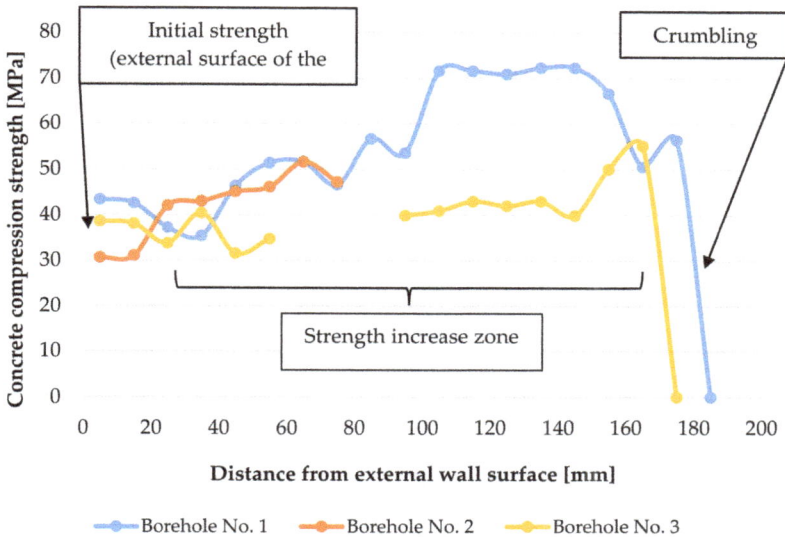

Figure 11. Change of concrete strength across the tank wall thickness in three boreholes.

Values of compressive strength of the samples taken from the tank walls show an increase in strength from 30–43 N/mm^2 in the cross sections near the external, unexposed wall layers to 55 and 72 N/mm^2 in the internal (exposed) cross sections of the walls. For the tested samples, this gives an increase in strength of 44% to 83% from its initial strength measured in the cross sections near the unexposed side of the structure. Sulphuric acid ions react with the cement compounds and the formation of gypsum leads to an increase in volume [17,18,50]. No destructive expansion of concrete takes place at this stage, and a reasonable conclusion can be drawn that the filling of pores and spaces in concrete by the calcium sulphate dihydrate causes a significant increase in the strength of the concrete, as has been shown in Figure 11. The compressive strength of the concrete samples decreases suddenly in the inner cross sections of the walls, when the salt crystallization pressure exceeds concrete tensile strength. In this stage of corrosion gypsum stone reacts with tricalcium aluminates (C3A) and hydrated calcium sulfoaluminate (monosulphate) and forms the final chemical product Candlot's salt what is associated with spalling and cracking in the surface zone of tested elements, and has been described in [51–53]. The completed tests confirmed that the drop of compressive strength took place only near internal cross-sections of the tested samples, within a distance not greater than 10 mm from their exposed surface. In the future, it is planned to carry out measurements of ultrasonic wave velocity and strength distribution in the tank walls at the height at which they are immersed in the corrosive substance and to compare them with the results presented in this article.

4. Conclusions

Based on the case study and its analysis presented in this article, the following conclusions can be made:

- Ultrasound testing methodology allowed determination of the distribution of strength as a function of depth of concrete elements under sulphate attack.
- The compressive strength of the concrete exposed to sulphate attack from one side is variable across its depth.
- The experimentally tested distribution of compressive strength at the depth of the elements showed an upward trend in the entire cross section towards the surface subject to corrosion.
- A decrease of strength appears only in the destroyed, crumbled zone of the concrete structure. The destroyed zone of tested elements did not exceed a depth of 10 mm from the surface exposed to sulphates attack.
- In the presented research, the difference in concrete strength between cross sections near the exposed and unexposed sides varied from an increase of 44% (borehole No. 3) to 83% (borehole No. 1).
- The performed tests indicate that gases may be a more corrosive environment, especially with high humidity, than liquids, therefore the coefficients of diffusion resistance or other permeability parameters, e.g., $g/m^2/24$ hours, are important parameters characterizing anticorrosive coatings. Chemical resistance to various acids, alkalis or other compounds is tested for the specific aggressive compound at a given level.
- In the tested tank, sulphur-containing gas (hydrogen sulphide) easily penetrated through the thin bituminic layer, based on the measurements at a thickness of 0.97 mm, and in contact with cement and lime formed sulphates, considerable quantities of which were found in the concrete.
- Since the lagging thickness had decreased already down to 2 mm, danger exists not only for the concrete but also for steel. The rate of concrete corrosion in the tank is probably influenced by the concentration of the gases above the liquid. If the tank had a properly built gravity ventilation, the concrete damage process would be much slower, because relative air humidity in the tank would also be much lower with effectively running ventilation.

Author Contributions: Conceptualization, B.S. and T.K.; Methodology, T.K. and B.S.; Validation, B.S., Investigation, B.S and T.K.; Resources, B.S. and T.K.; Writing-original draft preparation, B.S. and T.K.; Writing-review and editing, B.S. and T.K.; Visualization, B.S. and T.K.; Supervision, B.S.

Funding: This research received no external funding.

Conflicts of Interest: The authors declare no conflict of interest.

References

1. Feldman, R.F. Porestructure, Permeability and Diffusivity as Related to Durability. In *Eighth International Congress on the Chemistry of Cement*; Bertrand Brasil: Rio de Janeiro, Brazil, 1986; Volume 1, Theme 4.
2. Gruner, M. *Corrosion and Protection of Concrete*; PWN: Warsaw, Poland, 1983. (In Polish)
3. Flaga, K. The Role of the Tightness of the Aggregate Skeleton in the Design of a Concrete Mix. *Inz. Bud.* **1984**, *7*, 14–16. (In Polish)
4. Chodor, L. Repair and Protection of Reinforced Concrete. |Chodor-Projekt|. Available online: http://chodor-projekt. net/encyclopedia/naprawa-i-ochrona-zelbetu/ (accessed on 15 July 2018).
5. Maj, M.; Ubysz, A. Cracked reinforced concrete walls of chimneys, silos and cooling towers as result of using formworks. In *MATEC Web of Conferences 146, Proceedings of Building Defects 2017, České Budějovice, Czech Republic, November 23–24, 2017*; Šenitková, I.J., Ed.; EDP Sciences: Les Ulis, France, 2018; Volume 02002, pp. 1–8. [CrossRef]
6. Trapko, T.; Musiał, M.P. Failure of pillar of sports and entertainment hall structure. In *MATEC Web of Conferences 146, Proceedings of Building Defects 2017, České Budějovice, Czech Republic, November 23–24, 2017*; Šenitková, I.J., Ed.; EDP Sciences: Les Ulis, France, 2018; Volume 02002, pp. 1–8. [CrossRef]

7. Wells, T.; Melchers, R.E.; Bond, P. Factors involved in the long term corrosion of concrete sewers. In Proceedings of the 49th Annual Conference of the Australasian Corrosion Association, Corrosion and Prevention, Coffs Harbour, Australia, 15–19 November 2009.

8. Sun, C.; Chen, J.; Zhu, J.; Zhang, M.; Ye, J. A new diffusion model of sulfate ions in concrete. *Constr. Build. Mater.* **2013**, *39*, 39–45. [CrossRef]

9. Bonakdar, A.; Mobasher, B.; Chawla, N. Diffusivity and micro-hardness of blended cement materials exposed to external sulfate attack. *Cem. Concr. Compos.* **2012**, *34*, 76–85. [CrossRef]

10. Idiart, A.E.; L'opez, C.M.; Carol, I. Chemo-mechanical analysis of concrete cracking and degradation due to external sulfate attack: A meso-scale model. *Cem. Concr. Compos.* **2011**, *33*, 411–423. [CrossRef]

11. Lorente, S.; Yssorche-Cubaynes, M.-P.; Auger, J. Sulfate transfer through concrete: Migration and diffusion results. *Cem. Concr. Compos.* **2011**, *33*, 735–741. [CrossRef]

12. Condor, J.; Asghari, K.; Unatrakarn, D. Experimental results of diffusion coefficient of sulfate ions in cement type 10 and class G. *Energy Procedia* **2011**, *4*, 5267–5274. [CrossRef]

13. Roziere, E.; Loukili, A.; Hachem, R.; Grondin, F. Durability of concrete exposed to leaching and external sulphate attacks. *Cem. Concr. Res.* **2009**, *39*, 1188–1198. [CrossRef]

14. Santhanam, M.; Cohen, M.D.; Olek, J. Modeling the effects of solution temperature and concentration during sulfate attack on cement mortars. *Cem. Concr. Res.* **2002**, *32*, 585–592. [CrossRef]

15. Pommersheim, J.M.; Clifton, J.R. Expansion of cementitious materials exposed to sulfate solutions, scientific basis for nuclear waste management. *Mater. Res. Soc.* **1994**, *333*, 363–368. [CrossRef]

16. Parande, A.K.; Ramsamy, P.L.; Ethirajan, S.; Rao, C.R.K.; Palanisamy, N. Deterioration of reinforced concrete in sewer environments. *Inst. Civ. Eng. -Munic. Eng.* **2006**, *159*, 11–20. [CrossRef]

17. Basista, M.; Weglewski, W. Chemically-assisted damage of concrete: A model of expansion under external sulfate attack. *Int. J. Damage Mech.* **2008**, *18*, 155–175. [CrossRef]

18. Basista, M.; Weglewski, W. Micromechanical modelling of sulphate corrosion in concrete: Influence of ettringite forming reaction. *Theor. Appl. Mech.* **2008**, *35*, 29–52. [CrossRef]

19. Pommersheim, J.; Clifton, J.R. *Sulphate Attack of Cementitious Materials: Volumetric Relations and Expansion*; National Institute of Standards and Technology: Gaithersburg, MD, USA, 1994; pp. 1–19.

20. Skalny, J.; Marchand, J.; Odler, I. *Sulphate Attack on Concrete*; Spon Press: London, UK, 2002.

21. Zhou, Y.; Tian, H.; Sui, L.; Xing, F.; Han, N. Strength Deterioration of Concrete in Sulfate Environment: An Experimental Study and Theoretical Modeling. *Adv. Mater. Sci. Eng.* **2015**, *2015*, 951209. [CrossRef]

22. Shi, F.; Wang, J.H. Performance degradation of cube attacked by sulfate. *Concrete* **2013**, *3*, 52–53.

23. Du, J.M.; Liang, Y.N.; Zhang, F.J. *Mechanism and Performance Degradation of Underground Structure Attacked by Sulfate*; China Railway Publishing House: Beijing, China, 2011.

24. Neville, A.M. *Properties of Concrete*; Polski Cement Sp. z o.o.: Cracow, Poland, 2000. (In Polish)

25. Petersons, N. Should standard cube test specimens be replaced by test specimens taken from structures? *Mater. Struct.* **1968**, *1*, 425–435. [CrossRef]

26. Stawiski, B. The heterogeneity of mechanical properties of concrete in formed constructions horizontally. *Arch. Civ. Mech. Eng.* **2012**, *12*, 90–94. [CrossRef]

27. Jasinski, R.; Drobiec, Ł.; Mazur, W. Validation of Selected Non-Destructive Methods for Determining the Compressive Strength of Masonry Units Made of Autoclaved Aerated Concrete. *Materials* **2019**, *12*, 389. [CrossRef]

28. Bogas, J.A.; Gomes, M.G.; Gomes, A. Compressive strength evaluation of structural lightweight concrete by non-destructive ultrasonic pulse velocity method. *Ultrasonics* **2013**, *53*, 962–972. [CrossRef]

29. Brunarski, L. Estimation of concrete strength in construction. *Build. Res. Inst. Quat.* **1998**, *2–3*, 28–45.

30. Anugonda, P.; Wiehn, J.S.; Turner, J.A. Diffusion of ultrasound in concrete. *Ultrasonics* **2001**, *39*, 429–435. [CrossRef]

31. Sansalone, M.; Streett, W.B. *Impact-Echo Nondestructive Evaluation of Concrete and Masonry*; Bullbrier Press: Ithaca, NY, USA, 1997.

32. Breysse, D. Nondestructive evaluation of concrete strength: An historical review and a new perspective by combining NDT methods. *Constr. Build. Mater.* **2012**, *33*, 139–163. [CrossRef]

33. Facaoaru, I. Contribution à i'étude de la relation entre la résistance du béton à la compression et de la vitesse de propagation longitudinale des ultrasons. *RILEM* **1961**, *22*, 125–154.

34. Leshchinsky, A. Non-destructive methods instead of specimens and cores, quality control of concrete structures. In Proceedings of the Second International RILEM/CEB Symposium, Belgium, 12–14 June 1991; pp. 377–386.

35. Bungey, J.H. The validity of ultrasonic pulse velocity testing of in-place concrete for strength. *NDT Int.* **1980**, *13*, 296–300. [CrossRef]

36. Szpetulski, J. Testing of compressive strength of concrete in construction. *Constr. Rev.* **2016**, *3*, 21–24. (In Polish)

37. Gudra, T.; Stawiski, B. Non-destructive strength characterization of concrete using surface waves. *NDT Int.* **2000**, *33*, 1–6. [CrossRef]

38. Komlos, K.; Popovics, S.; Nurnbergerova, T.; Babal, B.; Popovics, J.S. Ultrasonic Pulse Velocity Test of Concrete Properties as Specified in Various Standards. *Cem. Concr. Compos.* **1996**, *18*, 357–364. [CrossRef]

39. PN-B-06261. *Non-Destructive Testing of Structures*; Ultrasound method of testing compressive strength of concrete; Polish Committee for Standardization: Warsaw, Poland, 1974.

40. EN 12504-4. *Testing Concrete-Part 4: Determination of Ultrasonic Pulse Velocity*; European Committee for Standardization: Brussels, Belgium, 2004.

41. EN 13791. *Assessment of In-Situ Compressive Strength in Structures and Precast Concrete Components*; Polish Committee for Standardization: Brussels, Belgium, 2004.

42. Stawiski, B.; Kania, T. Determination of the influence of cylindrical samples dimensions on the evaluation of concrete and wall mortar strength using ultrasound method. *Procedia Eng.* **2013**, *57*, 1078–1085.

43. Alam, B.; Afzal, S.; Akbar, J.; Ashraf, M.; Shahzada, K.; Shabab, M.E. Mitigating Sulphate Attack in High Performance Concrete. *Int. J. Adv. Struct. Geotech. Eng.* **2013**, *2*, 11–15.

44. Genovés, V.; Vargas, F.; Gosálbez, J.; Carrión, A.; Borrachero, M.V.; Payá, J. Ultrasonic and impact spectroscopy monitoring on internal sulphate attack of cement-based materials. *Mater. Des.* **2017**, *125*, 46–54. [CrossRef]

45. Cumming, S.R. *Non-Destructive Testing to Monitor Concrete Deterioration Caused by Sulfate Attack*; University of Florida: Gainesville, FL, USA, 2004.

46. Kocjan, R. *Analytical Chemistry*; Qualitative analysis, Quantitative analysis, Instrumental analysis; PZWL: Warsawa, Poland, 2015. (In Polish)

47. Blumenthal, P.L.; Guernsey, S.C. *The Determination of Sulfur as Barium Sulfate*; Research Bulletin No. 26; AES: Ames, IA, USA, 1915.

48. Reid, J.M.; Czerewko, M.A.; Cripps, J.C. *Sulfate Specification for Structural Backfills*; Report TRL447; TRL: Berkshire, UK, 2005.

49. St. John, T.W. Quantifying acid-soluble sulfates in geological materials: A comparative study of the British Standard gravimetric method with ICP-OES/AES. In Proceedings of the 19th International Conference on Soil Mechanics and Geotechnical Engineering, Seoul, Korea, 15 September 2017.

50. Piasta, W.; Marczewska, J.; Jaworska, M. Some aspects and mechanisms of sulphate attack. *Struct. Environ.* **2014**, *6*, 19–24.

51. Brown, W.; Taylor, H. The role of ettringite in external sulfate attack. *Mater. Sci. Concr.* **1999**, *5*, 73–98.

52. Yu, C.; Sun, W.; Scrivener, K. Mechanism of expansion of mortars immersed in sodium sulfate solutions. *Cem. Concr. Res.* **2012**, *43*, 105–111.

53. Whitaker, M.; Black, L. Current knowledge of external sulfate attack. *Adv. Cem. Res.* **2015**, *27*, 1–14. [CrossRef]

materials

MDPI

Article

Identification of the Degree of Degradation of Fibre-Cement Boards Exposed to Fire by Means of the Acoustic Emission Method and Artificial Neural Networks

Krzysztof Schabowicz [1], **Tomasz Gorzelańczyk** [1,*] **and Mateusz Szymków** [1]

Faculty of Civil Engineering, Wrocław University of Science and Technology, Wybrzeże Wyspiańskiego 27,
50-370 Wrocław, Poland; krzysztof.schabowicz@pwr.edu.pl (K.S.); mat.szymkow@gmail.com (M.S.)
* Correspondence: tomasz.gorzelanczyk@pwr.edu.pl; Tel.: +48-71-320-3742

Received: 27 December 2018; Accepted: 18 February 2019; Published: 21 February 2019

check for
updates

Abstract: This paper presents the results of research aimed at identifying the degree of degradation of fibre-cement boards exposed to fire. The fibre-cement board samples were initially exposed to fire at various durations in the range of 1–15 min. The samples were then subjected to three-point bending and were investigated using the acoustic emission method. Artificial neural networks (ANNs) were employed to analyse the results yielded by the acoustic emission method. Fire was found to have a degrading effect on the fibres contained in the boards. As the length of exposure to fire increased, the fibres underwent gradual degradation, which was reflected in a decrease in the number of acoustic emission (AE) events recognised by the artificial neural networks as accompanying the breaking of the fibres during the three-point bending of the sample. It was shown that it is not sufficient to determine the degree of degradation of fibre-cement boards solely on the basis of bending strength (*MOR*).

Keywords: fibre-cement boards; non-destructive testing; acoustic emission; degree of degradation

---

1. Introduction

The fibre-cement board is a building material that has been used in construction since the beginning of the last century. The Czech engineer Ludwig Hatschek developed and patented the technology for the production of this composite material. The first boards contained asbestos fibres. After asbestos had been found to be carcinogenic, they were replaced with cellulose fibres and synthetic fibres [1]. Being composed of cement, cellulose fibres, synthetic fibres and various innovative additives and admixtures, the currently produced fibre-cement boards are a completely different building product. The added components and fillers of fibre-cement boards are: limestone powder, mica, perlite, kaolin, microspheres and recycled materials [2,3]. As a result, fibre-cement boards continue to be innovative products consistent with the principles of sustainable development [4]. Fibre-cement boards are used in construction mainly for ventilated façade cladding [5], as shown in Figure 1. In the course of their service life, fibre-cement boards are exposed to variable environmental factors, chemical aggressivity (acid rains), and physical aggressivity (ultraviolet radiation). Moreover, fibre-cement boards are exposed to accidental operational factors, mainly the high temperature that a building fire generates.

Figure 1. Exemplary uses of fibre-cement boards as ventilated façade cladding: (**a**) University building in Łódz, Poland, (**b**) University building in Wrocław, Poland.

After a building fire occurs, it is necessary to determine the impact of the high temperatures on the façade cladding of the building and the neighbouring buildings with regard to their further service life. This means that identifying the degree of degradation of fibre-cement boards caused by the impact of fire is vital from both the scientific and practical point of view. Most of the research on fibre-cement boards to date has been limited to determining their standard physicomechanical properties, the effect of operational factors, such as soak–dry cycles, freeze–thaw cycles, heating, raining and high temperatures, and the effect of the use of various types of fibres and production processes, solely through bending (*MOR*) tests [6]. In the literature on the subject, one can find only a few nondestructive tests carried out on fibre-cement boards, limited to the imperfections arising during production [7–10]. The effect of fire is one of the most destructive accidental factors for many building products, especially the composite ones containing reinforcement in the form of fibres of various kinds, particularly cellulose fibres, considering that at a temperature above 200 °C cellulose fibres undergo pyrolysis. Thus, the action of fire critically affects the durability of the whole composite [6]. In order to verify this hypothesis, experiments were conducted during which fibre-cement board samples were exposed to fire at a temperature of about 400 °C at various durations in the range of 1–15 min. This time range and the temperature of 400 °C had been experimentally determined through many preliminary trials. The aim of the experimental research is to observe the advance of the degradation of the fibres due to fire and the associated high temperature. Preliminary tests conducted at the temperature of 500 °C very quickly, i.e., after 1–2 min, resulted in the complete destruction of the board, which no longer could be subjected to the bending test. A reduction in the temperature to 250–300 °C would considerably lengthen the duration of the tests. Preliminary investigations showed that, at a temperature of 250 °C, the fibres in the boards would be completely destroyed after 2–3 h. Therefore, in the authors' opinion, the choice of the temperature of 400 °C was optimal for conducting experiments aimed at identifying the degree of degradation of the fibres contained in fibre-cement boards under the high temperature generated by fire.

After they had been exposed to fire, the board samples were subjected to three-point bending, during which the acoustic emission (AE) was recorded, and the results were then analysed by means of artificial neural networks. Research carried out by the authors has shown that the identification of the degree of degradation of fibre-cement boards exposed to fire, based on only bending strength (*MOR*), is inadequate [11,12]. Thanks to the use of the acoustic emission method, the degree of degradation can be determined on the basis of not only the mechanical parameters, but also the acoustic phenomena that can occur in fibre-cement boards. From the recorded AE signals, model (reference) characteristics of the acoustic spectra that accompany the breaking of the cement matrix and the fibres during bending were derived. Under the continuing exposure to fire, the fibres in the boards would undergo gradual degradation, which was reflected in a decrease in the counts of AE events identified as accompanying fibre breaking. This phenomenon is described in more detail in this paper.

2. Survey of the Literature

Most of the research on fibre-cement boards to date has been devoted to the effect of operational factors [13–15] and the effect of high temperatures, investigated by testing the physicomechanical parameters of the boards, mainly their bending strength (*MOR*). The paper by Ardanuy et al. [6] presents the results of research on, inter alia, the effect of high temperature on fibre-cement boards, but only with reference to bending strength *MOR*. Li et al. [16] studied the effect of high temperatures on composites produced using the extrusion method, but also solely on the basis of their mechanical properties. The nondestructive investigations of fibre-cement boards have been mainly limited to the detection of imperfections arising at the production stage. Papers by Drelich et al. [8] and Schabowicz and Gorzelańczyk [17] presented the possibility of using Lamb waves in a noncontact ultrasonic scanner to detect defects in fibre-cement boards at the production stage. Paper [18] by Stark describes a method of detecting delaminations in composite elements by means of a moving ultrasonic probe. An ultrasonic device and a way of detecting delaminations are described in work [7] by Dębowski, Lewandowski, Mackiewicz, Ranachowski and Schabowicz. In works by Berkowski et al. [19], Hoła and Schabowicz [20] and Davis et al. [21] it was proposed to use the impact-echo method jointly with the impulse response method to identify delaminations in concrete members. Since the impulse response method is used to test elements thicker than 100 mm, it is not suitable for testing fibre-cement boards. Preliminary research had shown that striking such a board with a hammer could damage it, which is another reason why the impulse response method is not suitable for fibre-cement boards. Also, the impact echo method is not used to test fibre-cement boards. The drawback of this method is that multiple reflections of waves cause disturbances, making the interpretation of the obtained image difficult [19]. Therefore, it is not recommended to use this method for fibre-cement boards thicker than about 8 mm.

In the literature, there is little information on the use of other nondestructive methods for testing fibre-cement boards. The research described in the works of Chady et al. [9] and Chady and Schabowicz [22] showed the terahertz (T-Ray) method to be suitable for testing fibre-cement boards. Terahertz signals have a very similar character to that of ultrasonic signals, but their interpretation is more complicated. In work [23], the microtomography method was used to identify delaminations and low-density regions in fibre-cement boards. The test results indicate that this method clearly reveals differences in the microstructure of the boards. Therefore, the microtomography method can be a useful tool for testing the structure of fibre-cement boards in which defects can arise due to production errors. However, this method can be used only for small-sized boards. It should be noted that, so far, few cases of testing fibre-cement boards by means of an acoustic emission have been reported in the literature. Ranachowski and Schabowicz et al. [23] carried out pilot studies of fibre-cement boards produced using the extrusion method, including boards subjected to the temperature of 230 °C for 2 h, in which the acoustic emission method was used to determine the effect of cellulose fibres on the strength of the boards and where attempts were made to distinguish between the AE events emitted by the fibres and the cement matrix. The results of this research confirmed the suitability of this method for testing fibre-cement boards. In paper [11] by Gorzelańczyk et al., it was proposed to use the acoustic emission method to investigate the impact of high temperature on fibre-cement boards. It should be noted that the effects of high temperatures on concrete, and the interdependences involved in this process, have been widely described using the acoustic emission method; examples are the works [24,25] by Ranachowski. One should also note that the acoustic emission method is often used to test thin materials, for example steel and polymeric composites, and even to test brittle food products [26,27]. During acoustic emission measurements a lot of data are recorded, which need to be properly analysed and interpreted. For this purpose, it can be highly effective to combine the acoustic emission method with artificial intelligence, including artificial neural networks (ANNs). In the literature, ANNs are used to analyse and recognise the signals registered during the failure of various materials [28]. In paper [29], Schabowicz used ANNs to analyse the results of nondestructive tests carried out on concrete. Łazarska et al. [30] and Woźniak et al. [31] successfully used the acoustic emission method and artificial neural networks to analyse its results. Rucka and Wilde [32,33] successfully used the ultrasonic method to investigate damage to concrete structures.

Considering the above information, it was decided that the acoustic emission method combined with artificial neural networks would be proper for identifying the degree of degradation of fibre-cement boards exposed to fire.

3. Strength Tests

The effect of fire was investigated for two series of fibre-cement boards, designated respectively A and B. The basic parameters of the two series are specified in Table 1.

Table 1. The tested fibre-cement boards series A and B and their basic parameters.

Series Name	Board Thickness e (mm)	Board Colour	Application	Board Bulk Density ρ (g/cm^3)	Bending Strength MOR (MPa)
A	8.0	natural	exterior	1.60	25
B	8.0	pigmented	exterior	1.60	30

The effect of fire on the fibre-cement boards was investigated by applying a flame generating a temperature of about 400 °C to the surface of the board, for about 1–15 min, every 2.5 min. The reference samples under an air-dry condition were designated as A and B. The designations of the other series of boards are listed in Table 2.

Table 2. A list of fibre-cement board series and test cases with the adopted sample designations.

Series Name/Test Case	SERIES A	SERIES B
Air-dry condition (reference board)	A_1	B_1
Exposure to the temperature of 400 °C: 1 min	A_2	B_2
Exposure to the temperature of 400 °C: 2.5 min	A_3	B_3
Exposure to the temperature of 400 °C: 5 min	A_4	B_4
Exposure to the temperature of 400 °C: 7.5 min	A_5	B_5
Exposure to the temperature of 400 °C: 10 min	A_6	B_6
Exposure to the temperature of 400 °C: 12.5 min	A_7	B_7
Exposure to the temperature of 400 °C: 15 min	A_8	B_8

Since not the whole fire separation area, but the fibre-cement board alone was the object of the test, on the basis of experiments the duration of exposure to fire was limited to 15 min. A stand for exposing the samples to fire at the temperature of 400 °C is shown in Figure 2.

Figure 2. The stand for exposing samples to fire at the temperature of 400 °C.

In order to identify the effect of the high temperature generated by fire on the fibre-cement boards, acoustic emission investigations were carried out during three-point bending. Before the investigations, the samples were exposed to fire, then they were placed in a strength testing machine and subjected to three-point bending. In the course of the three-point bending, the breaking force F, the strain ε and acoustic emission signals were registered. The tested samples are shown in Figure 3, while the stand for three-point bending tests and the set of equipment for acoustic emission measurements are shown in Figure 4.

(a)   (b)

Figure 3. The tested fibre-cement board samples: (a) board A, (b) board B.

Three-point bending strength testing   Close-up of strength testing machine tip and tested

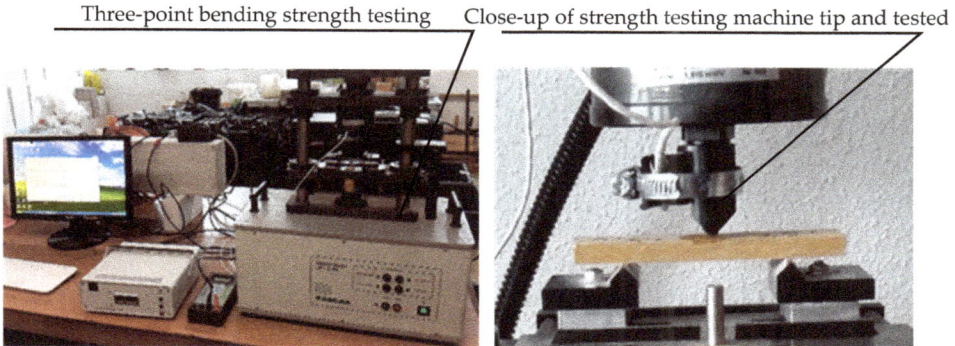

Figure 4. The test stand for the acoustic emission measurements, and a fibre-cement board during a test.

The trace of flexural strength σ_m, bending strength *MOR*, notch toughness W_f, limit of proportionality (*LOP*) and strain ε were analysed. Bending strength *MOR* was calculated from the standard formula [34]:

$$MOR = \frac{3Fl_s}{2b\,e^2} \tag{1}$$

where:

    F is the loading force (N);
    l_s is the length of the support span (mm);
    B is the specimen width (mm); and
    e is the specimen thickness (mm).

Notch toughness was calculated from the following formula given in [35]:

$$W_f = \frac{1}{S} \int_{F0}^{F0,4max} F\,da \tag{2}$$

where:

    F is the loading force (N);

S is the specimen cross-section (m2); and

a is the specimen deflection under the loading roller (m).

Figure 5 shows diagrams of the σ−ε dependence under bending for the samples of all the tested fibre-cement boards.

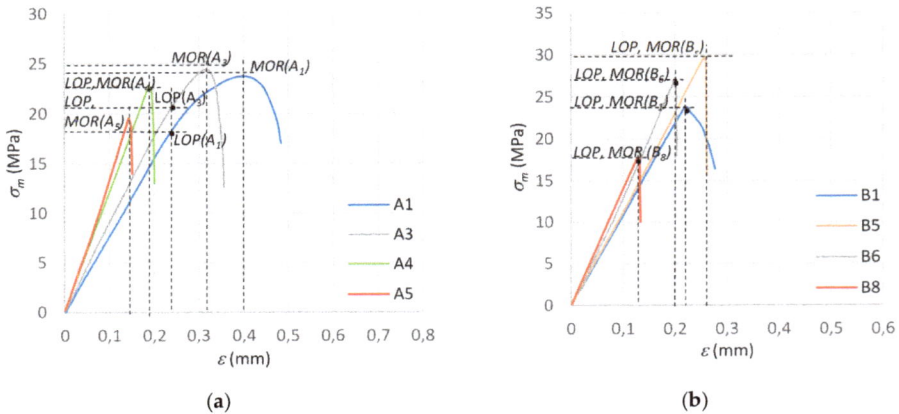

Figure 5. σ−ε dependence under bending for fibre-cement boards: (**a**) series A, (**b**) series B. *LOP*, limit of proportionality.

The results presented in Figure 5 show that, under high temperature, the bending strength *MOR* initially increases by 3–15% and then sharply decreases in comparison with that of the reference fibre-cement boards. In the initial phase of exposure to fire, especially in the time interval of 1–2.5 min, the sample would dry, but no fibres would be destroyed yet. It should be noted that the reference samples were under an air-dry condition and their bulk moisture content amounted to 6–8%. One can suppose that the dampness of the fibre-cement boards affects the bending strength *MOR* mainly as a result of the weakening of the bonds between the crystals in the structural lattice of the cement matrix. The weakening is due to the fact that, as the material's dampness increases, the bonds partially dissolve, whereby the bending strength slightly decreases. The decrease in bending strength *MOR* caused by dampness is partially reversible; i.e., after it is dried, the material regains most of its lost strength and so ultimately its strength is close to the initial one. In the case of the B_1–B_8 series samples, no significant changes in the σ−ε dependence diagrams for the reference boards and the ones exposed to fire were observed, except for the changes in bending strength values. In the case of the A series boards, the σ−ε dependence changed markedly with the time of their exposure to fire. On this basis, for the boards of series A, one can determine the effect of high temperature, reflected not only in the initial increase in bending strength *MOR*, followed by its decrease, but also in changes in the plot of σ−ε dependence. For the series A boards, it was also observed that as the time of their exposure to the temperature of 400 °C is extended, the structure of the fibre-cement board becomes stiffer and more brittle. For this series, as the time of exposure to the temperature generated by fire is extended, the extent of the nonlinear increase in bending stress is reduced until the bending strength *MOR* comes to be level with the proportionality limit *LOP*. In the case of the reference boards, the *MOR* and *LOP* values were clearly separated.

Figure 6 shows the obtained values of notch toughness W_f of the tested fibre-cement boards.

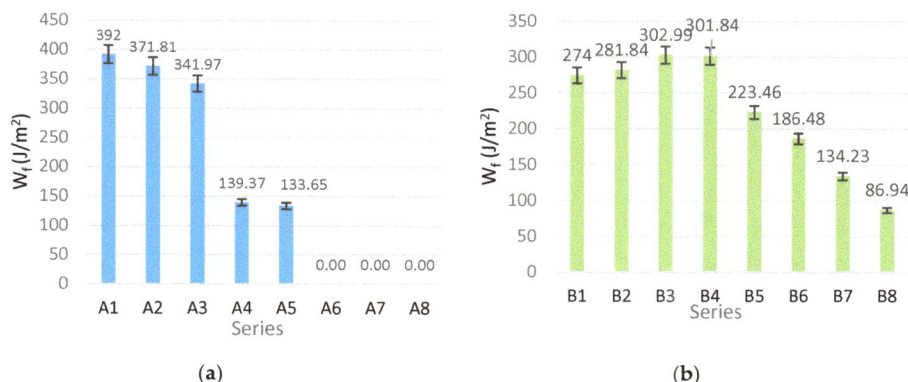

Figure 6. The values of notch toughness W_f for air-dry samples (case 1) and samples exposed to the temperature of 400 °C (cases 2–8): (**a**) series A, (**b**) series B.

An analysis of the results presented in Figure 6 showed that under the impact of high temperature, the notch toughness W_f of the boards of series B initially increases by 2–10% and then markedly decreases. For the boards of series A, the notch toughness W_f decreases with the time of exposure to fire. Hence, one can conclude that, as a result of the high temperature, the fibre-cement board matrix stiffens and becomes more brittle. The investigations carried out for the boards of series B showed that the identification of the degree of degradation on the basis of bending strength *MOR*, notch toughness W_f, proportionality limit *LOP* and $\sigma-\varepsilon$ dependence is inadequate. The results obtained for the fibre-cement boards of series A indicate that destructive changes took place in their structure. However, the fact that the bending strength *MOR* increased in the initial phase of the exposure to the high temperature is worth noting. The above results show that the identification of the degree of degradation solely on the basis of mechanical parameters, such as *MOR* and W_f, is insufficient to determine the destructive effect of high temperature on fibre-cement boards in the initial phase of their exposure to fire.

4. Tests Using the Acoustic Emission Method and Artificial Neural Networks

The next step in the investigations of the degree of degradation caused by fire consists in analysing the AE signals registered during three-point bending, such as the events rate N_{ev}, the events sum ΣN_{ev}, the events energy E_{ev} and the frequency distribution of the AE signal. For AE measurements, a broadband sensor with a frequency band of 5–500 kHz and an AE signal analyser (made by IPPT PAN, Warsaw, Poland), were used [35]. In the AE analyser the signal from the sensor is amplified and prefiltered to remove the acoustic background from the surroundings of the monitored element. Then, the signal is converted into a digital form. Further processing of digital records was carried out using audio file analysis software (made by IPPT PAN, Warsaw, Poland). Figure 7 shows exemplary values of events sum ΣN_{ev} for the boards of series A1–A6 and B1–B8. No three-point bending test was carried out for cases A6-A8 because the fibre-cement boards had been completely destroyed by fire.

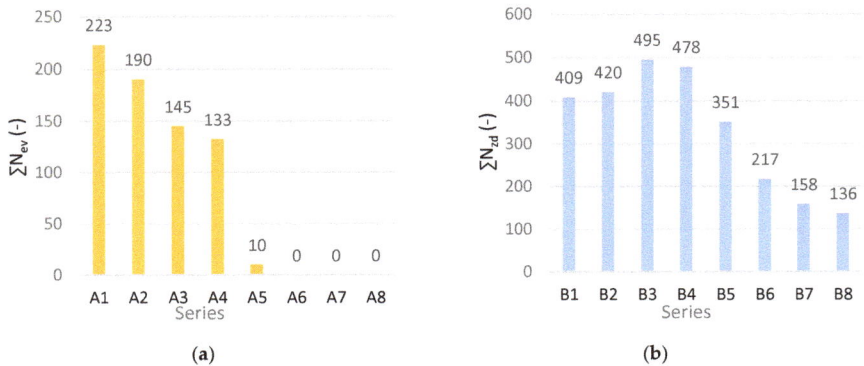

Figure 7. Exemplary registered values of events sum $\sum N_{ev}$ for boards of series A under an air-dry condition (case 1) and exposed to the temperature of 400 °C (cases 2–8): (**a**) series A, (**b**) series B.

In order to more closely analyse the course of the bending test and how it was affected by the degrading factor, in the form of the high temperature of 400 °C lasting from 1 to 15 min, the dependence between events rate $\sum N_{ev}$ and bending stress σ_m over time for selected cases is presented in Figure 8.

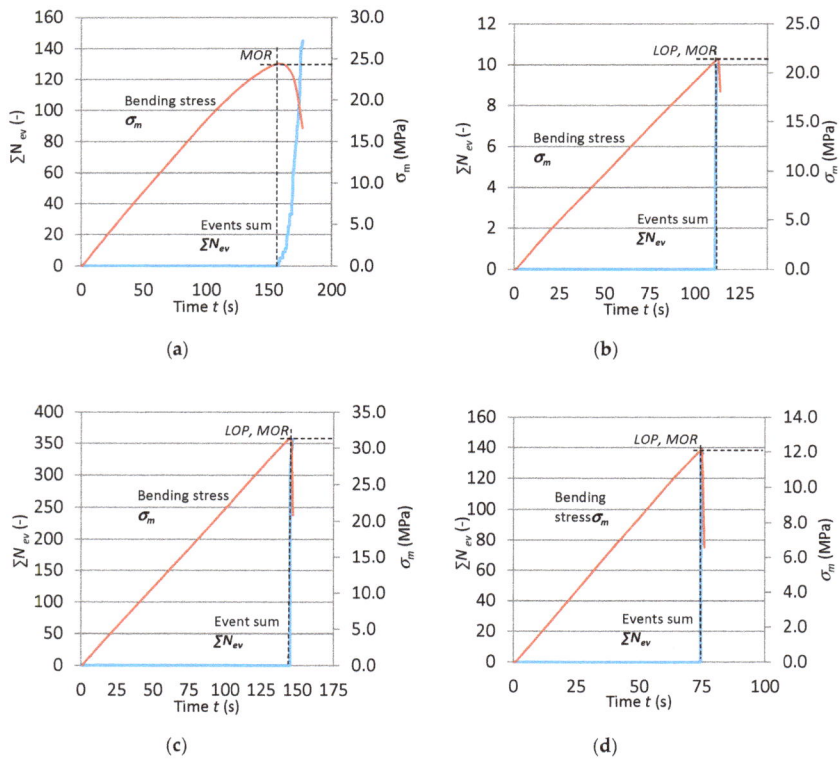

Figure 8. Dependence between σ_m and $\sum N_{ev}$ as function of time t for fibre-cement boards: (**a**) series A3, (**b**) series A5, (**c**) series B5, (**d**) series B8.

An analysis of the obtained results showed that AE signals were registered only after the stress corresponding to bending strength *MOR* had been exceeded. The analysis of the course of the three-point bending test, based on AE events sum $\sum N_{ev}$, confirmed the changes taking place in the fibre-cement boards. Marked qualitative changes in the registered events in comparison with the reference boards were visible. At the high temperature of about 400 °C, the sample is initially dried, whereby the strength of the board and the acoustic activity of the AE events registered during bending slightly increase. An analysis of the AE descriptors indicates that the brief effect of the high temperature of 400 °C proved to be significant for the fibre-cement boards. The graphs for series and A and B show a noticeable decrease in the registered events and a change in the path of events sum $\sum N_{ev}$. Besides the decrease in the value of events sum $\sum N_{ev}$ due to the exposure to the high temperature of 400 °C for 7.5 min in the case of series A_5 and for 12.5 min in the case of series B_7, it was also noticed that all the registered events were within one 1.5 second time interval. Therefore, it can be concluded that the events originated from a single fracture of the cement matrix. This indicates that the exposure to the high temperature of 400 °C for about 7.5 min was more destructive for the fibre-cement boards of series A. Because of the destructive effect of the high temperature of about 400 °C, the registered AE event counts declined.

By analysing the acoustic activity traces in the time-frequency system during the bending of the boards of series A and B exposed to the temperature of 400 °C for over 7.5 min, model (reference) acoustic spectra for the breaking of the cement matrix devoid of fibres were selected. A model characteristic of the fibre breaking acoustic spectrum was selected from the spectra read for the boards of series A_1 to B_1 under an air-dry condition, with a similar characteristic repeating itself in the frequency range of 10–24 kHz, clearly standing out against the characteristic of the cement matrix. The characteristic of the background acoustic spectrum originating from the strength tester was determined on the basis of the initial bending phase by averaging the characteristics obtained from all of the tested fibre-cement boards of series A and B. The selected spectral characteristics of fibre-breaking are understood to be the signal accompanying the breaking of the cement matrix together with the fibres, whereas the model matrix spectral characteristic is understood to be the signal accompanying the breaking of the matrix alone. The selected model (reference) acoustic spectrum characteristics were recorded in 80 intervals at every 0.5 kHz. Figure 9 shows the record of the model characteristics of the acoustic spectrum accompanying the breaking of, respectively, the cement matrix and the fibres, and of the background acoustic spectrum.

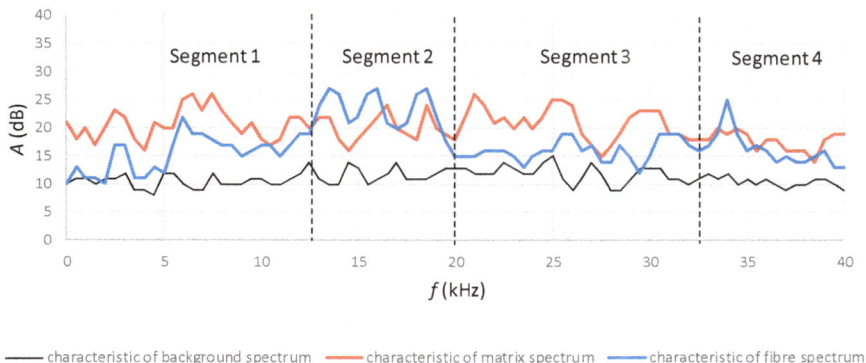

Figure 9. The characteristic of the background, cement matrix and fibre acoustic spectrum.

According to Figure 9, the acoustic activity of the background is in the range of 10–15 dB. The characteristic of the cement matrix acoustic spectrum reached the acoustic activity of 25 dB in the range of 5–10 kHz (segment 1) and 20–32 kHz (segment 3). In the case of the fibres, the activity was above 25 dB in the frequency range of 12–18 dB (segment 2) and 32–38 kHz (segment 4). The above reference spectra for the cement matrix, the fibres and the background were implemented in artificial neural networks (ANNs) for training and testing. A unidirectional multilayered ANN structure with error backpropagation with momentum was adopted. Eight appropriate learning sequences were adopted in an iterative manner to achieve optimal compatibility of the learned ANN with the training pattern. After the ANN had been trained on the input data, its mapping correctness was verified using the training data and the testing data. For this purpose, two pairs of input data were fed, i.e., the data used for training the ANN, to check its ability to reproduce the reference spectra, and the one used for testing the ANN, to check its ability to identify the reference spectral characteristics originating from the fibres and the cement matrix during the bending test. For the eight performed training sequences, ANN compliance was obtained with the training standard at 0.995. Then, a record of the ANN output in the form of recognised acoustic spectra for, respectively, fibre breaking, matrix breaking and the background was obtained.

Figure 10 shows the results of the acoustic spectrum pattern recognition for the cement matrix and the fibres. The results are marked on the record of events rate N_{ev} and flexural stress σ_m versus time. The graphs are for the reference samples of series A_1 and the samples of series A_3–A_5 exposed to fire. In order to better distinguish between the recognised acoustic spectra, the matrix patterns were marked green while the ones for the fibres were marked in violet.

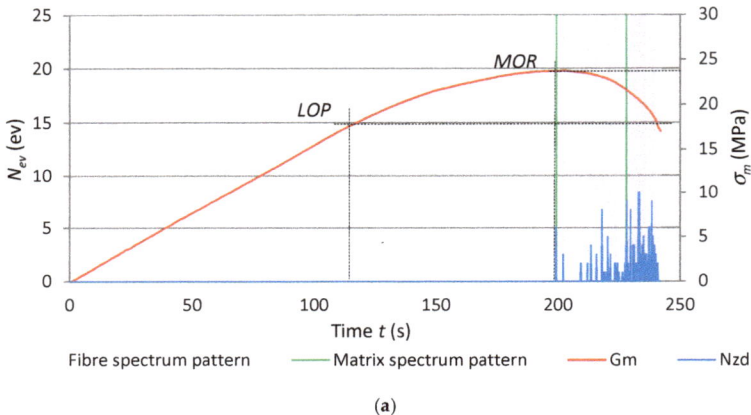

(a)

Figure 10. *Cont.*

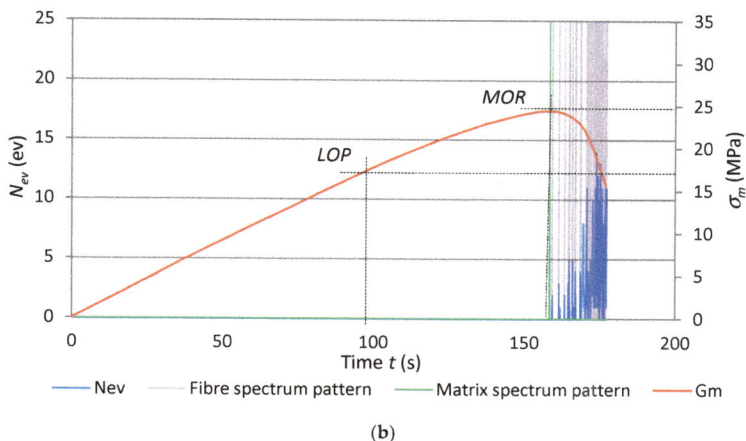

(b)

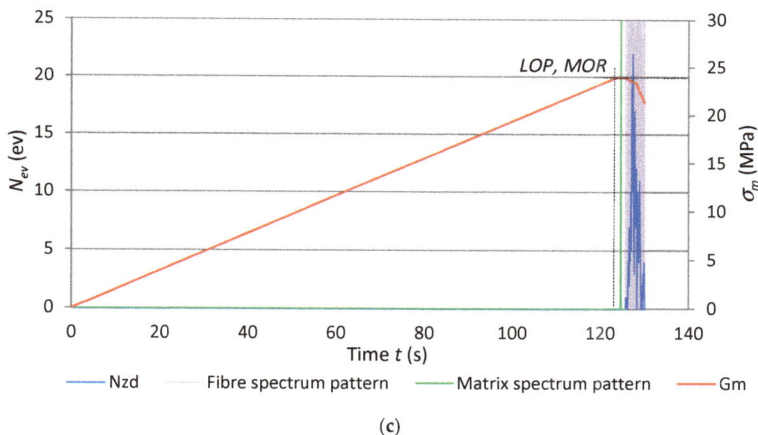

(c)

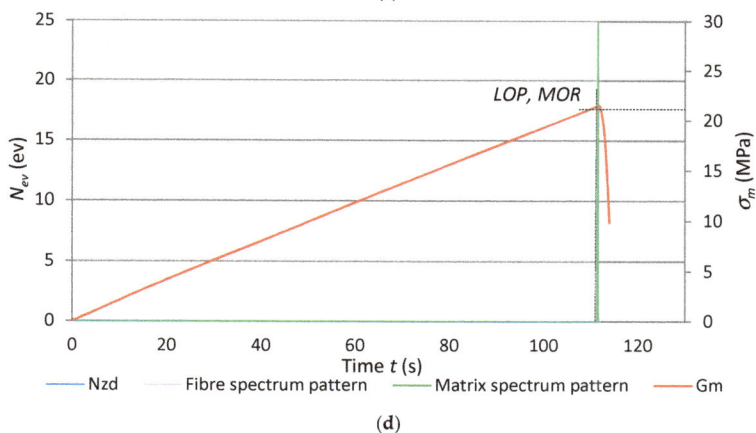

(d)

Figure 10. The events rate N_{ev} and flexural stress σ_m versus time for exposure to fire, with marked recognised reference spectral characteristics: (**a**) series A_1, (**b**) series A_3, (**c**) series A_4, (**d**) series A_5.

An analysis of the results presented in Figure 10 showed that the registered acoustic events repeatedly predominate after the bending strength *MOR* is exceeded. The recognised events result from the breaking of the fibres and the cement matrix. An event originating from cement matrix breaking initiates subsequent events originating from fibre breaking. The matrix fractures when the bending strength *MOR* is reached. Subsequently, events originating from the breaking of the fibres build up until the board's cross-section fractures. In Figure 10c, which shows the recognised events for the boards of series A_4 exposed to the high temperature for 5 min, one can clearly see that the drop in the counts of acoustic events originating from the fibres subjected to the high temperature is smaller. Similarly to the case of the boards of series A_3, the fracture of the cement matrix initiates the breaking of the fibres. The graph of flexural stress σ_m has no segment characterised by the linear increase of stress relative to strain. The limit of proportionality coincides with the bending strength *MOR*.

For the series A_5 sample, exposed to the high temperature of 400 °C for 7.5 min, only events recognised as originating from the breaking of the matrix were registered within the time of 0.1 s. All of the fibres in this board were destroyed. The samples of series A_6–A_8 exposed to fire for over 7.5 min underwent complete destruction, whereby no further tests could be carried out. Table 3 lists the events recognised as accompanying fibre breaking and cement matrix breaking for the fibre-cement boards of series A_2–A_5.

Table 3. A list of events recognised as accompanying fibre breaking and cement matrix breaking for boards of series A_1–A_8.

Series	Events Sum $\sum N_{ev}$	Sum of Recognised Events $\sum N_{ev,r}$	Sum of Events Ascribed to Fibre Breaking $\sum N_{ev,f}$	Sum of Events Ascribed to Matrix Breaking $\sum N_{ev,m}$
A_1	223	203	195	8
A_2	190	183	169	14
A_3	145	144	70	70
A_4	133	130	45	85
A_5	92	92	0	92

Figure 11 shows diagrams for samples of series B1, B5 and B8.

(a)

(b)

(c)

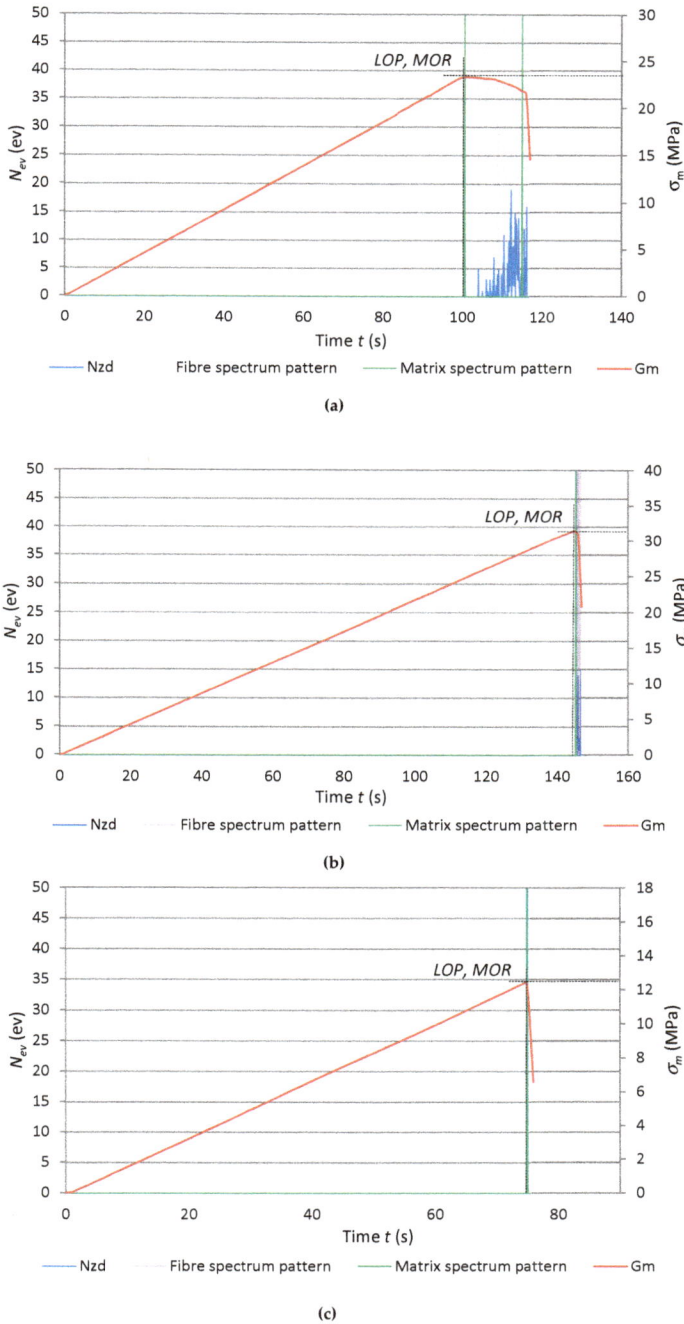

Figure 11. The events rate N_{ev} and flexural stress σ_m versus time for exposure to fire, with marked recognised reference spectral characteristics: (**a**) series B_1, (**b**) series B_5, (**c**) series B_8.

In Figure 11, one can see a decline in the counts of events recognized for the fibres and a considerably shorter time segment in which events were registered. It should be noted that, in the case of the fibre-cement boards of series B, distinct changes in the registered AE descriptors occurred only when the boards were exposed to the high temperature for 7.5 min. Moreover, the proportionality limit of the boards of series B was equal to their bending strength *MOR*. In Figure 11c, which shows the results for series B_8 exposed to the high temperature of 400 °C for 15 min, one can see that the recognised events originate from cement matrix breaking, but there are no recognised events originating from the fibres.

Table 4 lists the events recognised as accompanying fibre breaking and cement matrix breaking for boards of series B_1–B_8.

Table 4. A list of events recognised as accompanying fibre breaking and cement matrix breaking for boards of series B_1–B_8.

Series	Events Sum $\sum N_{ev}$	Sum of Recognised Events $\sum N_{ev,r}$	Sum of Events Ascribed to Fibre Breaking $\sum N_{ev,f}$	Sum of Events Ascribed to Matrix Cracking $\sum N_{ev,m}$
B_1	409	405	325	80
B_2	420	413	319	94
B_3	495	484	336	148
B_4	478	475	315	160
B_5	351	349	198	151
B_6	217	207	54	153
B_7	158	158	0	158
B_8	114	114	0	114

The above investigations have shown that it is insufficient to determine the degree of degradation solely on the basis of the value of bending strength *MOR*. As part of the investigations, the degree of degradation was determined not only on the basis of bending strength *MOR*, but also using the number of the registered AE events recognised (by artificial neural networks) as accompanying fibre breaking, and the dependence between notch toughness W_f and the events recognised as originating from fibre breaking. In this way, the degree of degradation caused by the high temperature could be unequivocally identified. As the number of events recognised as fibre breaking decreases, the notch toughness W_f also commensurately decreases. The sharp drop in notch toughness W_f is connected with the large number of fibres contained in fibre-cement boards and their rapid degradation under high temperature. It was experimentally found that the coefficient U_i determines the significance of the degradation of fibre-cement boards caused by high temperature. The coefficient U_i is expressed by the formula:

$$U_i = \frac{W_{fi}}{W_{f,ref}} \tag{3}$$

where: U_i is a value of the notch toughness coefficient common for samples exposed to high temperature and for the reference samples, W_{fi} is the notch toughness of a fibre-cement board exposed to high temperature, and $W_{f,ref}$ is the notch toughness of the reference board.

For the notch toughness coefficient U_i, its average value U_{av} and the lower and upper limit of the confidence interval were determined. The confidence level was assumed to amount to 95% of the average value expressed by the formulas:

$$U_L = U_{av} - 0.75 \times \bar{\bar{s}} \tag{4}$$

$$U_H = U_{av} + 0.75 \times \bar{\bar{s}} \tag{5}$$

where: U_L is the lower limit of the confidence interval (the confidence amounting to 95% of the average value of the notch toughness coefficient U), U_H is the upper limit of the confidence interval (the confidence amounting to 95% of the average value of notch toughness coefficient U), U_{av} is the average value of the notch toughness coefficient, and $\bar{s}$ is the standard deviation.

$U_H = 0.85$ was adopted as the boundary value of the upper limit of the confidence interval.

Hence, insignificant degradation is characterised by an insignificant decrease in registered events recognised as accompanying fibre breaking or its absence, in comparison with the reference fibre-cement boards. An insignificant decrease in events recognised as accompanying fibre breaking is such a decrease for which the notch toughness W_f satisfies the condition: $U_L > 0.85$. At the same time, the bending strength MOR condition, defined as $R_L > 0.75$, must be satisfied. Significant degradation is characterised by a significant drop in registered events recognised as accompanying fibre breaking, in comparison with the reference board. A significant drop in registered events recognised as accompanying fibre breaking is such a drop for which the notch toughness W_f satisfies condition $U_L < 0.85$. Simultaneously, the bending strength MOR condition, defined as $R_L > 0.75$, must be satisfied. Critical degradation is characterised by the total absence of or a significant drop in events recognised as accompanying fibre breaking, in comparison with the reference board, and when the fibre-cement boards do not satisfy the bending strength MOR condition, defined as $R_L < 0.75$.

Tables 5 and 6 show the degrees of degradation identified for the tested samples of the fibre-cement boards of, respectively, series A and B.

Table 5. The identified degrees of degradation of the tested fibre-cement boards of series A.

Series	R_L (-)	U_L (-)	Degree of Degradation
A_2	0.95	0.92	Insignificant degradation
A_3	1.00	0.79	Significant degradation
A_4	0.80	0.35	Significant degradation
A_5	0.7	–	Critical degradation

Table 6. The identified degrees of degradation of the tested fibre-cement boards of series B.

Series	R_L (-)	U_L (-)	Degree of Degradation
B_2	0.96	0.99	Insignificant degradation
B_3	0.94	1.00	Insignificant degradation
B_4	1.00	1.00	Insignificant degradation
B_5	1.00	0.81	Significant degradation
B_6	0.91	0.68	Significant degradation
B_7	0.56	–	Critical degradation
B_8	0.54	–	Critical degradation

5. Conclusions

High temperatures by nature have a degrading effect on most building products. The latter's resistance to high temperatures is measured by the length of time during which they preserve the properties required by the standards. The tests carried out on the fibre-cement boards of series A and B showed that they differed in the length of time during which they were resistant to fire. It should be emphasised that the tests have shown that it not sufficient to determine the resistance to high temperature solely on the basis of bending strength MOR. An analysis of the events registered during the bending test and their assignment to the signals accompanying the breaking of the fibres or the cement matrix broaden the range of investigations whereby one can correctly determine the high-temperature resistance of fibre-cement boards. Thanks to the use of the nondestructive methods and artificial neural networks, the degree of degradation of the fibre-cement boards exposed to the high temperature generated by fire was successfully determined. The following three degrees of degradation: insignificant degradation, significant degradation and critical degradation, were defined on the basis of the test results. Owing to the formulation of the dependence between the notch toughness W_f and the events originating from fibre breaking, registered during three-point bending, the degree of degradation of the fibre-cement boards exposed to high temperatures could be properly parameterised. The dependence between the notch toughness W_f and the counts ($\sum N_{ev,f}$) of recognised events accompanying fibre breaking during the bending of the fibre-cement boards was used for this purpose.

Author Contributions: K.S. conceived of and designed the experimental work and completed the experiments; T.G. analysed the test results and performed editing; and M.S. prepared the specimens.

Funding: This research received no external funding.

Conflicts of Interest: The authors declare no conflict of interest.

References

1. Schabowicz, K.; Gorzelańczyk, T. Fabrication of fibre cement boards. In *The Fabrication, Testing and Application of Fibre Cement Boards*, 1st ed.; Ranachowski, Z., Schabowicz, K., Eds.; Cambridge Scholars Publishing: Newcastle upon Tyne, UK, 2018; pp. 7–39. ISBN 978-1-5276-6.
2. Bentchikou, M.; Guidoum, A.; Scrivener, K.; Silhadi, K.; Hanini, S. Effect of recycled cellulose fibres on the properties of lightweight cement composite matrix. *Constr. Build. Mater.* **2012**, *34*, 451–456. [CrossRef]
3. Savastano, H.; Warden, P.G.; Coutts, R.S.P. Microstructure and mechanical properties of waste fibre–cement composites. *Cem. Concr. Compos.* **2005**, *27*, 583–592. [CrossRef]
4. Coutts, R.S.P. A Review of Australian Research into Natural Fibre Cement Composites. *Cem. Concr. Compos.* **2005**, *27*, 518–526. [CrossRef]
5. Schabowicz, K.; Szymków, M. Ventilated facades made of fibre-cement boards. *Materiały Budowlane* **2016**, *4*, 112–114. (In Polish) [CrossRef]
6. Ardanuy, M.; Claramunt, J.; Toledo Filho, R.D. Cellulosic Fibre Reinforced Cement-Based Composites: A Review of Recent Research. *Constr. Build. Mater.* **2015**, *79*, 115–128. [CrossRef]
7. Dębowski, T.; Lewandowski, M.; Mackiewicz, S.; Ranachowski, Z.; Schabowicz, K. Ultrasonic tests of fibre-cement boards. *Przegląd Spawalnictwa* **2016**, *10*, 69–71. (In Polish) [CrossRef]
8. Drelich, R.; Gorzelanczyk, T.; Pakuła, M.; Schabowicz, K. Automated control of cellulose fibre cement boards with a non-contact ultrasound scanner. *Autom. Constr.* **2015**, *57*, 55–63. [CrossRef]
9. Chady, T.; Schabowicz, K.; Szymków, M. Automated multisource electromagnetic inspection of fibre-cement boards. *Autom. Constr.* **2018**, *94*, 383–394. [CrossRef]
10. Schabowicz, K.; Jóźwiak-Niedźwiedzka, D.; Ranachowski, Z.; Kudela, S.; Dvorak, T. Microstructural characterization of cellulose fibres in reinforced cement boards. *Arch. Civ. Mech. Eng.* **2018**, *4*, 1068–1078. [CrossRef]
11. Gorzelańczyk, T.; Schabowicz, K.; Szymków, M. Non-destructive testing of fibre-cement boards, using acoustic emission. *Przegląd Spawalnictwa* **2016**, *88*, 35–38. (In Polish) [CrossRef]
12. Adamczak-Bugno, A.; Gorzelańczyk, T.; Krampikowska, A.; Szymków, M. Non-destructive testing of the structure of fibre-cement materials by means of a scanning electron microscope. *Badania Nieniszczące i Diagnostyka* **2017**, *3*, 20–23. (In Polish)
13. Claramunt, J.; Ardanuy, M.; García-Hortal, J.A. Effect of drying and rewetting cycles on the structure and physicochemical characteristics of softwood fibres for reinforcement of cementitious composites. *Carbohydr. Polym.* **2010**, *79*, 200–205. [CrossRef]
14. Mohr, B.J.; Nanko, H.; Kurtis, K.E. Durability of kraft pulp fibre-cement composites to wet/dry cycling. *Cem. Concr. Compos.* **2005**, *27*, 435–448. [CrossRef]
15. Pizzol, V.D.; Mendes, L.M.; Savastano, H.; Frías, M.; Davila, F.J.; Cincotto, M.A.; John, V.M.; Tonoli, G.H.D. Mineralogical and microstructural changes promoted by accelerated carbonation and ageing cycles of hybrid fibre–cement composites. *Constr. Build. Mater.* **2014**, *68*, 750–756. [CrossRef]
16. Li, Z.; Zhou, X.; Bin, S. Fibre-Cement extrudates with perlite subjected to high temperatures. *J. Mater. Civ. Eng.* **2004**, *3*, 221–229. [CrossRef]
17. Schabowicz, K.; Gorzelańczyk, T. A non-destructive methodology for the testing of fibre cement boards by means of a non-contact ultrasound scanner. *Constr. Build. Mater.* **2016**, *102*, 200–207. [CrossRef]
18. Stark, W. Non-destructive evaluation (NDE) of composites: Using ultrasound to monitor the curing of composites. In *Non-Destructive Evaluation (NDE) of Polymer Matrix Composites. Techniques and Applications*, 1st ed.; Karbhari, V.M., Ed.; Woodhead Publishing Limited: Cambridge, UK, 2013; pp. 136–181. ISBN 978-0-85709-344-8.

19. Berkowski, P.; Dmochowski, G.; Grosel, J.; Schabowicz, K.; Wójcicki, Z. Analysis of failure conditions for a dynamically loaded composite floor system of an industrial building. *J. Civ. Eng. Manag.* **2013**, *19*, 529–541. [CrossRef]

20. Hoła, J.; Schabowicz, K. State-of-the-art non-destructive methods for diagnostic testing of building structures—Anticipated development trends. *Arch. Civ. Mech. Eng.* **2010**, *10*, 5–18. [CrossRef]

21. Davis, A.; Hertlein, B.; Lim, K.; Michols, K. Impact-echo and impulse response stress wave methods: Advantages and limitations for the evaluation of highway pavement concrete overlays. In Proceedings of the Conference on Nondestructive Evaluation of Bridges and Highways, Scottsdale, AZ, USA, 4 December 1996; pp. 88–96. [CrossRef]

22. Chady, T.; Schabowicz, K. Non-destructive testing of fibre-cement boards, using terahertz spectroscopy in time domain. *Badania Nieniszczące i Diagnostyka* **2016**, *1–2*, 62–66. (In Polish)

23. Schabowicz, K.; Ranachowski, Z.; Józwiak-Niedźwiedzka, D.; Radzik, Ł.; Kudela, S.; Dvorak, T. Application of X-ray microtomography to quality assessment of fibre cement boards. *Constr. Build. Mater.* **2016**, *110*, 182–188. [CrossRef]

24. Ranachowski, Z. The Application of neural networks to classify the acoustic emission waveforms emitted by the concrete under thermal stress. *Arch. Acoust.* **1996**, *21*, 89–98.

25. Ranachowski, Z.; Józwiak-Niedźwiedzka, D.; Brandt, A.M.; Dębowski, T. Application of acoustic emission method to determine critical stress in fibre reinforced mortar beams. *Arch. Acoust.* **2012**, *37*, 261–268. [CrossRef]

26. Jawaid, M.; Khalil, H.P.S.A. Cellulosic/synthetic fibre reinforced polymer hybrid composites: A review. *Carbohydr. Polym.* **2011**, *86*, 1–18. [CrossRef]

27. Marzec, A.; Lewicki, P.; Ranachowski, Z.; Debowski, T. The influence of moisture content on spectral characteristic of acoustic signals emitted by flat bread samples. In Proceedings of the AMAS Course on Nondestructive Testing of Materials and Structures, Centre of Excellence for Advanced Materials and Structures, Warszawa, Poland, 20–22 May 2002; Deputat, J., Ranachowski, Z., Eds.; pp. 127–135.

28. Yuki, H.; Homma, K. Estimation of acoustic emission source waveform of fracture using a neural network. *NDT E Int.* **1996**, *29*, 21–25. [CrossRef]

29. Schabowicz, K. Neural networks in the NDT identification of the strength of concrete. *Arch. Civ. Eng.* **2005**, *51*, 371–382.

30. Łazarska, M.; Woźniak, T.; Ranachowski, Z.; Trafarski, A.; Domek, G. Analysis of acoustic emission signals at austempering of steels using neural networks. *Met. Mater. Int.* **2017**, *23*, 426–433. [CrossRef]

31. Woźniak, T.Z.; Ranachowski, Z.; Ranachowski, P.; Ozgowicz, W.; Trafarski, A. The application of neural networks for studying phase transformation by the method of acoustic emission in bearing steel. *Arch. Metal. Mater.* **2014**, *59*, 1705–1712. [CrossRef]

32. Rucka, M.; Wilde, K. Experimental study on ultrasonic monitoring of splitting failure in reinforced concrete. *J. Nondestruct. Eval.* **2013**, *32*, 372–383. [CrossRef]

33. Rucka, M.; Wilde, K. Ultrasound monitoring for evaluation of damage in reinforced concrete. *Bull. Pol. Acad. Sci. Tech. Sci.* **2015**, *63*, 65–75. [CrossRef]

34. EN 12467. *Cellulose Fibre Cement Flat Sheets. PRODUCT Specification and Test Methods*; Standards Australia International Ltd: Strathfield, Australia, 2000.

35. Ranachowski, Z.; Schabowicz, K. The contribution of fibre reinforcement system to the overall toughness of cellulose fibre concrete panels. *Constr. Build. Mater.* **2017**, *156*, 1028–1034. [CrossRef]

materials

MDPI

Article

Identification of the Destruction Process in Quasi Brittle Concrete with Dispersed Fibers Based on Acoustic Emission and Sound Spectrum

Dominik Logoń[iD]

Faculty of Civil Engineering, Wrocław University of Science and Technology, 50-377 Wrocław, Poland;
dominik.logon@pwr.edu.pl

Received: 20 June 2019; Accepted: 9 July 2019; Published: 15 July 2019

check for
updates

Abstract: The paper presents the identification of the destruction process in a quasi-brittle composite based on acoustic emission and the sound spectrum. The tests were conducted on a quasi-brittle composite. The sample was made from ordinary concrete with dispersed polypropylene fibers. The possibility of identifying the destruction process based on the acoustic emission and sound spectrum was confirmed and the ability to identify the destruction process was demonstrated. It was noted that in order to recognize the failure mechanisms accurately, it is necessary to first identify them separately. Three- and two-dimensional spectra were used to identify the destruction process. The three-dimensional spectrum provides additional information, enabling a better recognition of changes in the structure of the samples on the basis of the analysis of sound intensity, amplitudes, and frequencies. The paper shows the possibility of constructing quasi-brittle composites to limit the risk of catastrophic destruction processes and the possibility of identifying those processes with the use of acoustic emission at different stages of destruction.

Keywords: acoustic emission AE; acoustic spectrum; quasi brittle cement composites; destruction process

---

1. Introduction

The application of acoustic emission (AE) measurements in determining the cracks, maximum load, and failure of reinforcement in cement composites has been widely presented in the literature.

The continuous AE evaluation in composites was earlier reported [1–3] and this technique has been applied to determine crack propagation in the fracture process in cement composites with and without reinforcement [4,5]. The acoustic emission (AE) events sum was also recorded for easier recognition of the first crack and crack propagation process [6–8].

It was also noticed that at the preliminary stage of degradation, the damage of the concrete elements was possible to detect with the application of the AE method [9,10]. The effectiveness of acoustic emission (AE) measurements in determining the critical stress of cement composites was tested [11], which enables the accurate definition of the elastic range corresponding to Hook's law. Previously conducted tests have shown that AE is a good method for crack formation monitoring in mechanically loaded specimens [12–18] and has been successfully used to monitor structures [19,20]. Most of the papers have used AE to identify the destruction process of materials in structures [21–28] including crack orientation [29–31]. The AE method is still used and improved for the purpose of the identification of failure processes [32–34].

Previous works, however, have not focused on the correlation between AE and the individual failure processes of each of the different composite components based on the sound spectrum. These papers [10,11] showed that for the accurate recognition of composite failure processes, the AE (and the

AE events sum) recording should be expanded to include the analysis of each sound separately and the analysis of the range of sounds corresponding to a given mechanical effect with the use of acoustic spectrum. The acoustic spectrum should be correlated with the load-deflection curve and with other acoustic effects, which enables the identification of the failure process (of the structure or the applied reinforcement) [10,11,18]. The quasi-brittle ESD cement composites (ESD—elastic range, strengthening, deflection control) are characterized by a higher load and absorbed energy in the elastic range when compared to the sample without reinforcement (E/E_0) (Figure 2). Additionally, those composites are distinguished by a highly deflected structure damaged with macrocracks, multicracking effects, and the ability to carry additional stress in the strengthening area. Moreover, in the deflection control area, the samples' ability to carry stress is higher than in the elastic range area. This paper focuses on determining the relation between the acoustic and mechanical ESD effects, in other words, reinforcement breaking, pull-out, macrocracks, and microcracking with the use of space spectrum. In [11], it was noted that in order to assess the destruction process, the analysis of a single signal and the AE events sum with the acoustic spectrum was required (each kind of the mechanical effect results in a different acoustic spectrum). In order to conduct a more in-depth analysis of the composite destruction process, what should be taken into account when interpreting the acoustic spectrum is not only the range of signals corresponding to a given mechanical effect (in a wide range of frequencies corresponding to the sound intensity), but also a single signal in a very small range of frequencies.

It was confirmed that there is a possibility of correlation between AE and the failure process in ESD composites. That correlation enables a determination of the stage of damage in cement composites increasing the safety in the use of the composite and the decision of whether or not the damaged composite can be repaired.

2. Materials and Methods

The materials for the concrete (matrix—sample without reinforcement) consisted of: Portland cement CEM I 42.5R—368.7 kg/m^3, silica fume 73.75 kg/m^3, fly ash 73.75 kg/m^3, sand and coarse aggregates 0/16 mm–1640 kg/m^3, superplasticizer (SP), tap water 188.6 kg/m^3, w/c = 0.51.

The ESD concrete was reinforced with polypropylene fibers (curved/wave), minimum tensile strength 490 MPa, E = 3.5 GPa, equivalent diameter d = 0.8–1.2 mm, l = 54 mm. The reinforcement was randomly dispersed V_f = 1.5%.

Concrete was mixed in the concrete mixer and then used to mold samples. Beams (600 mm × 150 mm × 150 mm) were cast in slabs and then cured in water at 20 ± 2 °C. After 180 days of ageing, beams were prepared for the bending test (Figure 1). The samples were not notched.

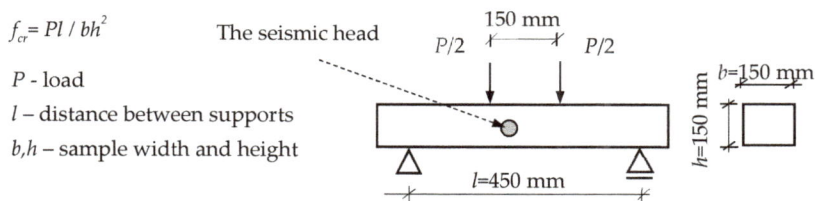

$f_{cr} = Pl / bh^2$

The seismic head

P - load

l – distance between supports

b,h – sample width and height

150 mm

$P/2$ $P/2$

b=150 mm

h=150 mm

l=450 mm

Figure 1. Four-point bending test.

Acoustic emission effects were recorded in order to monitor the progress of the fracture process in correlation with the load-deflection curve. The crosshead displacement was continuous and the rate was 1 mm/min. A seismic head HY919 (Spy Electronics Ltd.) was used to record the acoustic emission effects in the range from 0.2–20 kHz. The head was placed on the side in the central part of the loaded beams (Figure 1). The acoustic emission effects were presented as a 2D and 3D acoustic spectrum (amplitude of the frequency depending on sound intensity). The mechanical effects of the ESD composites were correlated with the recorded acoustic spectrum effects. The 2D sound spectrum was achieved with the use of the Audacity program (free digital audio editor) and the 3D spectrum using SpectraPLUS-SC (Pioneer Hill Software LLC, USA).

Figure 2 presents the mechanical effects of the ESD (Eng. elastic range, strengthening, deflection control) cement composites with the corresponding acoustic effects and compiled acoustic spectra with various amplitudes corresponding to different mechanical effects (reinforcement breaking, pull-out, macrocracks, and microcracking).

The ESD reinforcement effect is presented by characteristic points f_x(F_x-load, ε_x-deflection, W_x-work) and areas A_X under the load-deflection curve.

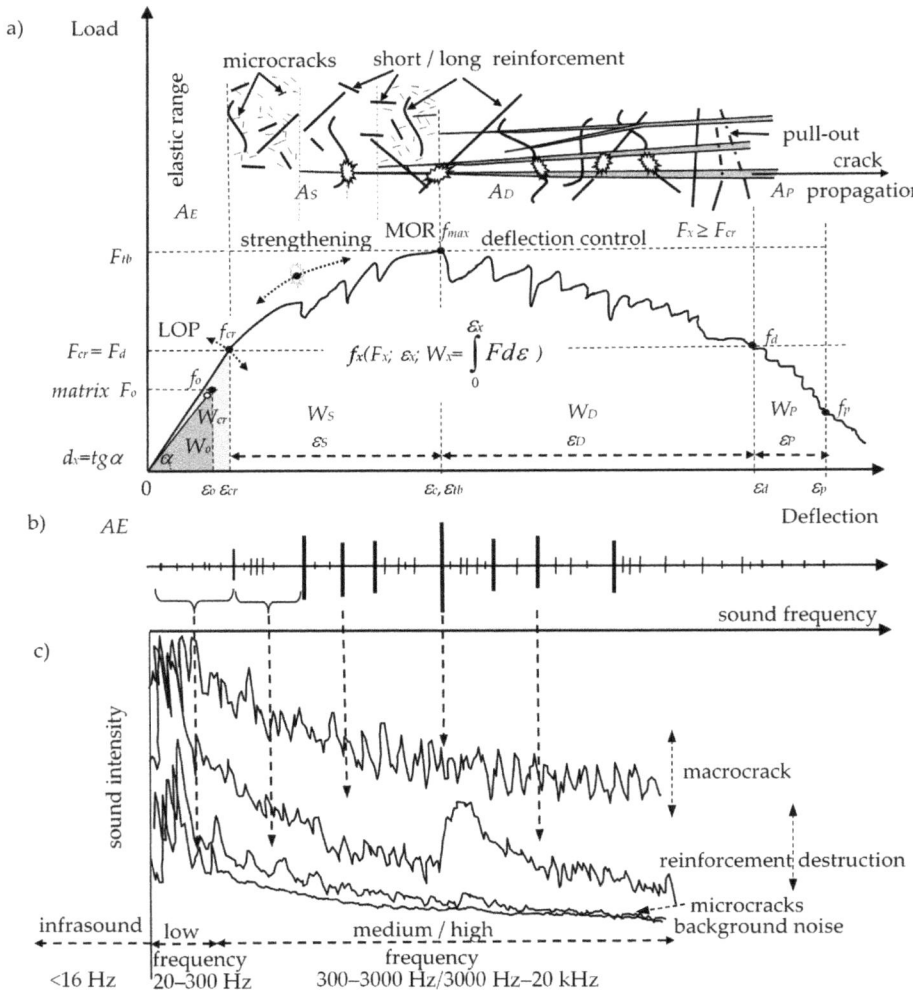

Figure 2. ESD composite: (**a**) load-deflection curve, (**b**) AE—acoustic emission effects, (**c**) 2D acoustic spectrum (frequency amplitude depending on sound intensity) [11].

3. Results

Figure 3 presents the testing area for the four-point bending test with the AE acoustic emission measurements. Subsequent pictures show the characteristic stages of the ESD concrete failure process.

Figure 3b indicates a crack occurring at the f_{cr} point, Figure 3c shows the multicracking (micro- and macrocracks), Figure 3d shows the progressing crack propagation, and Figure 3e shows the sample after the completed test.

The load-deflection curve of the ESD composite and matrix (concrete without the dispersed reinforcement) is presented in Figure 4a. Above the curves, there are the results of the AE measurement with characteristic failure process events.

The ESD effects in the quasi-brittle composite were described with the use of the formula defining any points on the load-deflection curve f_x (load; deflection; absorbed energy). This formula enables the description and assessment of the ESD effects in the elastic range, strengthening, deflection control, and propagation areas. The matrix is characterized by f_{max} (Table 1).

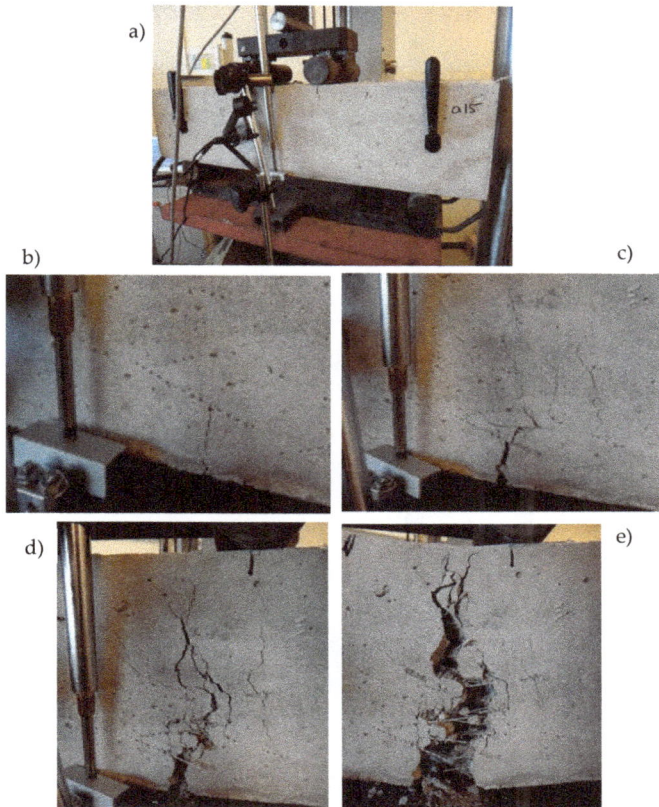

Figure 3. Four-point bending test: (**a**) sample before the test, (**b**) first crack at f_{cr} point, (**c**) multicracking (micro- and macrocracks), (**d**) destruction - propagation process, (**e**) view after the test.

Table 1. Mechanical properties of the matrix (concrete without reinforcement) and ESD composite.

Composite		Load F [N]	Deflection ε [mm]	Work W [kJ]	Ratio	Load	Deflection	Work
matrix	f_{max}	32.9	1.42	23.0	-	-	-	-
ESD	f_{cr}	38.9	1.87	36.3	$A_{E/Ematrix}$	1.2	1.3	1.6
	f_{tb}	49.7	3.46	106.7	$A_{S/E}$	0.3	0.9	1.9
	f_d	38.9	4.06	133.3	$A_{D/E}$	-	0.4	0.7

For the ESD composite, the following results were achieved: f_{cr}, f_{max}, f_d, (Table 1, Figure 4). Comparing the elastic range area of the ESD composite and the matrix, an x-time improvement was achieved for load, deflection, and absorbed energy $A_E/A_{Ematrix}$. $A_{S/E}$ and $A_{D/E}$ are the comparison of the strengthening A_S and deflection control A_D areas to the elastic range A_E. The amount of absorbed energy in the strengthening area was considerably larger than in the deflection control area. The failure process propagation range following the deflection control area was not important in the ESD composites and was omitted.

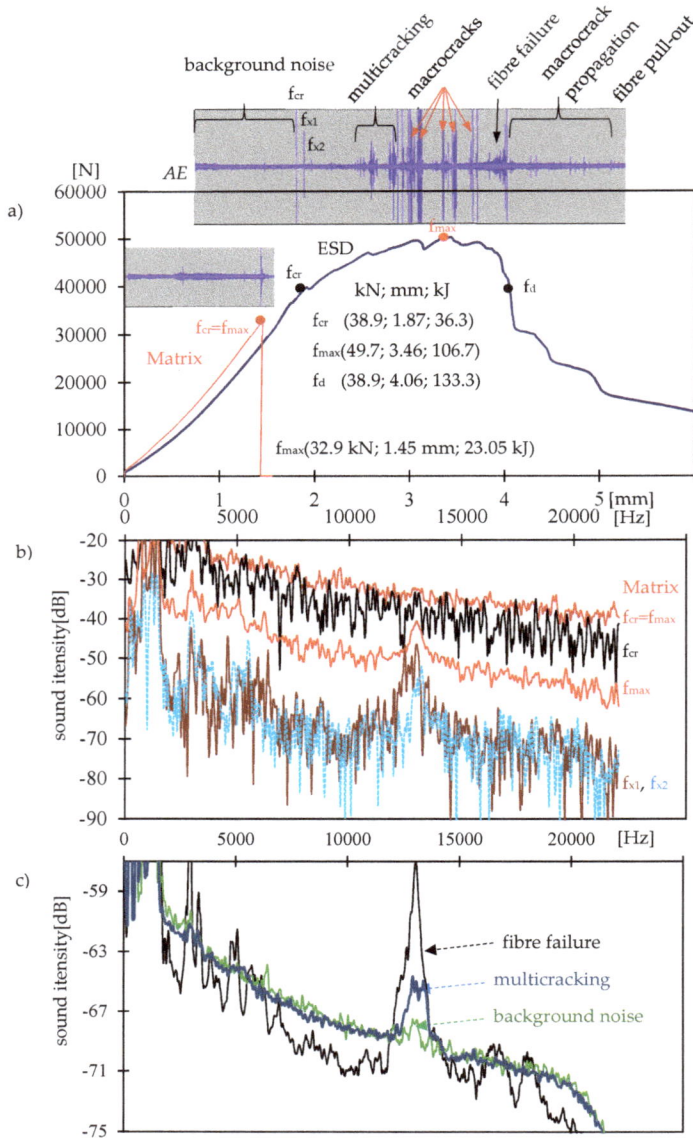

Figure 4. Matrix and ESD composite: (**a**) load-deflection curve, (**b**) 2D spectra of the matrix and ESD composite, (**c**) 2D spectrum of the ESD composite.

Two-dimensional (2D) spectra of the matrix and ESD composite in the frequency range of 0–22 kHz are presented in Figure 4. Figure 4b shows the matrix spectrum for a crack $f_{cr} = f_{max}$, additional spectra of the ESD sample for the first crack, f_{cr}, and subsequent cracks, f_{x1}, f_{x2}, and f_{max}. Figure 4c presents the ESD concrete spectra compared to the spectra of background noise, multicracking, and the fiber failure process. Three-dimensional (3D) spectra of the ESD composite are presented in Figures 5 and 6.

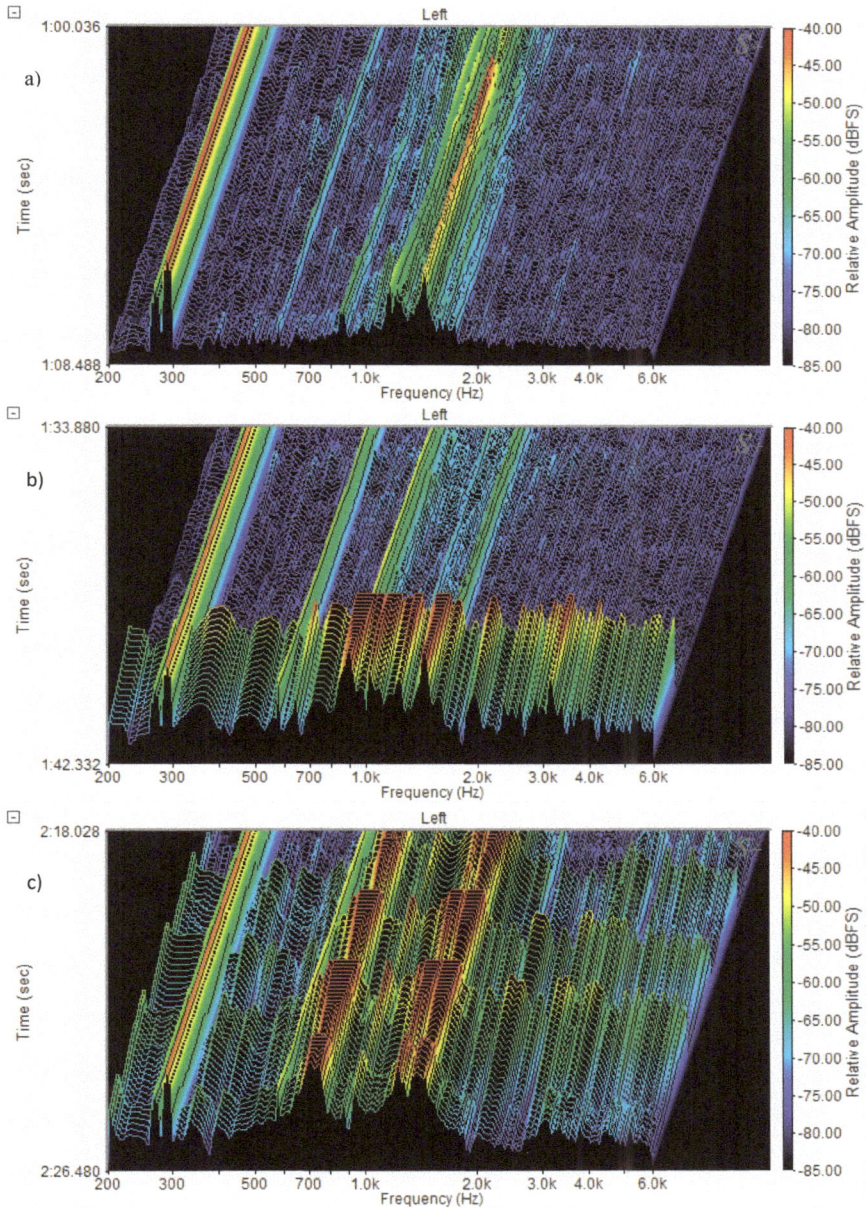

Figure 5. 3D spectrum of ESD composite: (**a**) background noise; (**b**) first crack, f_{cr}; and (**c**) multicracking.

The spectrum frequency range was limited to 200–6000 Hz as the greatest changes were observed within this frequency range in the sound spectra connected with the failure process. Figure 5a shows the background noise spectrum for the ESD composite recorded during the test, Figure 5b presents the first crack, f_{cr}, and Figure 5c shows the multicracking. Figure 6a displays the macrocrack spectrum, Figure 6b shows the reinforcement destruction, and Figure 6c presents the fiber pull-out effect.

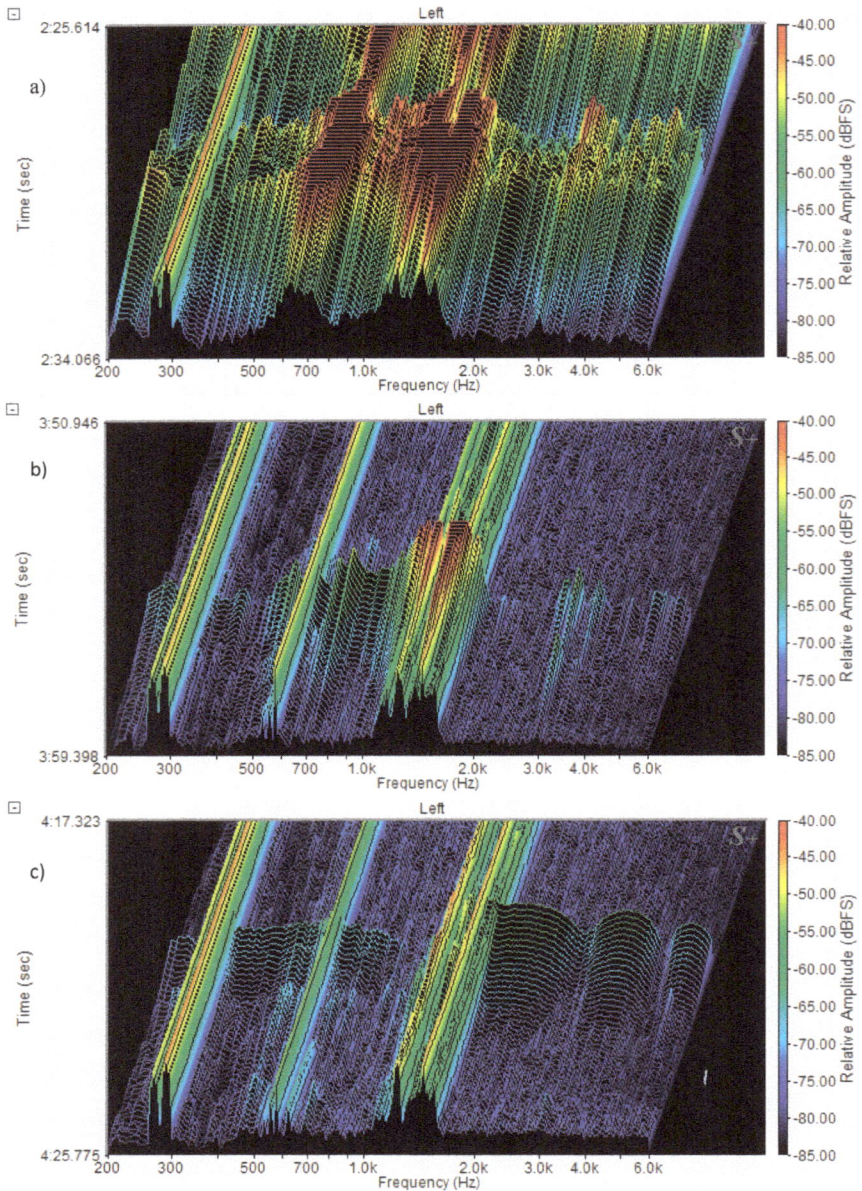

Figure 6. 3D spectra of the ESD composite: (**a**) macrocrack, (**b**) fiber destruction, (**c**) fiber pull-out.

4. Discussion

The existing provisions in the ASTM 1018 standard concerning the identification of characteristic points LOP, MOR, and ASTM indices I_5, I_{10}, I_{15} [35–37] have been extended by adding the possibility of describing any area or point f_x (load, deflection, energy). The introduced f_{cr}, f_{max}, and f_d points enable a precise description of the elastic range, strengthening, deflection control, and propagation areas as well as their comparison with one another with respect to the same sample or different samples.

The obtained results indicate that the ESD composite achieved the best effects in the strengthening area As. The improvement of properties in the elastic range AE was not good enough, which resulted from a low elasticity module of the fibers, causing a more significant deflection in this area. The deflection control range in the A_D area could be improved by increasing the fiber-matrix bond of the dispersed reinforcement. That effect may be achieved by increasing the strength of the composite or modifying the surface and geometry of the fibers.

2D and 3D spectra in the lowest frequency range did not record well (due to the head's measurement range from 0.2 to 20 kHz) and were not taken into account in the interpretation of the results (2D spectrum 0–0.2 kHz). The 3D spectrum frequency range was limited to 200–6000 Hz. Within that frequency range, the greatest changes were observed in the sound spectra connected with the failure process.

Concrete without reinforcement (matrix) is characterized by a catastrophic destruction process. The appearing crack causes a destruction—breaking in halves—of the sample. The sound spectrum (within the range from −20 to −40 dB) corresponding to that process was positioned the highest when compared to other spectra characterizing various destruction processes, as shown in Figure 4a and is characterized by a small range of amplitudes. The background noise spectrum of the matrix has not been presented here, but was similar to the background noise spectrum of the ESD composite.

The ESD composite in the elastic range showed 2D and 3D spectra of background noise located low and within the range of the lowest amplitudes, as seen in Figures 4c and 5a. The sound corresponding to f_{cr} was characterized by significant sound intensity, and the corresponding spectrum was located high, immediately below curve f_{cr} for the matrix, and the range of amplitudes was much larger (Figures 4a and 5b).

Subsequent cracks, f_{x1} and f_{x2}, were situated at the level of background noise spectrum, but with the greatest amplitude range (Figure 4b).

The sound intensity of the multicracking was similar to background noise spectrum, but with a slightly greater range of amplitudes and significant amplitudes in a narrow frequency range (Figures 4c and 5c).

Macrocracks showed the highest sound intensity. What is worth noting is the fact that the corresponding spectra were not characterized by the greatest amplitude range. The spectra were located the highest (Figure 4b).

The fiber pull-out process was characterized by a small range of amplitudes with a wide range of wavelengths (Figure 6c).

The sound spectrum corresponding to fiber failure was positioned low. The lower position of that spectrum when compared to that of the background noise may result from the manner of determining that spectrum. The background noise spectrum was determined with respect to a long period of time before the first crack and refers to a number of background noises in that period, whereas the fiber failure spectrum refers to a single signal. The sound spectrum for fiber failure was characterized by great amplitudes with a strong spike at 12–15 kHz. The average sound intensities of the fiber failure and the background noise were on a comparable level. The analysis of the background noise spectrum for a single sound in a short period of time resulted in a slight decrease in the sound intensity, but the amplitudes did not change significantly.

Frequency spike 12–15 kHz is an interesting, recurring correlation that may be used in the future for the identification of the failure process (Figure 4b,c). It is worth noting that that spike did not occur in the case of a catastrophic fracture f_{cr} (in a sample without reinforcement) in a short period of time. Frequency spike 12–15 kHz occurs in the case of defects generating acoustic effects that last for a longer period of time such as fiber failure, fiber pull out, and micro- and macrocracks that are blocked (stop propagating) or propagate slowly.

The conducted tests confirmed the possibility of identifying the failure process in traditional and ESD cement composites. The analysis of data showed that the 3D spectrum provided better general information for the identification of the failure process at each stage of the process, whereas the 2D spectrum enabled a more precise characterization of each of the sound spectra (sound intensity, amplitudes, frequency range) and their correlation with each of the failure processes. The simultaneous occurrence of failure processes makes their identification difficult. In order to differentiate them accurately, it is necessary to separately identify the sounds that do not overlap.

Summarizing the conducted tests, it can be stated that the analysis of 2D and 3D spectra is a good method of controlling the failure processes in the ESD cement composites. It increases the safety in the use of construction elements and enables correct decision-making in whether and how they should be repaired.

5. Conclusions

The research has allowed for the following conclusions to be formulated conclusions:

(1) The 3D sound spectrum is a good tool for the observation and identification of failure processes in cement composites.

(2) It has been noticed that the 2D spectrum enables a more precise identification and description of the sound spectra (sound intensity, amplitudes, frequency range) corresponding to different failure processes.

(3) It has been suggested that for the analysis and identification of failure processes, both the 2D and 3D spectra should be used at the same time in a wide frequency range.

(4) The development of ESD cement composites and the identification of failure processes with the use of AE and 2D and 3D spectra enables the control of failure processes (particularly useful in seismic areas or during natural disasters) or decisions of whether and how damaged cement composites should be repaired.

Funding: This research received no external funding.

Conflicts of Interest: The authors declare no conflict of interest.

References

1. Brandt, A.M. Fiber reinforced cement-based (FRC) composites after over 40 years of development in building and civil engineering. *Compos. Struct.* **2008**, *86*, 3–9, ISSN: 0263-8223. [CrossRef]
2. Kucharska, L.; Brandt, A.M. Pitch-based carbon fibre reinforced cement composites. In *Materials for the New Millenium, Proceedings of the Materials Engineering Conference ASCE, Washington, DC, USA, 10–14 November 1996*; Chong, K.P., Ed.; American Society of Civil Engineers: New York, NY, USA, 1996; Volume 1, pp. 1271–1280.
3. Yuyama, S.; Ohtsu, M. Acoustic Emission evaluation in concrete. In *Acoustic Emission-Beyond the Millennium*; Kishi, T., Ohtsu, M., Yuyama, S., Eds.; Elsevier: Amsterdam, The Netherlands; Tokyo, Japan, 2000; pp. 187–213.
4. Landis, E.; Ballion, L. Experiments to relate acoustic energy to fracture energy of concrete. *J. Eng. Mech.* **2002**, *128*, 698–702. [CrossRef]
5. Ouyang, C.S.; Landis, E.; Shh, S.P. Damage assessment in concrete using quantitative acoustic emission. *J. Eng. Mech.* **1991**, *117*, 2681–2698. [CrossRef]
6. Ranchowski, Z.; Jóźwiak-Niedźwiecka, D.; Brandt, A.M.; Dębowski, T. Application of acoustic emission method to determine critical stress in fibre reinforced mortar beams. *Arch. Acoust.* **2012**, *37*, 261–268. [CrossRef]
7. Kim, B.; Weiss, W.J. Using acoustic emission to quantify damage in restrained fiber-reinforced cement mortars. *Cem. Concr. Res.* **2003**, *33*, 207–214. [CrossRef]

8. Shieldsa, Y.; Garboczib, E.; Weissc, J.; Farnam, Y. Freeze-thaw crack determination in cementitious materials using 3DX-ray computed tomography and acoustic emission. *Cem. Concr. Comp.* **2018**, *89*, 120–129. [CrossRef]

9. Hoła, J. Acoustic Emission investigation of failure of high-strength concrete. *Arch. Acoust.* **1999**, *24*, 233–244.

10. Logoń, D. Monitoring of microcracking effect and crack propagation in cement composites (HPFRC) using the acoustic emission (AE). In Proceedings of the The 7th Youth Symposium on Experimental Solid Mechanics (YSESM '08), Wojcieszyce, Poland, 14–17 May 2008.

11. Logoń, D. The application of acoustic emission to diagnose the destruction process in FSD cement composites. In *Proceedings of the International Symposium on Brittle Matrix Composites (BMC-11), Warsaw, Poland, 28–30 September 2015*; Brandt, A.M., Ed.; Institute of Fundamental Technological Research: Warsaw, Poland, 2015; pp. 299–308.

12. Schabowicz, K.; Gorzelańczyk, T.; Szymków, M. Identification of the degree of degradation of fibre-cement boards exposed to fire by means of the acoustic emission method and artificial neural networks. *Materials* **2019**, *12*, 656. [CrossRef] [PubMed]

13. Chen, B.; Juanyu Liu, J. Experimental study on AE characteristics of free-point-bending concrete beams. *Cem. Concr. Res.* **2004**, *34*, 391–397. [CrossRef]

14. Reinhardt, H.W.; Weiler, B.; Grosse, C. Nondestructive testing of steel fibre reinforced concrete. In Proceedings of the Brittle Matrix Composites (BMC-6), Warsaw, Poland, 9–11 October 2000; pp. 17–32.

15. Granger, S.; Pijaudier, G.; Loukili, A.; Marlot, D.; Lenain, J.C. Monitoring of cracking and healing in an ultra high performance cementitious material using the time reversal technique. *Cem. Concr. Res.* **2009**, *39*, 296–302. [CrossRef]

16. Ohtsu, M. The history and development of acoustic emission in concrete engineering. *Mag. Concr. Res.* **1996**, *48*, 321–330. [CrossRef]

17. Ranachowski, Z.; Schabowicz, K. The contribution of fibre reinforcement system to the overall toughness of cellulose fibre concrete panels. *Constr. Build. Mater.* **2017**, *156*, 1028–1034. [CrossRef]

18. Logoń, D. FSD cement composites as a substitute for continuous reinforcement. In *Proceedings of the Eleventh International Symposium on Brittle Matrix Composites (BMC-11), Warsaw, Poland, 28–30 September 2015*; Brandt, A.M., Ed.; Institute of Fundamental Technological Research: Warsaw, Poland, 2015; pp. 251–260.

19. Parmar, D. *Non-Destructive Bridge Testing With Advanced Micro-II Digital AE system. Final Report*; Hampton University, Eastern Seaboard Intermodal Transportation Applications Center (ESITAC): Hampton, VA, USA, 2011.

20. Ono, K.; Gołaski, L.; Gębski, P. Diagnostic of reinforced concrete bridges by acoustic emission. *J. Acoust. Emiss.* **2002**, *20*, 83–98.

21. Paul, S.C.; Pirskawetz, S.; Zijl, G.P.A.G.; Schmidt, W. Acoustic emission for characterising the crack propagation in strain-hardening cement-based composites (SHCC). *Cem. Concr. Res.* **2015**, *69*, 19–24. [CrossRef]

22. Watanab, K.; Niwa, J.; Iwanami, M.; Yokota, H. Localized failure of concrete in compression identified by AE method. *Constr. Build. Mater.* **2004**, *18*, 189–196. [CrossRef]

23. Soulioti, D.; Barkoula, N.M.; Paipetis, A.; Matikas, T.E.; Shiotani, T.; Aggelis, D.G. Acoustic emission behavior of steel fibre reinforced concrete under bending. *Constr. Build. Mater.* **2009**, *23*, 3532–3536. [CrossRef]

24. Šimonová, H.; Topolář, L.; Schmid, P.; Keršner, Z.; Rovnaník, P. Effect of carbon nanotubes in metakaolin-based geopolymer mortars on fracture toughness parameters and acoustic emission signals. In Proceedings of the International Symposium on Brittle Matrix Composites (BMC 11), Warsaw, Poland, 28–30 September 2015; pp. 261–288.

25. Shahidan, S.; Rhys Pulin, R.; Bunnori, N.M.; Holford, K.M. Damage classification in reinforced concrete beam by acoustic emission signal analysis. *Constr. Build. Mater.* **2013**, *45*, 78–86. [CrossRef]

26. Aggelis, D.G.; Mpalaskas, A.C.; Matikas, T.E. Investigation of different modes in cement-based materials by acoustic emission. *Cem. Concr. Res.* **2013**, *48*, 1–8. [CrossRef]

27. Elaqra, H.; Godin, N.; Peix, G.; R'Mili, M.; Fantozzi, G. Damage evolution analysis in mortar, during compressive loading using acoustic emission and X-ray tomography: Effects of the sand/cement ratio. *Cem. Concr. Res.* **2007**, *37*, 703–713. [CrossRef]

28. Shiotani, T.; Li, Z.; Yuyama, S.; Ohtsu, M. Application of the AE Improved b-Value to Quantitative Evaluation of Fracture Process in Concrete Materials. *J. Acoust. Emiss.* **2004**, *19*, 118–133.

29. Ohtsu, M. Determination of crack orientation by acoustic emission. *Mater. Eval.* **1987**, *45*, 1070–1075.

30. Ohno, K.; Ohtsu, M. Crack classification in concrete based on acoustic emission. *Constr. Build. Mater.* **2010**, *24*, 2339–2346. [CrossRef]

31. Ohtsu, M. Elastic wave methods for NDE in concrete based on generalized theory of acoustic emission. *Constr. Build. Mater.* **2016**, *122*, 845–855. [CrossRef]

32. Van Steen, C.; Verstrynge, E.; Wevers, M.; Vandewalle, L. Assessing the bond behaviour of corroded smooth and ribbed rebars with acoustic emission monitoring. *Cem. Concr. Res.* **2019**, *120*, 176–186. [CrossRef]

33. Tsangouri, E.; Michels, L.; El Kadi, M.; Tysmans, T.G.; Aggelis, D. A fundamental investigation of textile reinforced cementitious composites tensile response by Acoustic Emission. *Cem. Concr. Res.* **2019**, *123*, 105776. [CrossRef]

34. Kumar Das, A.; Suthar, D.; Leung, C.K.Y. Machine learning based crack mode classification from unlabeled acoustic emission waveform features. *Cem. Concr. Res.* **2019**, *121*, 42–57.

35. ASTM 1018. *Standard Test Method for Flexural Toughness and First Crack Strength of Fiber–Reinforced Concrete*; ASTM 1018: West Conshohocken, PA, USA, 1992; Volume 04.02.

36. EN 14651. Test method for metallic fibre concrete. In *Measuring the Flexural Tensile Strength (Limit of Proportionality (LOP), Residual)*; European Committee for Standardization (CEN): Brussels, Belgium, 2005.

37. Japan Concrete Institute Standard (JCI). *Method of Test for Bending Moment-curvature Curve of Fiber-reinforced Cementitious Composites, S-003-2007*; Japanese Concrete Institute Standard Committee: Tokyo, Japan, 2007.

materials

MDPI

Article

Inverse Contrast in Non-Destructive Materials Research by Using Active Thermography

Paweł Noszczyk * and **Henryk Nowak**

Department of Building Physics and Computer Design Methods, Faculty of Civil Engineering,
Wroclaw University of Science and Technology, 50-370 Wrocław, Poland; henryk.nowak@pwr.edu.pl
* Correspondence: pawel.noszczyk@pwr.edu.pl; Tel.: +48-071-320-3203

Received: 11 January 2019; Accepted: 1 March 2019; Published: 12 March 2019

check for
updates

Abstract: Background: it is undesirable for defects to occur in building partitions and units. There is a need to develop and improve research techniques for locating such defects, especially non-destructive techniques for active thermography. The aim of the experiment was to explore the possibility of using active thermography for testing large-sized building units (with high heat capacity) in order to locate material inclusions. Methods: as part of the experiment, two building partition models—one made of gypsum board (GB) and another made of oriented strand board (OSB)—were built. Three material inclusions (styrofoam, granite, and steel), considerably differing in their thermal parameters, were placed in each of the partitions. A 7.2 kW infrared radiator was used for thermally exciting (heating) the investigated element for 30 min. The distribution of the temperature field was studied on both sides of the partition for a few hours. Results: using the proposed investigative method, one can detect defects in building partitions under at least 22 mm of thick cladding. At a later cooling down phase, inverse temperature contrasts were found to occur—the defects, which at the beginning of cooling down were visible as warmer areas, at a later phase of cooling down are perceived as cooler areas, and vice versa (on the same front surface). In the transmission mode, the defects are always visible as areas warmer than defect-free areas. Moreover, a quantitative (defect location depth) analysis with an accuracy of up to 10% was carried out using the Echo Defect Shape method. Conclusions: active thermography can be used in construction for non-destructive materials testing. When the recording of thermograms is conducted for an appropriate length of time, inverse contrasts can be observed (on the same front surface).

Keywords: thermovision; active thermography; thermal contrast; defect detection; location of inclusions; non-destructive testing; materials research; building partition

1. Introduction

The material structure of a building unit determines its physical parameters (e.g., its resistance to various external impacts). In the case of existing building structures, one can use either destructive, semi-destructive, or non-destructive tests to assess their structural components. If it is technically possible, it is desirable to carry out non-destructive tests, which do not adversely affect the tested member. A wide range of non-destructive tests is used in construction to investigate the material structure of individual structural components. The applications of most tests are described in the extensive literature on this subject [1–4]. Thermography, which is also referred to as thermal (infrared) imaging, has been widely used in construction for non-destructive testing purposes for many years [5–9]. It is anticipated that the number of its applications and the quality of thermal imaging surveys will increase in the nearest future [10,11]. A special kind of thermography is active thermography, which consists of thermally stimulation of the tested element and registering its thermal response to the set controlled excitation over time [12]. This method has been successfully applied in materials research to investigate thin elements (up to a few millimeters thick) [13–15]. The most commonly used active thermography methods are: lock in thermography (LT), stepped thermography (ST), and phase thermography (PT) [16–18]. Stepped thermography consists of thermally exciting (using an external heat source) the tested element with a continuous heat pulse and recording thermograms as the element cools down. There occur two main phases in this method: heating up (or cooling down) and cooling down (or heating up) of the investigated surface. According to the relevant European standard [19] in step of thermography, an energy source (e.g., a halogen or induction lamp) is switched on or/and off at a particular time for excitation. Contrary to pulse thermography, the thermal signature of the defects or of the rear side of the layer already appears during excitation. The image sequence may be analyzed in the time domain or in the frequency domain. Since it is more difficult to detect inclusions in large-sized and massive members, they are less often investigated using this method [20–48]. No experiment in which the time of recording thermograms is considerably extended until a thermal equilibrium between the tested element and the environment is reached. This was reported in the literature on the subject. As part of the present research, elements with a large thermal mass were tested and thermograms were recorded simultaneously in the transmission mode and the reflection mode for three defects made of different materials. A difference in defect detection depending on the distance from which the tested surface was heated up was observed. In most cases, the occurrence of material inclusions in building units is undesirable since it adversely affects the durability of the whole structure. It is vital to locate such material inclusions in order to ensure the safe serviceability of civil structures and their durability. For this reason, an attempt was made to employ active thermography to investigate building partition models and find out if modelled material inclusions in large-sized (a few centimeters thick) members could be located in this way. In the course of the experiment, a new kind of thermal contrast, which has not been described in the literature, was noticed. Building partition models made of oriented strand board (OSB) and gypsum board (GB) were investigated. Material inclusions made of steel, granite, and XPS (extruded polystyrene), i.e., materials markedly differing in their thermal parameters (Table 1), were modelled in each of the two partition models. The tested element was excited by a long-duration heat pulse. The distribution of the temperature field on the surface was registered simultaneously in the reflection mode (the thermal imaging camera situated on the thermal excitation side) and in the transmission mode (the camera situated on the opposite—relative to the thermal excitation—side of the partition). In this way, the material inclusions inside the investigated building partition model were located and identified. Thanks to the recording of the thermograms of the investigated surface over a longer time, an inverse contrast phenomenon (described further in this paper) was observed. In the earlier publications [49,50], the inverse contrast phenomenon was mentioned in connection with the testing of thin (a few mm) elements of a small size (up to 10–20 cm). In addition, in the paper [51], an inverse contrast occurring on different sides of the tested element (the front and rear side of the material sample) was mentioned. In the present research, the inverse contrast phenomenon occurs on one (the same) side of the tested building partition after a long observation time. The principal aim of this research was to examine the effect of the type of partition materials and the type of material inclusions for the possibility of detecting the latter in massive building partitions, i.e., elements with a large thermal capacity.

2. Materials and Methods

Two building partition models, consisting of four 1250 × 1250 mm (oriented strand and gypsum) boards bolted together, were investigated. The surface (outer) boards were 22 mm thick, while the inner boards were 10 mm thick. The overall thickness of the partition model amounted to 64 mm. Materials markedly differing in their thermal parameters, i.e., steel, granite, and styrofoam (XPS), were placed inside the model (inserted into the 10 mm thick boards). The inclusions were 200 × 100 × 20 mm in size and spaced from one another at a distance greater than their largest dimension in order to eliminate the influence of one inclusion on another during heat conduction in the course of the experiment. This distance was estimated on the basis of previously carried out numerical studies of heat conduction in the partition model. A precise geometrical model of the tested element, which shows the arrangement of the material inclusions inside the partition, is presented in Figure 1.

Two thermal imaging cameras are placed on the two sides of the partition model. A thermal excitation source and an air temperature and humidity sensor were used in the experiment. A Flir P65 camera with an infrared (IR) detector resolution of 320 × 240 pixels (the camera on the unheated side of the partition—the transmission mode) and an Optris PI400 camera with an IR detector resolution of 382 × 288 pixels (the camera on the side heated by the infrared radiator—the reflection mode) were used in the experiment. The thermal sensitivity of the two infrared cameras was below 80 mK. A Fobo infrared radiator with a total power of 7.2 kW (six lamps, each with a power of 1.2 kW) was used for thermal excitation. In order to prevent the tested element's edges from cooling (which would disturb the temperature field there), a Styrofoam band was placed around the partition model. A schematic of the test stand is shown in Figure 2.

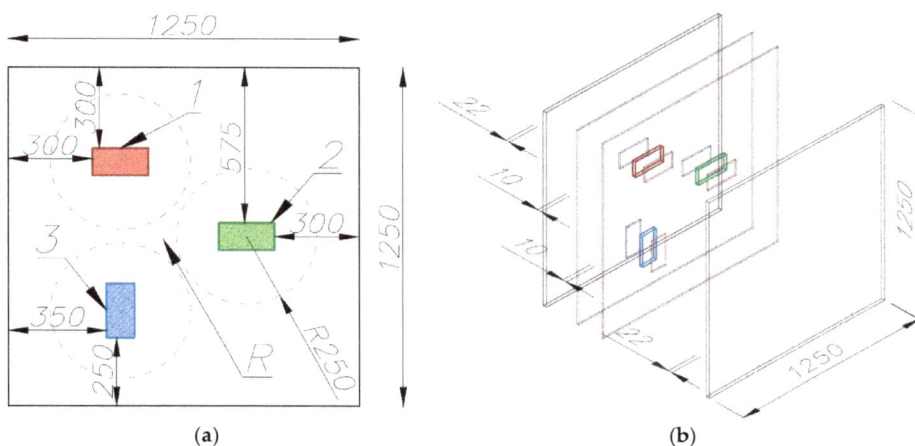

(a)   (b)

Figure 1. Geometrical structure of the partition model (given units in mm): (**a**) arrangement of material inclusions, internal structure. (**b**) View of the 3D partition model where inner boards with holes and inclusions are marked blue and red, while the outer boards are marked violet and grey. Designations in figures: 1—XPS inclusion, 2—granite inclusion, 3—steel inclusion, and R—homogenous area without inclusion.

(a)

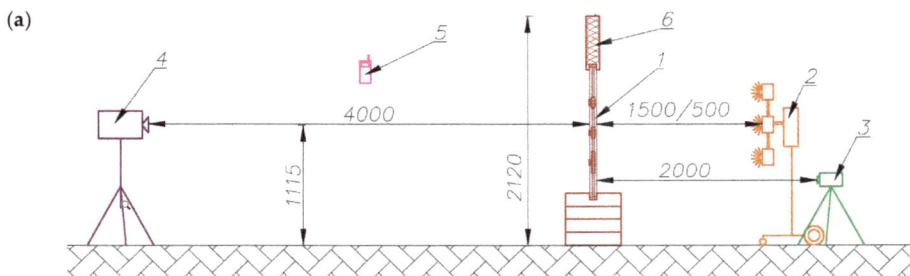

Figure 2. *Cont.*

(b)

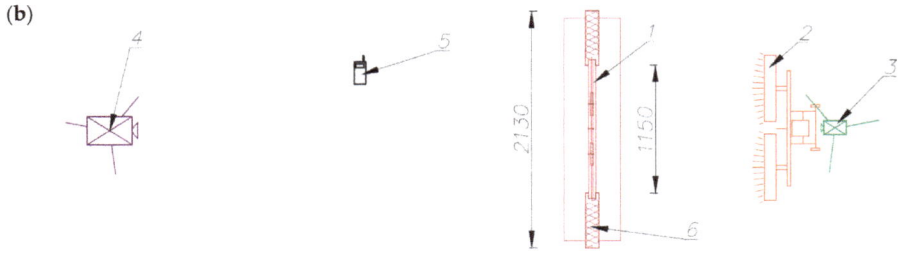

Figure 2. Schematic of test stand (given units in mm): (**a**) side view and (**b**) top view. Designations in the figure: 1—tested partition model, 2—infrared radiator, 3—thermal imaging camera Optris PI400, 4—thermal imaging camera Flir P65, 5—air temperature and relative humidity sensor, 6—Styrofoam band.

As part of the experiment, two building partition models, one made of OSB and the other of GB, were investigated. The experiment consisted in thermally exciting the partition with a 30-minute long continuous heat pulse. The heating was effected at two different infrared radiator distances from the excited surface, i.e., 1500 and 500 mm. The heating of the investigated surface was followed by its cooling during which the temperature field distribution was cyclically measured every 20 s on both sides of the tested element (the transmission mode and the reflection mode). The thermal imaging measurement was conducted for up to eight hours, counting from switching the radiator off. A simple analysis based on the obtained thermograms was carried out. For this purpose, the absolute contrast was calculated from the formula below.

$$C_a(t) = T_p(t) - T_{pj}(t) \quad [°C] \tag{1}$$

where:

$T_p(t)$—the temperature on a surface point of the cross-section with a defect [°C],

$T_{pj}(t)$—the temperature on a surface point of the homogeneous cross section without a defect [°C].

Measuring points in the thermograms were situated in the center of the area with a defect and at the center of the tested element in the defect-free cross-section. The thermogram analysis software "in-point reading" function was used to read temperature values. The size of the in-point function area was 3×3 pixels.

Various building materials were used in the experiment. Besides the kind of thermal excitation and the tested element geometry, the thermal parameters of the materials used also influenced the distribution of the temperature field. The thermal parameters of the materials used in the experiment are specified (on the basis of various literature sources) in Table 1.

Table 1. Thermal parameters of building materials used in the experiment (values based on various literature sources).

Type of Material	Bulk Density	Specific Heat	Heat Capacity	Thermal Conductivity	Thermal Diffusivity
[–]	ϱ_{vol} [kg·m^{-3}]	c_w [J·kg^{-1}·K^{-1}]	C_{vol} [J·m^{-3}·K^{-1}]	λ [W·m^{-1}·K^{-1}]	a [m^2·h^{-1}]
Styrofoam	30	1460	0.04×10^6	0.033	0.75×10^{-6}
Granite	2600	920	2.39×10^6	2.80	1.17×10^{-6}
Steel	7900	500	3.95×10^6	17.0	4.30×10^{-6}
GB	1000	1000	1.00×10^6	0.23	0.23×10^{-6}
OSB	650	1700	1.11×10^6	0.13	0.12×10^{-6}

The thermal parameters of the tested materials affected the test results the most. The homogeneous materials and the material inclusions were appropriately matched in order to compare the influence of the differences between the thermal parameters on the temperature field distribution recorded during the experiment.

3. Results

3.1. Thermograms

The obtained temperature field distribution results are presented as thermograms recorded at characteristic instants. Figure 3 shows the results for the measurement performed in the reflection mode with the heating from a distance of 0.5 m for the homogeneous OSB material (row I) and the GB material (row II), and in the transmission mode at the infrared radiator-surface distance of 1.5 m for the GB partition model (row III) and the OSB partition model (row IV) as well as for the measurement performed in the reflection mode with heating from a distance of 1.5 m for the GB partition (row V) and the OSB partition (row VI). The thermogram recording time, counting from the beginning of the element cooling down, is specified above each of the columns (A–D). The numbers from 1 to 18 indicate the temperature reading positions at the center of the cross section with a defect, while "R" designates the temperature reading position at the defect-free cross section.

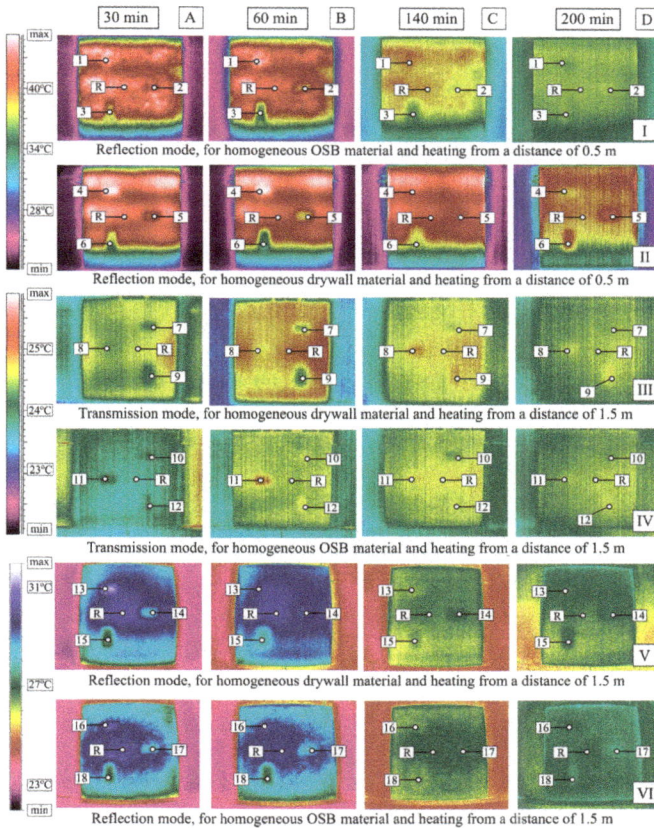

Figure 3. Thermograms at selected instants (time from the beginning of the tested element cooling down is specified above the figure). Numbers from 1 to 18 indicate the temperature reading position at the center of the cross section with a defect, while "R" designates the temperature reading position at the defect-free cross section.

3.2. Absolute Contrasts

On the basis of the obtained temperature field distribution on the investigated surface over time, the absolute contrasts were calculated from Formula (1). The absolute contrast values were calculated in the thermogram points designated with numbers from 1 to 18—the cross sections with defects—whereas "R" designates the temperature value in the defect-free cross section (see Figure 3). Figures 4–9 correspond to the next rows, according to Figure 3 (rows I–VI), for the tested partition model surfaces. The characteristic point numbers 1–18 and the cross section designations A–D are the same in Figures 3 and 4. The thick trend line was plotted as a 6th degree polynomial.

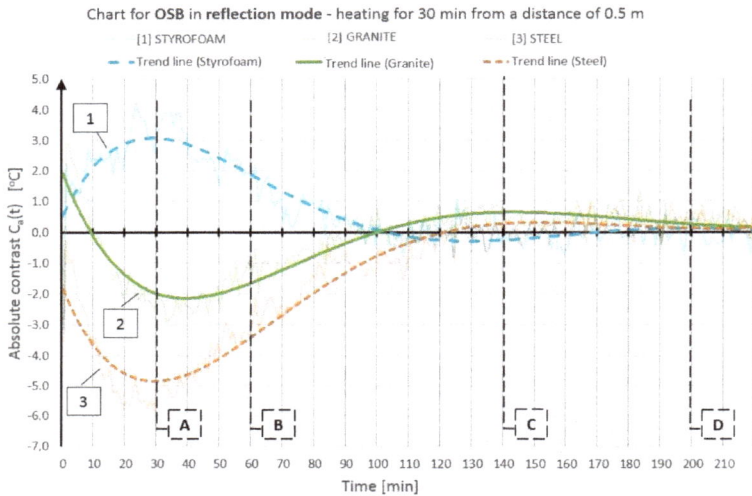

Figure 4. Diagram of the absolute contrast for measurement in a reflection mode for the OSB partition model with heating from a distance of 0.5 m (based on thermograms in Figure 3, row I).

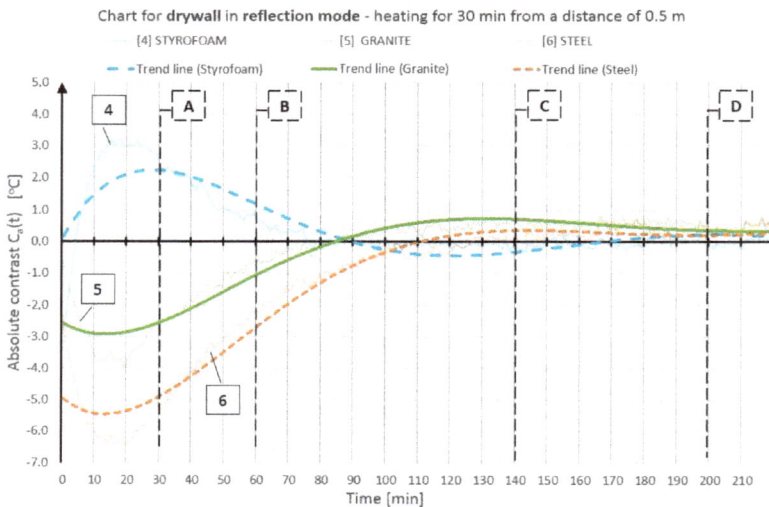

Figure 5. Diagram of absolute contrast for measuring the reflection mode for the GB partition model with heating from a distance of 0.5 m (based on thermograms in Figure 3, row II).

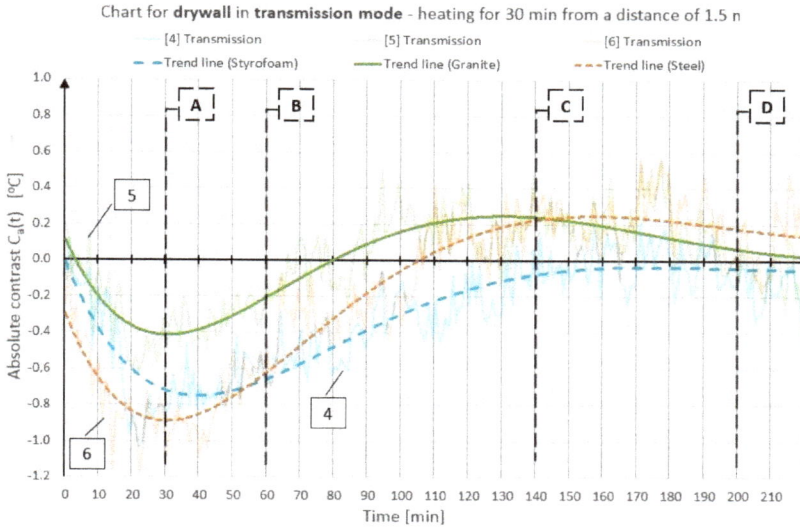

Figure 6. Diagram of absolute contrast for measuring in the transmission mode for the GB partition model with heating from a distance of 1.5 m (based on thermograms in Figure 3, row III).

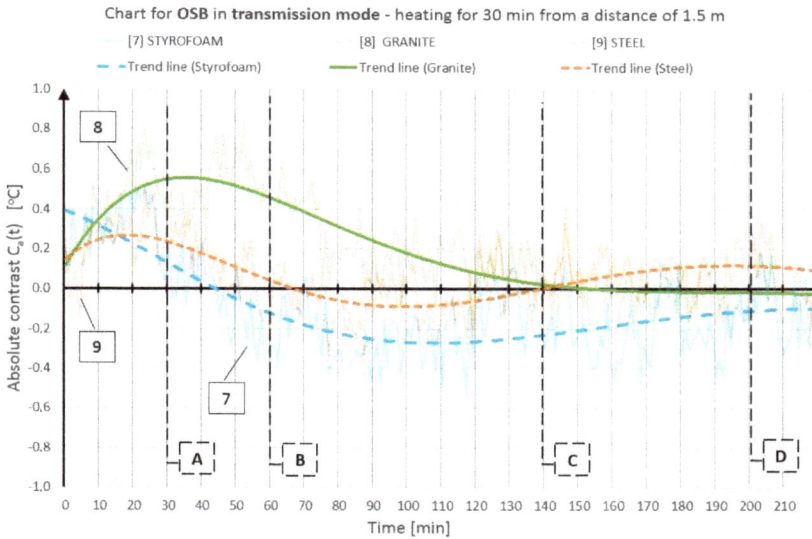

Figure 7. Diagram of absolute contrast for measuring in the transmission mode for the OSB partition model with heating from a distance of 1.5 m (based on thermograms in Figure 3, row IV).

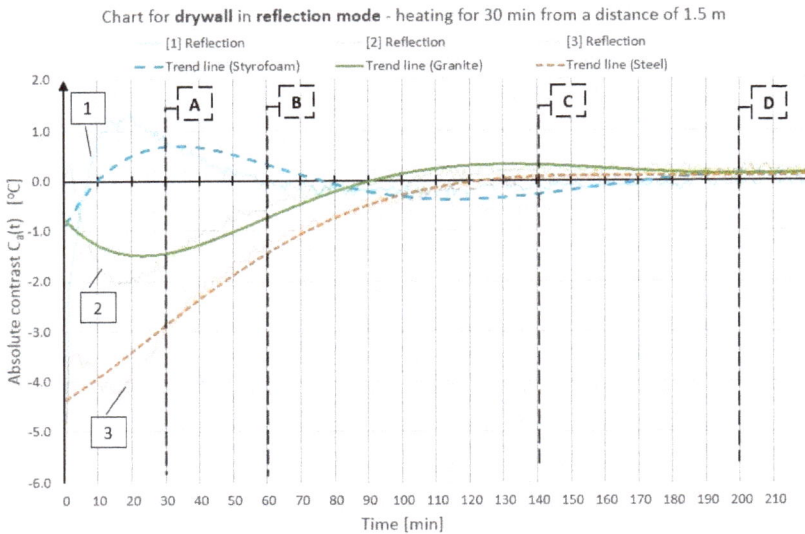

Figure 8. Diagram of absolute contrast for measuring in the reflection mode for a GB partition model with heating from a distance of 1.5 m (based on thermograms in Figure 3, row V).

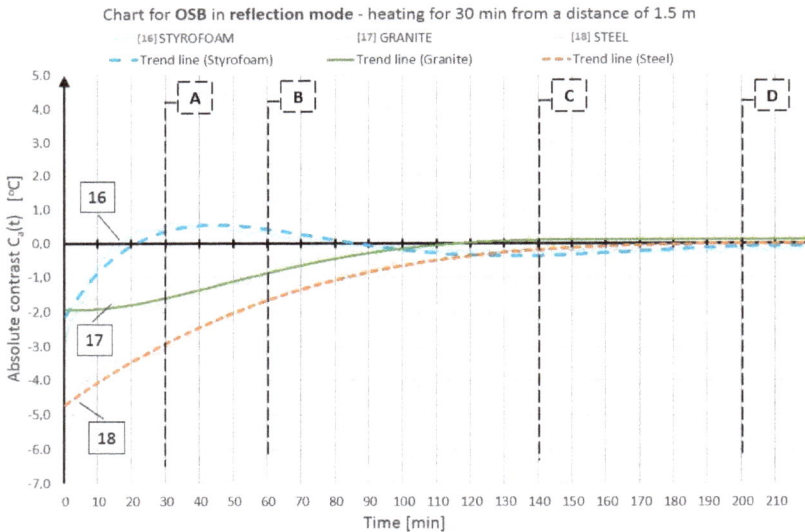

Figure 9. Diagram of absolute contrast for measuring in a reflection mode for the OSB partition model with heating from a distance of 1.5 m (based on thermograms in Figure 3, row VI).

In order to illustrate better and compare the differences in absolute contrast values for the particular types of material inclusions, the obtained data are presented in individual diagrams for styrofoam (Figure 10), granite (Figure 11), and steel (Figure 12).

Chart for the inclusion of **styrofoam** - heating for 30 min (R-reflective mode; T-transmission mode)

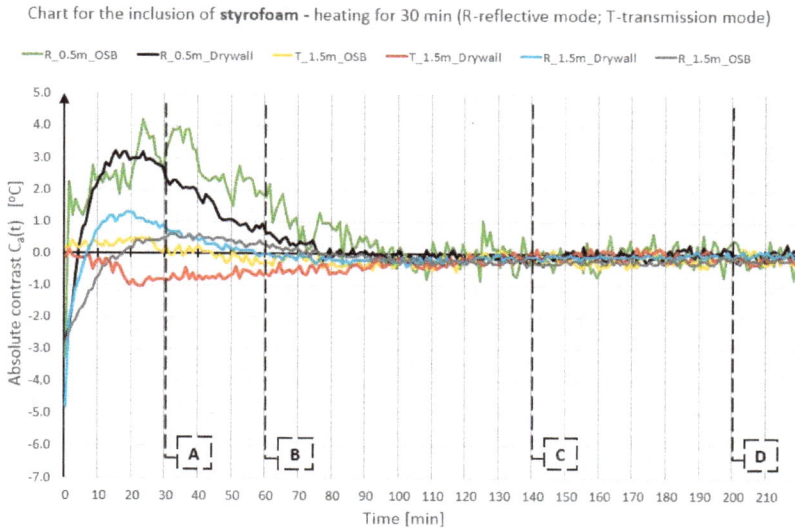

Figure 10. Comparison of absolute contrasts in a reflection mode and in a transmission mode for material inclusion made of Styrofoam.

Chart for the inclusion of **granite** - heating for 30 min (R-reflective mode; T-transmission mode)

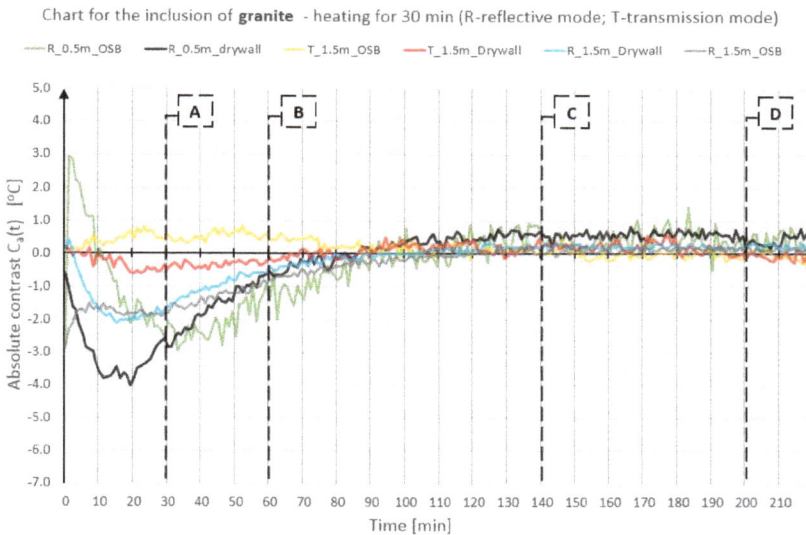

Figure 11. Comparison of absolute contrasts in a reflection mode and in a transmission mode for material inclusion made of granite.

Chart for the inclusion of **steel** - heating for 30 min (R-reflective mode; T-transmission mode)

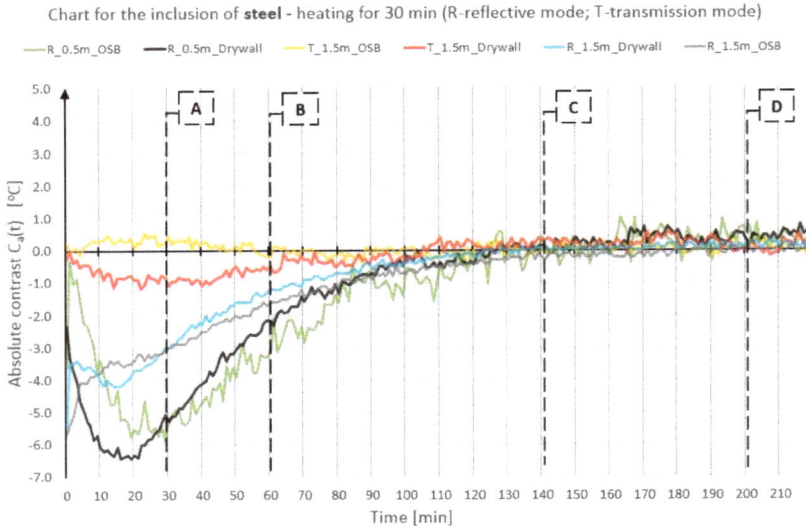

Figure 12. Comparison of absolute contrasts in a reflection mode and in a transmission mode for material inclusion made of steel.

4. Discussion

4.1. Absolute Contrasts and Absolute Inverse Contrasts

The inclusions incorporated into the OSB and GB partition models, which is invisible from the outside, can be located using the active thermography method. The different thermal properties of the inclusions in comparison with those of the basic material in which these inclusions occur cause disturbances in the temperature field distribution visible in the thermograms. The temperature is uniform in those places where there are no inclusions. The only non-uniformities, which can occur there, are due to the non-uniform heating of the surface (when the surface is heated for up to 60 min from a small distance—Figure 3, rows I and II). In the cross sections with inclusions, one can observe warmer or cooler areas in comparison with the homogeneous cross sections. In the case of measurement in the reflection mode, for both the partition materials (OSB and GB), the Styrofoam inclusion was visible as an area with a higher temperature, whereas the granite inclusion and the steel inclusion were observed as areas with a temperature lower than that of the areas without an inclusion. This dependence was observed during heating from both the distance of 0.5 m and 1.5 m. In the case of heating from the distance of 0.5 m for 30 min, the investigated surfaces was heated up to a temperature of about 90 °C, whereas, for heating from the distance of 1.5 m, the temperature on the surface did not exceed 66 °C. In both cases, the inclusions were clearly visible and locatable, but at surface temperatures, when higher than 90 °C, the investigative method is no longer non-destructive since the excessively high temperature damages the structure of the tested materials. Moreover, in order to reach such a high temperature of the surface, the heat source must be placed close to this surface, which results in the non-uniform heating of the latter, consequently, and reduces the visibility of the inner inclusions. When the surface is excited more intensively, the obtained maximum absolute thermal contrasts range from about 3 °C to over 6 °C (Figures 4 and 5). At less intensive heating of the surface, the maximum temperature contrasts range from about 1 °C to over 4 °C (Figures 8 and 9). As one can see in Figure 3, regardless of the degree to which the surface is heated up, at the 30th min from the beginning of the cooling down phase, all the material inclusions are clearly visible and locatable. Hence, one can conclude that, for this geometrical configuration of the building partition and the material inclusions, it suffices to heat up the surface to a temperature of about 60 °C (using the 30-min long continuous heating by an infrared radiator) in order to locate defects situated under a 22-mm thick layer of the basic material. With regard to the thermal parameters of the materials (Table 1), the two basic materials (OSB and GB) have similar heat capacity, but GB is characterized by a higher thermal diffusivity at nearly twice the value. This means that heat will propagate faster in the partition model made of GB. Styrofoam has a very low heat capacity in comparison with that of the partition material, whereby this cross section heats up very quickly and is visible as an area with a temperature higher than that of the defect-free area. By contrast, granite and steel have a heat capacity much higher than that of GB and OSB, which means that the cross section containing such inclusions heats up more slowly than the homogeneous area around the inclusions. The temperature distribution described above is visible in the reflection mode for both strong and weak heating, up to approximately the 90th min from the beginning of cooling down. After this time, inverse contrasts were observed, i.e., the areas with inclusions, which, at the beginning are visible as cooler areas, and, in the second phase of cooling down (after approximately the 90th min), begin to be visible as warmer areas and vice versa. This phenomenon is visible due to the very long recording of the cooling down of the tested elements. It occurs as a result of the difference in heat capacity between the inclusion and the basic material: steel and granite characterized by the high heat capacity cool down much slower than the partitions made of GB and OSB, respectively. In the case of Styrofoam, due to its lower heat capacity, the cross section with a defect cools down quicker than the homogeneous cross section without a defect. In the experiment, the inverse contrast phenomenon can be best observed for the steel inclusion and the granite inclusion when the GB partition model is heated from a distance of 0.5 m (Figure 3 row II). The change of the absolute contrasts over time is shown in Figures 4–9. The sharpest contrasts (the most clearly visible areas with an inclusion) would occur from approximately the 15th to approximately the 40th min since switching the heat source off. For the excitation from a distance of 0.5 m (for both the OSB partition (Figure 4) and the GB partition (Figure 5)) the contrast reached about +4 °C while the cross sections with the steel defect and the granite defect induced a contrast amounting to about −6 °C and about −4 °C, respectively (the minus sign indicates that the defect is visible as a cooler area, whereas the plus sign represents an area warmer than the surrounding area without a defect). In the case of heating from a distance of 1.5 m, the absolute contrast was lower by about 3 °C for the Styrofoam and by about 2 °C for the steel and the granite. Contrasts were also observed during the recording of thermograms in the transmission mode (Figure 3, rows III and IV). The values of the contrasts were considerably lower and amount to ±1 °C (Figures 6 and 7). Even at such low contrasts, however, the defects could be located up to the 60th min from the beginning of the cooling of the partition model.

Absolute contrasts occurring as the element cools down are marked in the diagrams (Figures 13–15). In the initial cooling down phase, a normal absolute contrast (positive for material inclusions with low heat capacity and negative for material inclusions with high heat capacity in comparison with the basic partition material) occurs. In the next cooling down phase, inversion of the contrasts occurs. This is referred to as an inverse absolute contrast. The inverse absolute contrasts are much smaller than the normal absolute contrasts, but they still allow the location of defects within the tested partition, even a few hours after the thermal stimulation of the surface has ended. In the adopted geometric partition model, an interesting dependence was observed. In the case of materials with high heat capacity (granite and steel), the contrasts have the same sign in both the reflection mode and the transmission mode, whereas, for defects with low heat capacity (Styrofoam), the contrast is positive in the reflection mode and negative in the transmission mode. Regardless of heat capacity, in the transmission mode, the defects were always perceived as areas warmer than the homogeneous areas without defects.

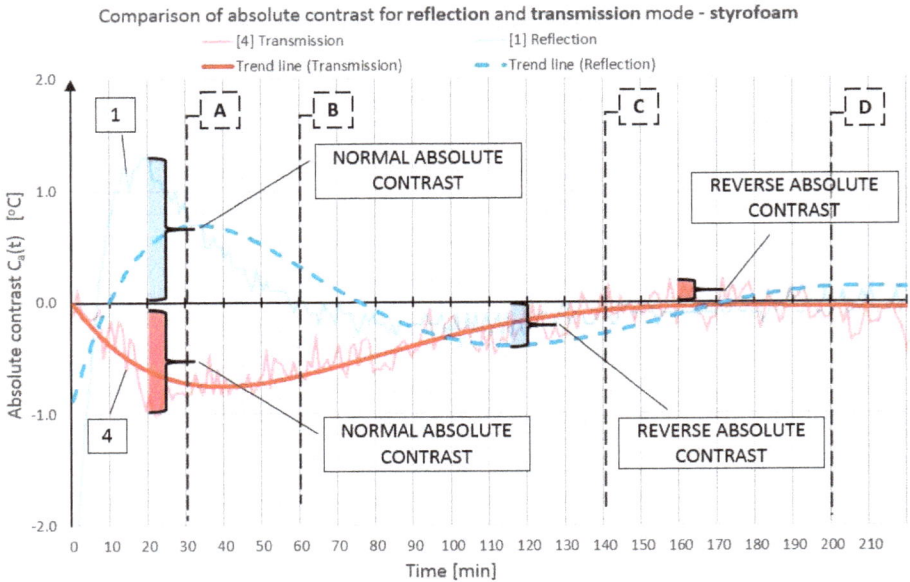

Figure 13. Diagram of absolute contrast in a reflection mode and in a transmission mode for Styrofoam inclusion, with marked places where normal absolute contrast and inverse absolute contrast occur (heating from a distance of 1.5 m).

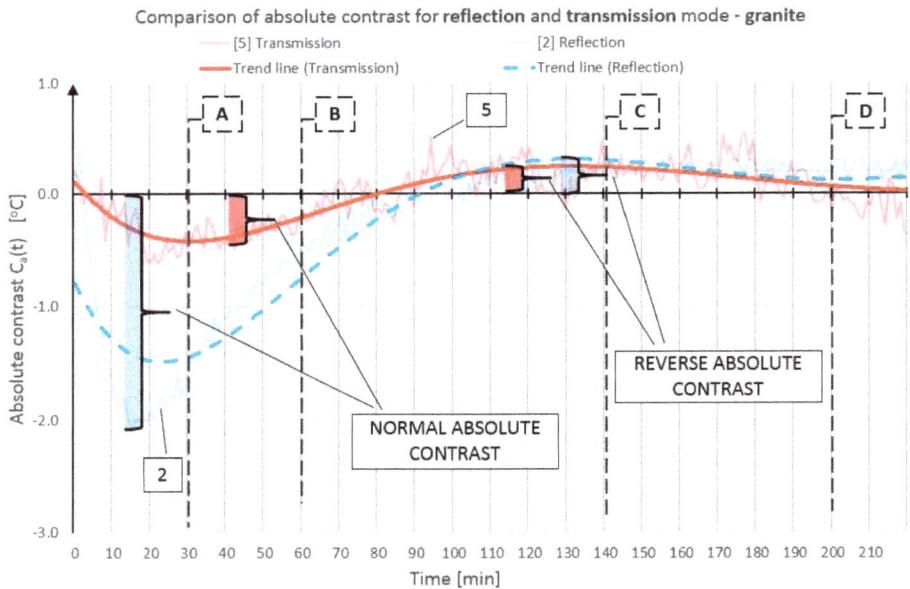

Figure 14. Diagram of absolute contrast in a reflection mode and in a transmission mode for granite inclusion, with marked places where normal absolute contrast and inverse absolute contrast occur (heating from a distance of 1.5 m).

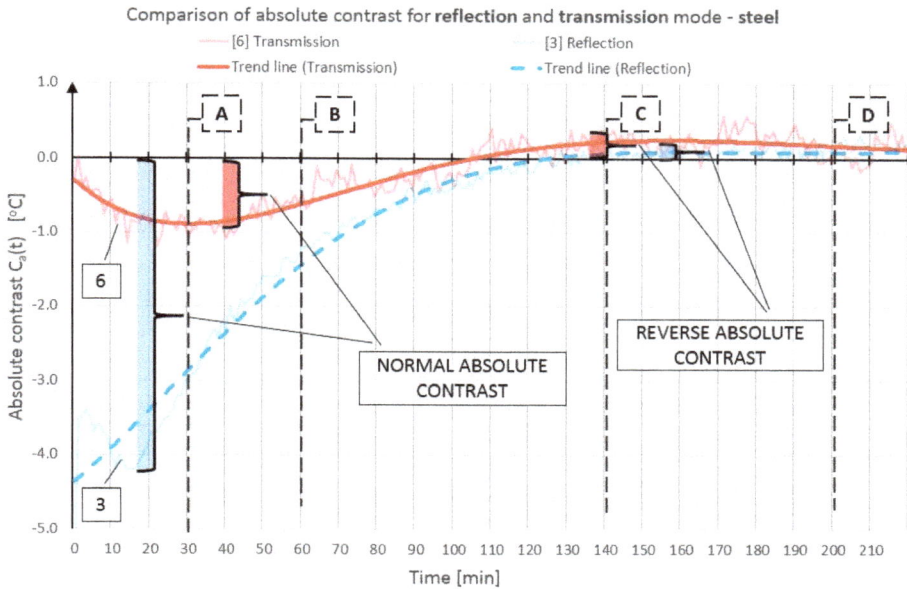

Figure 15. Diagram of absolute contrast in a reflection mode and in a transmission mode for steel inclusion, with marked places where normal absolute contrast and inverse absolute contrast occur (heating from a distance of 1.5 m).

4.2. Defect Location Depth

In addition, other parameters are used in defect detection. In References [16,17], it was noted that parameters "m" and R^2 could be used for this purpose.

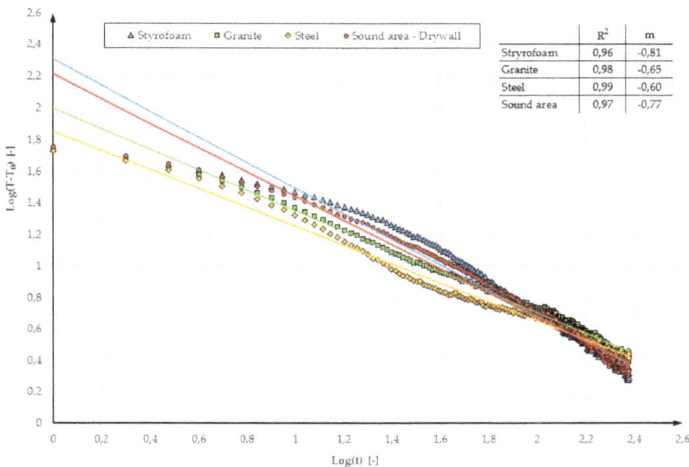

Figure 16. Log–log graph of the cooling phase and R^2 and m values (stimulated period of 30 min) for the OSB partition model with heating from a distance of 0.5 m.

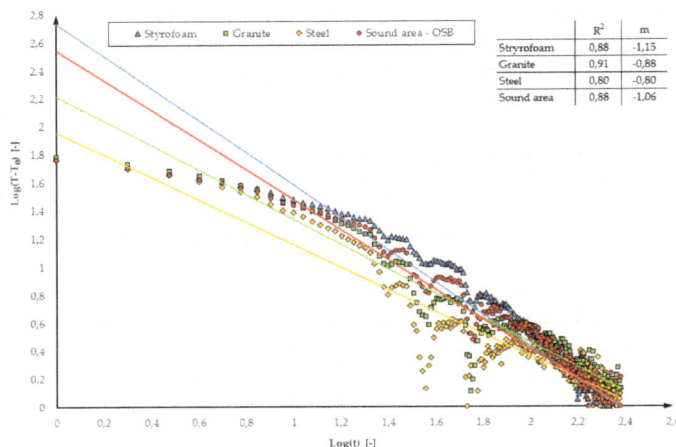

Figure 17. Log–log graph of the cooling phase and R^2 and m values (stimulated period 30 min) for the GB partition model with heating from a distance of 0.5 m.

In the above publications, parameter "m" is defined as the inclination and R^2 as a cooling curve determination coefficient. The cooling curve plot is a Log-Log graph. Time axis X represents Log(t) while axis Y represents the contrast logarithm Log(T − T$_0$), where T is the temperature varying over time and T$_0$ is the temperature measured before the stimulation. For these assumptions, Figures 16 and 17 showed the determined values of parameters "m" and R^2 for the Log-Log graph. In Reference [16], when testing a thin plate (16.2 mm) containing defects with a diameter of a few mm, located at a depth of a few mm (the first cycle of testing using the lock-in method was analysed), the determined R^2 parameter values were in the range of 0.92 to 0.98 (for the selected defects) and the deflections ranged from −0.43 to −0.76. One can notice that both when small elements with a low thermal capacity and massive partitions with a high thermal capacity are tested, similar temperature-time dependences are obtained in the characteristic cross sections with and without defects. A better fit of the cooling line was obtained for the partition model made of gypsum board (R^2 ranging from 0.96 to 0.99). A better cooling line fit means a greater possibility of locating defects on the thermograms. This dependence is visible in Figure 3 where defects in the GB partition are more distinct and their edges are "sharper" in comparison with the partition model made of OSB.

The echo defect shape (EDS) method can be used to determine the depth at which a defect is located in the tested material. This method is described in References [49,50] in more detail where the depth of defect location is calculated using Equation (2).

$$d = \sqrt{-\frac{\lambda \cdot t}{\rho \cdot c_w} \cdot \ln(C_{rel}(t))} \quad [\text{m}] \tag{2}$$

where:

λ—the thermal conductivity of the tested material (above the defect) [W·m^{-1}·K^{-1}],
t—the time elapsed since the end of heating [s],
ρ—the bulk density of the tested material (above the defect) [kg·m^{-3}],
c_w—the specific heat of the tested material (above the defect) [J·kg^{-1}·K^{-1}],
$C_{rel}(t)$—the relative contrast in relation to the reference area (the cross section with the defect) [-].

The relative contrast is a ratio of the absolute contrast to the temperature in a surface point of the homogenous cross section without a defect. When calculating "d", it is highly important to select a proper relative contrast, which must be high enough to reduce the measurement noise, but still be below the maximum values [50]. In Reference [49], it is recommended to adopt $C_{rel}(t)$ equal to 0.025 if the thermal noise is in the range of 0.01–0.015. In Reference [50], this value is optimized and selecting the value of 0.07 is recommended. It sometimes happens that the relative contrast does not reach the value of 0.07 during the test. For testing elements with a large thermal mass by means of a long thermal pulse, the authors optimized $C_{rel}(t)$ value selection to the value that occurs 3 min before the first peak of this contrast. The relative contrast and defect location depth versus time dependence (in the GB partition model heated from a distance of 0.5 m) is shown in Figures 18–20.

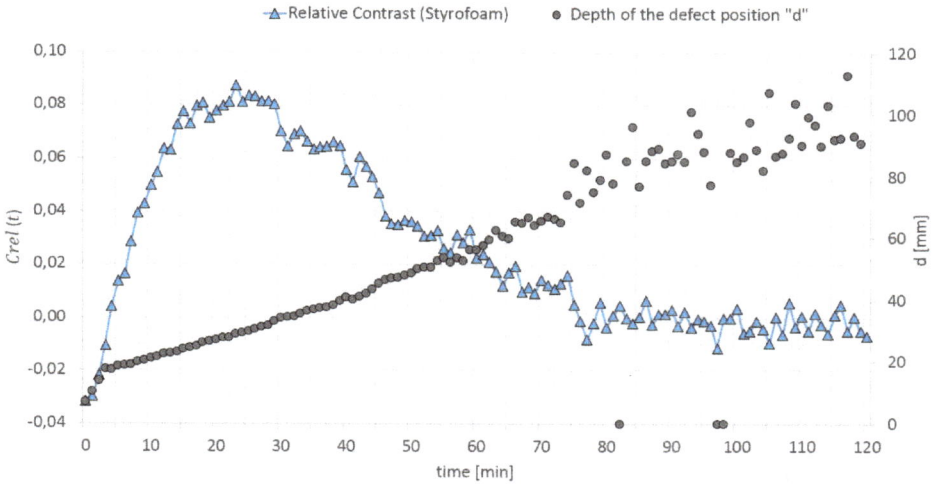

Figure 18. Relative contrast and defect location depth versus time (Styrofoam).

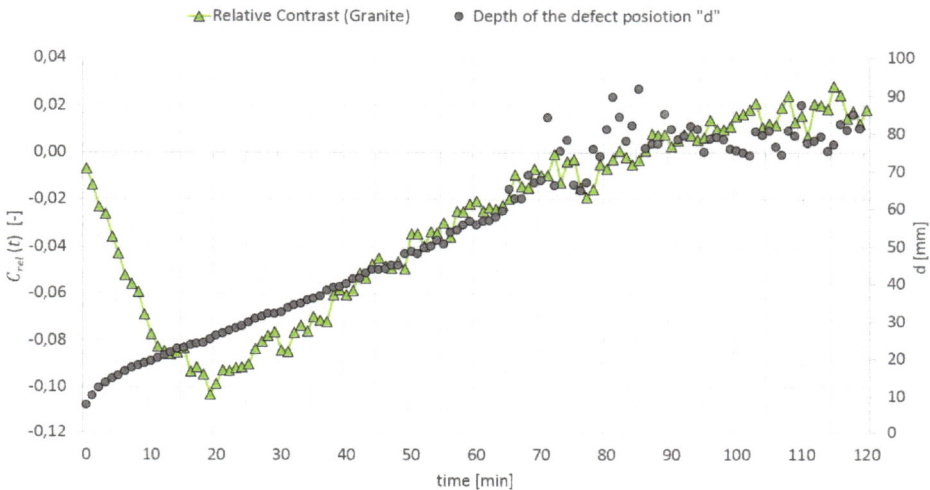

Figure 19. Relative contrast and defect location depth versus time (Granite).

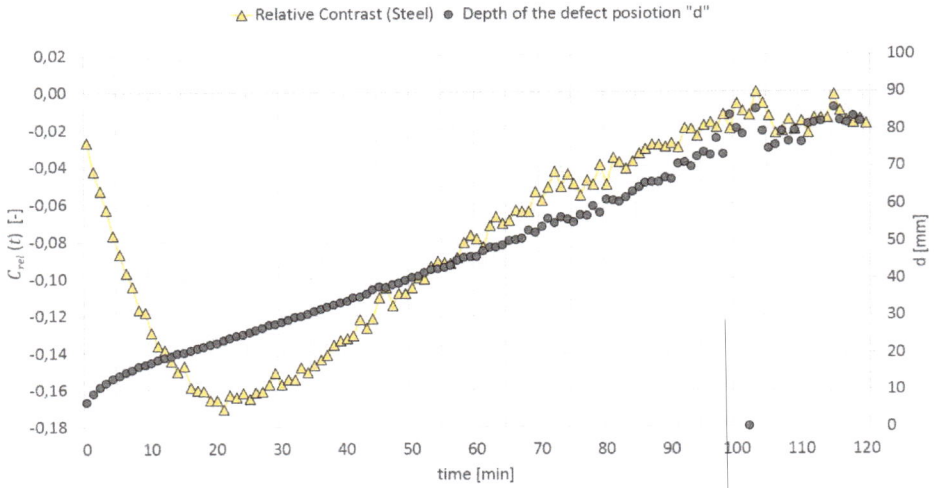

Figure 20. Relative contrast and defect location depth versus time (Steel).

For the above assumptions, the defect location depth was calculated for two partition models thermally stimulated from different distances. The GB and OSB values contained in Table 1 were used in the calculations. The results are shown in Table 2. The actual depth at which the defects are located amounts to 22 mm. For the presented investigative methodology, the calculated values do not exceed the relative error of 10% (except for the Styrofoam defect in the OSB model, heated from the distance of 1.5 m. In this case, the relative error amounts to 23%).

Table 2. Defect location depths calculated using the EDS method.

Defect	OSB, Heating from Distance 0.5 m	OSB, Heating from Distance 1.5 m	GK, Heating from Distance 0.5 m	GK, Heating from Distance 1.5 m
–	d [mm]	d [mm]	d [mm]	d [mm]
Styrofoam	21.67	27.14	22.51	22.93
Granite	24.15	20.63	22.89	21.92
Steel	19.97	22.36	21.32	19.51

4.3. Dimensions of the Defect

The measurement by means of the Flir P65 camera with a 24° × 18° lens was carried out from a distance of 4.0 m, which means that the size of a single pixel in the thermogram is 5.2 mm. For the Optris PI400 thermal imaging camera with a 60° × 45° lens and measurement taken from a distance of 2.0 m, the pixel size is 4.6 mm. Using the thermograms, one can easily assess qualitatively the location of the material inclusions. A disturbance in the temperature field distribution indicates the place where the inclusions occur. On the basis of the thermograms, one can also carry out a simple quantitative assessment and estimate the dimensions of the defects inside the tested partition. The procedure for estimating the width and height of a defect is schematically shown in Figure 16. First, one should select the thermogram in which the defect is most clearly visible. The best thermogram is the one in which the thermal contrast between the area with the defect and the defect-free area as well as the potential edges of the defect is the sharpest (most distinct). Then, virtual lines representing the predicted edge of the defect are assumed. In the program for analyzing thermograms, one can read the x, y coordinates of the particular lines and, on this basis, calculate the potential dimensions (in pixels) of the perceived defect. Lastly, knowing the size of a single pixel, one can estimate the size of the material inclusion.

Figure 21 shows how the defect made of granite is estimated using thermograms obtained from the Optris thermal imaging camera. The estimated dimensions are 211.6 mm × 124.2 mm, while the actual dimensions of the defect are 200 mm × 100 mm. It can be assumed that the relative estimation error does not exceed 6% for the width and 25% for the height. If the above procedure is used, the obtained dimensions will always exceed the actual dimensions of the defect since the assumed area delimiting the place where defects occur is an area of the temperature field disturbance. This disturbance always decays from the defect's edge outwards instead of inwards. Quantitative procedures for estimating defect dimensions by using pulsed thermography can be found, for example, in References [52,53].

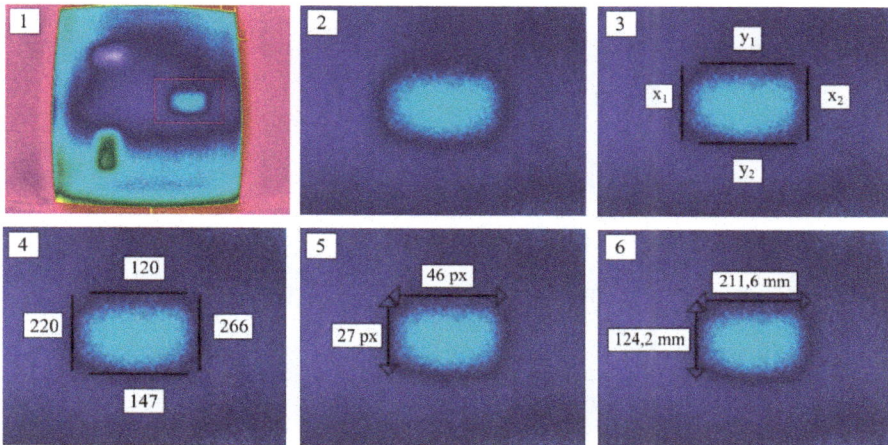

Figure 21. Procedure for estimating the defect size on the basis of the known size of a single pixel (4.6 mm).

5. Conclusions

From the experiment carried out on the building partition with modeled inclusions of various materials as well as from the analysis of the obtained thermograms and the calculated absolute contrasts, the following conclusions can be drawn (for the experimental geometry of the partition and the inclusions).

1. Using the non-destructive active thermography technique, one can locate material inclusions in building partitions (elements with a large thermal capacity), situated at a depth of at least 22 mm below the surface of cladding made of wood or gypsum;
2. In order to locate defects in a building partition, it is sufficient to heat its surface to about 60 °C whereby the test does not lose its non-destructive character;
3. The highest temperature contrasts between the cross sections with defects and the cross sections without defects (the inclusions are most clearly visible then) arise at the time interval from the 15th to the 40th min since the beginning of cooling down;
4. During testing in the reflection mode, inclusions with a considerably higher heat capacity than that of the basic materials are visible as areas whose temperature is lower than that of the areas without inclusions;
5. During testing in the reflection mode, inclusions with a considerably lower heat capacity than that of the basic materials are visible as areas whose temperature is higher than that of the areas without inclusions;

6. During testing in the transmission mode, the areas with defects are visible as areas whose temperature is higher than that of the areas without inclusions, regardless of the heat capacity of the material inclusions;

7. Approximately 90 min after the heating source is switched off, the second phase of cooling down takes place. In this phase, an inverse absolute contrast appears—the sign of the contrast changes and the inclusions initially visible as a warmer area are perceived as cooler areas, and vice versa (this phenomenon occurs on the same side of the partition);

8. Using the Echo Defect Shape method, one can successfully (with an error below 10%) determine the depth at which a defect is located in the building partition;

9. Knowing the specifications of the thermal imaging camera lens (the size [in mm] of 1 pixel in the thermogram), one can estimate the size of the defects present inside the partition by indicating the place in the thermogram where the defect's edges are likely situated.

The experiment shows that active thermography has great potential for materials testing in construction. It has been demonstrated that defect detection can be successfully conducted over large areas of massive members (large thermal capacity), using an appropriately powerful heat source and a long heat pulse. By recording the distribution of the temperature field during the cooling down of the tested partition over a long time, inverse contrasts (on the same front side of the partition) were observed. As part of further research, inverse heat conduction problems should be more broadly solved and the depth at which inclusions occur, as well as their thermal parameters, should be estimated on the basis of the obtained test results (thermograms). The further research will deal with applying inverse contrasts to improve the defect detection method (quantitative analysis). It is anticipated that the proposed investigative technique will be increasingly used for the non-destructive testing of materials, especially large-sized building units.

Author Contributions: Conceptualization, P.N. and H.N. Methodology, H.N. Validation, P.N. Formal Analysis, P.N. and H.N. Investigation, P.N. Resources, P.N. Data Curation, P.N. Writing-Original Draft Preparation, P.N. and H.N. Writing-Review & Editing, H.N. Visualization, P.N. and H.N. Supervision, H.N. Project Administration, H.N.

Acknowledgments: The authors thank Prof. Jacek Kasperski—the Head of the Department of Construction and Flow Machines at the Wrocław University of Science and Technology—for lending the thermographic camera used in the experiment.

Conflicts of Interest: The authors declare no conflict of interest.

References

1. Maierhofer, C.; Reinhardt, H.W.; Dobmann, G. *Non-Destructive Evaluation of Reinforced Concrete Structures—Non-Destructive Testing Methods*; Woodhead Publishing: Cambridge, UK, 2010; Volume 2, ISBN 978-18-4569-950-5.

2. Hoła, J.; Bień, J.; Sadowski, Ł.; Schabowicz, K. Non-destructive and semi-destructive diagnostics of concrete structures in assessment of their durability. *Bull. Pol. Acad. Sci.* **2015**, *63*, 87–96. [CrossRef]

3. Lüthi, T. *Non-Destructive Evaluation Methods. EMPA—Swiss Federal Laboratories for Materials Science and Technology*; Self-Published: Dübendorf, Switzerland, 2013.

4. Balageas, D.; Maldague, X.; Burleigh, D.; Vavilow, V.P.; Oswald-Tranta, B.; Roche, J.M.; Pradere, C.; Carlomagno, G.M. Thermal (IR) and Other NDT Techniques for Improved Material Inspection. *J. Nondestruct. Eval.* **2016**, *35*, 18. [CrossRef]

5. Nowak, H. *Application of Infrared Thermography in Construction*; Wrocław University of Science and Technology Publishing House: Wrocław, Poland, 2012; ISBN 978-83-7493-676-7. (In Polish)

6. Barreira, E.; Freitas, V.P. Evaluation of building materials using infrared thermography. *Constr. Build. Mater.* **2007**, *21*, 218–224. [CrossRef]

7. Barreira, E.; Almeida, R.M.S.F.; Delgado, J.M.P.Q. Infrared thermography for assessing moisture related phenomena in building components. *Constr. Build. Mater.* **2016**, *110*, 251–269. [CrossRef]

8. Wróbel, A.; Kisilewicz, T. Detection of thermal bridges—Aims, possibilities and conditions. In Proceedings of the QIRT 2008 Proceedings—9-th Quantitative InfraRed Thermography—International Conference, Krakow, Poland, 2–5 July 2008; pp. 227–232, ISBN 978-83-908655-1-5.

9. Kisilewicz, T.; Wróbel, A. Quantitative infrared wall inspection. In Proceedings of the 10-th Edition of the Quantitative InfraRed Thermography—International Conference, Québec-City, QC, Canada, 27–30 June 2010; pp. 589–594, ISBN 978-2-9809199-1-6.

10. Khodayar, F.; Sojasi, S.; Maldague, X. Infrared thermography and NDT: 2050 horizon. *Quant. Infrared Thermogr. J.* **2016**, *13*, 210–231. [CrossRef]

11. Vavilov, V. Thermal NDT: Historical milestones, state-of-the-art and trends. *Quant. Infrared Thermogr. J.* **2014**, *11*, 66–83. [CrossRef]

12. Wu, D.; Karpen, W.; Busse, G. Lockin thermography for multiplex photothermal nondestructive evaluation. In Proceedings of the QIRT Conference, Paris, France, 7–9 July 1992.

13. Ibarra-Castanedo, C.; Piau, J.M.; Guilbert, S.; Avdelidis, N.P.; Genest, M.; Bendada, A.; Maldague, X. Comparative Study of Active Thermography Techniques for the Nondestructive Evaluation of Honeycomb Structures. *Res. Nondest. Eval.* **2009**, *20*, 1–31. [CrossRef]

14. Maierhofer, C.; Myrach, P.; Steinfurth, H.; Reischel, M.; Röllig, M. Development of standards for flash thermography and lock-in thermography. In Proceedings of the 14th International Conference on Quantitative InfraRed Thermography (QIRT14), Bordeaux, France, 7–11 July 2014.

15. Dudzik, S. *Determining the Depth of Material Defects by Means of Dynamic Active Thermography and Artificial Neural Networks*; Częstochowa University of Technology Publishing House: Częstochowa, Poland, 2013; ISBN 978-83-7193-572-5. (In Polish)

16. Palumbo, D.; Galietti, U. Damage Investigation in Composite Materials by Means of New Thermal Data Processing Procedures. *Strain* **2016**, *52*, 276–285. [CrossRef]

17. Palumbo, D.; Cavallo, P.; Galietti, U. An investigation of the stepped thermography technique for defects evaluation in GFRP materials. *NDT E Int.* **2019**, *102*, 254–263. [CrossRef]

18. Maldague, X. *Theory and Practice of Infrared Technology of Non-Destructive Testing*; John Wiley & Sons: New York, NY, USA, 2001; ISBN 978-0-471-18190-3.

19. *EN 17119:2018 Non-Destructive Testing—Thermographic Testing—Active Thermography*; European Standard: Brussels, Belgium, 2018.

20. Milovanović, B.; Banjad Pečur, I. Review of Active IR Thermography for Detection and Characterization of Defects in Reinforced Concrete. *J. Imaging* **2016**, *2*, 11. [CrossRef]

21. Arndt, R.W. Square pulse thermography in frequency domain as adaptation of pulsed phase thermography for qualitative and quantitative applications in cultural heritage and civil engineering. *Infrared Phys. Technol.* **2010**, *53*, 246–253. [CrossRef]

22. Ibarra-Castanedo, C.; Sfarra, S.; Klein, M.; Maldague, X. Solar loading thermography: Time-lapsed thermographic survey and advanced thermographic signal processing for the inspection of civil engineering and cultural heritage structures. *Infrared Phys. Technol.* **2017**, *82*, 56–74. [CrossRef]

23. Noszczyk, P.; Nowak, H. Active thermography as a state-of-the-art method of testing reinforced concrete units (in Polish). *J. Civ. Eng. Environ. Arch.* **2016**, *33*, 279–286. [CrossRef]

24. Maierhofer, C.; Brink, A.; Röllig, M.; Wiggenhauser, H. Quantitative impulse-thermography as non-destructive testing method in civil engineering—Experimental results and numerical simulations. *Constr. Build. Mater.* **2005**, *19*, 731–737. [CrossRef]

25. Maierhofer, C.; Arndt, R.; Röllig, M.; Rieck, C.; Walther, A.; Scheel, H.; Hillemeier, B. Application of impulse-thermography for non-destructive assessment of concrete structures. *Cem. Concrete Compos.* **2006**, *28*, 393–401. [CrossRef]

26. Weiser, M.; Röllig, M.; Arndt, R.; Erdmann, B. Development and test of a numerical model for pulse thermography in civil engineering. *Heat Mass Transf.* **2010**, *46*, 1419–1428. [CrossRef]

27. Maierhofer, C.; Brink, A.; Röllig, M.; Wiggenhauser, H. Transient thermography for structural investigation of concrete and composites in the near surface region. *Infrared Phys. Technol.* **2002**, *43*, 271–278. [CrossRef]

28. Maierhofer, C.; Wiggenhauser, H.; Brink, A.; Röllig, M. Quantitative numerical analysis of transient IR-experiments on buildings. *Infrared Phys. Technol.* **2004**, *46*, 173–180. [CrossRef]

29. Maierhofer, C.; Brink, A.; Röllig, M.; Wiggenhauser, H. Detection of shallow voids in concrete structures with impulse thermography and radar. *NDT E Int.* **2003**, *36*, 257–263. [CrossRef]

30. Khan, F.; Bolhassani, M.; Kontsos, A.; Hamid, A.; Bartoli, I. Modeling and experimental implementation of infrared thermography on concrete masonry structures. *Infrared Phys. Technol.* **2015**, *69*, 228–237. [CrossRef]

31. Donatelli, A.; Aversa, P.; Luprano, V.A.M. Set-up of an experimental procedure for the measurement of thermal transmittances via infrared thermography on lab-made prototype walls. *Infrared Phys. Technol.* **2016**, *79*, 135–143. [CrossRef]

32. Rumbayan, R.; Washer, G.A. Modeling of Environmental Effects on Thermal Detection of Subsurface Damage in Concrete. *Res. Nondest. Eval.* **2014**, *25*, 235–252. [CrossRef]

33. Sfarra, S.; Marcucci, E.; Ambrosini, D.; Paoletti, D. Infrared exploration of the architectural heritage: From passive infrared thermography to hybrid infrared thermography (HIRT) approach. *Mater. Constr.* **2016**, *66*, 94. [CrossRef]

34. Aggelis, D.G.; Kordatos, E.Z.; Strantza, M.; Soulioti, D.V.; Matikas, T.E. NDT approach for characterization of subsurface cracks in concrete. *Constr. Build. Mater.* **2011**, *25*, 3089–3097. [CrossRef]

35. Lai, W.W.-L.; Lee, K.-K.; Poon, C.-S. Validation of size estimation of debonds in external wall's composite finishes via passive Infrared thermography and a gradient algorithm. *Constr. Build. Mater.* **2015**, *7*, 113–124. [CrossRef]

36. Kurita, K.; Oyado, M.; Tanaka, H.; Tottori, S. Active infrared thermographic inspection technique for elevated concrete structures using remote heating system. *Infrared Phys. Technol.* **2009**, *52*, 208–213. [CrossRef]

37. Halabe, U.B.; Vasudevan, A.; Klinkhachorn, P.; GangaRao, H.V.S. Detection of subsurface defects in fiber reinforced polymer composite bridge decks using digital infrared thermography. *Nondestruct. Test. Eval.* **2007**, *22*, 155–175. [CrossRef]

38. Keo, S.A.; Brachelet, F.; Breaban, F.; Defer, D. Steel detection in reinforced concrete wall by microwave infrared thermography. *NDT E Int.* **2014**, *62*, 172–177. [CrossRef]

39. Cotič, P.; Kolarič, D.; Bosiljkov, V.B.; Bosiljkov, V.; Jagličić, Z. Determination of the applicability and limits of void and delamination detection in concrete structures using infrared thermography. *NDT E Int.* **2015**, *74*, 87–93. [CrossRef]

40. Scott, M.; Kruger, D. Infrared Thermography as a Diagnostic Tool for Subsurface Assessments of Concrete Structures. In Proceedings of the ITC User Conference, Stockholm, Sweden, 29–30 October 2014.

41. Scott, M.; Kruger, D. Effects of Solar Loading on the Limits of Predictability of Internal Delamination Defects in Concrete Using Infrared Thermography. In Proceedings of the Conference InfraMation, Orlando, FL, USA, 4–7 November 2013.

42. Brachelet, F.; Keo, S.; Defer, D.; Breaban, F. Detection of reinforcement bars in concrete slabs by infrared thermography and microwaves excitation. In Proceedings of the Conference QIRT, Bordeaux, France, 7–11 July 2014.

43. Brink, A.; Maierhofer, C.; Röllig, M.; Wiggenhauser, H. Application of quantitative impulse thermography for structural evaluation in civil engineering—Comparison of experimental results and numerical simulations. In Proceedings of the 6th Quantitative InfraRed Thermography Conference, Dubrovnik, Croatia, 24–27 September 2002.

44. Arndt, R.; Maierhofer, C.; Röllig, M.; Weritz, F.; Wiggenhauser, H. Structural Investigation of Concrete and Masonry Structures behind Plaster by means of Pulse Phase Thermography. In Proceedings of the 7th International Conference on Quantitative Infrared Thermography (QIRT), Brussels, Belgium, 5–8 July 2004.

45. Vavilov, V.P.; Pan, Y.; Moskovchenko, A.I.; Čapka, A. Modeling, detecting and evaluating water ingress in aviation honeycomb panels. *Quant. Infrared Thermogr. J.* **2017**, *14*, 206–217. [CrossRef]

46. Lai, W.L.; Poon, C.-S. Boundary and size estimation of debonds in external wall finishes of high-rise buildings using Infrared thermography. In Proceedings of the QIRT2012—11th international conference on quantitative infrared thermography, Naples, Italy, 11–14 June 2012.

47. Vavilov, V.P.; Burleigh, D.D. Review of pulsed thermal NDT: Physical principles, theory and data processing. *NDT E Int.* **2015**, *73*, 28–52. [CrossRef]

48. Szymanik, B.; Frankowski, P.K.; Chady, T.; John Chelliah, C.R.A. Detection and Inspection of Steel Bars in Reinforced Concrete Structures Using Active Infrared Thermography with Microwave Excitation and Eddy Current Sensors. *Sensors* **2016**, *16*, 234. [CrossRef]

49. Lugin, S.; Netzelmann, U. A defect shape reconstruction algorithm for pulsed thermography. *NDT E Int.* **2007**, *40*, 220–228. [CrossRef]

50. Richter, R.; Maierchofer, C.; Kreutzbruck, M. Numerical method of active thermography for the reconstruction of back wall geometry. *NDT E Int.* **2013**, *54*, 189–197. [CrossRef]
51. Vavilov, V. Infrared Techniques for Materials Analysis and Nondestructive Testing. In *Infrared Methodology and Technology*; Maldague, X.P.V., Ed.; Gordon and Breach Publishers: London, UK, 1994; pp. 230–309.
52. Balageas, D.L.; Roche, J.M.; Leroy, H. Comparative Assessment of Thermal NDT Data Processing Techniques for Carbon Fiber Reinforced Polymers. *Mater. Eval.* **2017**, *75*, 1019–1031.
53. Giorleo, G.; Meola, C. Comparison between pulsed and modulated thermography in glass-epoxy laminates. *NDT E Int.* **2002**, *35*, 287–292. [CrossRef]

materials

MDPI

Article

Investigation of Structural Degradation of Fiber Cement Boards Due to Thermal Impact

Zbigniew Ranachowski [1,*], Przemysław Ranachowski [1], Tomasz Dębowski [1], Tomasz Gorzelańczyk [2] and Krzysztof Schabowicz [2]

[1] Experimental Mechanics Division, Institute of Fundamental Technological Research, Polish Academy of Sciences, Pawińskiego 5B, 02-106 Warszawa, Poland; pranach@ippt.pan.pl (P.R.); tdebow@ippt.pan.pl (T.D.)
[2] Faculty of Civil Engineering, Wrocław University of Science and Technology, Wybrzeże Wyspiańskiego 27, 50-370 Wrocław, Poland; Tomasz.Gorzelanczyk@pwr.edu.pl (T.G.); krzysztof.schabowicz@pwr.edu.pl (K.S.)
[*] Correspondence: zranach@ippt.pan.pl; Tel.: +48-22-698-817870

Received: 21 February 2019; Accepted: 17 March 2019; Published: 21 March 2019

check for updates

Abstract: The aim of the present study was to investigate the degradation of the microstructure and mechanical properties of fiber cement board (FCB), which was exposed to environmental hazards, resulting in thermal impact on the microstructure of the board. The process of structural degradation was conducted under laboratory conditions by storing the FCB specimens in a dry, electric oven for 3 h at a temperature of 230 °C. Five sets of specimens, that differed in cement and fiber content, were tested. Due to the applied heating procedure, the process of carbonization and resulting embrittlement of the fibers was observed. The fiber reinforcement morphology and the mechanical properties of the investigated compositions were identified both before, and after, their carbonization. Visual light and scanning electron microscopy, X-ray micro tomography, flexural strength, and work of flexural test W_f measurements were used. A dedicated instrumentation set was prepared to determine the ultrasound testing (UT) longitudinal wave velocity c_L in all tested sets of specimens. The UT wave velocity c_L loss was observed in all cases of thermal treatment; however, that loss varied from 2% to 20%, depending on the FCB composition. The results obtained suggest a possible application of the UT method for an on-site assessment of the degradation processes occurring in fiber cement boards.

Keywords: cement-based composites; fiber cement boards; durability; ultrasound measurements

---

1. Introduction

Fiber cement board (FCB) is a versatile, green, and widely-applied building material. It acts as a substitute for natural wood and wood-based products, such as plywood or oriented strand boards (OSB). The properties of FCB, as a construction material, make it preferable for use as a ventilated, façade cladding for newly-built and renovated buildings, interior wall coverings, balcony balustrade panels, base course and chimney cladding, and enclosure soft-fit lining [1]. FCB can be applied to unfinished, painted, or simply-impregnated surfaces. Fiber cement components have been used in construction for over 100 years, mainly as roofing covers, in the form of corrugated plates or non-pressurized tubes. FCB façades are exposed to a variety of different environmental hazards. Adverse factors can include visual an ultraviolet light radiation, wind and ice-clod impacts, and thermal stresses evoked by temperature changes, etc. [2,3]. These hazards may result in board embrittlement, shrinkage, or bending. An example of a FCB façade, showing considerable damage after ten years of exposure to climate hazards, is presented in Figure 1. The fracture process that can develop in building materials is complex because the strains are not uniformly distributed during the fracture, particularly in regions where there are cracks. The facade boards are usually fixed to the wall-construction on their edgings, which exposes them to flexural stresses. The currently-applied fiber cement boards are designed to carry the mechanical load by the cellulose and polyvinyl alcohol (PVA) fiber reinforcements. The fibers reinforce the FCB component only when they are added in a specific quantity (5–10% wt.) and when they are uniformly dispersed throughout the cementitious matrix. A highly complex procedure is required to achieve this goal, as well as to avoid faults under efficient industrial conditions. Hatschek solved the problem by inventing a machine with a rotating sieve and a vat containing a diluted fiber slurry, Portland cement, and mineral components [4,5]. A thin film of FCB is formed on a moving belt, partially wrapped around the sieve, similar to the procedure used in paper sheet-making [6].

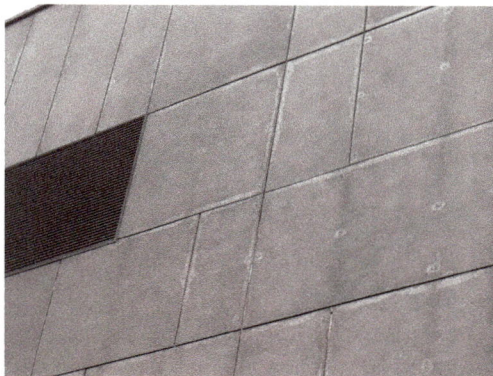

Figure 1. An example of a FCB façade showing considerable damage after ten years of exposure to climate hazards.

As the service performance of fiber cement boards may be affected by the improper function of reinforcements (i.e., damaged fibers, inhomogeneous concentrations, or poor quality fibers), several methods were proposed for testing the performance of the boards. These include:

- emitting and receiving the ultrasonic Lamb waves [7,8],
- the impact-echo method combined with the impulse-response method [9],
- the ultrasound (UT) longitudinal wave velocity c_L method [10,11],
- the acoustic emission method [12],
- X-ray micro tomography [13,14].

The UT wave velocity c_L method is one of the few methods that can be applied in situ to control the degradation processes in building façades. Some authors [15,16] have performed experiments that resulted in the degradation of fiber reinforcements by pyrolyzing the fibers; i.e., by exposing them to an elevated temperature for several hours. These authors reported that a loss of material elasticity can be observed when thermal treatment is conducted using temperatures exceeding 100 °C.

It is also possible to analyze the morphology and distribution of fibers in the cementitious matrix using microscopy, by applying different kinds of visible light during the testing process. A detailed description of this procedure can be found in [17]. Cellulose fibers are usually thicker than PVA fibers. They are beige-colored, like wood, while PVA fibers are pale and transparent. After thermal treatment, both types of fibers become brown, which is evidence of their structural dehydrogenation, so that what remains in the fibers is mostly dark-colored structural carbon.

The quality of the boards was evaluated using an exact measure to determine their mechanical toughness, understood here as the integrated product of applied stress and strain per unit of cross section of the investigated board, i.e., the work of flexural test W_f. The latter parameter can be determined as the work made over the deflection curve during the bending test [12]. In this study, the authors began testing with an initial force F_0 of 2 N and continued to break the fiber reinforcements so that the final decrease was registered at 40% of the maximum load—$F_{0.4\ MAX}$. Under these conditions, the flexural test W_f can be calculated by applying the formula:

$$W_f \frac{1}{S} \int_{F_0}^{F_{0.4\ MAX}} F\ da \qquad (1)$$

where S = specimen cross section; a = specimen deflection under the loading pin.

The loss of FCB elasticity can also be determined by applying the ultrasound testing (UT) method. In large objects with small thicknesses, like the flat boards, the following dependence combines UT longitudinal wave velocity c_L and the modulus of elasticity E:

$$c_L = \sqrt{\frac{E}{\rho(1 - v^2)}} \qquad (2)$$

where ρ = bulk material density; v = Poisson ratio.

2. Materials and Methods

In this study, five sets of specimens, made of five different types of FCB, were prepared for examination. These were labeled A, B, C, D and E. All of the specimens were fabricated by applying the Hatschek, or flow-on, forming method. Different matrix fillers and concentrations of fibers were determined for each set and resulted in different flexural strengths, which was measured by applying the standard EN 12467 three-point bending test [18]. The specimens of all FCB types, listed above, were stored in a dry, electric oven for 3 h at a temperature of 230 °C. The parameters of the heating procedure were chosen experimentally, after performing some preliminary tests to evoke considerable changes in the microstructure of the investigated materials. That treatment resulted in the decomposition (i.e., numerous broken chemical C–O bonds) of fiber reinforcements due to the process of carbonization. The carbonization process mostly concerns the linear chains of dehydrated glucose molecules, which are responsible for building up the cellulose fiber system.

The specimens, which underwent the high temperature treatment, were labeled A_T, B_T, C_T, D_T and E_T. Small pieces of each FCB type were prepared for microscopic observation, both before and after the elevated temperature treatment, and are presented in Figure 2. The mechanical properties of the specimens are shown in Table 1.

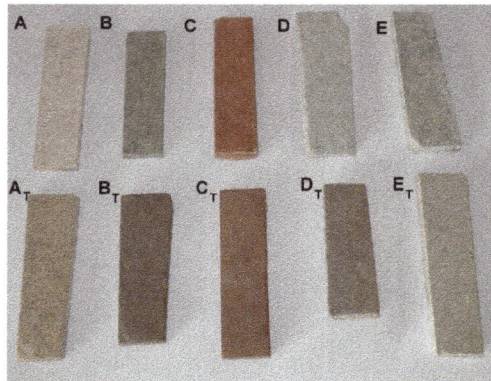

Figure 2. View of the small FCB (fiber cement board) pieces used for microscopic observation. Upper row: specimens of FCB types A, B, C, D and E before the elevated temperature treatment. Lower row: specimens of FCB types A, B, C, D and E after the elevated temperature treatment.

Table 1. Mechanical properties of the tested FCB compositions.

Board Symbol	Remarks	Board Thickness [mm]	Apparent Density [kg/m³]	Flexural Strength [MPa]
A	low cement content in the board matrix, approved for internal use	8.4	1000	14
B	low cement content in the board matrix, approved for internal use	7.4	1600	21
C	colored brown and approved for internal use	7.8	1700	23
D	approved for external use	8	1600	36
E	approved for internal use	9.3	1700	23

For detailed insight into each specimen's microstructure, the authors applied an X-ray microtomography (micro-CT) technique, which is described in more detail in [13]. A Nanotom 30, made by General Electric (Baker Hughes GE, Houston, TX, USA) was used in the investigation. The system included a micro-focal source of X-ray radiation, a movable table on which to place a specimen, and a flat FCB panel with a radiation detector, having a resolution of 2000 × 2000 pixels. The microstructure of each FCB sample was observed on the cross sections (tomograms) of the investigated specimens, using a grey scale convention that was directly related to the amount of the local radiation absorption of the materials. The grey scale covers a wide range of grey levels and is ordered from pure white, related to maximum absorption, to pure black, related to minimum absorption, respectively. Un-hydrated cement particles and aggregate grains are objects in the cement matrix that demonstrate the highest degree of absorption ability. The hydration products that make up the major components of the cement matrix present a slightly lower absorption ability. Next in line are the hydrated calcinates, which demonstrate an even lower absorption ability and, at the end of the scale, are the organic fibers (if present) and the regions of high porosity. To obtain the optimal X-ray penetration and absorption of the investigated specimens, the following parameters of the scanning procedure were set: X-ray lamp voltage—115 kV, lamp current—95 microamperes, and shot exposition time—750 ms. Scanning, performed by the authors in this study, resulted in a large set of tomograms (specimen cross sections), performed for every 5 μm of the specimen height.

The authors prepared a dedicated instrumentation set to determine the UT longitudinal wave velocity c_L in the boards made of fibrous materials. The Wave velocity c_L was determined using the UT material tester, which was capable of measuring the time of flight T of the elastic wave front, across a board of known thickness d, with the application of the formula $c_L = d/T$. The investigation was done using the ultrasonic material tester, UTC110, produced by Eurosonic (Vitrolles, France) [19]. A report in the related literature indicates that low-frequency ultrasound (50–200 kHz) was routinely used to characterize defects (a few centimeters in size) in the concrete structures on site. However, ultrasound at low-frequency ranges cannot be used to test fiber cement boards. Some experiments [11] have revealed that the sensitivity of the ultrasound parameters required to determine the structural properties of FCB is achieved when the ultrasound wavelength becomes comparable to the dimensions of the local delaminations and the lengths of the fiber inclusions. This wavelength λ remains in the following relation to the frequency f of the emitting source and the propagation velocity of the traveling ultrasonic longitudinal waveform:

$$\lambda = \frac{c_L}{f} \tag{3}$$

Thus, taking into account the propagation velocity of 1000–2000 m/s registered in the FCB, the authors recommend the application of a frequency of 1 MHz to achieve the propagation of wavelengths in the range of 1–2 mm. The instrumentation included an Olympus Videoscan [20] transmitting and receiving transducer, which emitted an ultrasonic beam measuring 19 mm in diameter, at a frequency of 1 MHz. The parameters were designed for coupling with low-density (i.e., 1000–2000 kg/m³) materials and, thus, exhibited a low-acoustic impedance of 10 MegaRayl. The contact between the rough surface of the FCB and the face of the transducer was achieved by using a 0.6 mm thick layer of polymer jelly interfacing foil (PM-4-12) produced by Olympus (Waltham, MA, USA) [20]. The custom-designed holder, with articulated joints and a compression spring, was prepared to ensure the correct coupling of the ultrasonic transducers to both surfaces of the investigated boards. A detailed view of the holder is presented in Figure 3.

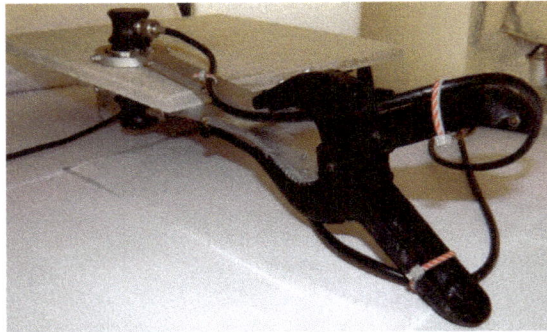

Figure 3. Detailed view of the custom-designed holder for correct coupling between the ultrasonic transducers and both sides of the rough surface board.

3. Results

3.1. Optical Light Microscopy

The morphology and surface views of all of the investigated compositions were analysed by visual light microscopy (AM4113ZTL 1.3 Megapixel Dino-Lite Digital Microscope with integral LED lighting, AnMo Electronics Corp. (Hsinchu, Taiwan). The magnified surface views of the investigated specimens are presented in Figure 4. Highly-diverse fiber distributions, due to the different fiber compositions, are noticeable in the micrographs.

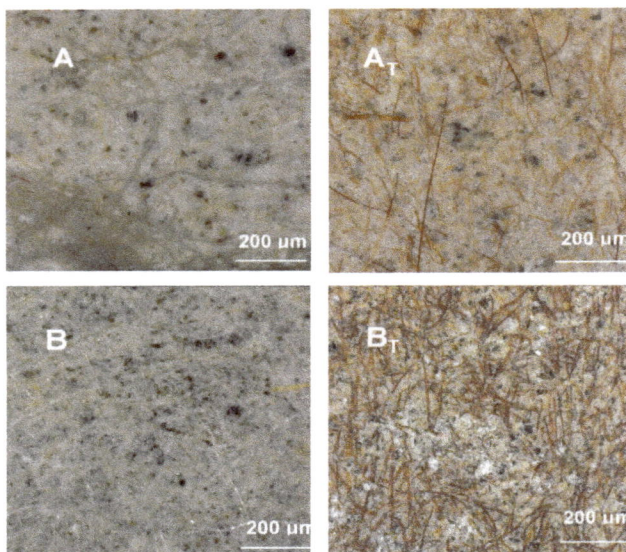

Figure 4. *Cont.*

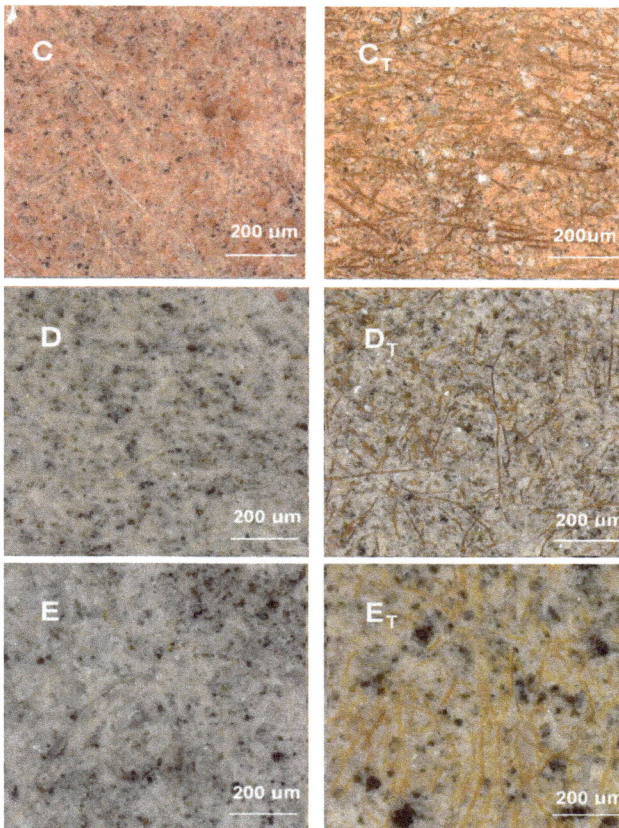

Figure 4. Micrographs of the surface views of the investigated specimens. **Left**: before the elevated temperature treatment; **Right**: after the elevated temperature treatment.

3.2. Scanning Electron Microscopy

A high-resolution environmental scanning electron microscope, Quanta 250 FEG, FEI (Hillsboro, OR, USA), was used in the investigations, along with an energy dispersive X-ray spectroscopy (EDS) analyzer. Figure 5 shows the exemplary SEM images for the tested fiber cement boards. Samples that were not exposed to a high temperature are shown on the left, while those that were exposed to a temperature of 230 °C are shown on the right. Exemplary elemental composition results for board C, which were obtained using the EDS analyzer, are shown in Figure 6.

An analysis of the images obtained from the scanning electron microscope and the EDS analyzer shows that the fiber cement boards, in the A–E series, have a compact macrostructure. Microscopic examinations revealed a fine-pore structure, with pores of up to 50 μm in size. Cavities and grooves, up to 500 μm wide, were visible in the fracture areas where the fibers had been pulled out. Cellulose fibers and, in some boards (B and D), PVA fibers, are clearly visible in the images. Various forms of hydrated calcium silicates of the C-S-H type occur. Both an "amorphous" phase and a phase built of strongly-adhering particles predominate. An analysis of the fiber composition showed that fiber elements and some cement elements were present. An analysis of the chemical composition of the matrix showed elements that are typical of cement. The surface of the fibers was covered with a thin layer of cement paste and hydration products. The fact that there are very few areas with a space between the fibers and the cement paste, indicates that the fiber-cement bond is strong.

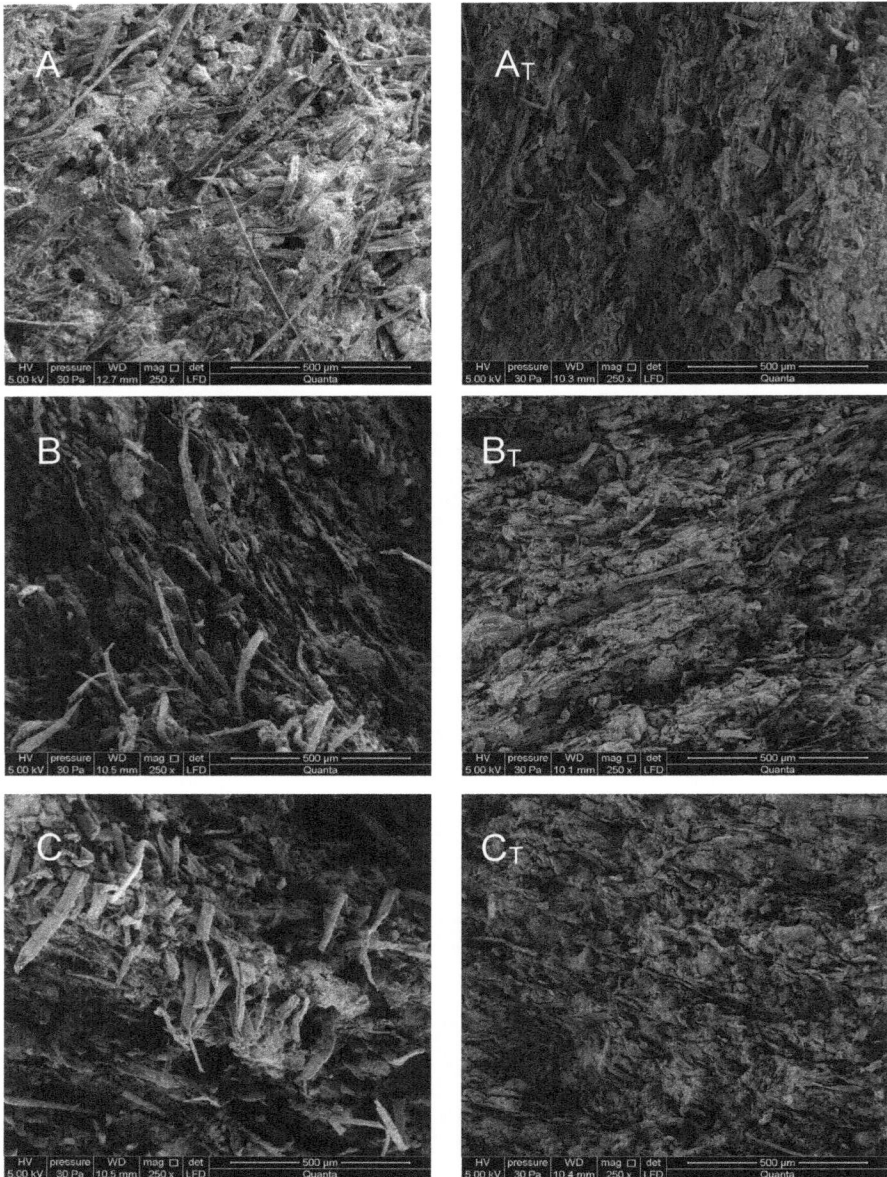

Figure 5. *Cont.*

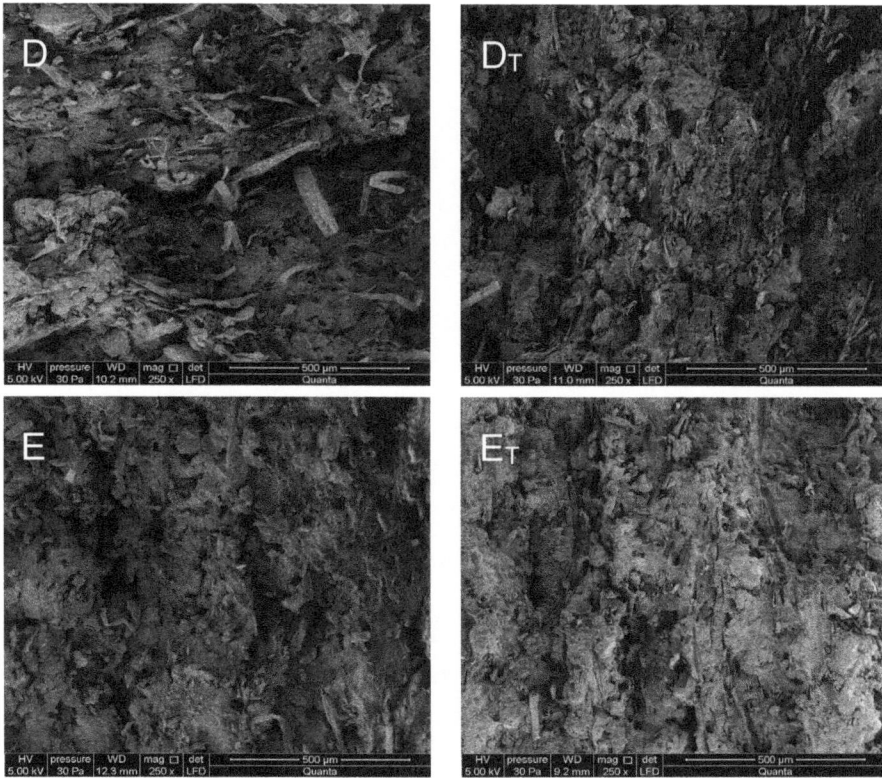

Figure 5. SEM images for boards (**A–E**) (left) and (**A$_T$–E$_T$**) (right).

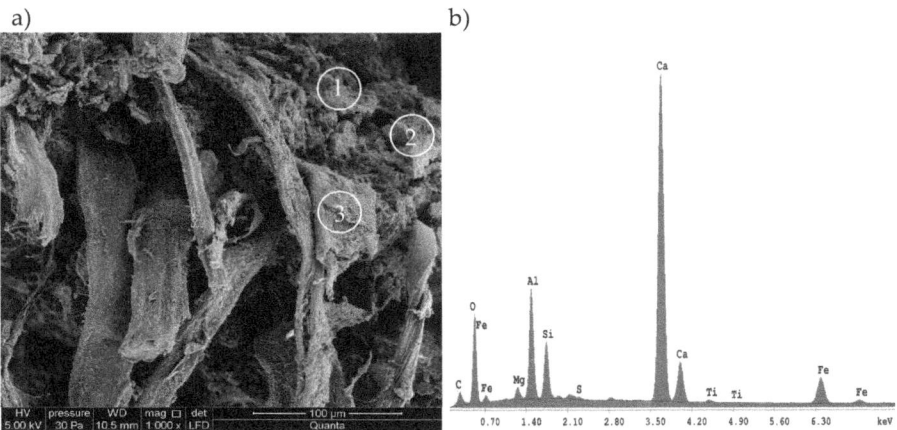

Figure 6. *Cont.*

c)   d)

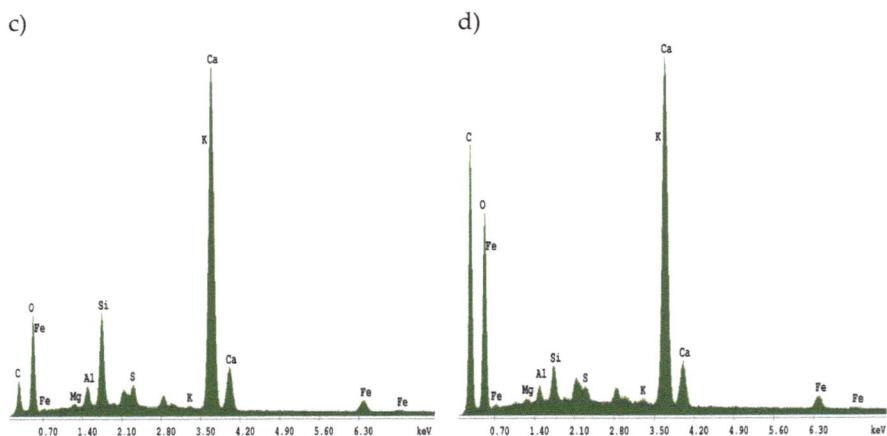

Figure 6. Results obtained using the energy dispersive X-ray spectroscopy (EDS) analyzer for board C: (a) areas of elemental composition analysis, (b) results of EDS in point 1, (c) results of EDS in point 2, (d) results of EDS in point 3.

A macroscopic analysis of the fiber cement boards in the A_T–E_T series, which were exposed to a temperature of 230 °C for 3 h, shows a clear change in the color of the samples. Examinations of the A_T–E_T fiber cement boards yielded consistent results. Most of the fibers in the boards were found to be burnt-out, or melted into the matrix, leaving cavities and grooves which were visible in all of the tested boards. The structure of the few remaining fibers was strongly degraded. Examination of the cement particles in further fractures revealed that burning-out also degrades their structure. The structure of the matrix was found to be more granular, showing many delaminations. Numerous caverns and grooves left by the pulled-out fibers, as well as the pulled-out cement particles, were observed.

3.3. X-ray Microtomography

Examples of virtually-cut, three-dimensional projections of $4 \times 4 \times 4$ mm^3 cubes, from specimens D, D_T and B_T, are presented in Figure 7. Specimen D, i.e., before the elevated temperature treatment, shows the regular microstructure without faults such as delaminations. The results of the elevated temperature impact is visible within the volume of specimen D_T, in the form of significant delaminations. The size and number of these delaminations are even more visible in the image of specimen B_T, which was made of a material with lower mechanical performance.

Another way to present information about the specimens, derived from the X-ray scanning procedure, is to determine the brightness distribution of all of the examined voxels (i.e., volumetric pixels). The magnitudes of brightness of the different microstructural elements were included in the following ranges: the area of voids and fibers: 0–50 arbitrary units, a.u., fillers: 50–140 a.u., and dense phases; i.e., un-hydrated cement and fine aggregate grains: 140–170 a.u. The greyscale brightness distribution (GBD) of all voxels belonging to the three specimens presented in Figure 7, are shown in Figure 8. It is worth noting that the occurrence of the delamination processes caused a shift of the affected GBD curves to the left, i.e., in the direction of a region of voids.

Figure 7. Examples of virtually-cut, three-dimensional projections of $4 \times 4 \times 4$ mm^3 cubes from specimens "D" (**left**), "D_T" (**center**) and "B_T" (**right**). Specimen "D" shows the regular microstructure without faults such as delaminations. The results of the elevated temperature impact is visible within the volume of specimen "D_T" and "B_T" in the form of significant delaminations.

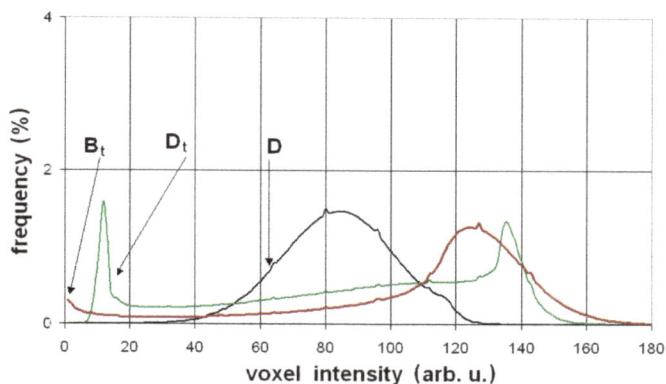

Figure 8. The greyscale brightness distribution (GBD) of all voxels belonging to the three examined specimens: "D", "D_T" and "B_T". The occurrence of the delamination processes caused a shift of the affected GBD curves to the left, i.e., in the direction of a region of voids.

3.4. Assessment of Ultrasound Wave Velocity c_L Loss and Presentation of the Results of Mechanical Tests

The results of the UT longitudinal wave velocity c_L tests are presented in Figure 9. For each 30×30 cm^2 FCB board (A, B, C, D, E, A_T, B_T, C_T, D_T and E_T), two series, of ten measurements each, were performed in order to determine the dispersion of the results. The measurements were done, randomly, at different locations, over the entire surface of the boards. The standard deviation of a single series of measurements was included in the range of 2–3% of the average value of the readings. The UT longitudinal wave velocity c_L loss was observed in all cases of thermal treatment; however, that loss varied from 2% to 20% depending on the FCB composition. It is worth mentioning that the time to perform ten UT measurements took approximately 5 min, suggesting that this may be a good method for in-situ application.

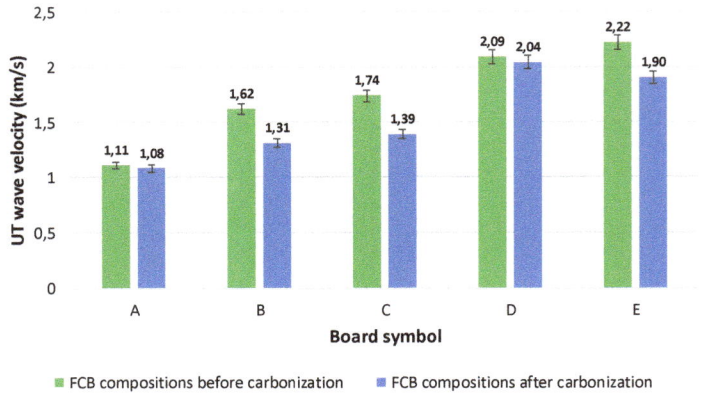

Figure 9. Results of the UT longitudinal wave velocity c_L tests, measured in all FCB compositions, before and after carbonization. The black marks depict standard deviations (2–3%), determined in populations of the measurements.

To determine the changes in the mechanical toughness of the specimens, the authors performed bending tests using three 30×30 cm^2 samples of each kind of board, following the requirements of the ISO 8336 [21]. The average results of the mechanical tests performed on the three samples are presented in Table 2.

Table 2. The results of the mechanical tests performed on the investigated FCB compositions.

Board Symbol	F_{MAX} in the State As-Delivered [N]	F_{MAX} after Thermal Treatment [N]	Work of Flexural Test W_f in the State As-Delivered [J/m^2]	Work of Flexural Test W_f after Thermal Treatment [J/m^2]
A	216	78	1330	29
B	250	82	681	32
C	475	115	848	190
D	330	384	1104	37
E	424	446	790	331

Figure 10 and Table 2 present the results of the mechanical tests. The blue curve shows the behavior of the specimens in an as-delivered state, and the red curve shows the behavior of the specimens with the pyrolyzed reinforcing fibers. Based on an analysis of the curves, the board specimens in an as-delivered state are capable of withstanding a load close to F_{MAX} for the time required to destroy the fiber reinforcement system. The energy required for that damage can be estimated as W_f, by applying Formula (1). The specimens that underwent the cellulose fiber carbonizing process demonstrated the brittle characteristics of the rupture, i.e., the break of the material cross section appeared immediately after reaching the critical F_{MAX} level. Work of flexural test, calculated for the carbonized specimens, equaled approximately 2–5% of the value estimated for the as-delivered state of the FCB specimens. It is also worth mentioning that, in the composition of boards with a low cement content, the brittle matrix broke under a relatively low loading force, while its internal fiber reinforcement system could withstand more stages of the damage process.

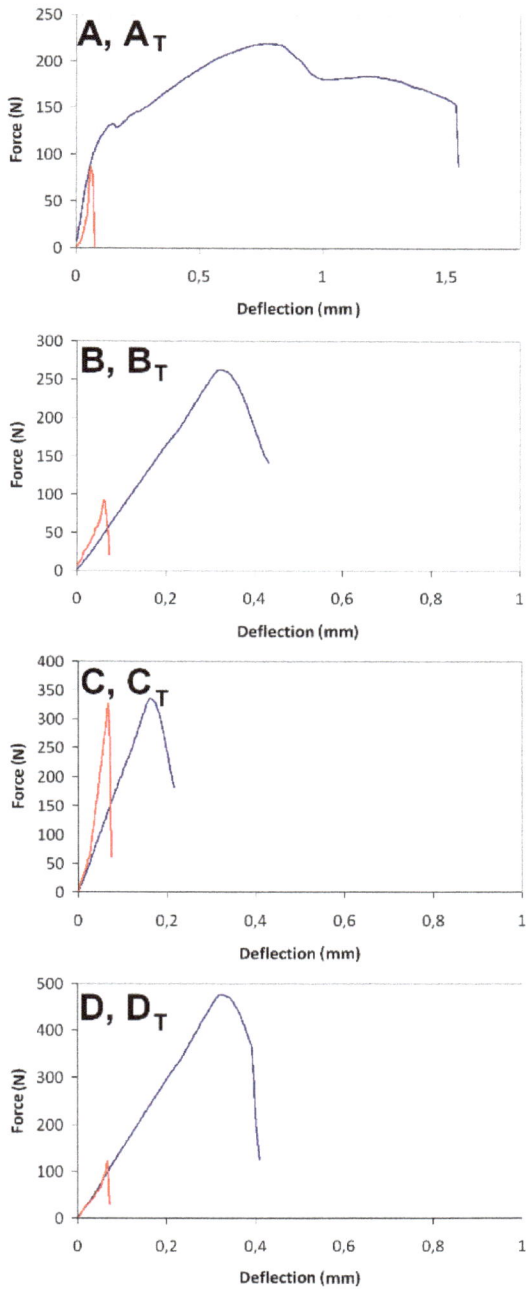

Figure 10. *Cont.*

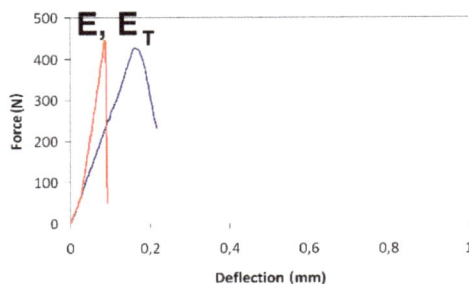

Figure 10. Typical load-deflection curves of the tested FCB compositions. The blue curves show the behavior of the specimen in an as-delivered state and the red curves show the behavior of the specimens with the carbonized reinforcing fibers.

4. Conclusions

The authors investigated five different compositions of fiber cement boards. Two of these compositions, A and B, contained a low amount of cement, which resulted in low flexural strength (14–21 MPa). The other three compositions contained more cement and their flexural strength was determined at the higher range of 23–36 MPa. The fibers applied in compositions A and D, having the best quality and proper length (approximately 3 mm), resulted in the highest value of the work of flexural test W_f, before carbonization. The carbonization process, designed in the laboratory to simulate the long exposure of FCBs to environmental hazards, significantly influenced the mechanical properties of all of the investigated compositions. The micrograph images of the carbonized specimens show the transition of the fibers from their original color into brown. The SEM examinations confirmed the marked changes in the structure that took place as a result of the exposure to a temperature of 230 °C for 3 h. In all of the tested fiber boards, most of the fibers were found to be burnt out, or melted into the matrix, leaving cavities and grooves. The structure of the few remaining fibers was highly degraded. The decrease of the W_f parameter was considerable for all of the tested compositions, as a result of the embrittlement of the fiber reinforcements. The delaminations within the microstructure of the specimens, due to the thermal treatment, was clearly visible in the three-dimensional projections obtained by applying the micro-CT technique. The delaminations also caused a shift in the affected GBD curves to the left, i.e., into the region signalling the presence of loose phases.

In the opinion of the authors, the ultrasound method has proven its applicability for testing the quality of fiber cement boards. The dedicated UT transducers, with low acoustic impedance and polymer jelly interface, were capable of achieving the required propagation of UT waves in order to determine their velocity in the investigated materials. The UT wave velocity c_L in compositions with low levels of flexural strength (A, B) was in the range of 1.1–1.6 km/s (1100–1600 m/s), whereas the wave velocity c_L in compositions with higher flexural strength (B, C, D, E) was in the range of 1.7–2.22 km/s (1700–2220 m/s). The decrease of wave velocity c_L after carbonization occurred in all tested compositions; however, its magnitude was diverse and was included in the range of 2–20%, in relative units. The lowest decrease of c_L occurred in the board made with the best quality components, i.e., the board intended for external use. All of these characteristics lead the authors to recommend the UT method as a useful tool for the on-site assessment of the degradation processes occurring in fiber cement boards.

Author Contributions: Z.R. analyzed the UT and micro-CT test results and performed paper editing; P.R. performed the microscopic observations; T.D. prepared the specimens; T.G. completed the experiments; K.S. conceived and designed the experimental work.

Funding: This research received no external funding.

Conflicts of Interest: The authors declare no conflict of interest.

References

1. Schabowicz, K.; Gorzelańczyk, T. Applications of fibre cement boards. In *The Fabrication, Testing and Application of Fibre Cement Boards*, 1st ed.; Ranachowski, Z., Schabowicz, K., Eds.; Cambridge Scholars Publishing: Newcastle upon Tyne, UK, 2018; pp. 107–121. ISBN 978-1-5276-6.
2. Mohr, B.J.; Nanko, H.; Kurtis, K.E. Durability of kraft pulp fibre-cement composites to wet/dry cycling. *Cement Concrete Compos.* **2005**, *27*, 435–448. [CrossRef]
3. Coutts, R.S.P. A Review of Australian Research into Natural Fibre Cement Composites. *Cement Concrete Compos.* **2005**, *27*, 518–526. [CrossRef]
4. Schabowicz, K.; Gorzelańczyk, T. Fabrication of fibre cement boards. In *The Fabrication, Testing and Application of Fibre Cement Boards*, 1st ed.; Ranachowski, Z., Schabowicz, K., Eds.; Cambridge Scholars Publishing: Newcastle upon Tyne, UK, 2018; pp. 7–39. ISBN 978-1-5276-6.
5. Bledzki, A.K.; Gassan, J. Composites reinforced with cellulose based fibres. *Prog. Polym. Sci.* **1999**, *24*, 221–274. [CrossRef]
6. Cooke, T. Formation of Films on Hatcheck Machines. Bonded Wood and Fibre Composites. Available online: www.fibrecementconsulting.com/publications/publications.htm (accessed on 16 November 2018).
7. Drelich, R.; Gorzelańczyk, T.; Pakuła, M.; Schabowicz, K. Automated control of cellulose fibre cement boards with a non-contact ultrasound scanner. *Autom. Constr.* **2015**, *57*, 55–63. [CrossRef]
8. Kaczmarek, M.; Piwakowski, B.; Drelich, R. Noncontact Ultrasonic Nondestructive Techniques: State of the Art and Their Use in Civil Engineering. *J. Infrastruc. Syst.* **2017**, *23*, 45–56. [CrossRef]
9. Mori, K.; Spagnoli, A.; Murakami, Y.; Kondo, G.; Torigoe, I. A new non-destructive testing method for defect detection in concrete. *NDT&E Int.* **2002**, *35*, 309–406.
10. McCann, D.M.; Forde, M.C. Review of NDT methods in the assessment of concrete and masonry structures. *NDT&E Int.* **2001**, *34*, 71–84.
11. Ranachowski, Z.; Schabowicz, K.; Gorzelańczyk, T.; Lewandowski, M.; Cacko, D.; Katz, T.; Dębowski, T. Investigation of Acoustic Properties of Fibre-Cement Boards. In Proceedings of the IEEE Joint Conference—ACOUSTICS, Ustka, Poland, 11–14 September 2018; pp. 275–279. [CrossRef]
12. Ranachowski, Z.; Schabowicz, K. The contribution of fiber reinforcement system to the overall toughness of cellulose fiber concrete panels. *Constr. Build. Mater.* **2017**, *156*, 1028–1034. [CrossRef]
13. Schabowicz, K.; Jóźwiak-Niedźwiedzka, D.; Ranachowski, Z.; Kudela, S., Jr.; Dvorak, T. Microstructural characterization of cellulose fibres in reinforced cement boards. *Arch. Civil Mech. Eng.* **2018**, *18*, 1068–1078. [CrossRef]
14. Ranachowski, Z.; Schabowicz, K.; Gorzelańczyk, T.; Kudela, S., Jr.; Dvorak, T. Visualization of Fibers and Voids Inside Industrial Fiber Concrete Boards. *Mater. Sci. Eng. Int. J.* **2018**, *1*, 00022. [CrossRef]
15. Ardanuy, M.; Claramunt, J.; Filho, R.D.T. Cellulosic fiber reinforced cement-based composites: A review of recent research. *Constr. Build. Mater.* **2015**, *79*, 115–128. [CrossRef]
16. Li, Z.; Zhou, X.; Shen, B. Fiber-Cement Extrudates with Perlite Subjected to High Temperatures. *J. Mater. Civil Eng.* **2004**, *16*, 221–229. [CrossRef]
17. Jarząbek, D. Application of advanced optical microscopy. In *The Fabrication, Testing and Application of Fibre Cement Boards*, 1st ed.; Ranachowski, Z., Schabowicz, K., Eds.; Cambridge Scholars Publishing: Newcastle upon Tyne, UK, 2018; pp. 89–106. ISBN 978-1-5276-6.
18. *BS EN 12467:2012 Fibre-Cement Flat Sheets—Product Specification and Test Methods*; British Standards Institution: London, UK, 31 July 2016.
19. MISTRAS Products & Systems. Available online: http://www.eurosonic.com/en/products/ut-solutions/utc-110.html (accessed on 19 November 2018).
20. Olympus Ultrasonic Transducers. Available online: http://www.olympus-ims.com/en/ultrasonic-transducers/contact-transducers/#! (accessed on 19 November 2018).
21. *ISO 8336:2009. Fibre-Cement Flat Sheets—Product Specification and Test Methods*; ISO Standards: Geneva, Switzerland, May 2009; Available online: https://www.iso.org/standard/45791.html (accessed on 20 March 2019).

materials

MDPI

Article

Machine Learning Techniques in Concrete Mix Design

Patryk Ziolkowski *[iD] and Maciej Niedostatkiewicz

Faculty of Civil and Environmental Engineering, Gdansk University of Technology, Gabriela Narutowicza 11/12, 80-233 Gdansk, Poland; mniedost@pg.edu.pl
* Correspondence: patziolk@pg.edu.pl; Tel.: +48-58-347-2385

Received: 3 March 2019; Accepted: 12 April 2019; Published: 17 April 2019

check for
updates

Abstract: Concrete mix design is a complex and multistage process in which we try to find the best composition of ingredients to create good performing concrete. In contemporary literature, as well as in state-of-the-art corporate practice, there are some methods of concrete mix design, from which the most popular are methods derived from The Three Equation Method. One of the most important features of concrete is compressive strength, which determines the concrete class. Predictable compressive strength of concrete is essential for concrete structure utilisation and is the main feature of its safety and durability. Recently, machine learning is gaining significant attention and future predictions for this technology are even more promising. Data mining on large sets of data attracts attention since machine learning algorithms have achieved a level in which they can recognise patterns which are difficult to recognise by human cognitive skills. In our paper, we would like to utilise state-of-the-art achievements in machine learning techniques for concrete mix design. In our research, we prepared an extensive database of concrete recipes with the according destructive laboratory tests, which we used to feed the selected optimal architecture of an artificial neural network. We have translated the architecture of the artificial neural network into a mathematical equation that can be used in practical applications.

Keywords: concrete; concrete mix design; concrete strength prediction; data mining; machine learning

1. Introduction

Concrete mix design is an essential and abstruse topic, which requires extensive knowledge of many expert issues. Obtaining concrete with appropriate strength, and other utility parameters, allows for the reliable use of the structure. The process of concrete hardening and hydration are irreversible. Therefore, any errors in the design of the concrete mix are incredibly costly for the investor, both at the construction stage and in the subsequent exploitation of the structure due to reduced durability. By definition, concrete mix is a mixture of cement, water, and coarse and fine aggregate, mostly enriched by additives and admixtures to improve some parameters, such as concrete strength, density, durability, or workability. The final product is in which the concrete mix is transformed into concrete. The concrete hardening is started by the cement hydration process, which is an exothermic chemical reaction between cement and water. Hydrated cement forms a tobermorite gel, hydroxide, and some secondary compounds that help with bonding between the fine and coarse aggregate. In the course of the hydration process, the hydration products gradually deposit on the original cement grains and fill the space occupied by water. The hydration process stops when there is no unreacted cement or the water molecules are retracted. The hardening of concrete continues further and ends around the twenty-eighth day, when the concrete reaches full compressive strength [1–3]. The necessary amount of water for full hydration of cement varies from 20% to 25% of its mass, without taking into account the water trapped in the pores [4,5]. According to Power's model, the water required to hydrate cement entirely is 42% by weight [6,7]. The issue of concrete mix design boils down to selecting the correct proportions of cement, fine and coarse aggregate, and water to produce concrete that has the specified properties. Progress in the design of concrete mixes is at a moderate level. The most popular method used to estimate the amount of main ingredients needed, in a little changed form, has been used for decades and consists of estimating the strength of concrete mortar for bending [8–10]. These methods have many disadvantages and are labour-intensive to use. We want to introduce a way to design a concrete mix based on a mathematical equation developed by the machine learning algorithm. In the following paper, we describe the adopted neural network architecture, which further will be feed by a large dataset of concrete mix recipes. Finally, we present a mathematical formula that allows estimating the compressive strength of concrete. The developed formula will evaluate the compressive strength of concrete based on four input variables, the amount of cement, water, fine, and coarse aggregate. The presented equation does not reflect the behaviour of the concrete perfectly and has boundary conditions. However, it is a step on the way to the introduction of machine learning techniques for concrete mix design. In its present form, it can be a tool for a rough estimation of the concrete class. In our further endeavours we would like to emphasise concrete mix design in terms of durability and service life estimation, then the influence of concrete admixtures, such as superplasticizers, would be essential [11–16].

2. The Contemporary Concrete Mix Design and Machine Learning Techniques

2.1. Concrete Mix Design in European Corporate Practice

The primary goal of concrete mix design is to estimate the proper quantitative composition and proportion of concrete mixture components. We should use a composition which allows us to achieve the best possible concrete performance. Concrete performance is characterised by several features, from which the most significant are compressive strength and durability. Both concrete strength and durability should play an essential role in the concrete mix design. The issue of durability is essential in the case of an aggressive environment [17–22]. Based on industry experience, we found out that, in European corporate engineering practice, there are a few most used methods for designing a concrete mix. These methods are the Bukowski method, the Eyman and Klaus method, and the Paszkowski method. All the solutions mentioned above are derived from the so-called "Three Equations Method", or Bolomey Method, which is a mixed experimental-analytical procedure [10,23,24]. It means that collected laboratory data should confirm that mathematical approach. We calculate a volume of required components by analytical measures and validate the results by destructive laboratory testing. In this method, we use a fundamental equation of strength, consistency, and tightness to determine the three searched values, as follows, the amount of aggregate, cement, and water, expressed in kilograms per cubic meter.

The first equation is the compressive strength equation or Bolomey formula (Equation (1)), which expresses the experimentally determined dependence of the compressive strength of hardened concrete on the grade of cement used, the type of aggregate used, and the water-cement ratio characterising the cement paste [23,24]. In this method, the concrete grade is assumed as input data.

$$f_{cm} = A_{1,2}\left(\frac{C}{W} \pm 0,5\right) \text{[MPa]},\tag{1}$$

where, f_{cm} is a medium compressive strength of concrete, expressed in kilograms. The value $A_{1,2}$ means coefficients, depending on the grade of cement and the type of aggregate; C is an amount of cement in 1 m^3 of concrete, expressed in kilograms; and W corresponds to the amount of water in 1 m^3 of concrete, expressed in kilograms. A second consistency Equation (2), is included in the water demand formula necessary to make a concrete mix with the required consistency.

$$W = C{\cdot}w_c + K{\cdot}w_k \text{ [dm}^3\text{]},\tag{2}$$

where W is the amount of water in 1 m^3 of concrete, expressed in kilograms; C corresponds to the amount of cement in 1 m^3 of concrete, expressed in kilograms; K means the amount of aggregate in 1 m^3 of concrete, expressed in kilograms; w_c is the cement water demand index in dm^3 per kilogram; and w_k is the aggregate water demand index in dm^3 per kilogram. The water-tightness of concrete Equation (3) is included in the simple volume formula, which indicates that a watertight concrete mix is obtained if the sum of the volume of the individual components is equal to the volume of the concrete mix.

$$\frac{C}{\rho_c} + \frac{K}{\rho_k} + W = 1000 \ [dm^3], \tag{3}$$

where W is the amount of water in 1 m^3 of concrete, expressed in kilograms, C corresponds to the amount of cement in 1 m^3 of concrete, expressed in kilograms, K means the amount of aggregate in 1 m^3 of concrete, expressed in kilograms; ρ_c is the cement density in kilograms per dm^3; and ρ_k is the aggregate density in kilograms per dm^3.

The system of equations presented above, with three unknowns variables, allows for calculating the sought amounts of cement (C), aggregate (K), and water (W) in one cubic meter of concrete mix. The system is valid, assuming that there are no air bubbles in the concrete. Another method used in the construction industry is "the double coating method" [25]. The methods above are ones that are used to determine the quantitative composition of the concrete mix. However, the actual process of creating a concrete mix is much broader, including the following steps: The first step is to determine the data needed to design the mix, such as the purpose of the concrete use, the compressive strength of the concrete, and the consistency of the concrete mix. Next, the qualitative characteristics of the components should be determined, namely the type and class of cement and the type and granularity of the aggregates. Subsequent steps include an examination of the properties of the adopted ingredients; a check of their compliance with the standard requirements; determining the characteristics of the components that will be needed to determine the composition of the concrete mix; and a projection of the aggregate pile. The successive step is the actual adoption of the design method and a calculation per unit of volume. The final stage is to make a trial sample and examine both the concrete mix and the hardened concrete with design assumptions [26].

2.2. Machine Learning Techniques

2.2.1. The Overall Concept of Machine Learning

Machine learning is an area of knowledge which is developing dynamically in recent times. This technology is a part science dealing with artificial intelligence and refers to scientific fields such as computer science, robotics, and statistics [27]. In practice, machine learning aims to use various state-of-the-art achievements in computer science to build upon a system that will be able to learn from data sets and, thus, seek patterns and relationships between variables and groups of variables, which would be challenging to conduct with conventional methods. Learning, in this case, can be considered as the instantiation of the sophisticated algorithm. One of the most popular methods of machine learning is artificial neural networks (ANN).

2.2.2. Artificial Neural Networks (ANN)

ANN are clusters of neurons, which are also its basic unit. We can consider an artificial neuron as a specific signal converter. The behaviour of artificial neurons, in a sense, imitates the behaviour of neurons in the human brain [28]. A primary example of ANN consist of three layers, called as follows:

- The input layer;
- The hidden layer;
- The output layer.

The input layer consists of input variables and combines them with neurons from the hidden layer. On the contrary, the output layer contains the target data to be obtained by the hidden layer [27]. Therefore, the whole process of learning happens in the hidden layer, where connections between neurons are sought. Vast numbers of neurons can build a complex model, which would be unattainable with simple architecture and so unobvious that it would be difficult to create a purely empirical formula. An essential thing that neural networks do is a search for patterns, which is why examples best teach neural networks. To teach a neural network how to solve a given problem, one must enter the input data into it using the first layer and put data in the output layer as a given target to which the network is to strive. Moreover, the input data can be adjusted by assigning weight to them, which can potentially represent the importance of a given variable. The weight control mechanism is also part of the neural network and is called the "learning rule". One artificial neuron has miserable problem-solving capabilities. Many neurons can be combined into more hidden layers, where layers pass the results to one another, looking to reach the target value [29,30].

2.3. Use of Machine Learning in Concrete Compressive Strength Prediction

Designing a concrete mix consists of selecting components and their amount to achieve specific parameters of the concrete. One of the most significant parameters for concrete performance is the compressive strength of concrete, which defines the class of concrete. Other important parameters that contribute to good concrete performance are durability and even the manufacturing process itself. Poor durability may contribute to lowering the service quality of building in time. With a wrong manufacturing process, for example, poor concrete care can cause excessive cracks and reduce concrete tightness [31].

The issue of machine learning applications, more precisely ANN, to predict the strength of concrete is present in the scientific discourse and is continuously evolving, making this topic very progressive.

The topic was first discussed in 1998 by Yeh et al. [32], which used linear regression and ANN to try to predict the strength of high-performance concrete using seven input variables. In the research, Yeh et al. used an extensive database, but in our opinion, they did not take into account the specificity of concrete and used samples in their database that were still in the maturing phase, even three days old, which, in our opinion, could seriously misrepresent the results.

Subsequently, the topic was taken up by Seung-Chang Lee [33], which used a modular network structure consisting of five ANN. In the presented solution, the author used the weighting technique of input neurons to improve the accuracy of predictions. To estimate the number of input neurons, he used the parameter condensation technique. The author concludes that the methods he uses, namely condensation and weighting techniques, are efficient in looking for the optimal performance network.

Another interesting approach in this matter is to use a neural-expert system, which was suggested further by Gupta et al. [34] to predict the compressive strength of high-performance concrete. The neural expert system architecture, in theory, allows for constructing the database automatically by learning from example inferences. In general, this architecture assumes the use of a multi-layered neural network, which is consequently trained with generalised backpropagation for interval training patterns. However, this may allow for the learning of patterns with irrelevant inputs and outputs. What is more, in the study by Gupta et al. [34], the input variables have very different input metrics and instead of the amounts of concrete mix components, the input variables refer to such parameters as curing time. In our opinion, the selected input parameters have no unambiguous effect on the strength of concrete and can imply false results. The topic of neural-expert systems was also undertaken by Dac-Khuong Bui et al. [35], which focused entirely on the practical application of the mentioned expert approach.

Fangming Deng et al. [36] practised deep learning architecture to predict the compressive strength of concrete. In this study their used recycled concrete with five input variables as follows, water-cement ratio, recycled coarse aggregate replacement ratio, recycled fine aggregate replacement ratio, and fly ash replacement. They used so-called deep features that refer to ratio rather than the individual amount of concrete mix components. We used a similar approach in our study by introducing feature scaling. To find out the proper prediction model they used a Softmax regression. In the results section of their paper, they state that the deep learning architecture they applied gives a higher efficiency, generalisation ability, and precision, in comparison with standard ANN. However, they do not present sufficient proofs to support their statements. Convolution networks are computationally expensive. This seems to be confirmed by a significantly lower number of samples (74 exactly) than in our study (741 records). However, such a small dataset might result in underfitting, which means that the model does not fit the data well enough to such an extent that it reduces the efficiency of the model. Moreover, Hosein Naderpour et al. [37] shows a comparable degree of precision between ANN and Deep Neural Networks (DNN).

3. Materials and Methods

3.1. Essentials

In our study, we want to implement machine learning for concrete mix design. Based on a large number of tested concrete mix recipes, we would like to build an ANN which will be able to estimate the compressive strength of the concrete mix. The ANN estimates the strength of the concrete based on the amount of the four main components of a concrete mix, more precisely cement, fine and coarse aggregate, and water. We translated the constructed ANN into the source code and simplified to one equation, defining the twenty-eight-day strength of concrete as a function of the four parameters. The equation can be used for concrete compressive strength estimation and can serve as a tool for a concrete mix recipe check. The practical application of this method in the concrete mix design process, required to adopt the approach, is presented in Figure 1.

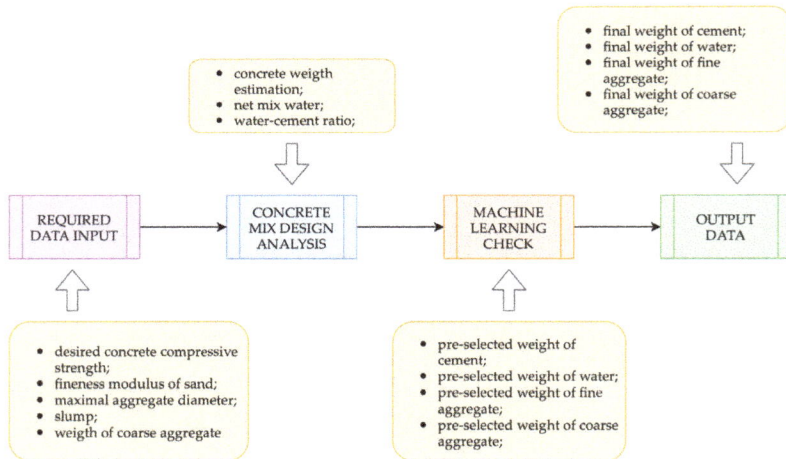

Figure 1. Block diagram of the practical application of machine learning in the concrete mix design.

It seems reasonable to set a boundary condition for this method. However, the ANN was trained on a limited number of samples so it may be difficult to predict how it will behave for amounts of material higher than in the considered ranges. It is essential to strictly control the water-cement ratio since the proper proportion is necessary for the full hydration of the cement. We have not analysed the influence of plasticisers.

3.2. The Database of Concrete Mix Recipes

In our research, we intend to teach the neural network the relationships between the number of individual components in a concrete mix and the compressive strength of concrete with a large number of examples. Thanks to this, the potential user of our solution will be able to design the right composition of ingredients and try to predict the compressive strength of concrete. To handle that task, we need a wide-ranging database containing a variety of concrete recipes with according data of their destructive laboratory tests. We prepared the database, which has many records from numerous sources, including literature, companies, institutions, and laboratories. The concrete mix recipes that we used for the analysis were designed for concrete structures of different dimensions, functions, and destinations. Therefore, there may be some differences between them, the sources of which we will not be able to predict. What is more, many of the recipes we have, besides the essential ingredients, have additives that have different functions. The most popular concrete additives are binding retardants, plasticisers, and workability boosters. The samples tested are standardised concrete cylinders with a diameter of 15 cm. Samples that were not cylindrical were converted into cylindrical ones according to valid norms [38]. The size of the aggregate in the dataset did not exceed 20 mm. The samples were made from normal Portland cement. We have carried out extensive consultations with experts and have adopted four components that have a significant impact on the compressive strength of concrete. The adopted input parameters are presented in Table 1.

Table 1. The parameters adopted in the dataset.

Parameter	Compressive Strength after 28 days	Cement	Water	Sand 0–2 mm	Aggregate above 2 mm
Codename	cs_28	cement	water	fine_aggregate	coarse_aggregate
Type	target	input	input	input	input
Description	The compressive strength of concrete 28 days after hydration. Considered as full strength.	The weight of cement added to the mixture	The weight of water added to the mixture	The weight of sand added to the mixture	The weight of aggregate, which have more than 2 mm, added to the mixture

We divided the parameters from Table 1 into two groups, the inputs and target, which characterise input and output variables, respectively. After initiating the cement hydration process, concrete strength grows, progressively over time, to full strength. In our deliberations, we adopted a general assumption that concrete achieves its designed compressive strength in twenty-eight days. Prior to the twenty-eighth day, the concrete has a partial strength, but it cannot be considered full strength. We assumed in our research that the concrete reached its full strength because a mixture is designed for such strength. We removed all records for concrete of lower ages from the base. Many factors have an indirect effect on the obtained concrete strength, which has not been included in the analysis, such as the curing process. We assumed that quality control was sufficient to produce full strength concrete. The minimum, maximum, and average values for every input variable are presented in Table 2.

Table 2. Ranges of input features of database input features.

Input Features	Minimum (kg/m^3)	Maximum (kg/m^3)	Average (kg/m^3)
Cement	86.00	540.00	278.00
Water	121.80	247.00	182.42
Fine aggregate (sand 0–2 mm)	372.00	1329.00	768.55
Coarse aggregate (aggregate above 2 mm)	597.00	1490.00	969.08

3.3. Results and Discussion

To carry out the simulation, we divided our set into three subsets, as follows: The training dataset, the selection dataset, and the testing dataset. The training dataset is used to create a neural network, the selection dataset is used to adjust parameters of the neural network, and the testing dataset is used to evaluate the efficiency of the network. The database has 741 records, but we had to exclude 79 records (10.7%) from the analysis as univariate outliners. The training dataset has 395 records (53.3%), the selection dataset has 133 records (17.9%), and the testing dataset has 134 records (18.1%). The scatter plots of a target variable versus the input variables are presented in Figure 2.

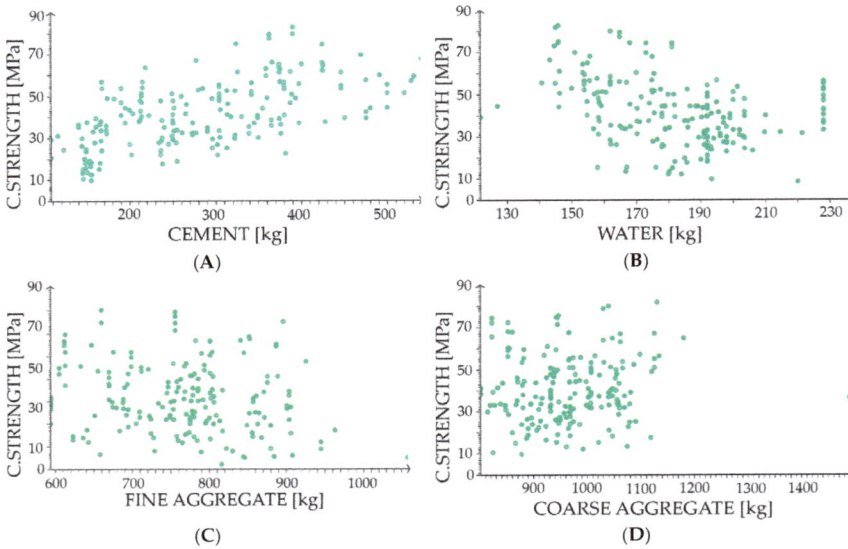

Figure 2. The scatter plots target versus input variables. The vertical axis is the full compressive strength of concrete expressed in megapascals. The horizontal axis is the amount of material in kilograms for cement and aggregate as well as in litres for water. (**A**) Cement; (**B**) water; (**C**) Fine aggregate (sand 0–2 mm); and (**D**) coarse aggregate (aggregate above 2 mm). The legend is in the left bottom corner.

Our neural network consists of four input variables, which refers to four principal components and generates one target output. The complexity of the model is expressed by the number of hidden layers, which in our case is three. The initial architecture that we prepared is shown in Figure 3, which consists of principal components (blue), perceptron neurons (red), and, because we used feature scaling, there are scaling and unscaling layers. The scaling and unscaling neurons are green and yellow, respectively.

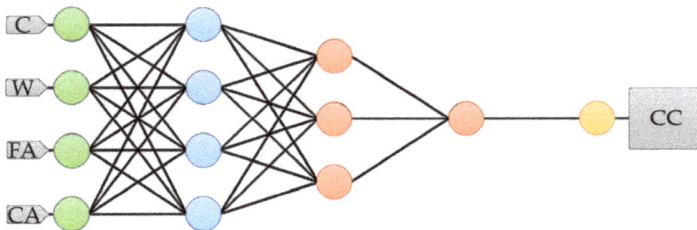

Figure 3. The initially used architecture of the ANN. The figure shows the network architecture, which includes the following parts, principal components (blue), perceptron neurons (red), and, because we use feature scaling, there are scaling and unscaling layers. The scaling and unscaling neurons are green and yellow, respectively. Abbreviations: C, cement; W, water; FA, fine aggregate; CA, coarse aggregate; CC, full compressive strength of concrete.

We want to point out that some input variables (cement, water, fine_aggregate, coarse_aggregate) correspond with some input neurons and target variable (cs_28) is associated with the output neuron. To obtain a proper training rate, we used the Broyden-Fletcher-Goldfarb-Shanno algorithm [39–44]. Then, to designate the quasi-Newton training direction step, we utilised the Brent method [45–48]. For the analysis, we calculated the linear correlation and determined a correlation matrix.

We have assessed the impact of individual variables on the final result, which is presented in Figure 4. We eliminated training input selectively and inspected the output results. An input contribution value 1.0 or lower than one denotes that the variable has less contribution to the results. Successively, a value higher than 1.0 means a more significant contribution. Our analysis indicates that the biggest contribution to the results have cement, which is in line with our assumptions that the water-cement ratio has the most significant impact on concrete strength. Literature findings also confirm that the cement content and type have a high influence on the compressive strength [49]. There are also other issues, including curing conditions and added admixture impact, that influence the compressive strength and concrete durability, especially an environments with a high risk of carbonation [50]. The detailed nature and the shape of the aggregate influence the workability and durability of concrete. The shape and texture of the aggregate affect the properties of fresh concrete more than hardened concrete [51]. Additionally, the grading or size distribution of aggregate is an important characteristic because it determines the paste requirement for workable concrete [52]. However, in our procedure, we did not make an exact distinction between the nature and shape of the aggregate. We only diversified the coarse and fine aggregates and sacrificed it for the sake of having larger data sample pools in these two categories. We also have not analysed the impact of environmental aggression and admixtures.

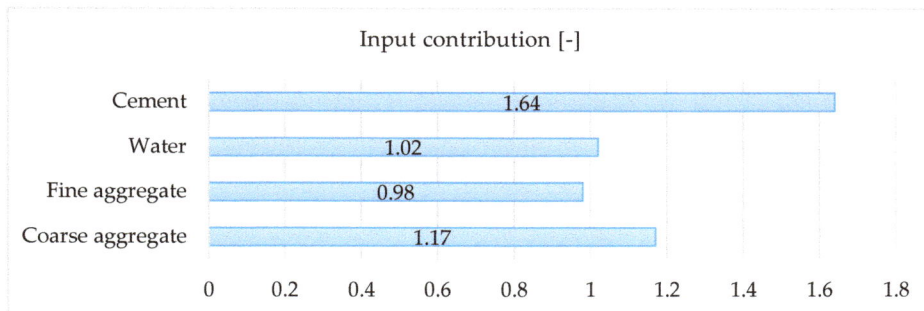

Figure 4. Input contribution.

We performed input selection by the growing inputs algorithm [53–56]. We found the optimal number of neurons by the order selection algorithm [57,58]. We carried out the output selection by the incremental order algorithm [59–61]. The loss history for the subsets used is presented in Figure 5.

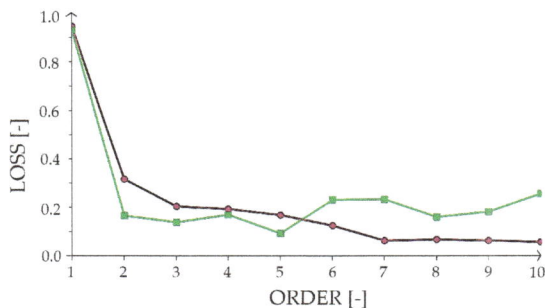

Figure 5. Incremental order algorithm performance. The chart presents a loss history, where the purple line is the training loss and the green one is the selection loss. The vertical axis is a loss and the horizontal axis is an order.

In Figure 6 we present a final architecture of the ANN, which consists of principal components (blue), perceptron neurons (red), and, because we used feature scaling there are scaling and unscaling layers. The scaling and unscaling neurons are green and yellow, respectively. We used a deep architecture with features scaling. Therefore it contains scaling and unscaling layers. Our final model, which is the most optimal for performing the given task, has four inputs, one output, and three hidden layers.

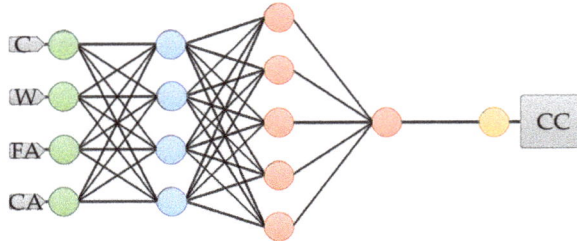

Figure 6. Finally used the architecture of ANN. The figure shows network architecture that includes following parts, principal components (blue), perceptron neurons (red), and, because we use feature scaling, there are scaling and unscaling layer. The scaling and unscaling neurons are green and yellow, respectively. Abbreviations: [C]—cement; [W]—water; [FA]—fine aggregate; [CA]—coarse aggregate; [CC]—full compressive strength of concrete.

In our study, we created an ANN which can be used for concrete mix design. The network targets the compressive strength of concrete with the four following input variables, cement, water, fine, and coarse aggregate. We can express our ANN by a mathematical Equation (4). The Equation (4) refers to the 28 day strength of concrete, which, as we mentioned, can be considered as full strength.

$$f_c^{full\ cs} = 0.5 \cdot \left(f_{ax}^{full\ cs} + 1.0 \right) \cdot (62.62184 + 240.13674) - 240.13674\ [\text{MPa}], \tag{4}$$

Auxiliary mathematical formulas are presented in Equations (5)–(14). The variables C, W, FA, CA used in Equations (15)–(18) mean cement, water, fine aggregate, and coarse aggregate, respectively. Units are expressed in kilograms per cubic meter.

$$f_{ax}^{full\ cs} = tanh(0.246033 - 0.959961 \cdot ax_{11} + 0.816467 \cdot ax_{12} + 0.526611 \cdot ax_{13} + 0.73407 \cdot ax_{14} - 0.270081 \cdot ax_{15}) \tag{5}$$

$$ax_{11} = tanh(0.15979 - 0.00384521 \cdot ax_{21} + 0.837402 \cdot ax_{22} - 0.148804 \cdot ax_{23} - 0.569336 \cdot ax_{24}) \tag{6}$$

$$ax_{12} = tanh(0.1297 + 0.820862 \cdot ax_{21} + 0.0808105 \cdot ax_{22} + 0.206116 \cdot ax_{23} - 0.601257 \cdot ax_{24}) \tag{7}$$

$$ax_{13} = tanh(0.262573 - 0.0964355 \cdot ax_{21} + 0.610413 \cdot ax_{22} - 0.380981 \cdot ax_{23} + 0.150024 \cdot ax_{24}) \tag{8}$$

$$ax_{14} = tanh(0.0534668 + 0.3526 \cdot ax_{21} + 0.929932 \cdot ax_{22} - 0.734924 \cdot ax_{23} - 0.415405 \cdot ax_{24}) \tag{9}$$

$$ax_{15} = tanh(-0.927551 - 0.993103 \cdot ax_{21} - 0.202698 \cdot ax_{22} - 0.719788 \cdot ax_{23} - 0.637817 \cdot ax_{24}) \tag{10}$$

$$ax_{21} = 0.984539 \cdot C_{ax} - 0.144211 \cdot W_{ax} - 0.0968967 \cdot FA_{ax} + 0.0222878 \cdot CA_{ax} \tag{11}$$

$$ax_{22} = -0.117904 \cdot C_{ax} - 0.946362 \cdot W_{ax} + 0.249228 \cdot FA_{ax} + 0.168472 \cdot CA_{ax} \tag{12}$$

$$ax_{23} = 0.0747692 \cdot C_{ax} - 0.00246874 \cdot W_{ax} + 0.576173 \cdot FA_{ax} - 0.813897 \cdot CA_{ax} \tag{13}$$

$$ax_{24} = 0.105786 \cdot C_{ax} + 0.28913 \cdot W_{ax} + 0.772348 \cdot FA_{ax} + 0.555601 \cdot CA_{ax} \tag{14}$$

$$C_{ax} = \frac{2 \cdot (C - 86)}{454} - 1 \tag{15}$$

$$W_{ax} = \frac{2 \cdot (W - 121.8)}{125.5} - 1 \tag{16}$$

$$FA_{ax} = \frac{2 \cdot (FA - 372)}{957} - 1 \tag{17}$$

$$CA_{ax} = \frac{2 \cdot (CA - 597)}{893} - 1 \tag{18}$$

We simplified the mathematical formula translated from the ANN source code and presented it in the form of Equation (19), $f_c^{full\ cs}$ with four variables C, W, FA, CA, which represent cement, water, fine aggregate, and coarse aggregate, respectively.

$$
\begin{aligned}
f_c^{full\ cs} =\ & 151.37929 \cdot tanh(0.24603 - 0.95996 \cdot tanh(-0.00077 \cdot C - 0.01524 \cdot W - \\
& 0.00066 \cdot FA - 0.00012 \cdot CA + 3.90060) + 0.816467 \cdot tanh(0.00331 \cdot C - 0.00588 \cdot \\
& W - 0.00085 \cdot FA - 0.00105 \cdot CA + 1.99855) + 0.526611 \cdot tanh(-0.00079 \cdot C - \\
& 0.00828 \cdot W + 0.00012 \cdot FA + 0.00111 \cdot CA + 0.78030) + 0.73407 \cdot tanh(0.00061 \cdot C - \\
& 0.01672 \cdot W - 0.00114 \cdot FA + 0.00119 \cdot CA + 2.67645) - 0.270081 \cdot tanh(-0.00474 \cdot \\
& C + 0.00243 \cdot W - 0.00180 \cdot FA + 0.00039 \cdot CA + 1.22858)\ ) - 88.757459\ [MPa],
\end{aligned}
\tag{19}
$$

To illustrate how the equation works we presented the charts of the output variable and the single input variable, while the other input variable is fixed. The charts are shown in Figure 7. It should be noted that, as presented in Figure 7, the output charts do not correspond to the combined correlation of the variables, but only show a trend of a given variable concerning the target variable. It also should be noted that the parameters give a different contribution to the final results, as we have shown in Figure 4.

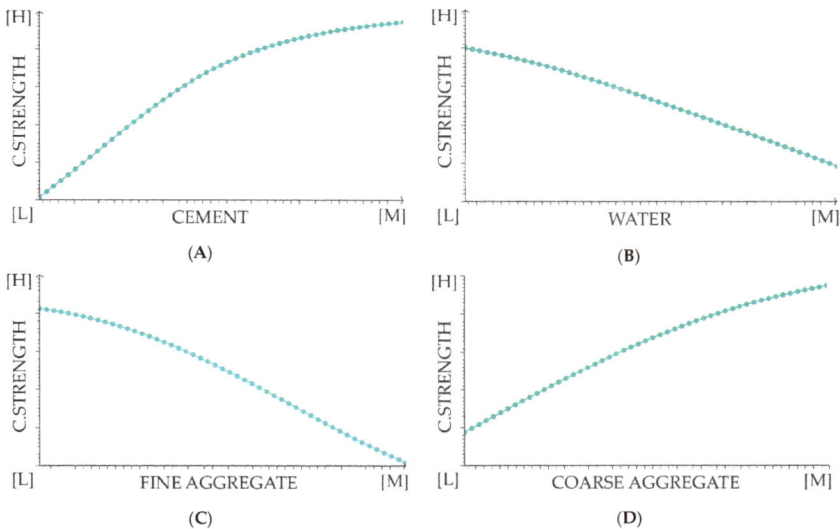

Figure 7. Output charts. The diagrams show variations in output variables for a single input variable, while the others are fixed. The vertical axis is the full compressive strength of concrete. The horizontal axis is the amount of material. (**A**) Cement; (**B**) water; (**C**) fine aggregate (sand 0–2 mm); (**D**) coarse aggregate (aggregate above 2 mm); [H], [M]—Higher/More; [L]—Lower/Less.

We compared the presented Formula (19) with a standard concrete mix design approach, based on the Bolomey design method. The comparison was prepared for 1 m³ of concrete designed for the concrete slab, with direct pouring, plastic slump, no special desired finishing, no special ambient conditions when casting, and negligible environmental aggression. To design a concrete mix, we used the following materials: Portland cement; network water; natural sand; limestone gravel 4/10 mm; and limestone gravel 10/20 mm. The tested recipes are presented in Table 3. The gradings and fitting curves for the designed recipes are shown in Figure 8. The comparison is presented in Figure 9.

Table 3. Tested concrete mix recipes.

Designed CC of Concrete Mix	Cement (kg/m^3)	Water (L/m^3)	Natural Sand (kg/m^3)	Limestone Gravel 4/10 (kg/m^3)	Limestone Gravel 10/20 (kg/m^3)
10	190.42	61.89	1089.78	531.24	676.11
20	249.52	81.09	991.52	541.32	661.89
30	308.62	94.81	901.23	554.15	651.75
40	367.71	101.25	821.33	570.82	646.97
50	426.81	107.97	745.14	583.21	641.99

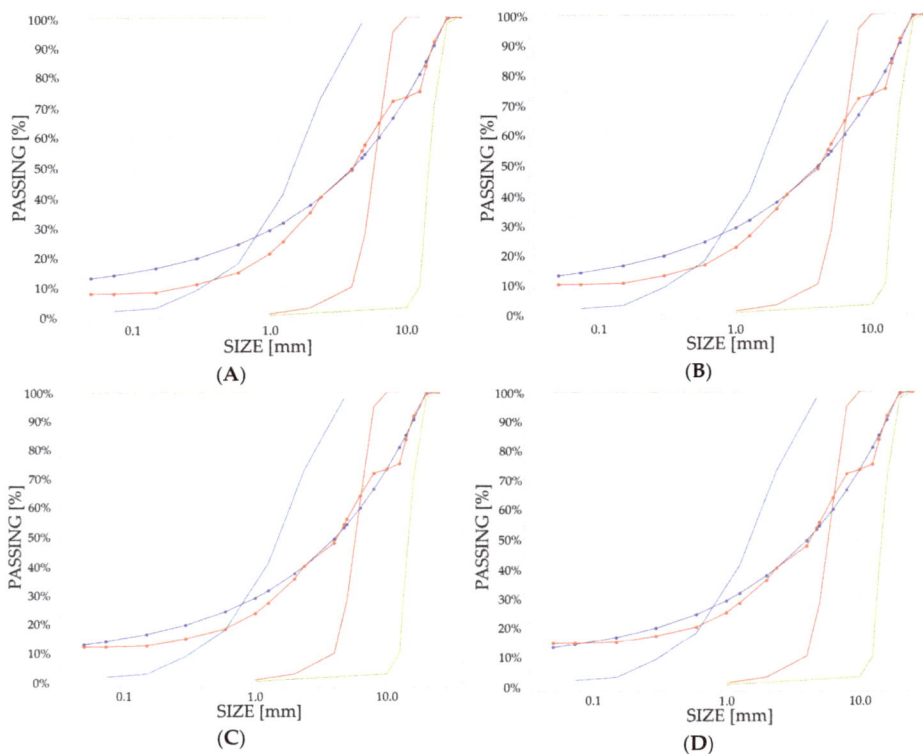

Figure 8. *Cont.*

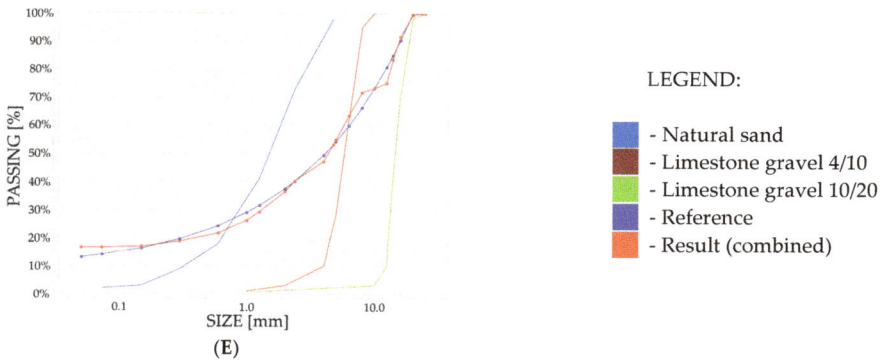

(E)

Figure 8. Gradings and fitting curves for designed concrete recipes: (**A**) 10 MPa; (**B**) 20 MPa; (**C**) 30 MPa; (**D**) 40 MPa; (**E**) 50 MPa; Legend in the left bottom corner.

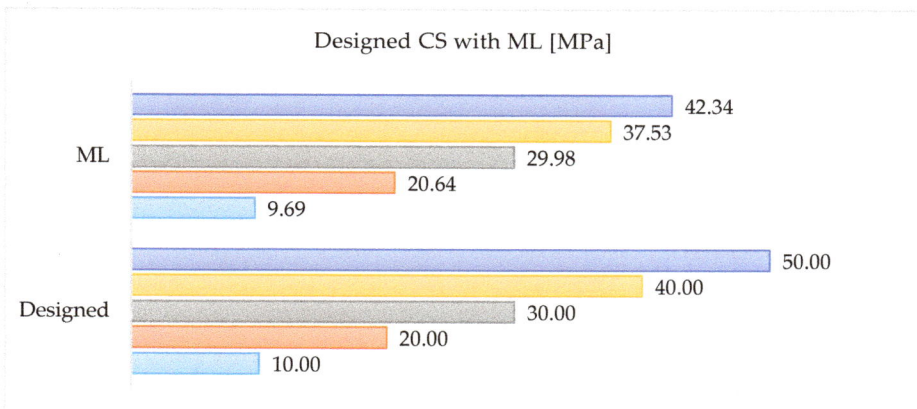

Figure 9. Comparison between designed Compressive Strength (CS) and calculated from Machine Learning (ML) formula.

We observed a low resilience of the ANN formula for recipes of high strength (50 MPa and above) concrete. It may be due to the smaller number of recipes on which we trained the ANN for these ranges. This behaviour of the ANN may be a sign of underfitting [27,28]. We must point out that the presented method is only an introduction to the broader use of machine learning in the design of concrete mixes and does not exhaust this topic. In particular, it does not take into account some crucial issues, such as durability and the technological process.

4. Conclusions

Our study is focusing on the application of machine learning in concrete mix design and building a practical tool that could be used in engineering practice. We designed the optimal ANN architecture and fed it with an extensive database of concrete mix recipes for the study. Every concrete mix recipe record has a corresponding laboratory destructive test. While building a neural network, the goal was to predict the compressive strength of the concrete resulting from a specific composition of concrete mix ingredients, or more precisely, what ratio of ingredients should be selected to obtain concrete with an adequate compressive strength. Our database has 741 records. We excluded 79 (10.7%) concrete samples from the dataset, as univariate outliners. The specificity of machine learning requires us to divide the database into three subsets, which we split up as follows: The training subset has 395 records (53.3%), the selection subset has 133 records (17.9%), and the testing subset has 134 records (18.1%). The initially adopted ANN model has four input variables, four principal components, four hidden neurons, and one target output. The suitable training rate and the step for the quasi-Newton training direction were calculated by the Broyden–Fletcher–Goldfarb–Shanno algorithm and the Brent method, respectively. Our input contribution analysis discloses that the most significant impact on the obtained results is the amount of cement that accurately points out the significance of the water-cement ratio to reach a higher concrete compressive strength. The finally adopted ANN model has four input variables, four principal components, six hidden neurons, and one target output. The pivotal point in making the machine learning techniques more applicable was a transformation of the ANN into an actual mathematical equation, which can be used in engineering practice. The initial conversion of the ANN into the mathematical formula had fifteen equations and required fourteen auxiliary variables. We simplified the expression into one general equation for the calculation of twenty-eight-day compressive strength of concrete. The equation we developed can be used as a rapid tool for concrete mix design check. The method allows checking the composition of four main concrete mix ingredients, cement, water, fine, and coarse aggregate, for achieving the desired concrete class. However, we would like to denote that the presented mathematical expression does not adequately reflect all the relationships between the components and have certain boundary conditions. We want to further develop the presented method. In the next step, to make this method more reliable, we would like to analyse the effect of admixtures and concrete durability.

Author Contributions: Conceptualisation, P.Z.; methodology, P.Z.; software, P.Z.; validation, P.Z.; formal analysis, P.Z.; investigation, P.Z.; resources, P.Z. and M.N.; data curation, P.Z.; writing—original draft preparation, P.Z.; writing—review and editing, P.Z.; visualisation, P.Z.; supervision, P.Z.; project administration, P.Z.; funding acquisition, M.N.

Funding: This research received no external funding.

Acknowledgments: The author wishes to acknowledge every institution which provided data and tools to conduct this study.

Conflicts of Interest: The authors declare no conflict of interest.

References

1. Marchon, D.; Flatt, R.J. Mechanisms of cement hydration. In *Science and Technology of Concrete Admixtures*; Elsevier: Amsterdam, The Netherlands, 2016; pp. 129–145.
2. Scrivener, K.; Snellings, R.; Lothenbach, B. *A Practical Guide to Microstructural Analysis of Cementitious Materials*; CRC Press: Boca Raton, FL, USA, 2018; ISBN 1498738672.
3. Kurdowski, W. *Cement and Concrete Chemistry*; Springer Science & Business: Berlin, Germany, 2014; ISBN 9400779453.
4. Young, J.F.; Mindess, S.; Darwin, D. *Concrete*; Prentice Hall: Upper Saddle River, NJ, USA, 2002; ISBN 0130646326.
5. Hover, K.C. The influence of water on the performance of concrete. *Constr. Build. Mater.* **2011**, *25*, 3003–3013. [CrossRef]
6. Jensen, O.M.; Hansen, P.F. Water-entrained cement-based materials: I. Principles and theoretical background. *Cem. Concr. Res.* **2001**, *31*, 647–654. [CrossRef]
7. Jensen, O.M.; Hansen, P.F. Water-entrained cement-based materials: II. Experimental observations. *Cem. Concr. Res.* **2002**, *32*, 973–978. [CrossRef]
8. Jiménez Fernández, C.G.; Barra Bizinotto, M.; Valls del Barrio, S.; Aponte Hernández, D.F.; Vázquez Ramonich, E. Durability of recycled aggregate concrete designed with the Equivalent Mortar Volume (EMV) method: Validation under the Spanish context and its adaptation to Bolomey methodology. *Mater. Construcción* **2014**, *64*, e006-1–e006-12. [CrossRef]
9. Abdelgader, H.S. How to design concrete produced by a two-stage concreting method. *Cem. Concr. Res.* **1999**, *29*, 331–337. [CrossRef]

10. Abdelgader, H.S.; Suleiman, R.E.; El-Baden, A.S.; Fahema, A.H.; Angelescu, N. Concrete Mix Proportioning using Three Equations Method (Laboratory Study). In Proceedings of the UKIERI Concrete Congress Innovations in Concrete Construction, Jalandhar, India, 5–8 March 2013.

11. Cartuxo, F.; De Brito, J.; Evangelista, L.; Jiménez, R.J.; Ledesma, F.E. Increased Durability of Concrete Made with Fine Recycled Concrete Aggregates Using Superplasticizers. *Materials* **2016**, *9*, 98. [CrossRef] [PubMed]

12. Serralheiro, M.I.; de Brito, J.; Silva, A. Methodology for service life prediction of architectural concrete facades. *Constr. Build. Mater.* **2017**, *133*, 261–274. [CrossRef]

13. Plank, J.; Sakai, E.; Miao, C.W.; Yu, C.; Hong, J.X. Chemical admixtures—Chemistry, applications and their impact on concrete microstructure and durability. *Cem. Concr. Res.* **2015**, *78*, 81–99. [CrossRef]

14. Huang, H.; Qian, C.; Zhao, F.; Qu, J.; Guo, J.; Danzinger, M. Improvement on microstructure of concrete by polycarboxylate superplasticizer (PCE) and its influence on durability of concrete. *Constr. Build. Mater.* **2016**, *110*, 293–299. [CrossRef]

15. Arredondo-Rea, S.P.; Corral-Higuera, R.; Gómez-Soberón, J.M.; Gámez-García, D.C.; Bernal-Camacho, J.M.; Rosas-Casarez, C.A.; Ungsson-Nieblas, M.J. Durability parameters of reinforced recycled aggregate concrete: Case study. *Appl. Sci.* **2019**, *9*, 617. [CrossRef]

16. Andrade, C.; Gulikers, J.; Marie-Victoire, E. *Service Life and Durability of Reinforced Concrete Structures*; Springer: Berlin, Germany, 2018; Volume 17, ISBN 3319902369.

17. Aïtcin, P.C. The durability characteristics of high performance concrete: A review. *Cem. Concr. Compos.* **2003**, *25*, 409–420. [CrossRef]

18. Ahmad, S. Reinforcement corrosion in concrete structures, its monitoring and service life prediction—A review. *Cem. Concr. Compos.* **2003**, *25*, 459–471. [CrossRef]

19. Schabowicz, K.; Gorzelańczyk, T.; Szymków, M. Identification of the Degree of Degradation of Fibre-Cement Boards Exposed to Fire by Means of the Acoustic Emission Method and Artificial Neural Networks. *Materials* **2019**, *12*, 656. [CrossRef]

20. Godycki-Ćwirko, T.; Nagrodzka-Godycka, K.; Wojdak, R. Reinforced concrete thin-wall dome after eighty years of operation in a marine climate environment. *Struct. Concr.* **2016**, *17*, 710–717. [CrossRef]

21. Schabowicz, K.; Gorzelańczyk, T.; Szymków, M. Identification of the degree of fibre-cement boards degradation under the influence of high temperature. *Autom. Constr.* **2019**, *101*, 190–198. [CrossRef]

22. Argiz, C.; Moragues, A.; Menéndez, E. Use of ground coal bottom ash as cement constituent in concretes exposed to chloride environments. *J. Clean. Prod.* **2018**, *170*, 25–33. [CrossRef]

23. Rajamane, N.P.; Ambily, P.S.; Nataraja, M.C.; Das, L. Discussion: Modified Bolomey equation for strength of lightweight concretes containing fly ash aggregates. *Mag. Concr. Res.* **2014**, *66*, 1286–1288. [CrossRef]

24. Zhang, X.; Deng, S.; Deng, X.; Qin, Y. Experimental research on regression coefficients in recycled concrete Bolomey formula. *J. Cent. South Univ. Technol.* **2007**, *14*, 314–317. [CrossRef]

25. Abdelgader, H.S.; Saud, A.F.; Othman, A.M.; Fahema, A.H.; El-Baden, A.S. Concrete mix design using the double-coating method. *Betonw. Fert. Plant Precast Technol.* **2014**, *80*, 66–74.

26. Rumman, R.; Bose, B.; Emon, M.A.B.; Manzur, T.; Rahman, M.M. An experimental study: Strength prediction model and statistical analysis of concrete mix design. In Proceedings of the International Conference on Advances in Civil Engineering, Istanbul, Turkey, 21–23 September 2016; pp. 21–23.

27. Pereira, F.C.; Borysov, S.S. Machine Learning Fundamentals. In *Pereira Big Data and Transport Analytics*; Antoniou, C., Dimitriou, L., Pereira, F., Eds.; Elsevier: Amsterdam, The Netherlands, 2019; Chapter 2; pp. 9–29, ISBN 978-0-12-812970-8.

28. Shobha, G.; Rangaswamy, S. Machine Learning. In *Computational Analysis and Understanding of Natural Languages: Principles, Methods and Applications*; Gudivada, V.N., Rao, C.R., Eds.; Elsevier: Amsterdam, The Netherlands, 2018; Volume 38, Chapter 8; pp. 197–228. ISBN 0169-7161.

29. Yang, Z.R.; Yang, Z. Artificial Neural Networks. In *Comprehensive Biomedical Physics*; Araghinejad, S., Ed.; Springer: Dordrecht, The Netherlands, 2014; Volume 6, pp. 1–17, ISBN 9780444536327.

30. Silva, I.N.; Hernane Spatti, D.; Andrade Flauzino, R.; Liboni, L.H.B.; dos Reis Alves, S.F. *Artificial Neural Networks*; Springer: Berlin, Germany, 2017; ISBN 978-3-319-43161-1.

31. Brown, C.J. *Design of Reinforced Concrete Structures*; John Wiley & Sons: New York, NY, USA, 2003; Volume 4, ISBN 9780470279274.

32. Yeh, I.-C. Modeling of Strength of High-Performance Concrete Using Artificial Neural Networks. *Cem. Concr. Res.* **1998**, *28*, 1797–1808. [CrossRef]

33. Lee, S. Prediction of concrete strength using artificial neural networks. *Eng. Struct.* **2003**, *25*, 849–857. [CrossRef]
34. Gupta, R.; Kewalramani, M.A.; Goel, A. Prediction of Concrete Strength Using Neural-Expert System. *J. Mater. Civ. Eng.* **2006**, *18*, 462–466. [CrossRef]
35. Bui, D.K.; Nguyen, T.; Chou, J.S.; Nguyen-Xuan, H.; Ngo, T.D. A modified firefly algorithm-artificial neural network expert system for predicting compressive and tensile strength of high-performance concrete. *Constr. Build. Mater.* **2018**, *180*, 320–333. [CrossRef]
36. Deng, F.; He, Y.; Zhou, S.; Yu, Y.; Cheng, H.; Wu, X. Compressive strength prediction of recycled concrete based on deep learning. *Constr. Build. Mater.* **2018**, *175*, 562–569. [CrossRef]
37. Naderpour, H.; Rafiean, A.H.; Fakharian, P. Compressive strength prediction of environmentally friendly concrete using artificial neural networks. *J. Build. Eng.* **2018**, *16*, 213–219. [CrossRef]
38. Toniolo, G.; di Prisco, M. *Reinforced Concrete Design to Eurocode 2*; Springer: Berlin, Germany, 2017; ISBN 978-3-319-52032-2.
39. Abdi, F.; Shakeri, F. A globally convergent BFGS method for pseudo-monotone variational inequality problems. *Optim. Methods Softw.* **2017**, *34*, 25–36. [CrossRef]
40. Andrei, N. An adaptive scaled BFGS method for unconstrained optimization. *Numer. Algorithms* **2018**, *77*, 413–432. [CrossRef]
41. Battiti, R.; Masulli, F. BFGS optimization for faster and automated supervised learning. In *International Neural Network Conference*; Springer: Berlin, Germany, 1990; pp. 757–760.
42. Berahas, A.S.; Nocedal, J.; Takác, M. A multi-batch l-bfgs method for machine learning. In *Advances in Neural Information Processing Systems*; Curran Associates Inc.: New York, NY, USA, 2016; pp. 1055–1063.
43. Hagan, M.T.; Menhaj, M.B. Training feedforward networks with the Marquardt algorithm. *IEEE Trans. Neural Netw.* **1994**, *5*, 989–993. [CrossRef] [PubMed]
44. Li, D.-H.; Fukushima, M. A modified BFGS method and its global convergence in nonconvex minimization. *J. Comput. Appl. Math.* **2001**, *129*, 15–35. [CrossRef]
45. Grabowska, K.; Szczuko, P. Ship resistance prediction with Artificial Neural Networks. In Proceedings of the Signal Processing: Algorithms, Architectures, Arrangements, and Applications (SPA), Poznan, Poland, 23–25 September 2015; IEEE: Piscataway, NJ, USA, 2015; pp. 168–173.
46. Le, D.; Huang, W.; Johnson, E. Neural network modeling of monthly salinity variations in oyster reef in Apalachicola Bay in response to freshwater inflow and winds. *Neural Comput. Appl.* **2018**, 1–11. [CrossRef]
47. Luenberger, D.G. *Introduction to Linear and Nonlinear Programming*; Pearson Education, Inc.: London, UK, 1973.
48. Yildizel, S.A.; Arslan, Y. Flexural strength estimation of basalt fiber reinforced fly-ash added gypsum based composites. *J. Eng. Res. Appl. Sci.* **2018**, *7*, 829–834.
49. Argiz, C.; Menéndez, E.; Sanjuán, M.A. Effect of mixes made of coal bottom ash and fly ash on the mechanical strength and porosity of Portland cement. *Mater. Construcción* **2013**, *63*, 49–64. [CrossRef]
50. Sanjuán, M.Á.; Estévez, E.; Argiz, C.; del Barrio, D. Effect of curing time on granulated blast-furnace slag cement mortars carbonation. *Cem. Concr. Compos.* **2018**, *90*, 257–265. [CrossRef]
51. Kumar Poloju, K. *Properties of Concrete as Influenced by Shape and Texture of Fine Aggregate*; American Journal of Applied Scientific Research: New York, NY, USA, 2017; Volume 3.
52. Chinchillas-Chinchillas, M.J.; Corral-Higuera, R.; Gómez-Soberón, J.M.; Arredondo-Rea, S.P.; Jorge, L.; Acuña-Aguero, O.H.; Rosas-Casarez, C.A. *Influence of the Shape of the Natural Aggregates, Recycled and Silica Fume on the Mechanical Properties of Pervious Concrete*; The Institute of Research Engineers and Doctors (IRED): New York, NY, USA, 2014; Volume 4, ISBN 9781632480064.
53. Mehnert, A.; Jackway, P. An improved seeded region growing algorithm. *Pattern Recognit. Lett.* **1997**, *18*, 1065–1071. [CrossRef]
54. Huang, G.-B.; Saratchandran, P.; Sundararajan, N. An efficient sequential learning algorithm for growing and pruning RBF (GAP-RBF) networks. *IEEE Trans. Syst. Man Cybern. Part B* **2004**, *34*, 2284–2292. [CrossRef]
55. Arora, P.; Varshney, S. Analysis of k-means and k-medoids algorithm for big data. *Procedia Comput. Sci.* **2016**, *78*, 507–512. [CrossRef]
56. Dariane, A.B.; Azimi, S. Forecasting streamflow by combination of a genetic input selection algorithm and wavelet transforms using ANFIS models. *Hydrol. Sci. J.* **2016**, *61*, 585–600. [CrossRef]
57. Tabakhi, S.; Moradi, P.; Akhlaghian, F. An unsupervised feature selection algorithm based on ant colony optimization. *Eng. Appl. Artif. Intell.* **2014**, *32*, 112–123. [CrossRef]

58. Song, Y.; Liang, J.; Lu, J.; Zhao, X. An efficient instance selection algorithm for k nearest neighbor regression. *Neurocomputing* **2017**, *251*, 26–34. [CrossRef]

59. Yu, H.; Reiner, P.D.; Xie, T.; Bartczak, T.; Wilamowski, B.M. An incremental design of radial basis function networks. *IEEE Trans. Neural Netw. Learn. Syst.* **2014**, *25*, 1793–1803. [CrossRef]

60. Iqbal, S.Z.; Gull, H.; Ahmed, J. Incremental Sorting Algorithm. In Proceedings of the Second International Conference on Computer and Electrical Engineering, Dubai, United Arab Emirates, 28–30 December 2009; pp. 378–381.

61. Gurbuzbalaban, M.; Ozdaglar, A.; Parrilo, P.A. On the convergence rate of incremental aggregated gradient algorithms. *SIAM J. Optim.* **2017**, *27*, 1035–1048. [CrossRef]

materials

MDPI

Article

Multi-Scale Structural Assessment of Cellulose Fibres Cement Boards Subjected to High Temperature Treatment

Tomasz Gorzelańczyk[ID], Michał Pachnicz *[ID], Adrian Różański[ID] and Krzysztof Schabowicz[ID]

Faculty of Civil Engineering, Wrocław University of Science and Technology, Wybrzeże Wyspiańskiego 27, 50-370 Wrocław, Poland
* Correspondence: michal.pachnicz@pwr.edu.pl

Received: 29 June 2019; Accepted: 31 July 2019; Published: 1 August 2019

check for updates

Abstract: The methodology of multi-scale structural assessment of the different cellulose fibre cement boards subjected to high temperature treatment was proposed. Two specimens were investigated: Board A (air-dry reference specimen) and Board B (exposed to a temperature of 230 °C for 3 h). At macroscale all considered samples were subjected to the three-point bending test. Next, two methodologically different microscopic techniques were used to identify evolution (caused by temperature treatment) of geometrical and mechanical morphology of boards. For that purpose, SEM imaging with EDS analysis and nanoindentation tests were utilized. High temperature was found to have a degrading effect on the fibres contained in the boards. Most of the fibres in the board were burnt-out, or melted into the matrix, leaving cavities and grooves which were visible in all of the tested boards. Nanoindentation tests revealed significant changes of mechanical properties caused by high temperature treatment: "global" decrease of the stiffness (characterized by nanoindentation modulus) and "local" decrease of hardness. The results observed at microscale are in a very good agreement with macroscale behaviour of considered composite. It was shown that it is not sufficient to determine the degree of degradation of fibre-cement boards solely on the basis of bending strength; advanced, microscale laboratory techniques can reveal intrinsic structural changes.

Keywords: cellulose fibre cement boards; microstructure; nanoindentation; SEM-EDS analysis; temperature

1. Introduction

Fibre-cement boards were invented by the Czech engineer Ludwik Hatschek over 100 years ago. They have been used in construction as siding, ceilings, floors, roofs and tile backer boards because they are damp-proof and nonflammable light-weight, strong and durable. Nowadays the cellulose fibre cement boards belong to a special class of fibre-reinforced cementitious composites and they consist of 50–70% of cement while the other components include: mineral fibres (usually cellulose) and fillers (limestone powder, kaolin, etc.). The mechanical properties, durability and microstructure of fibre-cement boards in the various environments are widely described in the literature [1,2]. The final properties of cellulose fibre cement composites depend, aside from the fibre and the matrix components, on the manufacturing process as well as on the internal microstructure.

Chady, Schabowicz et al. [3,4] proposed various non-destructive testing methods for evaluating fibre-cement boards as to the potential occurrence of heterogeneities or defects in them. Tonoli et al. [5] analyzed the effects of natural weathering on microstructure and mineral composition of cementitious roofing tiles reinforced with fique fibre. Savastano et al. [6] tested microstructure and mechanical properties of waste fibre-cement composites. At the same time the X-ray microtopography technique has been developed to study the microstructures for non-destructive characterization of the internal structure of various materials [7,8]. Cnudde et al. [9] were using the micro-CT method to determine the impregnation depth of water repellents and consolidants inside natural building stones. Li-Ping Guo et al. [10] investigated the effects of mineral admixtures on initial defects existing in high-performance concrete microstructures using a high-resolution X-ray micro-CT. Wang et al. [11] used this technique to produce the X-ray tomography images of porous metal fibre sintered sheet with 80% porosity. 3D information about the total porosity and the pore size distribution was obtained with the combination of micro-CT and home-made 3D software [12]. Schabowicz, Ranachowski at al. [13,14] successfully used the micro computed tomography (micro-CT) and SEM in the quality control system of cellulose fibre distribution in cement composites. Ranachowski and Gorzelańczyk et al. [15–17] investigate the degradation of the microstructure and mechanical properties of fibre cement board (FCB), which was exposed to environmental hazards, resulting in thermal impact on the microstructure of the board. Visual light and scanning electron microscopy, X-ray micro tomography, flexural strength, and work of flexural test W_f measurements were used.

From the literature review presented above, it is clear that until now the mechanical parameters of the microstructure of cellulose fibres cement boards have not been studied. It is however of primary importance to identify the mechanical morphology at microscale which directly affects the mechanical behavior of material at macroscale, i.e., at the scale of engineering applications. In the case of composite materials with a cement matrix, segmentation and characterization of components' mechanical properties is impossible with the use of classical macro scale experiments [18]. Therefore, advanced laboratory techniques that allow observation of the mechanical behavior of materials at different scales are used very often nowadays. An example of such technique is a classical nanoindentation test developed by Oliver and Pharr as a method to accurately calculate hardness and elastic modulus from the "load-displacement" curve [19].

The idea of nanoindentation is to determine the mechanical properties of composite components by observing their reaction to the point load followed by continuous unload of its surface. Initially, due to hardware limitations, the nanoindentation technique was mainly used in relatively homogeneous media, or layered materials with known thickness of individual layers [20,21]. Currently, better hardware capabilities have made it possible to use this technique for microheterogeneous materials, exhibiting different mechanical morphology depending on the observation scale [22,23]. Nanoindentation tests were successfully used to identify mechanical morphologies for different types of cementitious materials in the works [24–26].

This paper presents a multiscale approach for identification of internal structural changes of cellulose fibres cement boards subjected to high temperature treatment. Two advanced laboratory techniques are used, i.e., SEM imaging with EDS analysis and nanoindentation tests. The former technique provides microphotographs of the applied fibres together with the elemental composition of tested samples. The latter (nanoindentation) reveals the mechanical morphology of the material in terms of hardness and elastic moduli. All tests were carried out on two samples: reference material at air-dried state, and the one subjected to high temperature treatment. An evolution of geometrical and mechanical morphology of investigated boards is observed and discussed.

2. Materials and Methods

In this study, two sets of specimens were prepared for examination. These were labeled A and B. The specimens were fabricated by applying the Hatschek forming method. Air-dry reference specimens (not subjected to high temperature treatment) were denoted as A. The specimens, which underwent the high temperature treatment in an electric oven for 3 h at a temperature of 230 °C, were labeled as B. The parameters of the heating procedure were chosen experimentally, after performing some preliminary tests to evoke considerable changes in the microstructure of the investigated materials. The tested specimens were cut from the different fibre cement panels of 8 mm of thickness. Prior to the main research the panels were tested using the standard procedures to assess their performance.

Comparisons of tested panels are presented in Table 1 and the set of equipment for bending strength measurements is shown in Figure 1.

Table 1. Characteristic of the tested panels.

Symbol of the Board	A	B
Type of the board	fibre-cement, exterior	fibre-cement, exterior, after 3 h of burning (230 °C)
Thickness of board [mm]	8	8
Bending strength [MPa]	23.54 *	26.86 *
Density [kg/m³]	1600	1500
Photo of the board	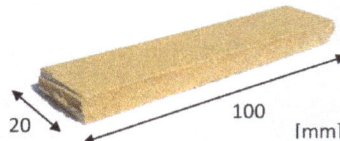	

* mean values calculated on the basis of ten independent measurements.

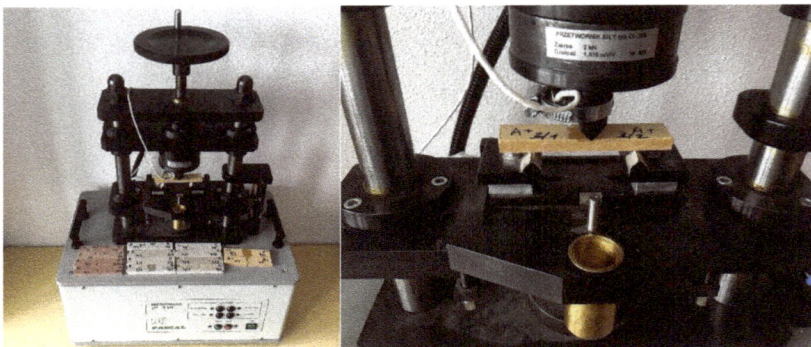

Figure 1. Test stand for bending strength measurements, and fibre-cement board during test.

Figure 2 shows an exemplary σ-ε curve of the tested fibre-cement boards under bending. The trace of flexural strength σ, bending strength *MOR*, limit of proportionality (*LOP*) and strain ε were analyzed. Bending strength *MOR* was calculated from the standard formula [27]:

$$MOR = \frac{3Fl_s}{2b\,e^2} \qquad (1)$$

where:

F is the loading force (N);
l_s is the length of the support span (mm);
b is the specimen width (mm); and
e is the specimen thickness (mm).

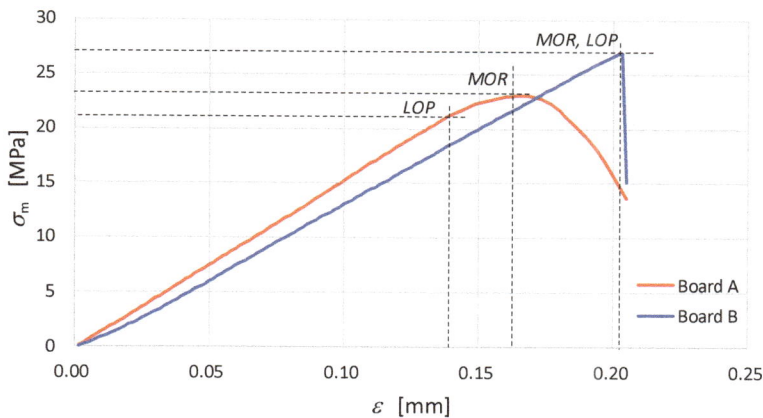

Figure 2. Diagrams of the σ-ε dependence under bending for fibre-cement boards.

In case of microscale experiments, first, the analysis was performed by SEM imaging with EDS elemental composition examination. For that purpose, the authors have prepared fractured specimens to produce the SEM micrographs applying the Quanta FEG-250 Scanning Electron Microscope (Hillsboro, OR, USA), FEI with an EDS analyser. The precondition was made at 50% of relative humidity and 22 °C to enable for different modes of decomposition of composite microstructure.

Next the boards were investigated in terms of nanoindentation approach. It is commonly known that the process of sample preparation for nanoindentation is of primary importance for getting reasonable results. In general, we can assume that the results from nanoindentation can only be as good as the sample used for testing. Hence, the aim of the preparation process was to obtain a satisfactory quality of: parallelism and roughness of surfaces, cleanness of the sample and sample tilt. We followed a common rule-of-thumb contained in the ISO standard for nanoindentation (ISO 14577) as well as requirements, concerning surface roughness criteria for cement paste nanoindentation, provided in [25]. Preparation of the samples for nanoindentation consisted of cutting fibre boards to required dimensions (approximately 1 × 1 cm), mounting specimens in the epoxy resin and thorough grinding and polishing process of the specimen surface. High-speed diamond saw Struers Labotom-5 (Copenhagen, Denmark) (Figure 3a), Struers CitoVac vacuum chamber (Copenhagen, Denmark) (Figure 3b) and Struers LaboPol-5 grinder (Copenhagen, Denmark) (Figure 3c) were used for the sample preparation. The photos of the samples after the preparation procedure are presented in Figure 4.

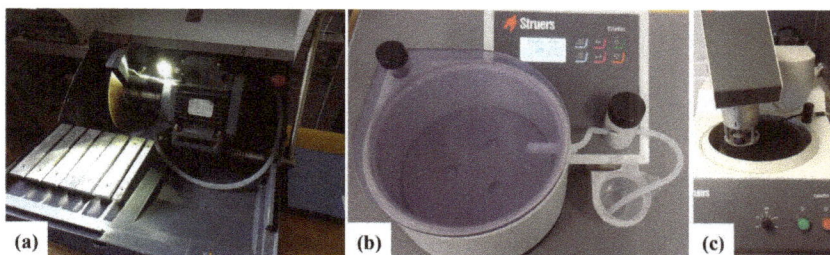

Figure 3. Equipment used for sample preparation: (a) Struers Labotom-5, (b) Struers CitoVac, (c) Struers LaboPol-5.

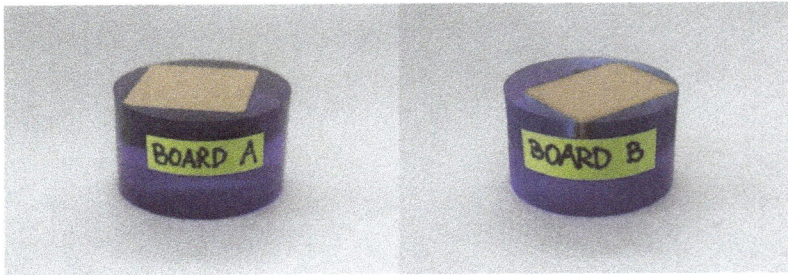

Figure 4. Samples prepared for the indentation.

Nanoindentation tests were performed using Nanoindenter CSM TTX-NHT (Neuchatel, Switzerland) (Figure 5) equipped with a diamond Berkovich tip (Poisson's ratio $v_i = 0.07$, elastic modulus $E_i = 1000$ GPa, $\beta = 1.034$).

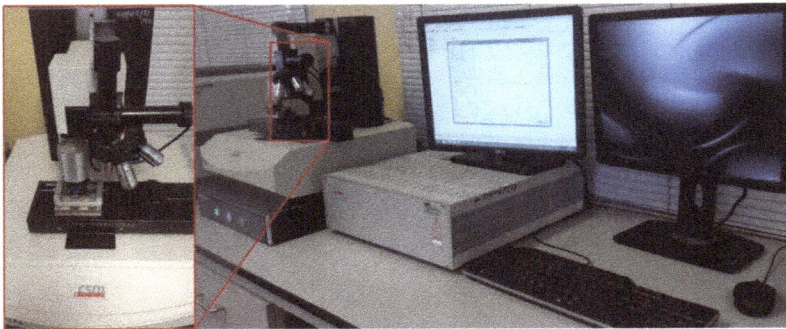

Figure 5. Nanoindentation test stand.

3. Results of Multiscale Approach

3.1. SEM Analysis

In this Section the results obtained with SEM imaging are presented. In particular, Figure 6 shows the microphotographs of the applied fibres. Note that left (right) panel of Figure 6 presents the microstructure of reference (subjected to high temperature) sample.

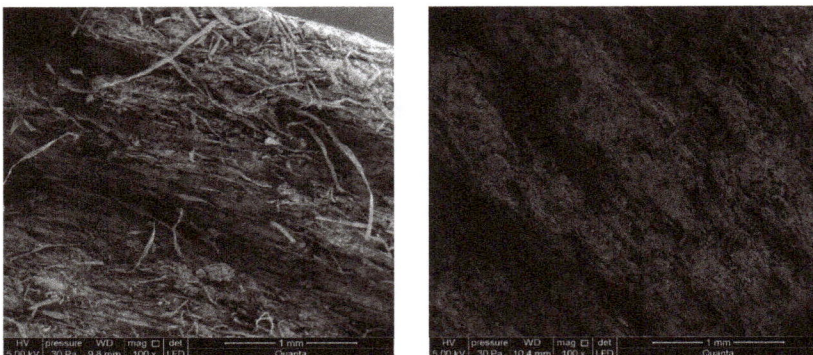

Figure 6. *Cont.*

Figure 6. The microscopic observations of the applied fibres, in scanning electron microscope (SEM) with scale bar in sequence 1 mm, 500 μm and 50 μm: (**a**) air-dry condition (reference board), (**b**) board exposure to temperature of 230 °C-3 h.

In Figure 7 elemental composition (results of EDS analysis) of tested samples is graphically presented. The investigation was carried out separately for matrix (Figure 7a), fibre of air-dry condition (reference board, Figure 7b) "fibre" of board exposed to temperature of 230 °C for 3 h (Figure 7c).

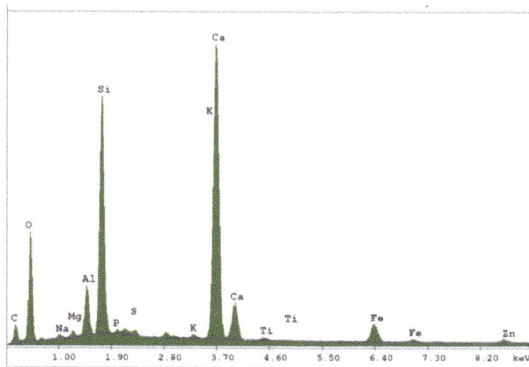

(**a**)

Figure 7. *Cont.*

(b)

(c)

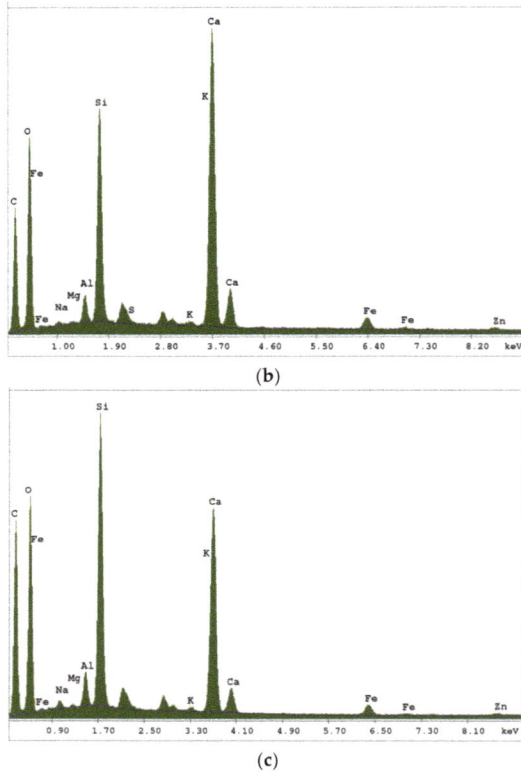

Figure 7. Composition analysis (EDS) of tested boards: (**a**) matrix, (**b**) fibre of air-dry condition (reference board) and (**c**) "fibre" of board exposure to temperature of 230 °C-3 h.

3.2. Nanoindentation

The assessment of mechanical parameters of the fibre cement boards was carried out using the method formulated for heterogeneous materials with a cement matrix. This technique, the Grid Indentation Technique (GIT), was introduced in [18]. For each specimen a regular grid of 220 tests was applied (Figure 8). The relevant grid parameters, such as the distance between indenter locations, the number of tests and the single test's maximum load were determined by trial nanoindentation tests in accordance with the recommendations presented in [18].

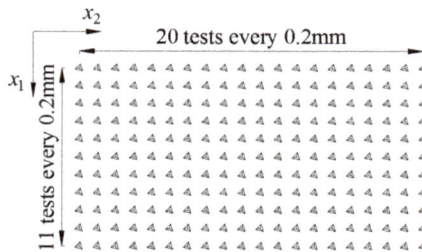

Figure 8. Applied nanoindentation grid.

In every grid point, a single standard nanoindentation test [19,28,29] was performed with the maximum load of 500 mN. The test runs as follows: continuous increase of force up to a fixed value F_{max}, then a short period in which the set maximum value of force is maintained, after that it is followed by unloading which is also carried out continuously (Figure 9a). During the test, the relation between the force F and the depth of penetration h is recorded. An example of *F-h* curve obtained for a single test is presented in Figure 9b. For every test two indentation parameters were calculated, namely hardness (H_{IT}) defined as follows:

$$H_{IT} = \frac{F_{max}}{A} \tag{2}$$

and the indentation modulus (M_{IT}):

$$M_{IT} = \frac{S}{2\beta} \frac{\sqrt{\pi}}{\sqrt{A}} \tag{3}$$

where, F_{max} is the maximum force of indentation, A is the projection of the contact area on the surface of the sample. This value is usually defined as a function of the maximum indentation depth h_{max} [18] and S is the initial slope of the unloading curve according to [18].

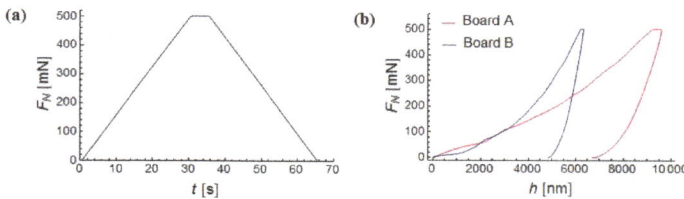

Figure 9. (**a**) Load function used for single nanoindentation, (**b**) and example of *F-h* curve.

As a consequence of performed tests, distribution of the mechanical parameters on the surface of the samples was evaluated. Contour maps of hardness and indentation modulus distribution are presented in Figures 10 and 11 for Board A and Board B, respectively.

All obtained values are presented in the form of histograms in Figure 12a,b. The results corresponding to the case of high temperature heating (230 °C-3 h, Board B) are displayed in grey. Red and blue colors are corresponding to the results obtained for reference samples (Board A). In particular, the red color represents the distribution of the indentation modulus, whereas the blue one represents the frequency of hardness values. The vertical dashed lines refer to the mean values averaged over 220 individual results. Furthermore, exact values of statistical measures, i.e., mean values μ and standard deviations σ, are summarized in Table 2.

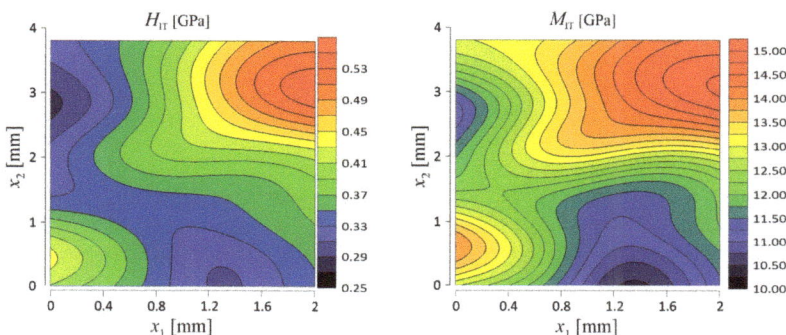

Figure 10. Distribution of hardness and indentation modulus for Board A (reference material).

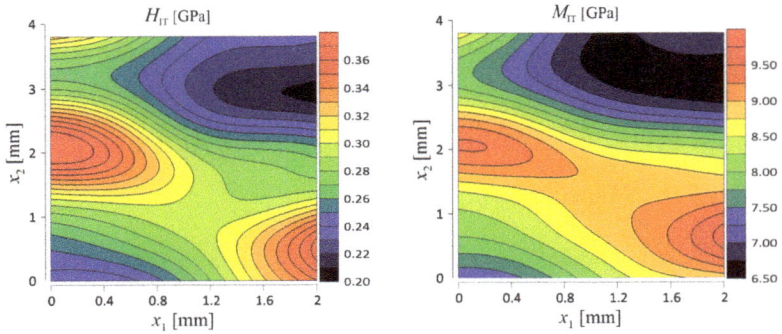

Figure 11. Distribution of hardness and indentation modulus for Board B.

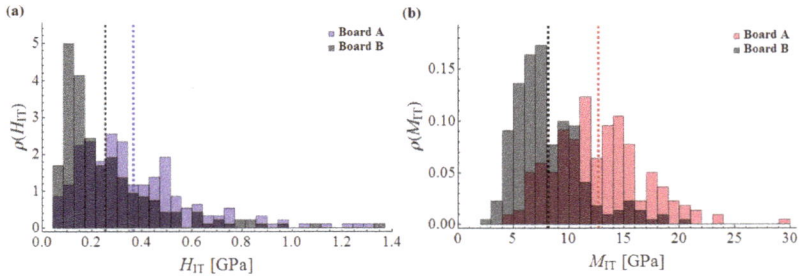

Figure 12. (a) Histograms of hardness; (b) histograms of indentation modulus.

Table 2. Summary of nanoindentation results.

Parameter		Specimen before High Temperature Treatment (Board A)	Specimen after High Temperature Treatment (Board B)
Mean value of hardness	$\mu^{H_{IT}}$ [GPa]	0.382	0.285
Standard deviation of hardness	$\sigma^{H_{IT}}$ [GPa]	0.226	0.201
Coefficient of variation * of hardness	c.o.v$^{H_{IT}}$	0.592	0.705
Mean value of indentation modulus	$\mu^{M_{IT}}$ [GPa]	12.686	8.147
Standard deviation of indentation modulus	$\sigma^{M_{IT}}$ [GPa]	4.093	3.138
Coefficient of variation * of indentation modulus	c.o.v$^{M_{IT}}$	0.318	0.385

* Coefficient of variation is defined as the ratio of the standard deviation to the mean value.

In Figures 13 and 14 the variation of mean values averaged along two independent directions, respectively for x_1 and x_2 (see coordinate system shown in Figure 8), is presented. The black color corresponds to the board subjected to heating and blue (hardness) and red (indentation modulus) colors are representing the results obtained for the reference sample.

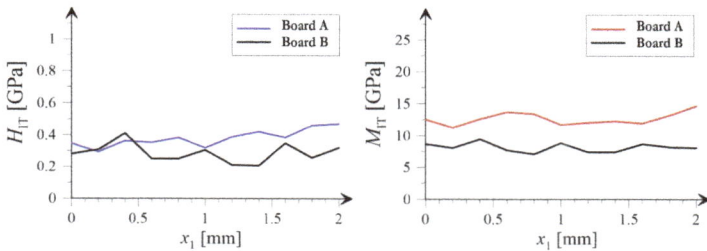

Figure 13. Average values of hardness (H_{IT}) and indentation modulus (M_{IT}) in the x_2 direction.

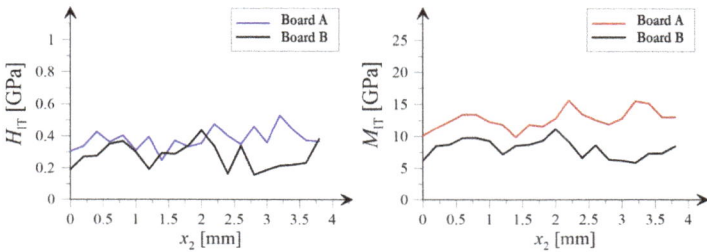

Figure 14. Average values of hardness (H_{IT}) and indentation modulus (M_{IT}) in the x_1 direction.

4. Discussion of Results

The results presented in Table 1 and Figure 2 show that, under high temperature, the bending strength *MOR* increases. The increase (in average sense) from 23.54 to 26.86 MPa is observed. For fibre-cement Board B, it was also observed that under the influence of a temperature of 230 °C, the structure of the Board Becomes more brittle; after peak value of stress, a sudden drop of strength is observed (Figure 2). An analysis of the classical macroscale results showed that for fibre-cement Board B the extent of the nonlinear increase in bending stress is reduced until the bending strength *MOR* comes to be level with the proportionality limit *LOP*. In the case of the reference board, the *MOR* and *LOP* values were clearly separated. It should be noted that the reference samples were under an air-dry condition and their bulk moisture content amounted to 6–8%.

Based on the macroscale results (bending strength) one can draw the conclusion that after exposure to the temperature (230 °C for 3 h) there is no degradation of the board, and even the strengthening effect is obtained. Nevertheless, advanced microscale laboratory techniques have revealed damaging and irreversible structural changes in both geometrical and mechanical morphology of microstructure.

An analysis of the images obtained from the scanning electron microscope and the EDS analyzer shows that the fibre cement Board A has a compact microstructure (Figure 6a). Microscopic examinations revealed a fine-pore structure, with pores of up to 50 µm in size. Cavities and grooves, up to 500 µm wide, were visible in the fracture areas where the fibres had been pulled out. Cellulose fibres and PVA fibres, are clearly visible in the images. Various forms of hydrated calcium silicates of the C-S-H type occur. Both an "amorphous" phase and a phase built of strongly-adhering particles predominate. An analysis of the fibre composition showed that fibre elements and some cement elements were present. An analysis of the chemical composition of the matrix showed elements that are typical of cement (Figure 7). The surface of the fibres was covered with a thin layer of cement paste and hydration products. The fact that there are very few areas with a space between the fibres and the cement paste, indicates that the fibre-cement bond is strong.

A microscopic analysis of the fibre cement Board B, which was exposed to a temperature of 230 °C for 3 h, shows a clear change in the colour of the samples (Figure 6b). Most of the fibres in the board were found to be burnt-out, or melted into the matrix, leaving cavities and grooves which were visible in all of the tested boards. The structure of the few remaining fibres was strongly degraded. An examination of the cement particles on the fracture surface revealed burning-out of their structure. The structure of the matrix was found to be more granular, showing many delaminations (Figure 6b). Numerous caverns and grooves left by the pulled-out fibres, as well as the pulled-out cement particles, were observed.

Noticeable changes of mechanical morphology of the analyzed boards due to the influence of high temperature is observed within the nanoindentation approach. The average value of the indentation modulus, being the measure of the elastic response of the material, after the exposure to the temperature of 230 °C for 3 hours, significantly decreased (see Figure 12b). The histogram for the Board B (grey colour) concentrates around lower values of modulus compared with the histogram for Board A. In addition, according to Table 2 the average value of the indentation modulus, after heating at the temperature of 230 °C, decreased from 12.686 GPa to 8.147 GPa which equals to an almost 50% drop in the elastic stiffness of examined material. Furthermore, as shown in the right panels of Figures 13 and 14, the mean value of indentation modulus is significantly smaller for Board B compared to the values obtained for Board A in the entire range of x_1 and x_2 values. These observations are in a very good agreement with SEM analysis. A decrease of elastic response of material's microstructure, its stiffness in fact, is mostly due to the changes in the microstructure geometry observed in SEM. The presence of cavities and grooves in Board B can be a direct cause of decrease in M_{IT} values. It is worth noticing that the changes of geometrical (SEM analysis) and mechanical (nanoindentation) morphology of Board B is a direct cause of the macroscale behavior; by observing Figure 2, one can simply notice that the stiffness of Board B is reduced compared to the one evaluated for Board A (the slope of σ-ε curve decreased after exposure to the temperature); and it was observed in case of all boards under bending.

Slightly different conclusions can be drawn for the hardness of the material. In general, the average hardness value, as a result of exposure to the temperature 230 °C, decreased, from 0.382 GPa for the Board A, to 0.285 GPa for the Board B (see Figure 12a and Table 2). On the other hand, observing the left panels of Figures 13 and 14 one can notice that "locally", i.e., for given ranges of x_1 or x_2 values, the decrease of H_{IT} is not observed. As shown in [30,31] it is the hardness H_{IT} which is the parameter determining the strength of the microstructure. This is due to fact that the strength is proportional to the hardness of individual microstructure components; the higher the hardness, the higher the strength of microstructure. Therefore the phenomenon observed e.g., in the left panels of Figure 13 (in the range of x_1 from 0 to 0.5 mm) and Figure 14 (in the range of x_2 from 0.5 to 2.0 mm), where no decrease in H_{IT} is observed, can justify the macroscale bending strength results. As shown in Table 1, the bending strength of Board B is slightly higher than the one evaluated for Board A; this is in the average sense since the values in Table 2 represent mean values. However, during individual bending tests we also observed the cases when the strength was not increased or even slightly decreased. This is the effect of the fact that "locally" hardness does not change under the influence of temperature, and hence, this can cause such macroscopic behavior of the boards.

It should also be mentioned that the standard deviations for both the indentation modulus and the hardness of the specimen subjected to temperature 230 °C slightly decreased. However, as mentioned above, the mean values of both properties decreased in a more evident manner, and hence the mechanical morphology of Board B seems to have more heterogeneous nature. This is clearly revealed by plots shown in Figures 13 and 14 as well as by c.o.v. (coefficient of variation) values summarized in Table 2. For Board B, the coefficient of variation of both hardness and indentation modulus increased.

5. Final Conclusions

Two methodologically different microscopic techniques were used to identify the evolution of geometrical and mechanical morphology of boards. The methodology of multi-scale structural assessment of the different cellulose fibre cement boards subjected to high temperature treatment was presented. For that purpose, two specimens were investigated: Board A (air-dry reference specimen) and Board B (exposed to a temperature of 230 °C for 3 h).

SEM examinations were carried out to get a better insight into the changes taking place in the structure of the tested boards. Significant changes take place in the structure of the boards, especially after the high temperature treatment in an electric oven for 3 h at a temperature of 230 °C. Most of the fibres in the board were burnt-out, or melted into the matrix, leaving cavities and grooves which were visible in all of the tested boards. The structure of the few remaining fibres was strongly degraded, as confirmed by the nanoindentation tests. Nanoindentation tests revealed significant changes of mechanical properties caused by high temperature treatment: "global" decrease of the stiffness (characterized by nanoindentation modulus) and "local" decrease of hardness. The results observed at microscale are in a very good agreement with macroscale behaviour of considered composite.

In the authors' opinion, the above findings are important for building practice because it was clearly shown that it is not sufficient to determine the degree of degradation of fibre-cement boards solely on the basis of bending strength. Based only on the macroscale results (bending strength) one can draw the conclusion that after exposure to the temperature (230 °C for 3 h) there is no degradation of the board. Moreover, in the average sense, some strengthening effect can also be observed. Microscale laboratory techniques adapted in this work, however, reveal damaging and irreversible microstructural changes of boards caused by the high temperature treatment.

It should be also noticed that the presented results are preliminary and starting a research cycle. Based on them, changes in mechanical parameters, especially of fibres in the fibre cement board, after exposed to a temperature of 230 °C for 3 h have been demonstrated. Currently, studies are being carried out to show the impact of high-temperature on the fibre cement board, but in a much shorter time. The authors hope to publish promising results soon.

Author Contributions: Conceptualization, K.S. and A.R.; Investigation, T.G. and M.P.; Methodology, T.G. and A.R.; Software, T.G. and M.P.; Supervision, K.S.; Visualization, T.G. and M.P.; Writing—original draft, K.S. and A.R.; Writing—review & editing, T.G. and M.P.

Funding: This research received no external funding.

Conflicts of Interest: The authors declare no conflict of interest.

References

1. Akhavan, A.; Catchmark, J.; Rajabipour, F. Ductility enhancement of autoclaved cellulose fiber reinforced cement boards manufactured using a laboratory method simulating the Hatschek process. *Constr. Build. Mater.* **2017**, *135*, 251–259. [CrossRef]
2. Ardanuy, M.; Claramunt, J.; Toledo Filho, R.D. Cellulosic fiber reinforced cement-based composites: A review of recent research. *Constr. Build. Mater.* **2015**, *79*, 115–128. [CrossRef]
3. Chady, T.; Schabowicz, K.; Szymków, M. Automated multisource electromagnetic inspection of fibre-cement boards. *Autom. Constr.* **2018**, *94*, 383–394. [CrossRef]
4. Schabowicz, K.; Gorzelańczyk, T. A nondestructive methodology for the testing of fibre cement boards by means of a non-contact ultrasound scanner. *Constr. Build. Mater.* **2016**, *102*, 200–207. [CrossRef]
5. Tonoli, G.H.D.; Santos, S.F.; Savastano, H.; Delvasto, S.; Mejía de Gutiérrez, R.; Del M. Lopez de Murphy, M. Effects of natural weathering on microstructure and mineral composition of cementitious roofing tiles reinforced with fique fibre. *Cem. Concr. Compos.* **2011**, *33*, 225–232. [CrossRef]
6. Savastano, H.; Warden, P.G.; Coutts, R.S.P. Microstructure and mechanical properties of waste fibre-cement composites. *Cem. Concr. Compos.* **2005**, *27*, 583–592. [CrossRef]
7. Schabowicz, K.; Ranachowski, Z.; Jóźwiak-Niedźwiedzka, D.; Radzik, Ł.; Kudela, S.; Dvorak, T. Application of X-ray microtomography to quality assessment of fibre cement boards. *Constr. Build. Mater.* **2016**, *110*, 182–188. [CrossRef]
8. Nowak, T.; Karolak, A.; Sobótka, M.; Wyjadłowski, M. Assessment of the Condition of Wharf Timber Sheet Wall Material by Means of Selected Non-Destructive Methods. *Materials* **2019**, *12*, 1532. [CrossRef] [PubMed]

9. Cnudde, V.; Cnudde, J.P.; Dupuis, C.; Jacobs, P.J.S. X-ray micro-CT used for the localization of water repellents and consolidants inside natural building stones. *Mater. Charact.* **2004**, *53*, 259–271. [CrossRef]

10. Guo, L.P.; Carpinteri, A.; Sun, W.; Qin, W.C. Measurement and analysis of defects in high-performance concrete with three-dimensional micro-computer tomography. *J. Southeast Univ.* **2009**, *25*, 83–88.

11. Wang, Q.; Huang, X.; Zhou, W.; Li, J. Three-dimensional reconstruction and morphologic characteristics of porous metal fiber sintered sheet. *Mater. Charact.* **2013**, *86*, 49–58. [CrossRef]

12. Liu, J.; Li, C.; Liu, J.; Cui, G.; Yang, Z. Study on 3D spatial distribution of steel fibers in fiber reinforced cementitious composites through micro-CT technique. *Constr. Build. Mater.* **2013**, *48*, 656–661. [CrossRef]

13. Ranachowski, Z.; Schabowicz, K. The contribution of fiber reinforcement system to the overall toughness of cellulose fiber concrete panels. *Constr. Build. Mater.* **2017**, *156*, 1028–1034. [CrossRef]

14. Schabowicz, K.; Jóźwiak-Niedźwiedzka, D.; Ranachowski, Z.; Kudela, S.; Dvorak, T. Microstructural characterization of cellulose fibres in reinforced cement boards. *Arch. Civ. Mech. Eng.* **2018**, *18*, 1068–1078. [CrossRef]

15. Ranachowski, Z.; Ranachowski, P.; Dębowski, T.; Gorzelańczyk, T.; Schabowicz, K. Investigation of Structural Degradation of Fiber Cement Boards Due to Thermal Impact. *Materials* **2019**, *12*, 944. [CrossRef] [PubMed]

16. Schabowicz, K.; Gorzelańczyk, T.; Szymków, M. Identification of the degree of fibre-cement boards degradation under the influence of high temperature. *Autom. Constr.* **2019**, *101*, 190–198. [CrossRef]

17. Schabowicz, K.; Gorzelańczyk, T.; Szymków, M. Identification of the Degree of Degradation of Fibre-Cement Boards Exposed to Fire by Means of the Acoustic Emission Method and Artificial Neural Networks. *Materials* **2019**, *12*, 656. [CrossRef]

18. Constantinides, G.; Ravi Chandran, K.S.; Ulm, F.J.; Van Vliet, K.J. Grid indentation analysis of composite microstructure and mechanics: Principles and validation. *Mater. Sci. Eng. A* **2006**, *430*, 189–202. [CrossRef]

19. Oliver, W.C.; Pharr, G.M. An improved technique for determining hardness and elastic modulus using load and displacement sensing indentation experiments. *J. Mater. Res.* **1992**, *7*, 1564–1583. [CrossRef]

20. Cheng, Y.T.; Cheng, C.M. Scaling, dimensional analysis, and indentation measurements. *Mater. Sci. Eng. R Rep.* **2004**, *44*, 91–149. [CrossRef]

21. Oliver, W.C.; Pharr, G.M. Measurement of hardness and elastic modulus by instrumented indentation: Advances in understanding and refinements to methodology. *J. Mater. Res.* **2004**, *19*, 3–20. [CrossRef]

22. Randall, N.X.; Vandamme, M.; Ulm, F.J. Nanoindentation analysis as a two-dimensional tool for mapping the mechanical properties of complex surfaces. *J. Mater. Res.* **2009**, *24*, 679–690. [CrossRef]

23. Giannakopoulos, A.E.; Suresh, S. Determination of elastoplastic properties by instrumented sharp indentation. *Scr. Mater.* **1999**, *40*, 1191–1198. [CrossRef]

24. Constantinides, G.; Ulm, F.J.; Van Vliet, K. On the use of nanoindentation for cementitious materials. *Mat. Struct.* **2003**, *36*, 191–196. [CrossRef]

25. Miller, M.; Bobko, C.; Vandamme, M.; Ulm, F.J. Surface roughness criteria for cement paste nanoindentation. *Cem. Concr. Res.* **2008**, *38*, 467–476. [CrossRef]

26. Ulm, F.J.; Vandamme, M.; Bobko, C.; Ortega, J.A.; Tai, K.; Ortiz, C. Statistical Indentation Techniques for Hydrated Nanocomposites: Concrete, Bone, and Shale. *J. Am. Ceram. Soc.* **2007**, *90*, 2677–2692. [CrossRef]

27. *Cellulose Fibre Cement Flat Sheets-Product Specification and Test Methods*; EN 12467; British Standards Institution: London, UK, 2018.

28. Rajczakowska, M.; Stefaniuk, D.; Łydżba, D. Microstructure Characterization by Means of X-ray Micro-CT and Nanoindentation Measurements. *Studi. Geotech. Mech.* **2015**, *37*, 75–84. [CrossRef]

29. Rajczakowska, M.; Łydżba, D. Durability of crystalline phase in concrete microstructure modified by the mineral powders: Evaluation by nanoindentation tests. *Studi. Geotech. Mech.* **2016**, *38*, 65–74. [CrossRef]

30. Ganneau, F.P.; Constantinides, G.; Ulm, F.J. Dual-indentation technique for the assessment of strength properties of cohesive-frictional materials. *Int. J. Solids Struct.* **2006**, *43*, 1727–1745. [CrossRef]

31. Cariou, S.; Ulm, F.J.; Dormieux, L. Hardness–packing density scaling relations for cohesive-frictional porous materials. *J. Mech. Phys. Solids* **2008**, *56*, 924–952. [CrossRef]

materials

MDPI

Article

Non-Destructive Assessment of Masonry Pillars using Ultrasonic Tomography

Monika Zielińska [1] and Magdalena Rucka [2,*]

[1] Department of Technical Fundamentals of Architectural Design, Faculty of Architecture,
Gdansk University of Technology, Narutowicza 11/12, 80-233 Gdansk, Poland; monika.zielinska@pg.edu.pl
[2] Department of Mechanics of Materials and Structures, Faculty of Civil and Environmental Engineering,
Gdansk University of Technology, Narutowicza 11/12, 80-233 Gdansk, Poland
* Correspondence: magdalena.rucka@pg.edu.pl or mrucka@pg.edu.pl; Tel.: +48-58-347-2497

Received: 21 November 2018; Accepted: 10 December 2018; Published: 13 December 2018

check for updates

Abstract: In this paper, a condition assessment of masonry pillars is presented. Non-destructive tests were performed on an intact pillar as well as three pillars with internal inclusions in the form of a hole, a steel bar grouted by gypsum mortar, and a steel bar grouted by cement mortar. The inspection utilized ultrasonic stress waves and the reconstruction of the velocity distribution was performed by means of computed tomography. The results showed the possibilities of tomographic imaging in characterizing the internal structure of pillars. Particular attention was paid to the assessment of the adhesive connection between a steel reinforcing bar, embedded inside pillars, and the surrounding pillar body.

Keywords: non-destructive testing; masonry structures; strengthening; ultrasonic tomography; adhesion assessment

1. Introduction

A significant part of engineering structures consists of masonry objects. Historic buildings are usually made of bricks or stones [1,2] while to make contemporary objects, both ceramic elements and blocks of autoclaved aerated concrete are used [3]. The technical condition of masonry structures requires a careful quality assessment and often intervention, enabling their further proper functioning, due to the influence of atmospheric conditions, excessive loads and processes of the natural ageing. Properly carried out works, aimed at strengthening, repairing or maintaining masonry, should be preceded by a precise diagnosis process. In general, two main diagnostic approaches are possible: invasive and non-invasive [4]. In destructive testing (DT), material samples acquired from an object are destroyed to evaluate their mechanical properties. Particularly important in the diagnostics of engineering structures are non-destructive testing (NDT) methods, because they do not violate the integrity of tested objects. In recent years, various non-destructive techniques dedicated to masonry structures have been developed, including ultrasonic methods, ground penetrating radar, thermography or acoustic emissions (e.g., [1,5–10]).

The basic methods for strengthening masonry structures are total or partial brick replacement and the introduction of new reinforcing elements. The brick replacing technique relies on substituting old, degraded material with new material. This method significantly affects the structure of the object, and it is not recommended for works carried out on objects of cultural heritage, where any interference with the historic substance should be avoided. The structural efficiency of masonry can be also increased by introducing new strengthening elements (e.g., [11,12]). Reinforced concrete elements, introduced in the form of columns and beams [13] or jackets [14] constitute a substitute supporting structure. Additional steel (or composite) elements are usually used as grouted anchors or tie-rods (e.g., [15–18]) as well as composite jackets made of fabric-reinforced cementitious matrix [19] or fiber-reinforced polymers [20,21]. If a replacement technique is used, the adhesion of an old wall to a new part is not as important, because usually the new wall has only the role of filling, and in this case, it is completely lightened. However, the adhesive connection is very important when steel elements in the form of bars are used, because their main purpose is to transfer stresses to the masonry. Incorrect anchoring of the steel elements, or damage developing at the interface between two materials, i.e., steel and mortar, may be the reason why the strengthening method does not fulfil its assumed function. Therefore, the non-destructive evaluation of the adhesive connection is crucial for properly conducting the repairing process.

In this work, an evaluation of masonry pillars is presented. Particular attention was paid to the condition assessment of the adhesive connection between a steel reinforcing bar embedded inside pillars and the surrounding pillar body. The research was carried out using ultrasound waves and tomography imaging. Ultrasound tomography is one of the more developed imaging methods, utilizing the properties of elastic waves. Many previous works concern Lamb-wave tomography for damage detection in metal or composite thin plates. Rao et al. [22] performed a study on the online corrosion monitoring of a steel plate and thickness reconstruction of the corrosion damage. Zhao et al. [23] investigated damage imaging in aluminum plates with an artificial thinning area. Leonard et al. [24] studied Lamb-wave tomography in both aluminum and composite plates, with defects of various sizes and thicknesses. A structural health monitoring system of composite plates with through-thickness holes was presented by Prasad et al. [25]. Ultrasound tomography was also widely used for defect imaging in concrete structures (e.g., [26,27]). Martin et al. [26] examined post-tensioned concrete beams. They identified both the location of ducts and the voiding in ducts. Chai et al. [28] developed attenuation tomography for visualizing defects in a concrete slab. Aggelis et al. [29] applied numerical simulations of wave propagation to investigate the possibility of detecting different types of inhomogeneities in concrete. Schabowicz and Suvorov [30] described an ultrasonic tomogram, equipped with a multi-element antenna array, and its application for the estimation of the thickness of concrete and the localization of flawed zones. Schabowicz [31] presented the one-sided non-destructive testing of concrete cubic specimens and a foundation concrete slab using ultrasonic tomography. Haach and Ramirez [32] analyzed different arrangements of transducers in a study on the detection of cylindrical polystyrene blocks in concrete prismatic specimens. Choi and Popovics [33] compared one-sided imaging with through-thickness tomography using the example of highly reinforced concrete elements with internal defects. Integrated ultrasonic tomography and a 3D computer vision technique was developed by Choi et al. [34]. They obtained volumetric internal images to detect defects within concrete. Chai et al. [27] identified the tendon duct filling, as well as honeycomb defects, in concrete specimens. In the case of masonry structures, ultrasonic tomography was used to assess the general condition or to evaluate the effectiveness of the conducted repair. Schuller et al. [35] performed a velocity reconstruction in a wall specimen made of bricks. Binda, Saisi, and Zanzi [36] examined the stone pillars of the temple of S. Nicolò l'Arena using sonic tests. They obtained maps of the velocity distribution that enabled the presence of the different building techniques applied during the erection of the pillars to be recognized. The detection of voids in laboratory masonry specimens was studied by Paasche et al. [7]. They applied ultrasonic tomography and compared the results with those of ground penetrating radar tomography. Pérez-Gracia et al. [37] determined the velocity distribution in columns of the Mallorca Cathedral. Santos-Assunçao et al. [9] evaluated laboratory models of masonry columns. Changes in the wave velocity in tomography maps were identified as damages or changes in the bricks and mortar. The literature on the assessment of reinforced masonry structures is rather limited.

This study presents a comprehensive condition assessment of masonry pillars. Experimental and numerical investigations were performed on an intact pillar as well as three pillars with internal inclusions in the form of a hole, a steel bar grouted by gypsum mortar, and a steel bar grouted by cement mortar. The inspection was conducted using ultrasonic waves and computed tomography. The investigations focused on the characterization of the internal structure of the pillars by velocity reconstruction in the examined cross-section. The influence of different numbers of pixels and different configurations of paths on tomographic velocity reconstruction was studied.

2. Theoretical Background of Ultrasonic Tomography

Ultrasonic tomography imaging allows the internal structure of an investigated object to be reconstructed by a set of projections through the sample in many different directions. A schematic diagram of ultrasound tomography is shown in Figure 1. At first, the tested specimen must be divided into small, geometrical cells, called pixels (marked with dotted lines in Figure 1a). Each pixel in the image represents a discrete area in the sample, and it is associated with an intensity value. The conducted calculations enable a quantitative description of each pixel, based on its physical characteristics, such as the wave propagation velocity. The number of pixels can be changed, enabling the study of the effect of spatial resolution on the quality of the obtained image.

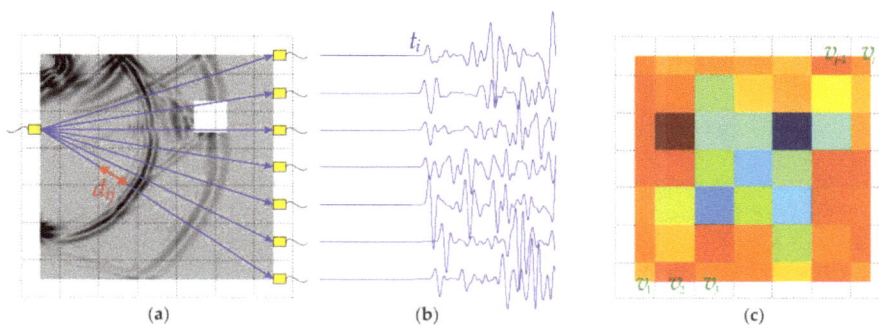

(a)   (b)   (c)

Figure 1. Schematic diagram of ultrasound tomography: (**a**) cross-section divided into pixels, with an indicated transmission point, receivers and a simulated wave field; (**b**) wave propagation signals; (**c**) tomographic image.

In this study, through-transmission tomography was applied (cf. [27,29,32–34]). The tomographic procedure is based on elastic waves passing through the tested specimen, from the transmitter (T) to receivers (R) (Figure 1b). The image reconstruction is performed based on the information obtained from the received pulses. Typically, this information is the time-of-flight (TOF), measured along many ray paths. Based on the known geometry of the specimen, the average velocity of wave propagation can be determined. This velocity depends on the mass density, modulus of elasticity and Poisson's ratio, characterizing the material inside the tested specimen. Any obstacle present in the path of the travelling wave results in a change of the propagation time. Defects, in the form of air voids or cracks on the wave path, cause a delay in the ray reaching the receiver. On the other hand, inclusions made of materials characterized by higher propagation velocities than the neighboring medium lead to an increase of the total velocity along all the rays passing through this inclusion.

The TOF between the transmitter and receiver can be represented by a line integral of the transition time distribution along the propagation way w:

$$t = \int_T^R dt = \int_T^R s\, dw = \int_T^R \frac{1}{v}\, dw,$$ (1)

where v denotes the average velocity, and s denotes the slowness, which is the inverse of the velocity. Therefore, by measuring the direct wave passing time, it is possible to determine the distribution of local velocities in the tested cross-section. The reconstitution of the velocity profile v_j for each cell in the plane of the transition of wave rays from the transmitter to the receiver can be performed based on the following formula:

$$t_i = \sum_{j=1}^{n} w_{ij} s_j, \quad i = 1, 2, 3, \ldots, m, \quad j = 1, 2, 3, \ldots, n, \tag{2}$$

where m is the number of measurement paths (rays), n is the number of cells (pixels), w_{ij} denotes the transition way of the i-ray through the j-pixel, t_i denotes the transition time of the P-wave, between the transmitter and receiver along the i-ray, and s_j is the slowness at pixel j, whose inverse is the velocity v_j, i.e., $v_j = 1/s_j$. It is assumed that the value of the velocity v_j in individual cells is constant.

Equation (2) can be expressed in matrix form, with known matrices t and w (representing the passage path through a given cell and the transition time along a given path, respectively) as well as the unknown matrix s (slowness):

$$
\begin{bmatrix} t_1 \\ t_2 \\ t_3 \\ \vdots \\ t_i \end{bmatrix}
=
\begin{bmatrix}
w_{11} & w_{12} & w_{13} & \cdots & w_{1j} \\
w_{21} & w_{22} & w_{23} & \cdots & w_{2j} \\
w_{31} & w_{32} & w_{33} & \cdots & w_{3j} \\
\vdots & \vdots & \vdots & \ddots & \vdots \\
w_{i1} & w_{i2} & w_{i3} & \cdots & w_{ij}
\end{bmatrix}
\begin{bmatrix} s_1 \\ s_2 \\ s_3 \\ \vdots \\ s_j \end{bmatrix}. \tag{3}
$$

In computed tomography, the above system of equations is usually overdetermined or underdetermined, so the problem, described by Equation (3), is ill posed. It can be efficiently solved using the algebraic reconstruction technique (ART) [38]. In the first step of the calculations, each cell is assigned the same slowness (inverse of the velocity value, equal to the average velocity of the wave propagation in the examined material). In this way, an initial image is created. Next, the iteration process is started, and corrections are calculated according to the relation [39]:

$$s_j^{(k)} = s_j^{(k-1)} + \frac{w_{ij} \Delta t_i}{\sum_{j=1}^{n} w_{ij}^2}, \tag{4}$$

where Δt_i is the difference between the time of the original projection and the rebuilding time. The system of equations is iteratively solved until the reconstructed travel time reaches the measured travel time, with an established error.

3. Materials and Methods

3.1. Description of Specimens

A masonry pillar was used as a testing object. The pillar was manufactured using solid bricks, with dimensions of 25 cm × 12 cm × 6.5 cm. Joints between bricks were 1 cm thick and were filled with gauged (cement-lime) mortar. Nine brick layers were used to build the specimen. The model had a length of 66.5 cm and cross-section dimensions of 38 cm × 38 cm. The geometry of the pillar and a view of the even and odd layers are shown in Figure 2.

The test specimens included one pristine pillar and three pillars with inclusions (Figure 3). The first pillar (#1) was prepared as a reference model. Pillars #2, #3 and #4 had a square hole through the entire height of the specimen. The hole, with dimensions of 5.3 cm × 5.3 cm, was situated at a distance of 7.7 cm from the pillar edges (Figure 2b). The hole in pillar #2 remained empty, while in the case of pillars #3 and #4, an anchor was mounted in it. The threaded steel bar had a diameter of 32 cm, and was grouted using two types of mortar. In pillar #3, the anchor was mounted using gypsum mortar to represent a weak bond, while in the case of pillar #4, cement mortar was used to represent a strong bond.

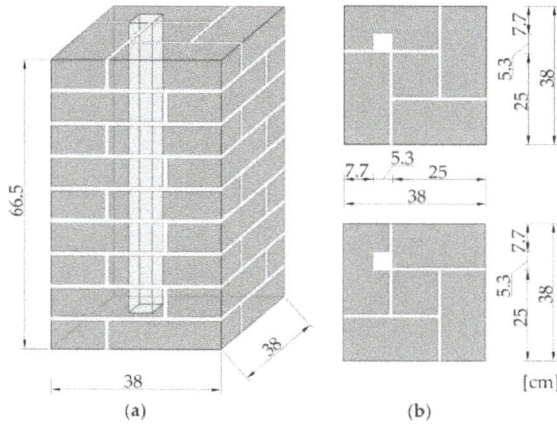

Figure 2. Geometry of the tested pillars: (**a**) 3D view; (**b**) plane view of even and odd layers.

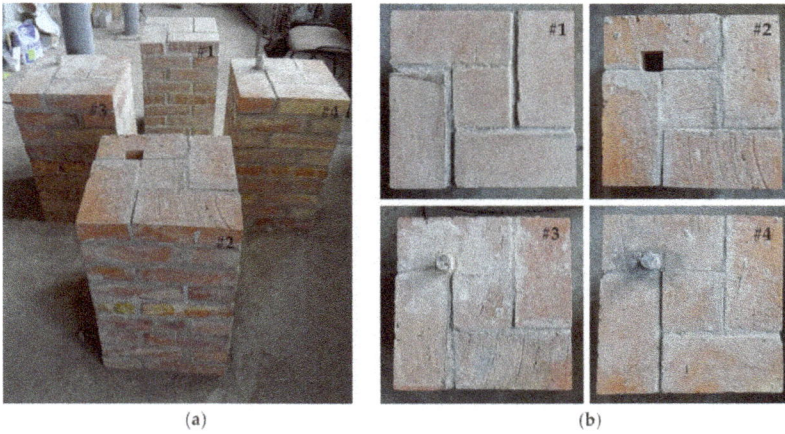

Figure 3. Photographs of laboratory models of the brick pillars: (**a**) overall view; (**b**) plane views of specimens #1 to #4.

3.2. Identification of the Material Parameters

The tests were carried out on a single brick as well as on cubes made of cement mortar, gypsum mortar and gauged mortar, with dimensions of $10 \times 10 \times 10$ cm^3. Each sample was measured and weighted, and then the mass density ρ was calculated. The Young's modulus E of brick and mortar was determined by a non-destructive approach using the ultrasonic pulse velocity (UPV) method ([40–42]). This approach was chosen because in this study a dynamic analysis was carried out in the ultrasonic frequency, which requires the identification of the dynamic modulus. In each sample, the transmission time of the P-wave was measured, and the P-wave velocity c_p was calculated. The Poisson's ratio ν was assumed, following Alberto et al. [43], as 0.15 for mortar and 0.2 for brick. Finally, the Young's modulus was calculated according to the following equation:

$$c_p = \sqrt{\frac{E(1-\nu)}{\rho(1+\nu)(1-2\nu)}}. \tag{5}$$

The results of the identified parameters are given in Table 1.

Table 1. Mechanical properties of the bricks and mortar used for the test specimens.

Material	Density [kg/m^3]	Young's Modulus [GPa]	Poisson's Ratio [-]
brick	1642.04	10.55	0.20
cement mortar	1719.10	8.53	0.15
gauged (cement lime) mortar	1799.90	8.20	0.15
gypsum mortar	1185.70	3.70	0.15

3.3. Experimental Setup

After 28 days, the masonry pillars were subjected to ultrasonic tests. The velocity of the stress waves was measured using the ultrasonic pulse velocity method. The PUNDIT PL-200 instrument (Proceq SA, Schwerzenbach, Switzerland) was utilized for measurements of the ultrasonic pulse velocity. Two exponential transducers of 54 kHz were used in the through transmission mode, and dry coupling was applied. The experimental setup is shown in Figure 4.

(a) (b)

Figure 4. Experimental setup: (**a**) testing in through transmission mode; (**b**) registration unit and exponential transducers.

Measurements were carried out on the middle (i.e., 5th) layer of bricks. Two experiments were conducted. In the first test (configuration #A), 7 measurement points were distributed at each pillar edge, while in the second test (configuration #B), measurements were taken at 13 points. The configurations of the measurement points are given in Figure 5. Transmitting points are denoted by T, and receiving points by R. The receiving transducer was set at a given point, and then the transmitting transducer was moved from the first to the last point on the opposite wall. There were 14 transmitting/receiving points for configuration #A and 23 transmitting/receiving points for configuration #B.

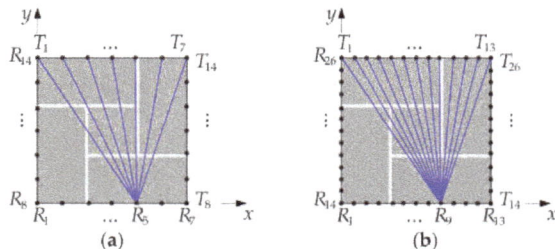

(a) (b)

Figure 5. Configuration of the measurement points: (**a**) configuration #A (transmission of waves at points T_1-T_{14} and sensing of waves at points R_1-R_{14}), (**b**) configuration #B (transmission of waves at points T_1-T_{26} and sensing of waves at points R_1-R_{26}).

3.4. FEM Modeling

Numerical analyses of the stress wave propagation were carried out using the finite element method (FEM) in the Abaqus/Explicit. A two-dimensional model of the fifth layer of the pillar was developed using four-node plane strain elements, with reduced integration (CPE4R). The size of all elements in the model was 1 mm × 1 mm. The material of the pillar was assumed to have linear elastic behavior. The material properties of the bricks and mortar were determined experimentally (Table 1), and for the steel, the following parameters were adopted: E = 200 GPa, ρ = 7850 kg/m^3, v = 0.3. The connections between the bricks and mortar as well as between the mortar and steel were modelled as a tie constraint. The boundary conditions were assumed to be free on all edges. The excitation signal was a one-cycle sine wave of 54 kHz frequency, modulated by the Hann window. The size of the integration step was 10^{-7} s. The output acceleration signals were recorded at the same points as in the experimental investigations (Figure 5).

3.5. Configuration of Pixels and Paths

The acquired experimental and numerical signals were processed to obtain tomography images. At first, the cross-section of the pillar was divided into 49 pixels, in the case of configuration #A, and 169 pixels, in the case of configuration #B. Data analysis was conducted in three stages, with different arrangements of pats (Figure 6). In the first stage, paths were assumed only between opposite transmitting/receiving points. There were 14 paths in configuration #A.1 and 26 paths in configuration #B.1. In the second stage, from each transmitting point, two or three paths were considered, giving 38 paths in configuration #A.2 and 74 paths in configuration #B.2. The third stage took into account seven paths from each transmitting point. As a result, 98 paths were traced in configuration #A.3 and 338 paths in configuration #B.3.

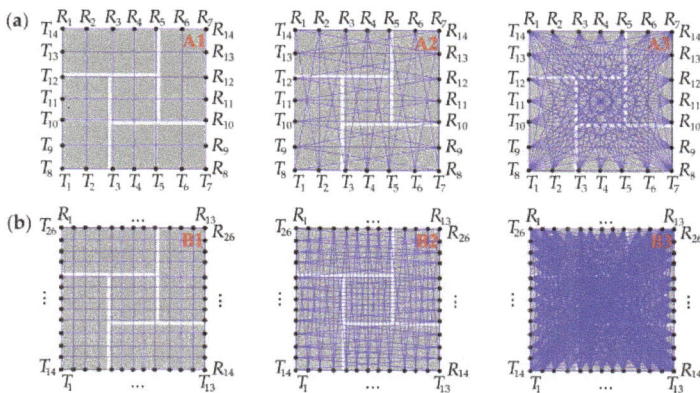

Figure 6. Configuration of paths: (**a**) configuration #A (#A1, #A2 and #A3); (**b**) configuration #B (#B1, #B2 and #B3).

4. Results and Discussion

Ultrasound tomography images, determined based on numerical FEM signals, are shown in Figures 7 and 8, with a division into 49 and 169 pixels, respectively (see Section 3.4). In general, the increase in the number of transmitters, as well as the increase in the number of paths, improved the resolution of the velocity maps and provided a more precise image of the internal structure of the pillars.

Tomograms for the intact pillar (#1) enabled the identification of the arrangement of joints (Figures 7a and 8a). Joints made of cement-lime mortar were characterized by a lower wave propagation velocity than the brick itself. The impact of the existence of joints on the tomographic images was visible, regardless of the number of transmitters used and the number of measurement paths. However, maps created based on the simplest configurations of paths (#A.1, #B.1) did not give satisfactory results regarding precise joint localization. The use of vertical and horizontal wave paths enabled only the identification of the directions of the mortar joints. More accurate tomographic images were obtained when a larger amount of measurement data was collected. Increasing the number of paths (configurations #A.2–A.3 and #B.2–B.3) allowed the position of joints in the considered cross-section and the exact brick arrangement pattern to be determined.

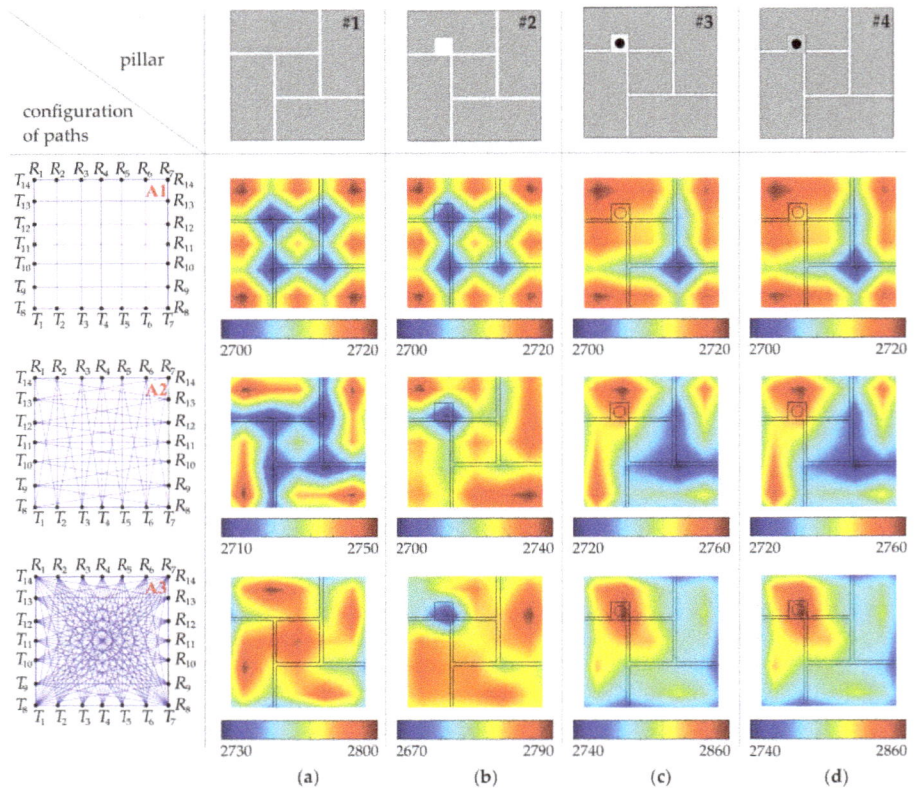

Figure 7. Tomographic velocity images (values in (m/s)), with a division into 49 pixels (configuration #A), obtained from FEM signals: (**a**) intact pillar; (**b**) pillar with the square hole; (**c**) pillar with the steel bar grouted by means of gypsum mortar; (**d**) pillar with the steel bar grouted by means of cement mortar.

Tests conducted for the pillar with the hole (pillar #2) allowed its position to be estimated, but only when a sufficiently large number of ray paths was used (Figures 7b and 8b). In the case of configurations of vertical and horizontal paths (configurations #A.1 and #B.1), the obtained tomographic images were very similar to those obtained for the intact pillar. To obtain the precise localization of the hole, the configuration, including all paths between the transmitting point and receiving points, were required. For such configurations (#A.3 and #B.3), a clear area, with a lower value of the velocity, can be observed. It was also possible to identify the arrangement of mortar joints, especially in the case of the image divided into 169 pixels.

The results for the pillars with the steel bar grouted by means of gypsum cement mortar are shown in Figure 7c,d (49 pixels) and Figure 8c,d (169 pixels). Tomographic velocity images of the simplest paths differed considerably between configuration #A.1 and #B.1. In the case of configuration #A.1, no path crossed the steel bar, therefore, it was not possible to localize the bar. However, since two paths passed directly along the joints, two straight lines with lower velocities appeared in the map, indicating the direction of the mortar joints. When 26 instead of 14 horizontal and vertical paths were used, two perpendicular paths crossed the steel bar. Therefore, for configuration #B.1, two longitudinal areas with a higher velocity appeared on the maps. These lines were the effect of blurring, which occurs in linear tomography. They indicated the presence of the bar but did not allow for its exact localization. The precise localization was achieved after considering more paths crossing the bar (configurations #A.3 and B.3). As a result, a distinct area with higher velocity values can be observed around the bar position. It should also be noted that the maps for pillars #3 and #4 are almost the same. This means that the main influence on the velocity distribution was the presence of the bar, and therefore, it was not possible to determine the type of mortar (gypsum or cement) used.

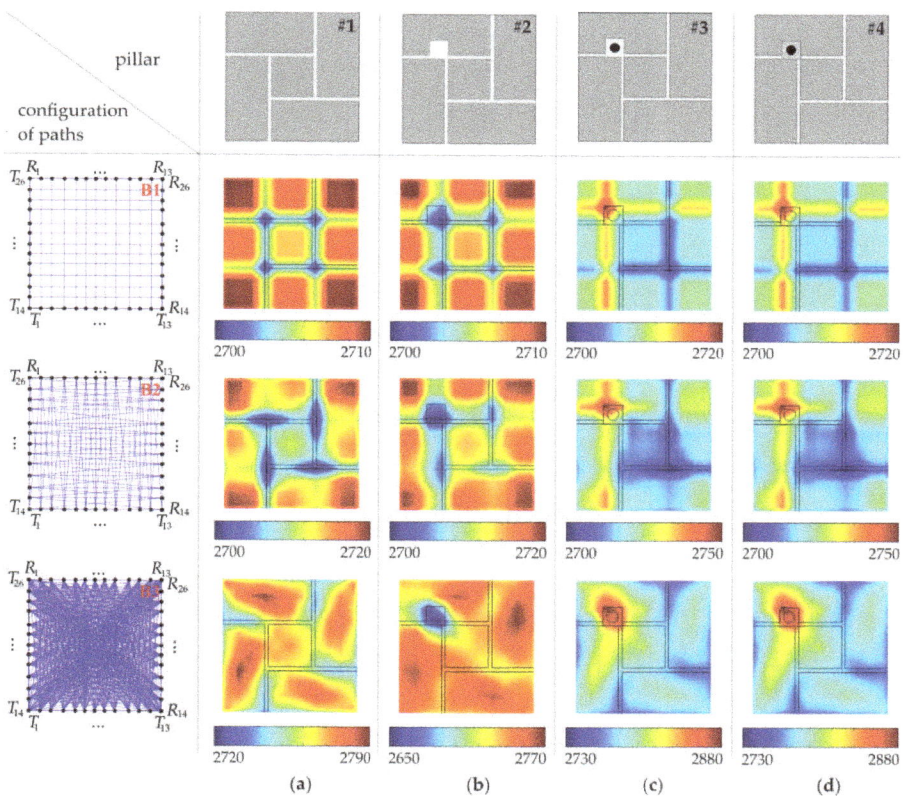

Figure 8. Tomographic velocity images (values in (m/s)), with a division into 169 pixels (configuration #B), obtained from FEM signals: (**a**) intact pillar; (**b**) pillar with the square hole; (**c**) pillar with the steel bar grouted by means of gypsum mortar; (**d**) pillar with the steel bar grouted by means of cement mortar.

The quantitative analysis of the wave propagation values through pillars along many different directions, is given in Tables 2 and 3. For each configuration of paths (#A.1–A.3 and #B.1–B.3), the velocity values were calculated along all rays, and then the minimum and maximum values as well as the mean value of the wave velocity were listed. Moreover, measures of variability, i.e., the standard deviation (SD) and coefficient of variation (CV), were calculated. The average values of the velocity of the particular arrangements were close to each other and ranged from 2719.18 m/s to 2745.48 m/s. The standard deviation ranged from 17.2 m/s to 29 m/s, and the coefficient of variation ranged from 0.63% to 1.07%. This indicates a small variation in the speed of wave propagation along the considered paths. Due to the idealized connections between mortar and bricks, implemented in FEM models, the velocity variations were caused only by internal inclusions.

Table 2. Numerical wave propagation velocities for configuration #A.

Pillar	Arrangement	v_{min} [m/s]	v_{max} [m/s]	$\Delta v = v_{max} - v_{min}$ [m/s]	v_{avg} [m/s]	SD [m/s]	CV [%]
	A1	2699.52	2750.93	51.41	2732.93	22.0	0.81
#1	A2	2699.52	2765.47	65.95	2736.24	21.0	0.77
	A3	2681.93	2767.02	85.09	2730.40	20.4	0.75
	A1	2692.63	2751.52	58.89	2733.03	23.2	0.85
#2	A2	2640.62	2766.05	125.44	2731.99	27.7	1.02
	A3	2640.62	2766.05	125.44	2721.84	28.8	1.06
	A1	2700.46	2753.27	52.81	2738.87	17.4	0.64
#3	A2	2700.46	2822.93	122.47	2742.47	22.0	0.80
	A3	2682.24	2825.21	142.97	2737.58	23.0	0.84
	A1	2700.46	2753.27	52.81	2738.87	17.4	0.64
#4	A2	2700.46	2822.93	122.47	2742.49	21.9	0.80
	A3	2682.24	2825.21	142.97	2737.58	23.0	0.84

Table 3. Numerical wave propagation velocities for configuration #B.

Pillar	Arrangement	v_{min} [m/s]	v_{max} [m/s]	$\Delta v = v_{max} - v_{min}$ [m/s]	v_{avg} [m/s]	SD [m/s]	CV [%]
	B1	2699.52	2750.93	51.41	2737.67	17.4	0.64
#1	B2	2693.96	2756.58	62.62	2737.61	17.2	0.63
	B3	2667.36	2771.10	103.74	2727.88	20.9	0.77
	B1	2692.63	2751.52	58.89	2735.22	18.8	0.69
#2	B2	2647.13	2757.36	110.23	2734.08	22.5	0.82
	B3	2618.33	2771.67	153.34	2719.18	29.0	1.07
	B1	2700.46	2810.41	109.95	2745.15	21.5	0.78
#3	B2	2696.75	2810.41	113.66	2745.48	21.4	0.78
	B3	2666.05	2825.21	159.15	2735.68	24.5	0.90
	B1	2700.46	2810.41	109.95	2745.15	21.5	0.78
#4	B2	2696.75	2810.41	113.66	2745.48	21.4	0.78
	B3	2666.05	2825.21	159.15	2735.68	24.5	0.90

The conducted research also allowed the impact of the number of joints along the wave path on the velocity of propagating ultrasonic waves to be determined. This effect was investigated for each of the four pillars (#1–#4), taking into account the results obtained for arrangement #B. While checking the influence of the number of joints on the wave velocities, paths crossing the hole or the bar as well as the paths along the pillar edges were omitted. The velocity of the wave passing through one joint was calculated as the mean obtained between points T_{10}-R_{10}, T_{11}-R_{11}, T_{12}-R_{12}, T_{15}-R_{15}, T_{16}-R_{16}, T_{17}-R_{17}. The velocity of the wave passing through two joints was calculated as the mean obtained between points T_6-R_6, T_7-R_7, T_8-R_8, T_{19}-R_{19}, T_{20}-R_{20}, T_{21}-R_{21}, and finally, for calculations concerning three joints, the data obtained between points T_1-R_{11} and T_3-R_{13} were used. The results of the obtained average propagation velocities are summarized in Table 4. The velocities obtained for each pillar are almost the same, because no considered path crossed internal inclusions. However, the difference between the velocity of the waves passing through one, two, and three joints was observed. The velocity value decreased with the increase of the number of joints on the wave path.

Table 4. The influence of the number of joints on the numerical wave propagation velocities.

Pillar	v_{avg} [m/s] 1 joint	v_{avg} [m/s] 2 joints	v_{avg} [m/s] 3 joints
#1	2747.70	2733.58	2726.77
#2	2747.19	2734.16	2727.26
#3	2747.45	2734.48	2727.47
#4	2747.45	2734.48	2727.47

The experimental tomographic images are shown in Figures 9 and 10, with a division into 49 and 169 pixels, respectively (see Section 3.4). Contrary to numerical maps, the quality of experimental images did not increase with the increase of the number of pixels or paths used. In real masonry elements, connections between mortar and bricks are not ideal. Air voids, pores and a lack of adhesion between a brick and mortar may occur. The heterogeneous nature of the tested specimens as well as the imperfections of the connections between bricks and mortar caused a strong dissipation of the energy of propagating waves and consequently affected the quality of the obtained tomography images.

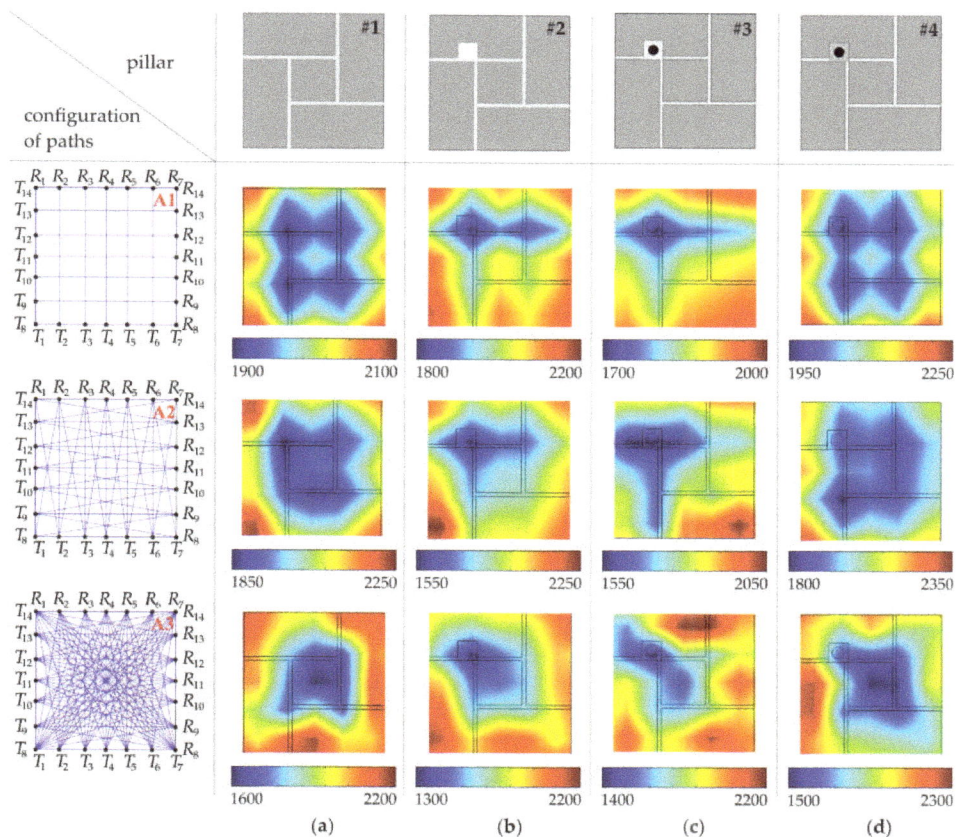

Figure 9. Tomographic velocity images (values in (m/s)), with a division into 49 pixels (configuration #A), obtained from experimental signals: (**a**) intact pillar; (**b**) pillar with the square hole; (**c**) pillar with the steel bar grouted by means of gypsum mortar; (**d**) pillar with the steel bar grouted by means of cement mortar.

Measurements made for the intact pillar (#1) using only vertical and horizontal paths revealed lower propagation velocities at joint intersections (Figures 9a and 10a). With the increase of the number of paths, for configurations #A.3 and #B.3 a zone with lower velocity values appeared in the central part of the pillar due to the accumulation of joints and paths crossing them.

Maps of the pillar with the hole are shown in Figures 9b and 10b. The hole unambiguously disturbed the wave transition. The location of the hole can be indicated as the area with lower velocity values. The identification of the position of the hole was clearest in the case of vertical and horizontal rays (#A.1 and #B.1), however, it was affected by the blurring effect. As in the case of the intact pillar, the accumulation of lower velocities can be seen for configurations #A.2, #A.3 and #B.2, #B.3.

The maps obtained for the pillar with the bar grouted by gypsum mortar (#3) revealed similar patterns as those for the pillar with the hole (Figures 9c and 10c). This may indicate poor adhesion between the bar and the gypsum mortar. On the other hand, the tomographic velocity images for the pillar with the bar grouted by cement mortar (#4) were very similar to those obtained for the intact pillar (Figures 9c and 10c). This means that the connection between the bar and gypsum mortar was of a good quality, and strong adhesion occurred.

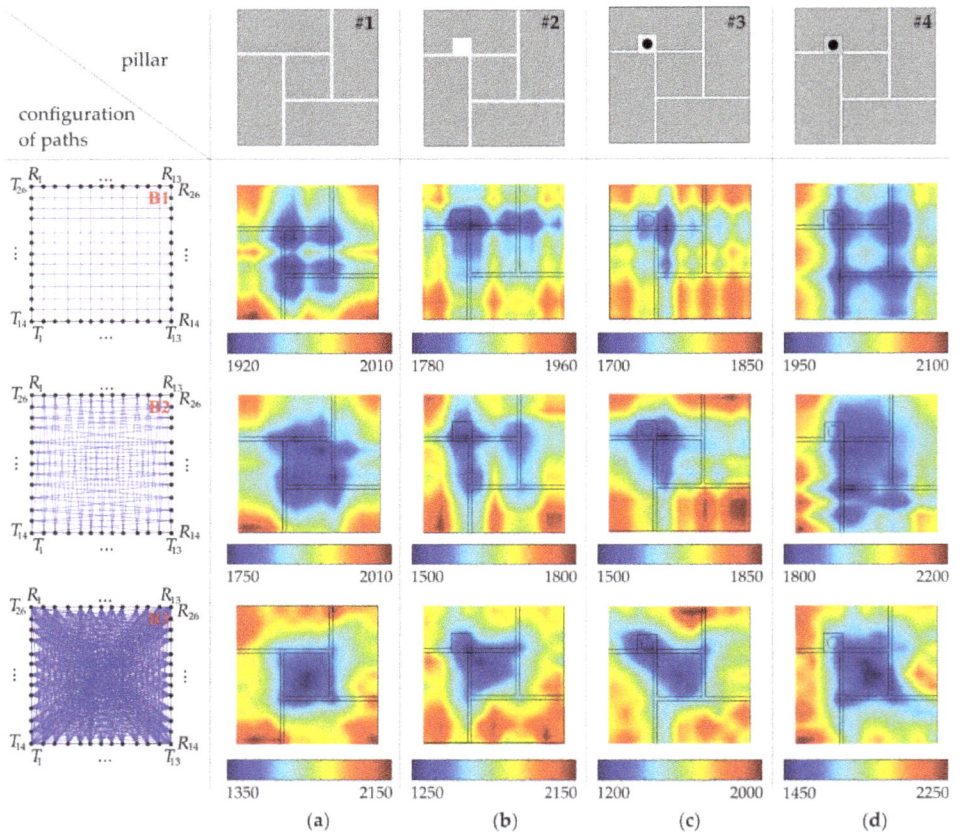

Figure 10. Tomographic velocity images (values in (m/s)), with a division into 169 pixels (configuration #B), obtained from experimental signals: (**a**) intact pillar; (**b**) pillar with the square hole; (**c**) pillar with the steel bar grouted by means of gypsum mortar; (**d**) pillar with the steel bar grouted by means of cement mortar.

The minimum, maximum and average velocities of ultrasonic waves, based on experimental tests for each configuration of paths (#A.1–A.3 and #B.1–B.3), are shown in Tables 5 and 6. The average values of the velocity for the particular pillars varied from 1635.34 m/s to 2163.11 m/s. The velocities for pillars #1 and #4 were higher than those for pillars #2 and #3, which proved weak adhesion in pillar #3 and strong adhesion in pillar #4. The standard deviation varied from 254.4 m/s to 465.7 m/s, and the coefficient of variation varied from 13.42% to 22.4%. These values were approximately 10–20 times larger than those obtained from the FEM simulations. At the same time, the experimental velocities appeared to be smaller than the numerical ones. This is the result of non-ideal connections between the joints and bricks in the real model of pillars.

Table 5. Experimental wave propagation velocities for configuration #A.

Pillar	Arrangement	v_{min} [m/s]	v_{max} [m/s]	$\Delta v = v_{max} - v_{min}$ [m/s]	v_{avg} [m/s]	SD [m/s]	CV [%]
	A1	1687.58	2651.26	963.68	2099.66	363.7	17.32
#1	A2	1687.58	2651.26	963.68	2104.60	296.1	14.07
	A3	1574.48	2651.26	1076.78	2030.73	258.4	12.72
	A1	1253.62	2908.28	1654.66	2100.92	463.0	22.04
#2	A2	1253.62	2908.28	1654.66	2042.58	395.5	19.36
	A3	1253.62	2908.28	1654.66	1927.77	322.5	16.73
	A1	1402.37	2615.69	1213.32	1936.08	304.9	15.75
#3	A2	1313.12	2615.69	1302.58	1876.67	310.2	16.53
	A3	1150.91	2615.69	1464.78	1731.45	291.4	16.83
	A1	1617.59	3068.45	1450.86	2163.11	465.7	21.53
#4	A2	1617.59	3068.45	1450.86	2107.37	365.2	17.33
	A3	1511.52	3068.45	1556.93	2009.67	316.0	15.72

Table 6. Experimental wave propagation velocities for configuration #B.

Pillar	Arrangement	v_{min} [m/s]	v_{max} [m/s]	$\Delta v = v_{max} - v_{min}$ [m/s]	v_{avg} [m/s]	SD [m/s]	CV [%]
	B1	1597.71	2651.26	1053.55	2014.90	330.1	16.38
#1	B2	1493.14	2651.26	1158.12	1969.88	283.3	14.38
	B3	1321.90	2651.26	1329.36	1896.30	254.4	13.42
	B1	1253.62	2908.28	1654.66	1887.48	422.8	22.40
#2	B2	1117.83	2908.28	1790.45	1807.67	362.3	20.04
	B3	1111.81	2908.28	1796.47	1722.46	311.5	18.08
	B1	1222.19	2615.69	1393.51	1814.67	314.8	17.35
#3	B2	1222.19	2615.69	1393.51	1748.82	314.9	18.01
	B3	1056.88	2615.69	1558.81	1635.34	277.0	16.94
	B1	1407.43	3068.45	1661.02	2055.87	424.4	20.64
#4	B2	1407.43	3068.45	1661.02	2001.88	378.7	18.92
	B3	1354.30	3068.45	1714.15	1921.07	315.8	16.44

The comparison of the propagation velocities, depending on the wave transition through one, two or three mortar joints, is shown in Table 7. The same wave paths as those in the numerical calculations were assumed. The velocities obtained for each pillar differed due to non-ideal connections between the mortar and bricks. Differences between the velocity of the waves passing through one, two, and three joints were also observed. As in the case of the FEM simulations, the velocity value decreased with the increase of the number of joints on the wave path.

Table 7. The influence of the number of joints on the experimental wave propagation velocities.

Pillar	v_{avg} [m/s] 1 joint	v_{avg} [m/s] 2 joints	v_{avg} [m/s] 3 joints
#1	2082.75	1727.25	1682.13
#2	1967.77	1657.43	1638.79
#3	1923.85	1578.54	1557.31
#4	2002.48	1800.36	1708.91

5. Conclusions

In this study, the ultrasonic tomography technique was applied to the non-destructive diagnostics of masonry pillars. Experimental and numerical investigations were performed on four laboratory specimens: one intact pillar and three pillars with inclusions. The conducted investigations focused on the tomographic velocity reconstruction and the assessment of the internal structure of the pillars.

The study of ultrasonic tomography applied to the assessment of masonry pillars led to the following conclusions:

1. The increase of the number of pixels and paths did not guarantee an improvement of the quality of the tomographic images. In numerical simulations made for the pillar models with idealized connections between the mortar and bricks, more accurate tomograms were obtained when a larger amount of measurement data was collected with a denser division in cells. This observation has not been confirmed in experimental tests conducted on real specimens, with connections between bricks and mortar influenced by air voids and non-ideal adhesion.

2. The change in the velocity value was observed depending on the number of joints through which the wave passed. The velocity decreased with the increase of the number of joints on the wave path.

3. Detection of the arrangement of joints in the cross-section was possible. The joints could be observed, based on numerical data, as line patterns with lower velocities. However, in experimental tests, the increase of the number of paths resulted in the appearance of a large zone, with lower velocity values in the central part of the pillar due to the accumulation of joints and paths crossing them.

4. The inclusion in the form of a hole was identified in both the numerical and experimental tests as an area with lower velocity values.

5. The inclusion in the form of an embedded bar was identified in the numerical data as an area with higher velocity values. This observation was not confirmed in experimental tests due to the existence of many factors that slowed the speed of the wave.

6. The experimental tests enabled the assessment of the quality of the adhesive connection between a steel reinforcing bar embedded inside pillars, and the surrounding pillar body. The tomograms obtained for the pillar with the bar grouted by gypsum mortar revealed similar patterns as those for the pillar with the hole, which indicated poor adhesion. The maps obtained for the pillar with the bar grouted by cement mortar were similar to those obtained for the intact pillar, which indicated strong adhesion.

To summarize, ultrasonic tomography appeared to be an effective technique for the reconstruction of the internal structure of brick pillars. The presented approach may be particularly useful in the diagnostics of applied strengthening and the assessment of the compatibility of reinforcement materials with brick structures, and it may become the basis for the strategic planning of repair procedures. Further investigations should consider the tomographic imaging of masonry structures, accessible from one or two sides, to develop efficient tomographic procedures based on a limited number of inputs and examined ray paths.

Author Contributions: Conceptualization, M.Z. and M.R.; Formal analysis, M.Z. and M.R.; Funding acquisition, M.Z.; Investigation, M.Z.; Methodology, M.Z. and M.R.; Software, M.Z.; Supervision, M.R.; Validation, M.R.; Visualization, M.Z.; Writing–original draft, M.Z.; Writing–review and editing, M.R.

Funding: The research work was carried out within the project No. 2017/27/N/ST8/02399, financed by the National Science Centre, Poland.

Acknowledgments: Abaqus calculations were carried out at the Academic Computer Centre in Gdańsk. The help of Karol Grębowski during the preparation of the pillars is gratefully acknowledged.

Conflicts of Interest: The authors declare no conflict of interest.

References

1. Binda, L.; Saisi, A.; Tiraboschi, C. Investigation procedures for the diagnosis of historic masonries. *Constr. Build. Mater.* **2000**, *14*, 199–233. [CrossRef]
2. Piroglu, F.; Ozakgul, K. Site investigation of masonry buildings damaged during the 23 October and 9 November 2011 Van Earthquakes in Turkey. *Nat. Hazards Earth Syst. Sci.* **2013**, *13*, 689–708. [CrossRef]
3. Jasiński, R.; Drobiec, Ł. Comparison Research of Bed Joints Construction and Bed Joints Reinforcement on Shear Parameters of AAC Masonry Walls. *J. Civ. Eng. Archit.* **2016**, *10*, 1329–1343. [CrossRef]
4. Vasanelli, E.; Sileo, M.; Leucci, G.; Calia, A.; Aiello, M.A.; Micelli, F. Mechanical characterization of building stones through DT and NDT tests: Research of correlations for the in situ analysis of ancient masonry. *Key Eng. Mater.* **2015**, *628*, 85–89. [CrossRef]
5. McCann, D.M.; Forde, M.C. Review of NDT methods in the assessment of concrete and masonry structures. *NDT E Int.* **2001**, *34*, 71–84. [CrossRef]
6. Schuller, M.P. Nondestructive testing and damage assessment of masonry structures. *Prog. Struct. Eng. Mater.* **2003**, 239–251. [CrossRef]
7. Paasche, H.; Wendrich, A.; Tronicke, J.; Trela, C. Detecting voids in masonry by cooperatively inverting P-wave and georadar traveltimes. *J. Geophys. Eng.* **2008**, *5*, 256–267. [CrossRef]
8. Bosiljkov, V.; Uranjek, M.; Žarnić, R.; Bokan-Bosiljkov, V. An integrated diagnostic approach for the assessment of historic masonry structures. *J. Cult. Herit.* **2010**, *11*, 239–249. [CrossRef]
9. Santos-Assunçao, S.; Perez-Gracia, V.; Caselles, O.; Clapes, J.; Salinas, V. Assessment of complex masonry structures with GPR compared to other non-destructive testing studies. *Remote Sens.* **2014**, *6*, 8220–8237. [CrossRef]
10. Khan, F.; Rajaram, S.; Vanniamparambil, P.A.; Bolhassani, M.; Hamid, A.; Kontsos, A.; Bartoli, I. Multi-sensing NDT for damage assessment of concrete masonry walls. *Struct. Control Heal. Monit.* **2015**, *22*, 449–462. [CrossRef]
11. Micelli, F.; Cascardi, A.; Marsano, M. Seismic strengthening of a theatre masonry building by using active FRP wires. In *Brick and Block Masonry: Proceedings of the 16th International Brick and Block Masonry Conference*; CRC Press: Padova, Italy, 2016; pp. 753–761.
12. La Mendola, L.; Lo Giudice, E.; Minafò, G. Experimental calibration of flat jacks for in-situ testing of masonry. *Int. J. Archit. Herit.* **2018**. [CrossRef]
13. Rucka, M.; Lachowicz, J.; Zielińska, M. GPR investigation of the strengthening system of a historic masonry tower. *J. Appl. Geophys.* **2016**, *131*, 94–102. [CrossRef]
14. Lachowicz, J.; Rucka, M. Diagnostics of pillars in St. Mary's Church (Gdańsk, Poland) using the GPR method. *Int. J. Archit. Herit.* **2018**, 1–11. [CrossRef]
15. Paganoni, S.; D'Ayala, D. Testing and design procedure for corner connections of masonry heritage buildings strengthened by metallic grouted anchors. *Eng. Struct.* **2014**, *70*, 278–293. [CrossRef]
16. Collini, L.; Fagiani, R.; Garziera, R.; Riabova, K.; Vanali, M. Load and effectiveness of the tie-rods of an ancient Dome: Technical and historical aspects. *J. Cult. Herit.* **2015**, *16*, 597–601. [CrossRef]
17. Ural, A.; Firat, F.K.; Tuğrulelçi, S.; Kara, M.E. Experimental and numerical study on effectiveness of various tie-rod systems in brick arches. *Eng. Struct.* **2016**, *110*, 209–221. [CrossRef]
18. Pisani, M.A. Theoretical approach to the evaluation of the load-carrying capacity of the tie rod anchor system in a masonry wall. *Eng. Struct.* **2016**, *124*, 85–95. [CrossRef]
19. Ombres, L.; Verre, S. Masonry columns strengthened with Steel Fabric Reinforced Cementitious Matrix (S-FRCM) jackets: Experimental and numerical analysis. *Measurement* **2018**, *127*, 238–245. [CrossRef]
20. Micelli, F.; Cascardi, A.; Aiello, M.A. A Study on FRP-Confined Concrete in Presence of Different Preload Levels. In Proceedings of the 9th International Conference on Fibre-Reinforced Polymer (FRP) Composites in Civil Engineering—CICE, Paris, France, 17–19 July 2018; pp. 493–499.

21. Ferrotto, M.F.; Fischer, O.; Cavaleri, L. A strategy for the finite element modeling of FRP-confined concrete columns subjected to preload. *Eng. Struct.* **2018**, *173*, 1054–1067. [CrossRef]
22. Rao, J.; Ratassepp, M.; Lisevych, D.; Hamzah Caffoor, M.; Fan, Z. On-Line Corrosion Monitoring of Plate Structures Based on Guided Wave Tomography Using Piezoelectric Sensors. *Sensors* **2017**, *17*, 2882. [CrossRef]
23. Zhao, X.; Royer, R.L.; Owens, S.E.; Rose, J.L. Ultrasonic Lamb wave tomography in structural health monitoring. *Smart Mater. Struct.* **2011**, *20*, 105002. [CrossRef]
24. Leonard, K.R.; Malyarenko, E.V.; Hinders, M.K. Ultrasonic Lamb wave tomography. *Inverse Probl.* **2002**, *18*, 1795–1808. [CrossRef]
25. Prasad, S.M.; Balasubramaniam, K.; Krishnamurthy, C.V. Structural health monitoring of composite structures using Lamb wave tomography. *Smart Mater. Struct.* **2004**, *13*, N73–N79. [CrossRef]
26. Martin, J.; Broughton, K.J.; Giannopolous, A.; Hardy, M.S.A.; Forde, M.C. Ultrasonic tomography of grouted duct post-tensioned reinforced concrete bridge beams. *NDT E Int.* **2001**, *34*, 107–113. [CrossRef]
27. Chai, H.K.; Liu, K.F.; Behnia, A.; Yoshikazu, K.; Shiotani, T. Development of a tomography technique for assessment of the material condition of concrete using optimized elastic wave parameters. *Materials* **2016**, *9*, 291. [CrossRef] [PubMed]
28. Chai, H.K.; Momoki, S.; Kobayashi, Y.; Aggelis, D.G.; Shiotani, T. Tomographic reconstruction for concrete using attenuation of ultrasound. *NDT E Int.* **2011**, *44*, 206–215. [CrossRef]
29. Aggelis, D.G.; Tsimpris, N.; Chai, H.K.; Shiotani, T.; Kobayashi, Y. Numerical simulation of elastic waves for visualization of defects. *Constr. Build. Mater.* **2011**, *25*, 1503–1512. [CrossRef]
30. Schabowicz, K.; Suvorov, V.A. Nondestructive testing of a bottom surface and construction of its profile by ultrasonic tomography. *Russ. J. Nondestruct. Test.* **2014**, *50*, 109–119. [CrossRef]
31. Schabowicz, K. Ultrasonic tomography—The latest nondestructive technique for testing concrete members—Description, test methodology, application example. *Arch. Civ. Mech. Eng.* **2014**, *14*, 295–303. [CrossRef]
32. Haach, V.G.; Ramirez, F.C. Qualitative assessment of concrete by ultrasound tomography. *Constr. Build. Mater.* **2016**, *119*, 61–70. [CrossRef]
33. Choi, H.; Popovics, J.S. NDE application of ultrasonic tomography to a full-scale concrete structure. *IEEE Trans. Ultrason. Ferroelectr. Freq. Control* **2015**, *62*, 1076–1085. [CrossRef] [PubMed]
34. Choi, H.; Ham, Y.; Popovics, J.S. Integrated visualization for reinforced concrete using ultrasonic tomography and image-based 3-D reconstruction. *Constr. Build. Mater.* **2016**, *123*, 384–393. [CrossRef]
35. Schullerl, M.; Berra, M.; Atkinson, R.; Binda, L. Acoustic tomography for evaluation of unreinforced masonry. *Constr. Build. Makrials* **1997**, *11*, 199–204. [CrossRef]
36. Binda, L.; Saisi, A.; Zanzi, L. Sonic tomography and flat-jack tests as complementary investigation procedures for the stone pillars of the temple of S. Nicolo 1'Arena (Italy). *NDT E Int.* **2003**, *36*, 215–227. [CrossRef]
37. Pérez-Gracia, V.; Caselles, J.O.; Clapés, J.; Martinez, G.; Osorio, R. Non-destructive analysis in cultural heritage buildings: Evaluating the Mallorca cathedral supporting structures. *NDT E Int.* **2013**, *59*, 40–47. [CrossRef]
38. Kak, A.C.; Slaney, M. *Principles of Computerized Tomographic Imaging*; The Instituite of Electrical and Electronics Engineers, Inc.: New York, NY, USA, 1988.
39. Oliveira, E.F.; Dantas, C.C.; Vasconcelos, D.A.A.; Cadiz, F. Comparison Among Tomographic Reconstruction Algorithms With a Limited Data. In Proceedings of the International Nuclear Atlantic Conference-INAC 2011, Belo Horizonte, Brazil, 24–28 October 2011.
40. Lu, X.; Sun, Q.; Feng, W.; Tian, J. Evaluation of dynamic modulus of elasticity of concrete using impact-echo method. *Constr. Build. Mater.* **2013**, *47*, 231–239. [CrossRef]
41. Węglewski, W.; Bochenek, K.; Basista, M.; Schubert, T.; Jehring, U.; Litniewski, J.; Mackiewicz, S. Comparative assessment of Young's modulus measurements of metal-ceramic composites using mechanical and non-destructive tests and micro-CT based computational modeling. *Comput. Mater. Sci.* **2013**, *77*, 19–30. [CrossRef]
42. Wolfs, R.J.M.; Bos, F.P.; Salet, T.A.M. Correlation between destructive compression tests and non-destructive ultrasonic measurements on early age 3D printed concrete. *Constr. Build. Mater.* **2018**, *181*, 447–454. [CrossRef]
43. Alberto, A.; Antonaci, P.; Valente, S. Damage analysis of brick-to-mortar interfaces. *Procedia Eng.* **2011**, *10*, 1151–1156. [CrossRef]

materials MDPI

Article

Non-Destructive Testing of a Sport Tribune under Synchronized Crowd-Induced Excitation Using Vibration Analysis

Karol Grębowski [1],*, Magdalena Rucka [2] and Krzysztof Wilde [2]

[1] Department of Technical Fundamentals of Architectural Design, Faculty of Architecture, Gdansk University of Technology, Narutowicza 11/12, 80-233 Gdansk, Poland
[2] Department of Mechanics of Materials and Structures, Faculty of Civil and Environmental Engineering, Gdansk University of Technology, Narutowicza 11/12, 80-233 Gdansk, Poland
* Correspondence: karol.grebowski@pg.edu.pl; Tel.: +48-58-347-1877

Received: 17 June 2019; Accepted: 2 July 2019; Published: 4 July 2019

check for updates

Abstract: This paper presents the concept of repairing the stand of a motorbike speedway stadium. The synchronized dancing of fans cheering during a meeting brought the stand into excessive resonance. The main goal of this research was to propose a method for the structural tuning of stadium stands. Non-destructive testing by vibration methods was conducted on a selected stand segment, the structure of which recurred on the remaining stadium segments. Through experiments, we determined the vibration forms throughout the stand, taking into account the dynamic impact of fans. Numerical analyses were performed on the 3-D finite element method (FEM) stadium model to identify the dynamic jump load function. The results obtained on the basis of sensitivity tests using the finite element method allowed the tuning of the stadium structure to successfully meet the requirements of the serviceability limit state.

Keywords: non-destructive testing; reinforced concrete grandstand stadium; vibration analysis; crowd-induced excitation; structural tuning

---

1. Introduction

Stadiums are sport objects that require a particular level of attention to ensure human security. During the last century, many tragedies have been caused mainly by negligence at the stage of designing the structure. Nowadays, many structural projects are sought to be optimized. The results are slender, more effective and economical objects; however, they are often sensitive to dynamic impacts.

Since 1902, more than 20 construction disasters of stadium stands have taken place around the world [1,2]. These range from the stand collapse at Ibrox Park stadium in Glasgow (Scotland) on April 5, 1902 during the Scotland vs. England match, which resulted in 26 dead and 550 injured supporters, to the tragedy that took place on November 26, 2007 at the Salvador stadium (Brazil) during the Bahia vs. Vila Nova match, which resulted in eight deaths and 150 injured due to the collapse of the stand.

Strong vibrations caused by moving people mainly occur in structures with low rigidity and mass. Stadiums are an example of this type of structure. High dynamic loads occur, caused by jumping fans cheering their teams. This type of load is often omitted during the structure design, because procedures in current standards or guidelines regarding the jump load are limited.

In recent years, many studies have been devoted to the dynamic impact of stadium structures. Development in sensor technology, through the supply of adapted vibration transducers (low-level vibration, high-level/shock vibration, near-static sensitivity, etc.) in conjunction with the dissemination of digital processing algorithms, has led to a notable increase in the study of vibrations caused by crowd-induced excitation in structures [3]. Modal identification results can be used as part of the vibration reduction methodologies [4]. In [5], it was found that crowd occupation can significantly alter the modal properties of a stadium, and that the changes vary according to crowd configuration. A large number of publications on structural response and stadium vibration resistance due to crowd-induced loads have shown that the dynamic response of stadiums depends not only on the basic mechanical characteristics, i.e., mass, stiffness and damping, but also on the nature of the load, which can become complex in the case of jumping supporters (e.g., [6]). In order to limit the vibrations of the structure, one of the countermeasures is frequency tuning a structure [7,8]. A comprehensive analysis of the literature on dynamic performance tests of existing stadium structures was conducted in [9]. It was concluded that the available knowledge on this subject is not yet sufficiently advanced and thus jump load is not currently included in most design standards. Therefore, the dynamic identification based on non-destructive testing methods is increasingly popular and effective in civil engineering research. Several literature studies have shown how non-destructive testing and structural health monitoring can be efficiently used to assess the durability of reinforced concrete structures [8–15].

An example of stand failure due to the dynamic impact of fans is the case of the Swiss Krono Arena motorbike speedway stadium in Zielona Góra (Poland). The new stand was built in 2009–2010. In July 2010, it was put into service. The stand is a reinforced concrete structure that is used in a very specific way. Speedway meetings are a sport discipline that arouse great emotions and interest among supporters. One of the basic forms of team cheering at the Swiss Krono Arena is the so-called "Labado dance". The dance is considered as almost a hymn among speedway fans. In this dance, fans put their hands on their neighbor's shoulders and in the rhythm of the animator's drumming, they jump simultaneously to the song "(...) we dance labado, labado, (...)". In 2010 it was noticed that the fans' dance causes an increase in the vibrations of the structure, which led to a discussion on the safety of using the new stand. Research on the assessment of the harmfulness of vibrations was carried out in 2010–2011 by a team from University of Zielona Góra, and in the years 2011–2012 by a team from Gdańsk University of Technology. The Labado dance appeared to be dangerous for the stand's structure, because by jumping fans caused a vertical periodic force, which when synchronized movements of a large number of people could lead to resonant vibrations, possibly ending with the collapse of the structure. A large number of people generating regular dynamic excitations could be able to destroy any bridge or reinforced concrete stadium stand, which is why this phenomenon is considered vandalism.

This paper presents non-destructive testing of a sport facility subjected to dynamic interactions in the form of jumping fans. Experimental tests were conducted in the field on the Swiss Krono Arena speedway stadium in Zielona Góra. Later on, numerical simulations were performed to determine the dynamic characteristics of the structure. Validation was carried out by comparing the numerical model with the results obtained in the field. After the validation of the numerical model, the identification of the jump load dynamic function was determined based on laboratory tests. Lastly, the stadium stand tuning model was re-tested in situ after structure strengthening.

2. Theoretical Background of the Experimental Modal Analysis

Modal analysis is a method of determining dynamic properties of structures (i.e., natural frequencies, modal shapes and damping coefficients) under vibrational excitation. The analysis involves the registration of vibrational signals and the application of different signal processing techniques. In general, two types of modal analysis can be distinguished: operational modal analysis (OMA) [16–19] and experimental modal analysis (EMA) [20–24]. Environmental excitations (e.g., wind, sea waves, micro-seismic vibrations, traffic loads, etc.) are used in the OMA technique, and only the structure's response is recorded. In the case of EMA, both excitation and response are measured. The most common approach in EMA is the impulse test, which is based on the excitation of the structure using a modal hammer and the harmonic test, in which electromechanical or piezoelectric actuators are usually applied (Figure 1).

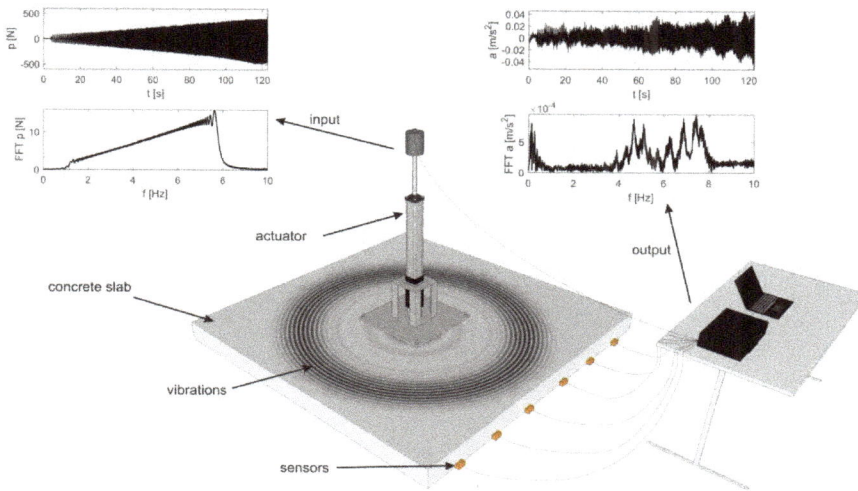

Figure 1. Scheme of the experimental modal analysis technique.

First, the input force $p(t)$ and output displacement $u(t)$ time signals are measured in EMA, and then they are transformed into the frequency domain by Fourier transform, resulting in $P(\omega)$ and $U(\omega)$ signals (Figure 1). The next step is the calculation of the frequency response function (FRF) in all measured points, which is a ratio of the j-th output and the k-th input:

$$H_{jk}(\omega) = \frac{U_j(\omega)}{P_j(\omega)}.$$ (1)

In the practical application of EMA, acceleration signals are often measured instead of displacement signals, as they are more convenient in analytical works. Then, the accelerance function can be defined as:

$$A_{jk}(\omega) = \frac{\ddot{U}_j(\omega)}{P_j(\omega)}.$$ (2)

Both the receptance and accelerance functions describe the same dynamic properties of the system and the relationship between them is given by:

$$A_{jk}(\omega) = -\omega^2 H_{jk}(\omega).$$ (3)

Based on the frequency response functions (FRFs), resonance frequencies, mode shapes and damping coefficients can be identified (e.g., [25,26]).

3. Experimental Dynamic Identification of the Grandstand

3.1. Description of the Structure

The Swiss Krono Arena motorbike speedway stadium is located in Zielona Góra (Poland). The grandstand was erected in 2009–2010 and its geometry is a circular sector with a center angle of 150° (Figure 2).

The stand consists of a reinforced concrete structure in which prefabricated under-seat beams rest on the main beams of reinforced concrete girders. The main girder is composed of two parts and it is supported on three reinforced concrete columns. The first part of the main girder (BR1) is an inclined simply supported beam, located in the bottom part of the stand. The second part of the girder (BR2), with a variable cross-sectional height, also works as a simply supported beam, and it is finished with a cantilever beam located in the upper part of the stand.

(a)   (b)

Figure 2. Swiss Krono Arena: (**a**) the view on the grandstand, (**b**) the location of the grandstand on the land development plan.

Both girders are connected to each other by an articulated joint. The main beams are based on columns via elastomeric pads. Steel anchors prevent horizontal displacement of the main beams against columns. The columns and the monolithic wall supporting the main girder are located on reinforced concrete strip foundations. The walls of additional objects are made of silicate brick, while the stairs structure is made as a prefabricated reinforced concrete element (Figure 3).

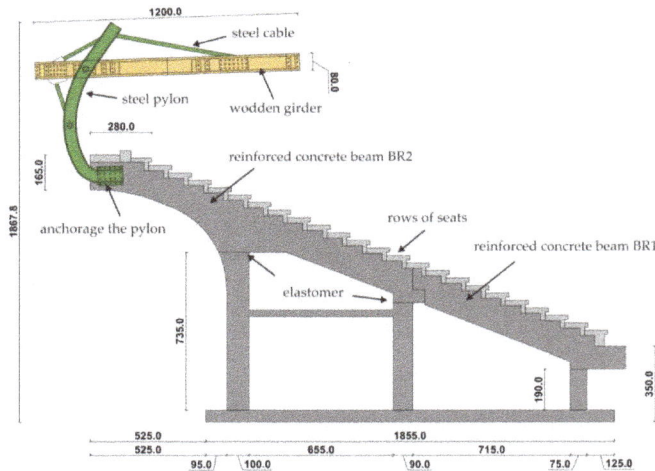

Figure 3. Cross-section of the stand (dimensions in cm).

The roof structure over the stand consists of three elements: steel columns, which are attached to the cantilever ends of the main load-bearing girder; the roof, which consists of wooden girders; and purlins, to which the covering trapezoidal sheeting and cables are attached. The cables connect the wooden roof structure with steel columns. The steel supporting columns of the roof structure are braced with steel braces. Roof bracings are also used in the wooden roof structure.

3.2. Test Procedure and Vibration Measurement

The measurements were conducted on a selected section of the stand, as shown in Figure 4. The experimental program included a sweep sine test and two types of people-induced vibrations, namely single jumps and synchronous Labado dancing.

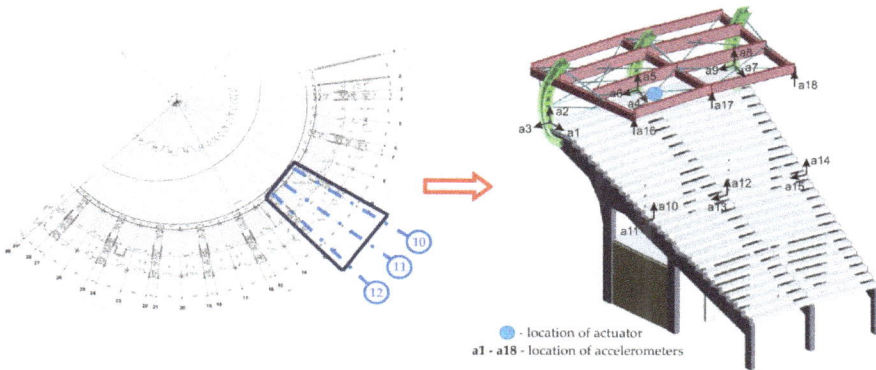

● - location of actuator
a1 - a18 - location of accelerometers

Figure 4. The stand section selected for field tests.

Triaxial piezoelectric accelerometers 356B18 (PCB Piezotronics, Inc., Depew, NY, USA) were used for the measurement of vibrations. They were attached to both the concrete part of the stand as well as to the roof structure. Vibrations signals were registered at nine points (Figure 4), in one, two and three directions. In total, 18 acceleration signals were acquired (a1 to a18). Data acquisition and signal conditioning were performed by the LMS SCADAS portable system (Siemens, Leuven, Belgium). The sampling frequency was set as 256 Hz.

In the first stage, the experimental modal analysis was conducted. The harmonic load was excited by means of the electromechanical actuator shown in Figure 5a. The excitation signal, created by an arbitrary signal generator, was a sweep sine of a smoothly adjustable frequency from 1 Hz to 8 Hz (Figure 5b).

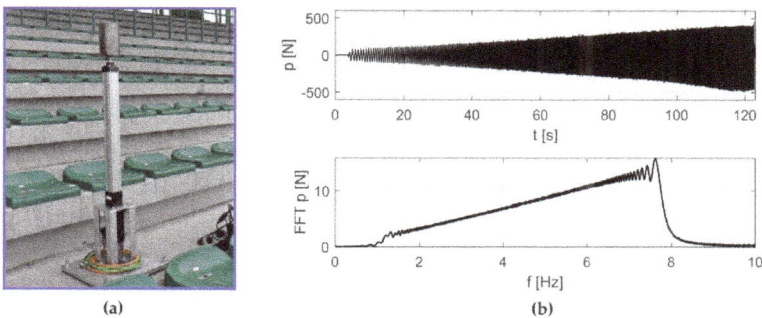

(a)   (b)

Figure 5. Experimental setup: (**a**) electromechanical actuator, (**b**) excitation signal in the time and frequency domains.

Dynamic parameters were determined based on the frequency response functions, according to procedure described in Section 2. The imaginary parts of the FRFs for signals registered on the girder and on the roof are given in Figure 6. Several peaks can be distinguished. The experiment allowed the identification of three natural frequencies in the range from 2 to 5 Hz (Figure 6). Their values were: 3.36 Hz, 3.92 Hz and 4.68 Hz. Experimentally determined mode shapes are presented in Figure 7. The directions and values of displacements for particular degrees of freedom are shown by means of vectors and numerical values. The shapes of modes of the girder and roof are similar, wherein the amplitude for the roof is much larger. The dominant vibration types for each identified form are vertical mode shapes. The main displacements of the girder and roof occurred as a result of vertical motion.

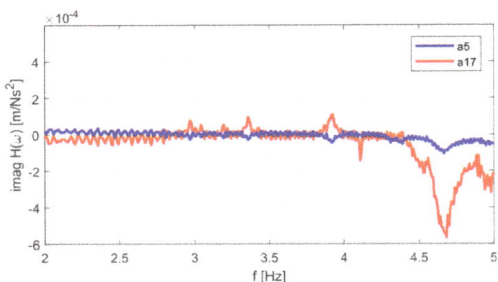

Figure 6. Imaginary parts of the frequency response function (FRF) for acceleration signals a5 (girder) and a17 (roof).

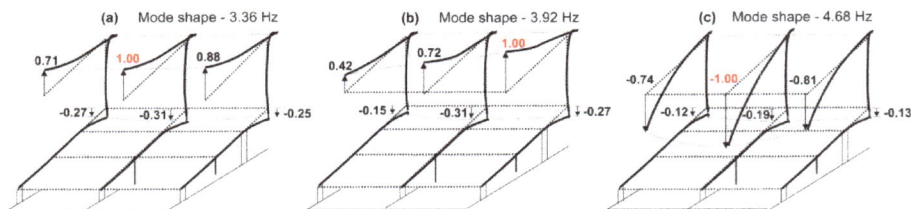

Figure 7. Identified mode shapes and natural vibrations: (**a**) 3.36 Hz, (**b**) 3.92 Hz, (**c**) 4.68 Hz.

The next step included the non-destructive evaluation of the stand throughout the analysis of the forced vibrations. The impact of the real dynamic forces was implemented thanks to the fans who, at the request of the club authorities, came to participate in the measurements. About 400–450 people participated in the research. For the purpose of the research, the fans presented two forms of active support: the so-called Scotland-type jumps and the Labado dance. By making the Scotland-type jumps, the fans performed single jumps to a drum beat. They started very slowly and finished with asynchronous jumps. In the Labado dance, almost all fans performed synchronous jumps with a constant frequency.

During the research, one of the most important tasks was to find the most unfavorable load combination. Different combinations of fans' arrangement on stadium stand rows were tested. One of the most unfavorable combinations, resulting in the greatest vibrations of the cantilever part of the stand, was selected for the purpose of the tests. Figure 8 presents the most unfavorable setting in which the dancing fans are located at the bottom of the stand (beam BR1) and on the upper part of the stand (four rows at the cantilever part of beam BR2).

Figure 8. Dynamic tests with fans: (a) experimental setup, (b) scheme showing the arrangement of people during tests (orange color—rows indicating the most unfavorable fan positions).

The results of the in situ tests are shown in Figures 9 and 10 for the Scotland-type jumps and the Labado dance, respectively. In the case of the Scotland-type jumps, many jump events are visible, with decreasing time between individual jumps. In the Fourier transformation results, two wide peaks are visible, the first around 2.5 Hz and the second around 4.5 Hz. The Labado dance, in which jumps were more regular, resulted in narrower peaks concentrated at 2.2 and 4.4 Hz.

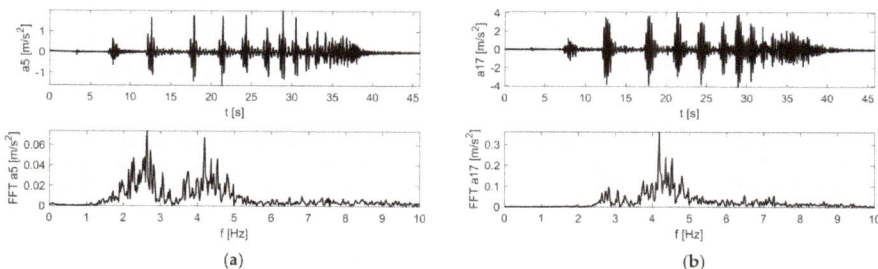

Figure 9. Accelerations registered during dynamic tests with fans (Scotland-type jumps) at: (a) girder, (b) roof.

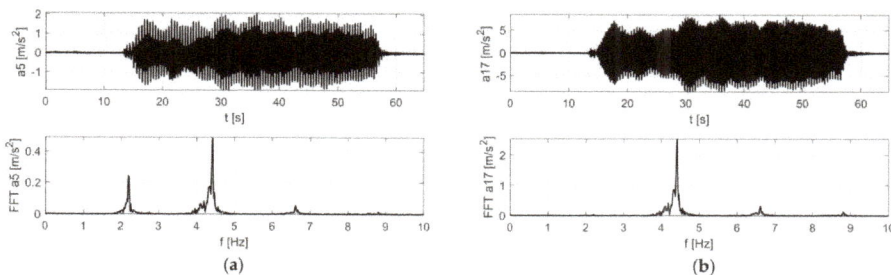

Figure 10. Accelerations registered during dynamic tests with fans (Labado dance) at: (a) girder, (b) roof.

Based on the obtained results, it was found that the frequency of vibrations of the end of the cantilever girder was equal to 2.2 Hz, coinciding with the frequency of the exciting force, which was caused by jumping supporters. The end of the roof structure vibrated at 4.4 Hz, which means that the roof vibrated twice as fast as the end of the girder.

4. Finite Element Method Modeling

4.1. Identification of the Jump Load Function

Experimental tests were conducted in the laboratory of the Gdańsk University of Technology in order to identify the jump load function. The test object was a composite plate with dimensions of 190 cm × 7.5 cm × 40 cm (Figure 11a). The plate was made of two sheets of poplar plywood of a thickness of 2 cm. Between them, C 20 class wood with a 3.5 cm thickness was inserted. Material parameters adopted for testing were: bending strength 20 MPa, stretching along fibers 12 MPa, compression along fibers 19 MPa, modulus of elasticity along fibers 10 GPa, average density 330 kg/m^3. The plate was placed on supports using elastomer pads with dimensions of 7 cm × 40 cm × 0.5 cm.

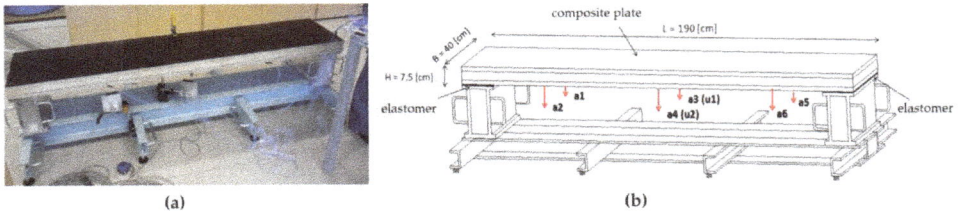

Figure 11. Experimental setup: (**a**) general view, (**b**) geometry of the tested plate and the location of points for measurement of accelerations (a1 to a5) and displacements (u1, u2).

The plate was subjected to dynamic tests. The acceleration measurements (a1 to a6) were performed at six points by means of the PCB accelerometers model 356A16, while displacement measurements were taken at two points (u1 to u2) by means of optoNCDT 1302 laser sensors, as shown in Figure 11b. The LMS SCADAS vibration measurement system was used to record the time histories. The research agenda included measurements of free vibrations as well as measurements of forced vibrations in the form of synchronized jumps performed by one, two and three people (Figure 12).

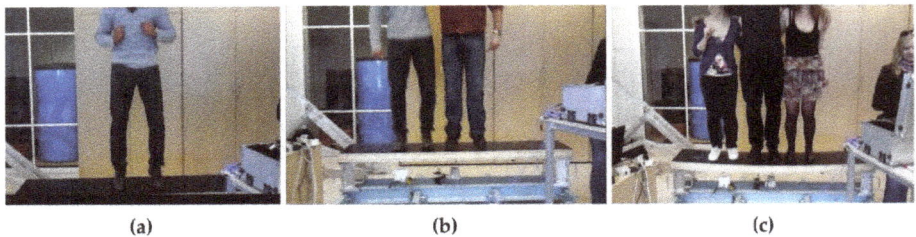

Figure 12. Implementation of excited vibrations in the form of rhythmical jumps of (**a**) one person, (**b**) two people, (**c**) three people.

The first natural frequency of the plate was identified using a typical impact test. Its value was 26 Hz. Figure 13 present the results of vibrations in the form of accelerations and displacements recorded during the jumps of three people. Apart from the jump frequency, further components of harmonically excited vibrations are visible in the Fourier transform diagrams. Eight harmonic components are visible on the Fourier transforms diagrams on acceleration, while the Fourier transforms of displacement signals show four harmonic components.

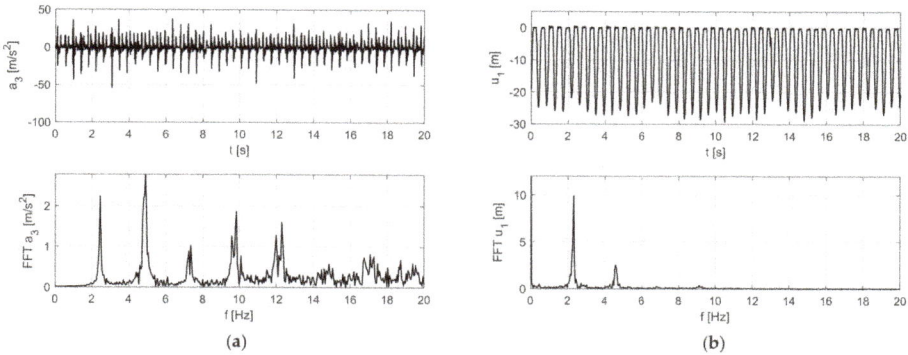

Figure 13. Results of vibrations in the form of (**a**) accelerations and (**b**) displacements recorded during the jumps of three people.

The dynamic impact function during the jumping was described by means of an impulse shaped in half of the sinus function period form. The loading cycle for the Labado dance includes the flight phase and the contact phase. On the basis of the measurement results, the duration of the periodic force full period was identified as 0.45005 s, which is composed of: the phase of load contact with the plate equal to 0.27003 s and the flight phase equal to 0.18002 s. The shape of a single dance cycle is shown in Figure 14a. The in situ tests indicated the jump frequency to be 2.2 Hz. The amplitude of the load function was determined as the average value of the weight of fans participating in the experiment, and was equal to approximately 0.78 kN.

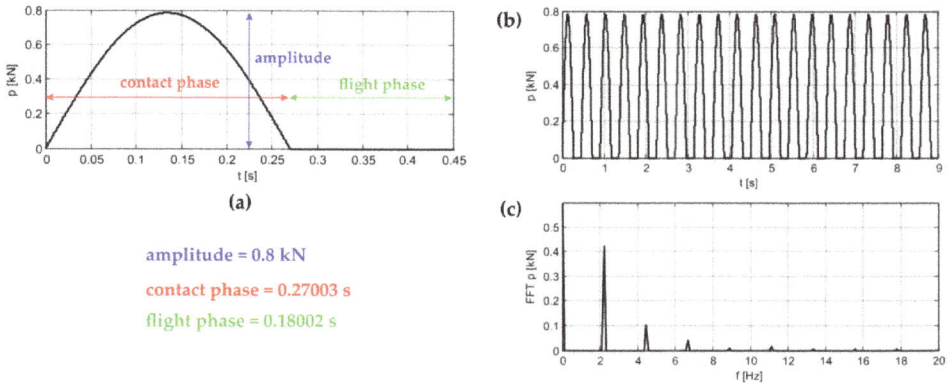

Figure 14. Labado function: (**a**) one jump, (**b**) 20 jumps, (**c**) Fourier transformation of the Labado function.

Figure 14b,c shows excitation over time during 20 rhythmic jumps and the corresponding Fourier transform. The Fourier transform graph proves that, in addition to the 2.2 Hz jump frequency, further harmonics are present in the spectrum. The occurrence of higher harmonics is characteristic of this periodic signal. Higher harmonic components are also present on transforms from the experimental data.

In order to validate the laboratory tests results, numerical simulations for the composite plate were performed. The calculated plate length was set as 183 cm. It was assumed that the plate works in a simply supported scheme. In the discretization process, the slab was divided into 504 solid elements in accordance with the results obtained during the convergence division analysis. Numerical calculations were performed for excitation caused by one, two and three people (Figure 15).

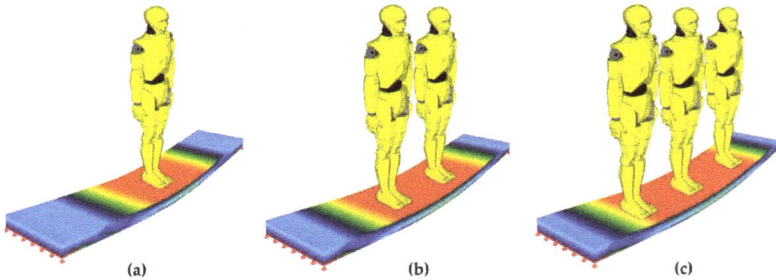

Figure 15. Numerical simulations of the implementation of excited vibrations in the form of rhythmic jumps performed by: (**a**) one person, (**b**) two people, (**c**) three people.

Figure 16 presents a comparison of the numerical and experimental results in the form of displacement in time, registered at the middle of the span, and their Fourier transforms for the cases of one and three people jumping. In the flight phase, plate vibrations occurred at a frequency of 26 Hz for both the experimental and numerical results. Slight phase shifts between the experimental and numerical graphs can be observed. The reason for the shifts lies in the fact that the real jumps were not perfectly repetitive. A very good experiment consistency of the numerical simulations was obtained, which proves the correctness of the assumed load model for the Labado dance. The excitation frequency was approximately 2.2 Hz during both the experimental and numerical tests. Later in the article, the obtained Labado dance load model is used to simulate vibrations and structure tuning of the stand at the Swiss Krono Arena speedway stadium.

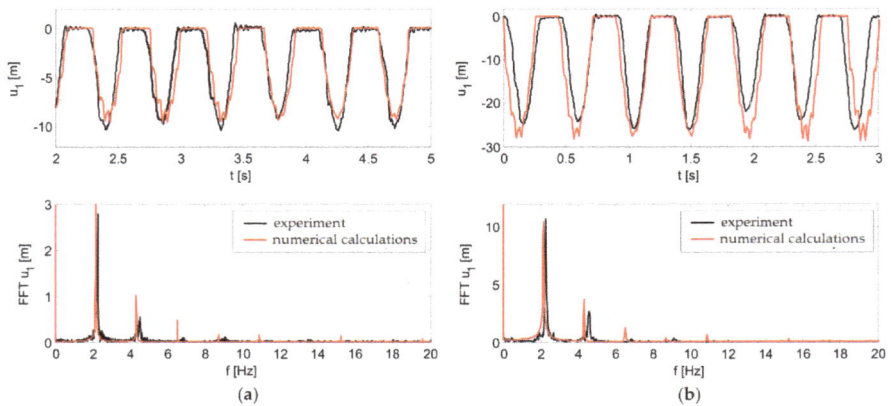

Figure 16. Numerical model validation results for jumps performed by: (**a**) one person, (**b**) three people.

4.2. Numerical Analysis

The creation of the 3-D FEM model of the stand at the Swiss Krono Arena stadium was implemented in two stages with the use of the commercial FEM software SOFiSTiK (SOFiSTiK AG, Oberschleißheim, Germany). In the first stage, models of the main structural parts of the stand were performed using solid elements (Figure 17b), while in the second stage they were performed with the use of beam elements (Figure 17a). The construction of beam models was verified based on the results of numerical analyses carried out on solid models, due to which a very high consistency was obtained between the results of the static and dynamic tests. The FEM calculation model of the entire prefabricated 3-D stadium contains approximately 3008 beam elements and 2001 nodes, in accordance with the results collected during the convergence analysis. Substitute cross-sections were applied in the beam model, in which the cross-section of concrete and steel was replaced with a representative cross-section of a homogeneous material. The under-seat beams of the stand were modeled as beam elements. The columns were articulately connected with the girders, and an articulated joint can also be found at the connection of the two main girders. Glued laminated timber girders were modeled as beam elements, and steel bracings and bar elements model the truss elements. In models that consider various stand strengthening methods, the tensile elements are modeled using tensile elements (cable) taking into account the prestressing force. Non-structural elements of significant weight (e.g., non-structural steel columns) were modeled as concentrated masses applied at the place of mounting. The material parameters taken for the purpose of the numerical analysis were adopted in accordance with the construction and structural design.

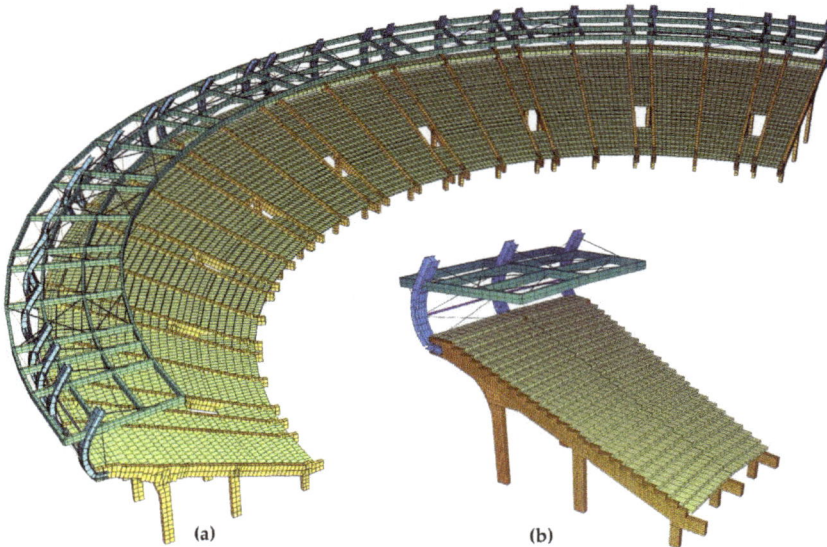

(a)   (b)

Figure 17. (a) Numerical model of the stadium, (b) numerical model of a single section of the stand.

The eigenvalues were calculated at the beginning. Thirty eigenvalues of the system were calculated. The total mass contributions in the dynamic response were more than 90% in directions: X (longitudinal), Y (transverse) and Z (vertical), which complied with the standard recommendations. The first three eigenvalues and their corresponding natural vibration frequencies determined on the basis of modal analysis are presented below (Figure 18).

The determined basic natural vibration frequency of the stand construction with a roof and without strengthening was used to simulate forced oscillations and to determine the degree of stiffness reduction of the concrete cross-section due to scratches. For the purpose of the analysis, the structure was loaded with its own weight and an evenly distributed load on the stands of 8.0 kN/m^2, in accordance with PN-EN-1991-1-1 [27]. It was assumed that this load changes harmonically according the to the relation $p(t) = p\sin(\omega t)$, where the amplitude $p = 8.0$ kN/m^2 and the circular frequency $\omega = 15.71$ rad/s (which corresponds to the frequency $f = 2.3$ Hz). It has been assumed that the load of such nature lasts $t = 60$ s, the same as in the experiment.

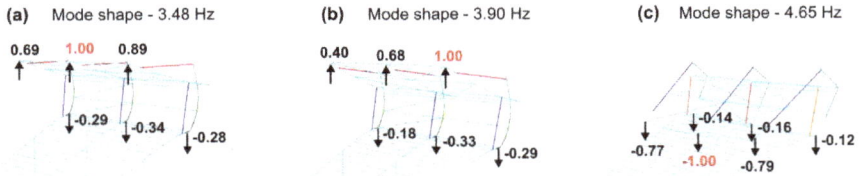

Figure 18. Mode shapes and frequencies of natural vibrations of a single stand section obtained from numerical simulations: (**a**) 3.48 Hz, (**b**) 3.90 Hz, (**c**) 4.65 Hz.

Figure 19 shows the stresses in the BR2 girder. Figure 19b shows equivalent von Mises stresses, Figure 19c presents the main tensile stresses and Figure 19d demonstrates the normal horizontal stresses in the girder plane (in the global coordinate system). The main tensile stress in the upper part of the BR2 beam was greater than the concrete tensile strength, i.e., $\sigma_1 = 8$ MPa $> \sigma_{cr} = f_{ctm} = 3.2$ MPa, which means that the reinforced concrete section becomes scratched.

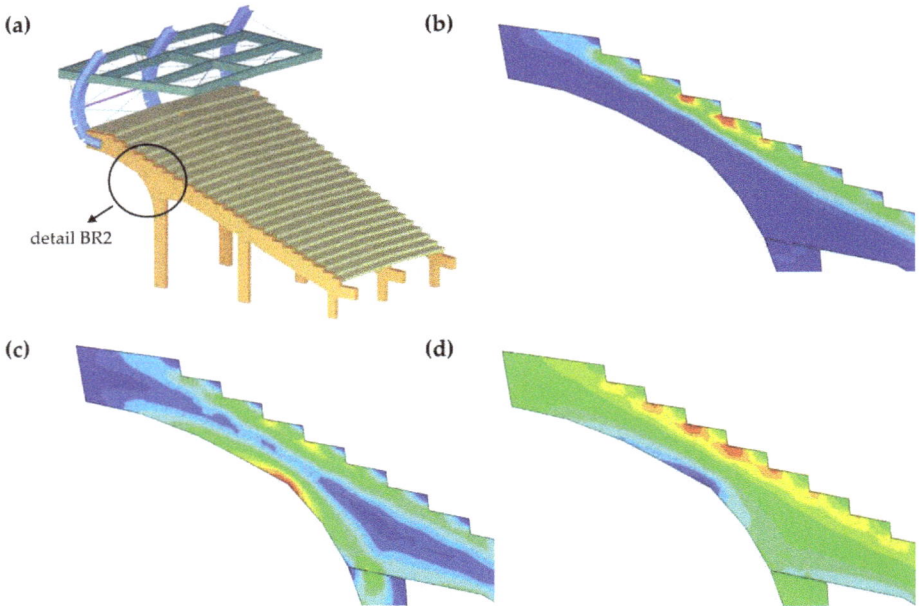

Figure 19. (**a**) Placement of the highest stress concentrations in the main girder of the stadium stand. (**b**) von Mises map of stresses in the BR2-reinforced concrete girder area (red color—maximum stress equal to 10 MPa). (**c**) Map of the main tensile stresses within the BR2-reinforced concrete girder (red color—maximum tensile stress equal to 8 MPa). (**d**) Map of the horizontal normal stress s11 within the BR2-reinforced concrete girder (red color—maximum tensile stress equal to 8 MPa, blue color—maximum compressive stress equal to 8 MPa).

The decrease in the stiffness of the scratched girder was equal to about 50%. The decrease of the stiffness of the reinforced concrete section after scratching was determined on the basis of the obtained stress distribution in the prefabricated beam supporting the stands. During the numerical simulation, the basic natural vibration frequencies of the structure were determined, taking into account the stiffness reduction of the reinforced concrete cross-section due to scratching.

In the further stage of the numerical analysis, the stadium structure was loaded with jumping fans. Each of the fans was replaced by a single concentrated force with a load value of 0.8 kN, as this load corresponded to the average human weight. A jump function determined during experimental research was assigned to every concentrated force. The duration time was equal to 40 s. The forces were set in the most unfavorable load combination, i.e., at the bottom of the stand (chairs on the whole BR1 beam width) and on the upper part of the stand (four rows above the cantilever section). Simulations of stand vibrations for the excitation form of non-synchronized Scotland-type jumps were carried out at the beginning (Figure 20). Afterwards, an analysis for the form of the synchronized Labado dance was performed (Figure 21).

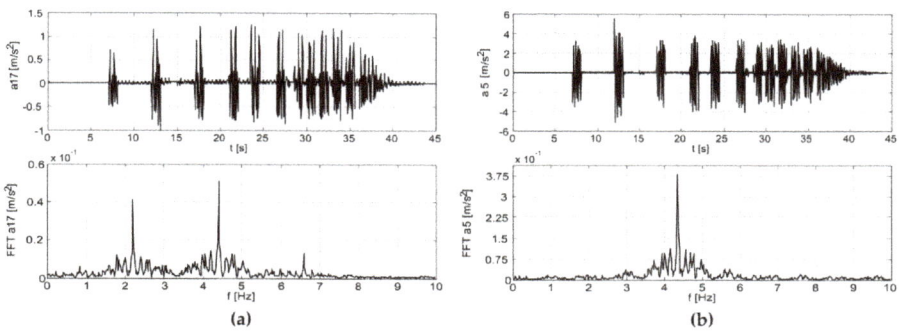

Figure 20. Results of numerical simulations from the excitation of vibrations by jumping fans (Scotland-type jumps): (**a**) girder, (**b**) roof.

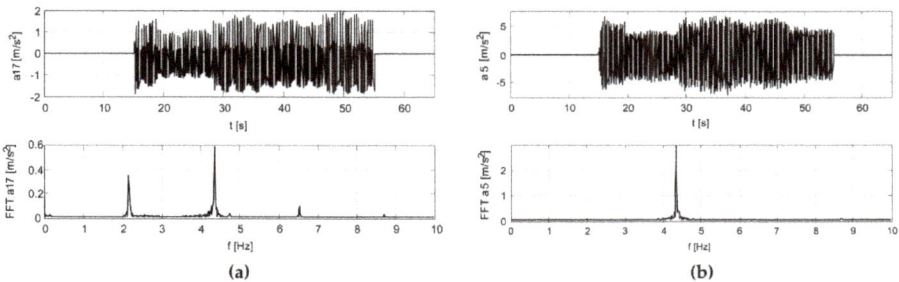

Figure 21. Results of numerical simulations from the excited vibrations caused by jumping fans (Labado dance): (**a**) girder, (**b**) roof.

On the basis of the results obtained during non-destructive tests and numerical simulations, it was concluded that as a result of dynamic interactions, the phenomenon of higher harmonic resonance occurred on the stand and the roof structure. The vibrations frequency at the end of the roof was equal to approximately 4.7 Hz, which coincides with the second harmonic of the excitation force, i.e., 4.4 Hz, caused by jumping fans. The form and dynamic characteristics of the Labado dance can be considered as vandalism. A large number of jumping fans with their synchronized movements drives the tribune structure's vibrations, and their regular dance could cause the collapse of the entire object. The results of non-destructive tests and numerical simulations showed a significant excess of the serviceability limit state.

5. Structural Tuning

An original method of stand structural tuning was proposed in response to the excess of the serviceability limit state. This method assumed the implementation of wall bracings stiffening the steel roofing columns, as well as roof bracings on the wooden roof structure (Figures 22 and 23). The main goal of the concept of bracing of all columns and roof girders was to significantly limit the free end of the wooden roof vibration amplitudes, and hence reduce the concrete cantilever displacements. The bracing cross-sections were the same as those of the bracings already installed in the stand structure. The bracings were made from Macalloy bar rods with a yield point of 520 MPa and tensile strength of 690 MPa.

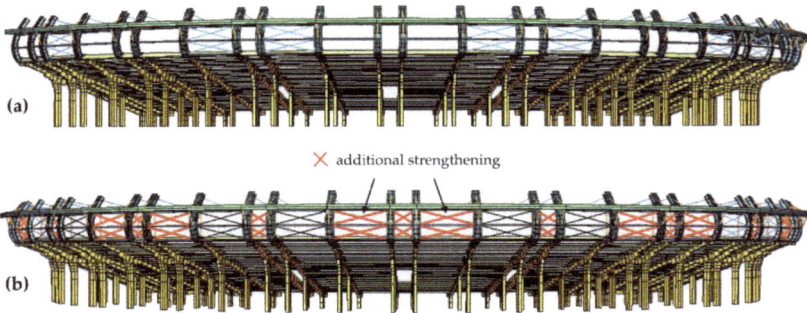

Figure 22. Additional wall bracings (**a**) before structural tuning, (**b**) after structural tuning.

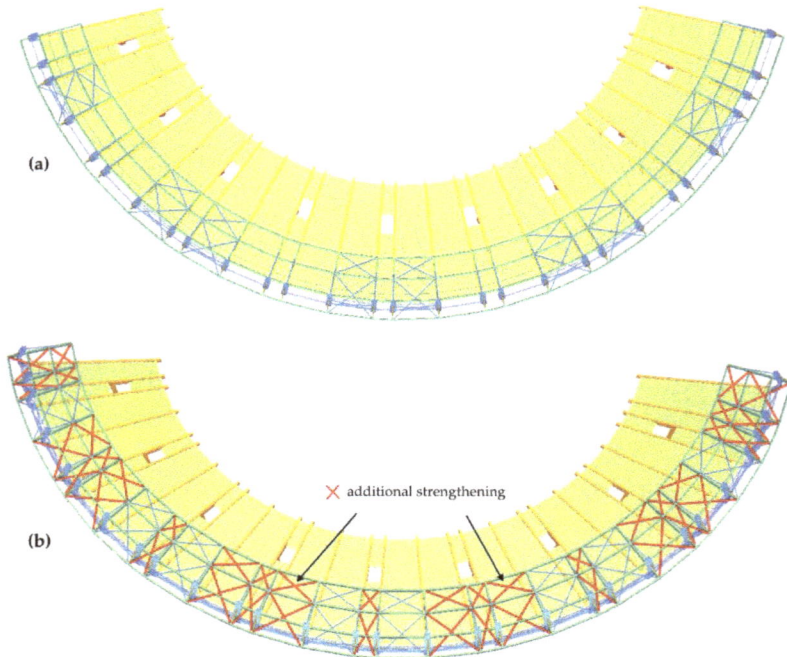

Figure 23. Additional roof bracings (**a**) before structural tuning, (**b**) after structural tuning.

The stand FEM model with additional roof bracings and vertical bracings was implemented in SOFiSTiK software. The beam stand model was made on the basis of the object design documentation (Figure 24). The basic model parameters were determined on the basis of detailed models with the use of solid elements, in which the location of reinforcement bars and details of beams support, such as elastomer pads and anchors, were taken into account. The models include the scratching of concrete elements by reducing the value of the concrete modulus of elasticity. The FEM model parameters were updated on the basis of the obtained measurement data.

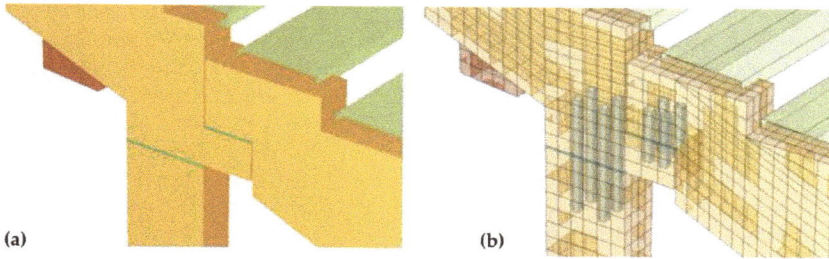

(a)   (b)

Figure 24. Detailed view of the stand's load-bearing beams: (**a**) support, (**b**) connection.

Figure 25 presents the numerical results obtained during the calculations for the load induced by excitation fans, for the purposes of post-strengthening the stadium structure. The exemplary first form of natural vibration frequencies determined numerically is shown in Figure 26. The dynamic analysis proved that in the range of 2.9 Hz to 5 Hz, there are at least 15 natural vibration frequencies. Such a large number of natural vibration frequencies is ordinary for a spatial structure made of repetitive elements. The strengthening target was to improve the spatial structure performance and reduce natural vibration frequencies within the range of 4.3–4.7 Hz, in order to eliminate the phenomenon of the beat and characteristic of the excitation of two very close natural vibration frequencies.

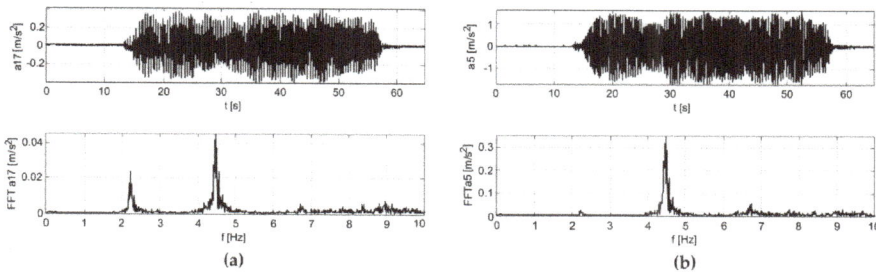

(a)   (b)

Figure 25. The results of numerical simulations after stand structural tuning from vibrational excitation by the jumping supporters: (**a**) girder, (**b**) roof.

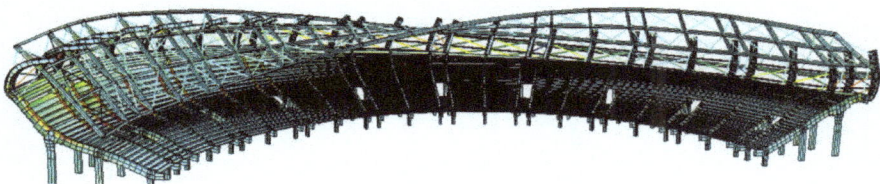

Figure 26. The first numerical form of natural vibrations of the stand after structural tuning.

Figure 27 shows the visualization of vibrations induced by synchronized dancing of the crowd in the stands, located in rows 22–25 throughout the stadium. The applied dynamic load was a periodic function consisting of a contact phase (sinusoidal impulse) and a flight phase (no load). The load function of a jumping crowd is typically not included in standard regulations. The integration of motion equations was carried out using the Newmark–Wilson method. Figure 27a shows a plane view of the relations of the object degrees of freedom vibration amplitudes in the case when the roof and vertical bracings were fixed in accordance with the design, i.e., every two fields. Figure 27b presents the nature of vibrations for bracings installed in each roofing field. The bracings installation resulted in the roof structure stiffening and in the change of its vibrational spatial character. The calculated amplitude of vibrations from the fans' synchronized dance on the whole structure in rows 22–25 was equal to 3.6 mm at the end of the concrete cantilever, and 5 mm at the free end of the roof. The strengthening method caused a slight reduction in the vibrations of the concrete cantilevers and a significant reduction in the vibration amplitudes of the free end of the roof. A requirement for the correct bracing performance during vibrations was the presence of an initially prestressing force. However, the reduction of concrete cantilevers vibration amplitudes was problematic, because they took over the main part of the dynamic forces generated by fans dancing directly on them.

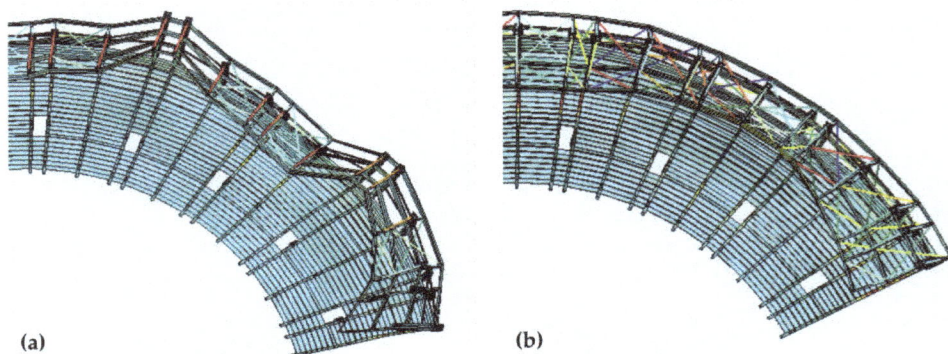

(a)   (b)

Figure 27. Top view of the stand's roof loaded with synchronous jumps of people showing the vibration amplitudes (**a**) before structural tuning, (**b**) after structural tuning.

Repeated in situ tests were carried out in order to verify the correctness of the original concept of stadium stand tuning. The results of measurements made after structural tuning are shown in Figures 28 and 29. It can be seen that that frequencies identified during crowd-induced vibrations are similar to the measurements before tuning (see Figures 9 and 10), because the same nature of the jumping load. It is important, however, that the natural frequencies of the structure were shifted after structural tuning (see Figure 28), decreasing resonance significantly. The individual resonance zones moved towards the higher frequencies.

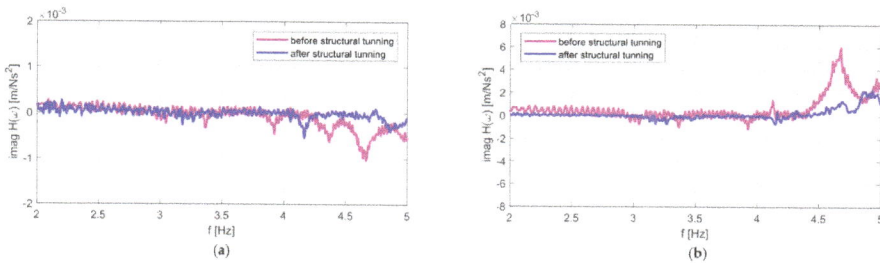

Figure 28. Experimental transition functions for the end of the concrete cantilever (**a**) before and (**b**) after stand structural tuning.

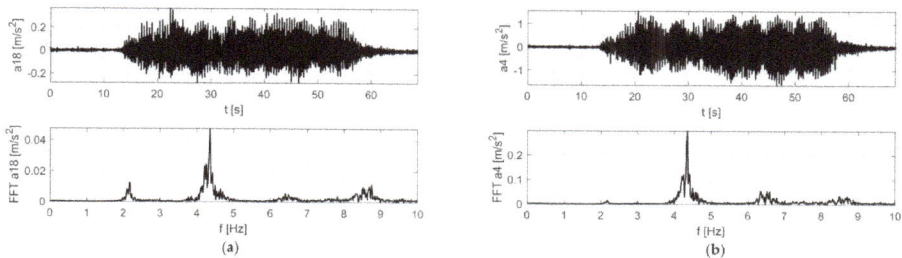

Figure 29. Results of in situ tests after the stand structural tuning from vibration excitation with the jumping supporters: (**a**) girder, (**b**) roof.

6. Conclusions

In this study, we adopted the vibration testing method for non-destructive diagnostics of a stadium stand. Experimental and numerical studies were conducted on the real object. The reason for undertaking the investigation was excessive stand vibrations, in particular vertical oscillations of the free end of the roof. On the basis of the conducted vibration tests, the following conclusions of the technical condition of the stadium stand were formulated:

(1) The key natural vibration frequency of 4.68 Hz coincided with the frequency of the higher harmonics of the excitation force (approximately 4.4 Hz).

(2) The reinforced concrete cross-section of the BR2 main girder was scratched. The calculated decrease in the stiffness after scratching was equal to about 50%. The inclusion of scratches reduced the calculated initial natural vibration frequencies of the structure by 5%–10% in comparison with the unscratched cross-section.

(3) The dynamic and synchronic nature of the Labado dance is accidental, but can constitute vandalism. Fans jumping to the rhythm of the animator's regular drumming results in increasing stand vibrations, causing the serviceability limit state to be exceeded. The stand vibrations can lead to material fatigue and, as a result, to its complete destruction.

(4) The proposal to solve the problem of overlapping frequencies causing resonance by changing the animator's drumming rhythm and using it for the needs of dynamic impact elimination in favor of the static impact of people on the stand structure was not accepted by club authorities nor by fans. This objection was explained by the long-term Ladabo dance tradition at the Swiss Krono Arena stadium. Any changes in the drumming rhythm would result in the loss of the cheering spirit among fans used to dancing the Labado.

(5) The proposed original concept of stadium structural tuning significantly reduced the amplitude of the stand structure vibrations. Individual resonance zones were moved towards higher frequencies after tuning, which allowed for safe use of the stadium.

(6)   There is a high probability of the occurrence of double natural vibrations frequencies or frequencies with close values in the case of spatial constructions with large dimensions and repeated construction segments. Therein, the phenomenon of the beat will appear, which may cause an increase in tribune elements displacement. The post-tuning FEM model of a structure does not indicate the occurrence of such a case. However, if the actual post-tuning stand structure had double frequencies in a range of up to 5 Hz, additional structure strengthening might be necessary.

(7)   The implementation of the architectural structure form resulted in a static scheme in which a roof structure cantilever steel column was mounted at the end of the stand's reinforced concrete cantilever. The girders working as cantilevers were fixed at the cantilever steel column end. This results in fragility under dynamic load applied to the cantilever part of the concrete beams. Sports facilities must ensure the freedom of supporters gathered in the stands and all possible dances and ways to cheer their favorite teams must be included in the design and implementation process.

(8)   The non-destructive tests and numerical simulations conducted in this paper are innovative and pioneering. Currently, no standard has been developed that allows for the determination of performance conditions and methods for testing the resistance of stadium stands to jump loads. The normative documents also lack a ready-to-use jump load function that could be used when designing a structure.

Finally, it can be concluded that the use of the vibration-based method as a non-destructive testing technique for massive structures such as stadiums enabled the creation of a jump function in order to study dance phenomena on the stands. Therefore, it was possible to prevent failures of and damage to the stadium structure, thus ensuring safety for fans supporting their teams. At the design stage, with the use of the created function, the dance influence as a negative factor affecting the stands can be taken into account, allowing the efficient elimination of potential errors and mistakes.

Author Contributions: Conceptualization and Methodology, K.W., M.R. and K.G.; Experimental Investigations, M.R. and K.W.; FEM Calculations, K.G.; Formal Analysis, K.G., M.R. and K.W.; Visualization, K.G. and M.R.; Writing—Original Draft Preparation, K.G.; Writing—Review and Editing, M.R.; Supervision, K.W.

Funding: This research received no external funding.

Acknowledgments: The authors would like to thank the Swiss Krono Arena for providing access to the structure and supplementary data.

Conflicts of Interest: The authors declare no conflict of interest.

References

1.   Melrose, A.; Hampton, P.; Manu, P. Safety at sports stadia. *Procedia Eng.* **2011**, *14*, 2205–2211. [CrossRef]
2.   Soomaroo, L.; Murray, V. Disasters at Mass Gatherings: Lessons from History. *PLOS Curr. Disasters* **2012**. [CrossRef] [PubMed]
3.   Proença, J.M.; Branco, F. Case studies of vibrations in structures. *Rev. Eur. génie Civ.* **2007**, *9*, 159–186. [CrossRef]
4.   Sachse, R.; Pavic, A.; Reynolds, P. Parametric study of modal properties of damped two-degree-of-freedom crowd–structure dynamic systems. *J. Sound Vib.* **2004**, *274*, 461–480. [CrossRef]
5.   Reynolds, P.; Pavic, A. Vibration Performance of a Large Cantilever Grandstand during an International Football Match. *J. Perform. Constr. Facil.* **2006**, *20*, 202–212. [CrossRef]
6.   Yang, Y. Comparison of bouncing loads provided by three different human structure interaction models. In Proceedings of the 2010 International Conference on Mechanic Automation and Control Engineering, Wuhan, China, 26–28 June 2010; pp. 804–807.
7.   Bachmann, H. Case Studies of Structures with Man-Induced Vibrations. *J. Struct. Eng.* **1992**, *118*, 631–647. [CrossRef]
8.   Santos, F.; Cismaşiu, C.; Cismaşiu, I.; Bedon, C. Dynamic Characterisation and Finite Element Updating of a RC Stadium Grandstand. *Building* **2018**, *8*, 141. [CrossRef]

Materials **2019**, *12*, 2148

9. Jones, C.; Reynolds, P.; Pavic, A. Vibration serviceability of stadia structures subjected to dynamic crowd loads: A literature review. *J. Sound Vib.* **2011**, *330*, 1531–1566. [CrossRef]
10. Gul, M.; Catbas, F.N. A Review of Structural Health Monitoring of a Football Stadium for Human Comfort and Structural Performance. *Struct. Congr.* **2013**, 2445–2454.
11. Ren, L.; Yuan, C.-L.; Li, H.-N.; Yi, T.-H. Structural Health Monitoring System Developed for Dalian Stadium. *Int. J. Struct. Stab. Dyn.* **2015**, *16*, 1640018. [CrossRef]
12. Di Lorenzo, E.; Manzato, S.; Peeters, B.; Marulo, F.; Desmet, W. Structural Health Monitoring strategies based on the estimation of modal parameters. *Procedia Eng.* **2017**, *199*, 3182–3187. [CrossRef]
13. Spencer, B.F.; Ruiz-Sandoval, M.E.; Kurata, N.; Ruiz-Sandoval, M.E. Smart sensing technology: opportunities and challenges. *Struct. Control. Heal. Monit.* **2004**, *11*, 349–368. [CrossRef]
14. Bedon, C.; Bergamo, E.; Izzi, M.; Noè, S. Prototyping and Validation of MEMS Accelerometers for Structural Health Monitoring—The Case Study of the Pietratagliata Cable-Stayed Bridge. *J. Sens. Actuator Networks* **2018**, *7*, 30. [CrossRef]
15. Hoła, J.; Bień, J.; Sadowski, Ł.; Schabowicz, K. Non-destructive and semi-destructive diagnostics of concrete structures in assessment of their durability. *Bull. Pol. Acad. Sci. Tech. Sci.* **2015**, *63*, 87–96. [CrossRef]
16. Brandt, A. A signal processing framework for operational modal analysis in time and frequency domain. *Mech. Syst. Signal Process.* **2019**, *115*, 380–393. [CrossRef]
17. Chen, G.-W.; Omenzetter, P.; Beskhyroun, S. Operational modal analysis of an eleven-span concrete bridge subjected to weak ambient excitations. *Eng. Struct.* **2017**, *151*, 839–860. [CrossRef]
18. Torres, W.; Almazán, J.L.; Sandoval, C.; Boroschek, R. Operational modal analysis and FE model updating of the Metropolitan Cathedral of Santiago, Chile. *Eng. Struct.* **2017**, *143*, 169–188. [CrossRef]
19. Idehara, S.J.; Júnior, M.D. Modal analysis of structures under non-stationary excitation. *Eng. Struct.* **2015**, *99*, 56–62. [CrossRef]
20. Jannifar, A.; Zubir, M.; Kazi, S. Development of a new driving impact system to be used in experimental modal analysis (EMA) under operational condition. *Sens. Actuators A Phys.* **2017**, *263*, 398–414. [CrossRef]
21. Prashant, S.W.; Chougule, V.; Mitra, A.C. Investigation on Modal Parameters of Rectangular Cantilever Beam Using Experimental Modal Analysis. *Mater. Today Proc.* **2015**, *2*, 2121–2130. [CrossRef]
22. Hu, S.-L.J.; Yang, W.-L.; Liu, F.-S.; Li, H.-J. Fundamental comparison of time-domain experimental modal analysis methods based on high- and first-order matrix models. *J. Sound Vib.* **2014**, *333*, 6869–6884. [CrossRef]
23. Clemente, P.; Marulo, F.; Lecce, L.; Bifulco, A. Experimental modal analysis of the Garigliano cable-stayed bridge. *Soil Dyn. Earthq. Eng.* **1998**, *17*, 485–493. [CrossRef]
24. Lin, R. Identification of modal parameters of unmeasured modes using multiple FRF modal analysis method. *Mech. Syst. Signal Process.* **2011**, *25*, 151–162. [CrossRef]
25. Rucka, M.; Wilde, K. Application of continuous wavelet transform in vibration based damage detection method for beams and plates. *J. Sound Vib.* **2006**, *297*, 536–550. [CrossRef]
26. Chroscielewski, J.; Miśkiewicz, M.; Pyrzowski, Ł.; Rucka, M.; Sobczyk, B.; Wilde, K. Modal properties identification of a novel sandwich footbridge—Comparison of measured dynamic response and FEA. *Compos. Part B Eng.* **2018**, *151*, 245–255. [CrossRef]
27. EN C. 1-1: Eurocode 1: *Actions on Structures–Part 1-1: General Actions–Densities, Self-Weight, Imposed Loads for Buildings*; European Committee for Standardization: Brussels, Belgium, 2002.

Article

Non-Destructive Testing of Technical Conditions of RC Industrial Tall Chimneys Subjected to High Temperature

Marek Maj [1],*, Andrzej Ubysz [1] , Hala Hammadeh [2] and Farzat Askifi [3]

[1]   Faculty of Civil Engineering, Wroclaw University of Science and Technology, Wroclaw 50-370, Poland; andrzej.ubysz@pwr.edu.pl
[2]   Faculty of Engineering, Middle East University, Amman 11831, Jordan; hhamm2131@gmail.com
[3]   Department of Structure Engineering, Faculty of Civil Engineering, Damascus University, Damascus PO Box 30621, Syria; frzataskifi@gmail.com
*   Correspondence: marek.maj@pwr.edu.pl; Tel.: +48-601-729-184

Received: 16 May 2019; Accepted: 18 June 2019; Published: 24 June 2019

check for updates

Abstract: Non-destructive tests of reinforced concrete chimneys, especially high ones, are an important element in assessing their condition, making it possible to forecast their safe life. Industrial chimneys are often exposed to the strong action of acidic substances, They are negatively exposed to the condensation of the flue gases. Condensate affects the inside of the thermal insulation and penetrates the chimney wall from the outside. This is one reason for the corrosion of concrete and reinforcing steel. Wet thermal insulation settles, and drastically reduces its insulating properties. This leads to an increase in temperature in the reinforced concrete chimney wall and creates additional large variations in temperature fields. This consequently causes a large increase in internal forces, which mainly increase tensile and shear stresses. This results in the appearance of additional cracks in the wall. The acid condensate penetrates these cracks, destroying the concrete cover and reinforcement. Thermographic studies are very helpful in monitoring the changes in temperature and consequently, the risk of concrete and reinforcement corrosion. This simple implication between changes in temperature of the chimney wall and increasing inner forces as shown in this article is particularly important when the chimney cannot be switched off due to the nature of the production process. Methods for interpreting the results of thermovision tests are presented to determine the safety and durability of industrial chimneys.

Keywords: nondestructive testing; thermography; monitoring of structures; reinforced concrete chimney; corrosion processes; service life of a structure

---

1. Introduction

Industrial reinforced concrete chimneys are often exposed to a chemically aggressive environment. The combustion gases conveyed via the chimney undergo condensation inside it or are dissolved in precipitation, becoming strongly acidic liquids. The concrete/acid contact results in the corrosion of the concrete and after the concrete cover is penetrated, in the corrosion of the steel also. The two corrosion processes result in the rapid degradation of the chimney structure.

As regards chimneys, one should bear in mind not only the high cost and the technological challenges involved in their construction but also that they perform an essential role in the production processes and so cannot be put out of service for repairs. For example, in steelworks and coking plants, the damping of the furnace from which the combustion gases are conveyed to a chimney results in the destruction of the whole power unit. Therefore, nondestructive tests are vital for both assessing the current technical condition of the chimney and monitoring the degradation processes over its whole service life [1].

2. Causes and Effects of Reduced Effectiveness of Chimney Thermal Insulation

One of the major causes of the degradation of the chimney's reinforced concrete (RC) shell is the too rapid fall in the temperature of the combustion gas as it flows through the chimney flue. The chimney wall consists of the following three layers:

- A reinforced concrete shaft;
- An internal wall (e.g., made of fire brick) constituting the chimney's inner lining which is very resistant to high temperatures;
- A mineral wool layer, placed between the two walls, serving as thermal insulation.

The durability of a chimney to a considerable degree depends on the quality and longevity of the thermal insulation. Since the chimney cannot be taken out of the production process, it is highly important to constantly monitor the condition of the insulation. Particularly suitable for this purpose are non-destructive testing methods. If the RC shell were removed in places to expose the insulation, the places could become corrosion centres and a thermal bridge could form.

When designing the geometry and thermal insulation of a chimney, one should consider the combustion gas inlet and outlet temperature, the amount and rate of flow of the flue gas and its condensation temperature. Significant internal forces can also arise from a temperature difference [2].

The above parameters are determined by

- The amount of the exhausted gas;
- Changes in flue gas temperature due to changes in production technology;
- The daily changes of the physicochemical parameters of the chimney environment, such as the wind velocity, the atmospheric pressure, the air humidity, etc.

An important factor is a reduction in the insulating power of the chimney walls caused by changes in the physical properties of the insulation due to, e.g., the mineral wool getting damp [3]. As a result, the temperature of the flue gas in the chimney's upper part decreases, and after the critical temperature is reached, this causes excessive flue gas condensation on the chimney lining.

As the condensate penetrates through cracks in the lining, it makes the mineral wool damp, whereby the latter loses its insulating properties. Moreover, the damp mineral wool sinks, and as a result, areas devoid of thermal insulation are created. As the coefficient of thermal conductivity decreases, the temperature of the chimney's inner wall falls further, and so does the temperature of the flue gas. As a result, more and more flue gas condenses on the chimney's inner wall, whereby the degradation processes in the chimney's RC shell intensify (Figures 1–3).

Figure 1. Deposition of condensation on outer surface chimney wall along construction joints.

Figure 2. Condensation degrading of chimney wall.

Figure 3. Condensation dripping down from a chimney upper ring.

As condensation drips down the inner wall, it penetrates via cracks to the mineral wool and outside to the reinforcement of the concrete wall, causing intense corrosion of the concrete and the reinforcing steel (Figure 4).

Figure 4. Condensation penetrating chimney crown wall.

3. The Idea of a Thermographic Survey of Chimney Thermal Insulation

Thermographic surveys [4–9] have been used to detect thermal bridges in residential buildings for many years. Thermographic surveys are conducted using a camera which can take thermograms. In the case of chimneys, only thermograms of the external surface of chimney's wall are taken. The air temperature considerably affects the accuracy of thermographic surveys. The latter is more precise. The larger the thermal contrast between the surface of the chimney shell and the ambient air temperature the better the thermogram. Therefore, the best period for evaluating the condition of the thermal insulation in chimneys is in winter.

Passive thermography is the optimal method for the thermographic surveying of chimneys. Based on the thermographic images and reference readings, one can determine the temperatures on the surface of the chimney shell [3,10]. In this method, the image is obtained for a set scale range. Knowing the temperature of the flow gas flowing through the chimney and the temperature on the outer surface of the chimney shell, one can determine the actual thermal transmittance coefficient and compare it with its calculated values specified in the design documents. Based on the relative temperature differences, one can determine the degree of damage to the thermal insulation. The measurements were made by a company specializing in thermovision. The cameras were calibrated before measurements.

However, there may be a few percentages of error due to the curvature of the chimney surface, non-homogeneity of the surface, etc.

4. Thermographic Surveys of Chimney Thermal Insulation

The monitoring of the condition of the chimney's insulation is an important element in the assessment of the durability of the chimney. Based on such monitoring, one can forecast the actual service lifespan of the structure and systematically eliminate the causes of chimney shell degradation. The condition of the thermal insulation of industrial chimneys is examined using the classic thermographic method [11,12]. It is mainly the structure's envelope which is examined in this way to detect places where there are gaps in the insulation, or there is no insulation. In the considered case, the thermograms showed that temperature differences on the chimney's outer surface occurred along practically its whole height and on its circumference (latitudinally). According to the results [13] of the thermographic surveys of the 120 m high chimney with the lower and upper diameter of respectively $D_d = 7.16$ m and $D_g = 4.48$ m, the temperature difference in the particular points of the surface amounted to

- As much as 21.8 °C in the chimney's upper part;
- 15.9 °C in the chimney's middle part;

- 10.5–11.9 °C (being more uniform) in the chimney's lower part (Figure 5).

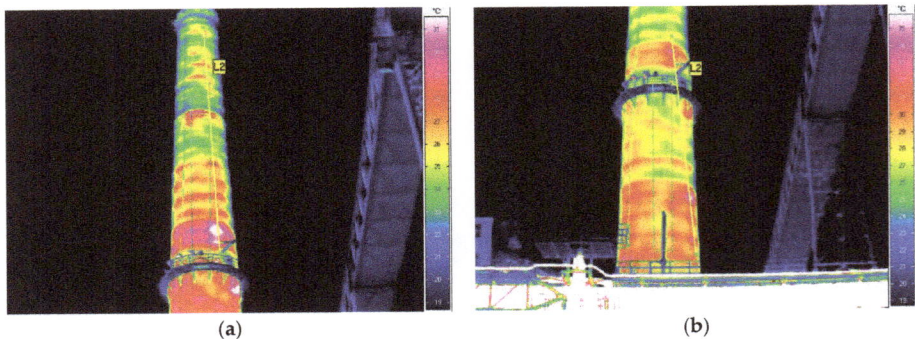

Figure 5. Chimney thermograms revealing insulation imperfections: (**a**) No thermal insulation more than two metres above and under platform; (**b**) Varied temperature distribution in chimney shaft.

The lower part of the chimney does not show such a high degree of degradation as its middle and upper parts. The thickness of the outer reinforced concrete shell, which is considerably greater in the lower part of the chimney, is one of the determining factors.

To nondestructively investigate the condition of the thermal insulation, an analysis of temperature values v_e on the outer surface of the chimney's shell was carried out. The parameters for the analysis had been determined based on the known thermophysical characteristics of the materials, the chimney's geometry (specified in the design documents), and temperature measurements. The following parameters of the layered chimney wall were assumed for the analysis:

- Thermal conductivity:

  - fire brick = 1.30 [W/(m K)],
  - mineral wool = 0.05 [W/(m K)],
  - reinforced concrete = 1.74 [W/(m K)],

- Heat-insulating layer thickness:

  - fire brick layer—11.4 cm,
  - mineral wool layer—13 cm,
  - reinforced-concrete chimney shell wall—24 cm,

- Temperature:

  - inside chimney 200 °C,
  - external 12.3 °C.

- Temperature values v_e on the outer surface of the chimney shell are:

  - for the wall with mineral wool: $v_e = 14.4$ °C,
  - for the wall without a mineral wall: $v_e = 32.4$ °C.

Calculations and in situ measurements indicate that in the places where the temperature of the reinforced-concrete shell is the highest there is no mineral wool or the wool there has very poor insulating properties. The measured minimum temperatures of the shell in well-insulated places on average amounted to 17 °C while the maximum temperatures in the same thermograms on average amounted to 33 °C. The variation coefficient for the temperature on the chimney surface ranges from

0.13 to 0.25. The average decrease in thermal performance in the particular chimney wall surface areas ranges from 20% to 90%.

5. Thermal and Static Load Analysis

As a result of the nonuniform temperature distribution, caused by damage to the thermal insulation, additional internal forces, such as bending moments, shear stresses, and annular tensile and compressive forces, arise. The values of the forces can be traced by studying real cases of damaged chimneys.

5.1. Adopted Assumptions

The following were assumed: the velocity of flow of the flue gas in the chimney: $v = 10$ m/s, the operating temperature of the flue gas at the chimney outlet: $t_{w2} = 180$ °C, and the operating temperature of the flue gas at the chimney inlet: $t_{w1} = 220$ °C. The inlet of the combustion gas takes place through a connecting flue pipe located below the ground level.

The external design temperature was assumed based on the national annex to standard [14]: in winter $T_{min} = -36$ °C and in summer $T_{max} = +40$ °C.

Since no detailed information was available, the emergency flue gas temperature was assumed as 20% higher than the typical one. The emergency temperature amounts to $t_{wa,1} = 220$ °C and $t_{wa,2} = 270$ °C, respectively, at the chimney outlet and inlet. The emergency temperature was used in the calculations.

5.2. Types of Thermal Effects in the Chimney

The linearly elastic models of chimney concrete walls are the simplest models that indicate the influence of thermal factors. The wall thickness of the reinforced concrete chimney is not large as in the case of silos [15] for hot materials. The model of linear temperature gradient used for these silos gives excessive bending moments. The wall of a reinforced concrete chimney subjected to repeated wind and thermal loads behaves linearly. Another problem is with cracks in the wall which reduce the stiffness of the chimney. Cracks in a chimney reduce the impact of bending moments but also expose the chimney to corrosion. The first load results from the difference between the temperatures on the surface of the chimney shaft. The load was calculated for an ideal situation, i.e., immediately after chimney erection—no degraded insulation, and for a situation when some of the insulation has degraded. The temperature load was introduced as the gradient of the temperatures in the chimney shaft.

The second load stems from the difference between the chimney operating temperature during chimney service life and the initial temperature, i.e., the temperature at which the chimney was erected.

5.3. Distribution of Temperature in Chimney Wall for Undamaged Insulation

The chimney wall consists of the following layers: the shaft, the thermal insulation, and the lining. The thicknesses and diameters were assumed according to Table 1. As the reference situation, the condition of the chimney immediately after its erection, i.e., with the continuous mineral wool and fibre brick insulation, was adopted. The following thermal conductivity coefficients were assumed: $\lambda_b = 1.74$ W/mK for the reinforced concrete wall, $\lambda_{iz} = 0.05$ W/mK for the insulation (mineral wool), and $\lambda_{sz} = 1.30$ W/mK for the fire brick.

Thermal transmittance coefficient k:

$$\frac{1}{k} = \frac{1}{\alpha_n} + \sum_i \left(\frac{t_i}{\lambda_i} \kappa_i \frac{r_z}{r_i}\right) + \frac{1}{\alpha_o}, \tag{1}$$

where $\alpha_n = 8 + v = 8 + 10 = 18$ W/m²K—the inflow coefficient (the zone of the inner surface of the lining), where v—the mean velocity of the flue gas;

$\alpha_o = 24$ W/m²K—the outflow coefficient (the outer surface of the shaft);

t_i—the thickness of layer i;
λ_i—the thermal conductivity coefficient of layer i;
r_i—the outside radius of layer i;
r_z—the outside radius of the shaft;
κ_i—correction coefficients taking into account wall curvature:
$\kappa_i = \left(\frac{r_z}{r_i}\right)^{0.57}$ where $i = b$ for the reinforced concrete shaft, $i = iz$ for the mineral wool, $i = sz$ for the fire brick.

Table 1. Geometric dimensions of the thermographically surveyed existing chimney.

Segment No.	H	Shaft Outside Radius	Chimney Wall Thickness	Shaft Inside Radius	Insulation Thickness	Lining Thickness
-	[m]	[m]	[m]	[m]	[m]	[m]
1	2.5	4.00	0.42	3.58	0.15	0.23
2	15	3.80	0.39	3.41	0.15	0.23
3	30	3.61	0.36	3.25	0.13	0.23
4	45	3.41	0.33	3.08	0.13	0.114
5	60	3.21	0.3	2.91	0.13	0.114
6	75	3.01	0.27	2.74	0.13	0.114
7	90	2.82	0.24	2.58	0.13	0.114
8	105	2.62	0.21	2.41	0.13	0.114
9	120	2.42	0.18	2.24	0.13	0.114

The temperature drop in a given layer is expressed by the formula

$$\Delta T_i = k\frac{t_i}{\lambda_i}\kappa_i\frac{r_z}{r_i}\Delta t, \tag{2}$$

where $\Delta t = t_w - t_z$.
For the inflow and outflow, the temperature drops are

$$\Delta T_n = k\cdot\frac{1}{\alpha_n}\Delta t \qquad \Delta T_o = k\cdot\frac{1}{\alpha_o}\Delta t. \tag{3}$$

The temperature at the boundary of each of the layers is

$$T_j = t_w - \frac{k}{\alpha_n}\cdot\Delta T - \sum_i \Delta T. \tag{4}$$

Figures 6 and 7 show temperature drops in the particular layers in the winter season and in the summer season. The largest temperature drop occurs in the full insulation layer, and it is about 10 times larger than in the reinforced concrete shaft. Because of the small thickness of the fire brick layer, the temperature drop in this layer is the smallest.

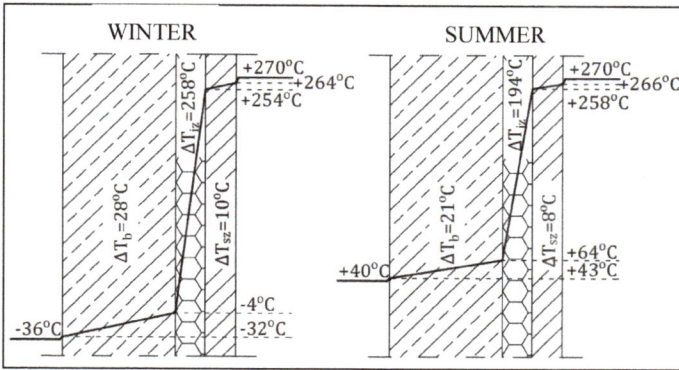

Figure 6. Temperature drops in particular layers for the winter and summer season.

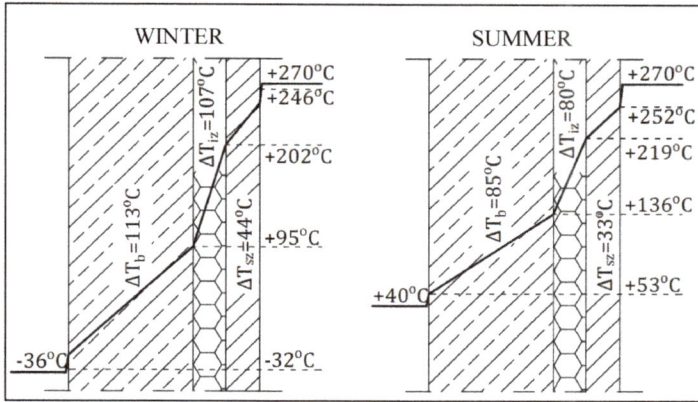

Figure 7. Temperature drops in particular layers of design chimney insulation in the first segment with damaged insulation.

5.4. Distribution of Temperature in Chimney Wall in Case of Insulation Discontinuity

Insulation discontinuities were assumed to occur in 2.5 high segments uniformly distributed along the whole height of the chimney. The mineral wool in the segments was assumed to lack the design insulating power.

The temperature drops in the particular layers for the decreased mineral wool insulating power are shown in Table 2. Since it is not possible to directly assess the degree to which the mineral wool's insulating power decreased, an approximate method was used. Calculations in which λ_{iz} was the unknown were carried out on the basis of the thermographic surveys of the existing chimney and the temperatures on its surface. The value of λ_{iz} was adjusted consistently with the actual temperatures on the surface of the investigated chimney. The results are presented in Table 3. The results apply to the 10-fold decrease in the insulating power of the mineral wool in the selected places ($\lambda_{iz} = 0.5$ W/mK).

The maximum temperature gradient in the reinforced concrete shell in the winter season increased by about 420% in comparison with the gradient calculated for the worst insulation case. For the summer season temperature, the gradient increased by about 405%. The temperature gradient which the insulation transfers is approximately equal to the gradient transferred by the chimney shaft under both the winter and summer temperature load. Such a large increase in temperature load can lead to the cracking of the reinforced concrete shell and to its damage. The temperature gradient values in the summer season and the temperature on the chimney's surface in summer and in winter and a figure (Figures 8–10) showing the temperature gradient values in each of the layers are presented below.

Table 2. Temperatures on the surface of the calculated chimney at an ambient temperature of 15 °C.

Seg. No.	H		$T_{sz,w}$	$T_{sz,z}$	$T_{iz,z}$	$T_{b,z}$	$T_{ext.}$		$T_{sz,w}$	$T_{sz,z}$	$T_{iz,z}$	$T_{b,z}$	$T_{ext.}$
-	[m]		[°C]	[°C]	[°C]	[°C]	[°C]		[°C]	[°C]	[°C]	[°C]	[°C]
1	10		250	213	124	30	15		265	256	41	19	15
2	20		246	208	118	30	15		261	252	39	19	15
3	30	Damaged Insulation	241	203	112	30	15	Full Insulation	257	248	38	19	15
4	40		234	192	118	32	15		251	240	42	20	15
5	50		230	187	111	33	15		247	236	40	20	15
6	60		223	175	113	35	15		241	227	44	21	15
7	70		218	169	106	35	15		237	223	41	21	15
8	80		212	160	103	37	15		232	216	42	22	15
9	90		207	153	95	37	15		228	212	39	22	15
10	100		204	150	94	37	15		224	208	38	21	15
11	110		200	148	92	36	15		220	204	38	21	15
12	120		193	142	89	35	15		212	197	37	21	15

Where $T_{sz,w}$—the temperature of the outside surface of the ceramic wall; $T_{sz,z}$—the temperature of the inside surface of the ceramic wall (contact with isolation); $T_{iz,z}$—the temperature of the inside surface of the concrete wall; $T_{b,z}$—the temperature of the outside surface of the concrete wall; T_{ext}—outside temperature.

Table 3. Differences between chimney operating temperature and chimney erection temperature.

Seg. No.	H	Winter		Summer	
-	[m]	Insulation		Insulation	
		Full	Damaged	Full	Damaged
		[°C]	[°C]	[°C]	[°C]
1	10	−28	28	44	86
2	20	−29	25	43	83
3	30	−30	22	42	81
4	40	−27	27	44	84
5	50	−28	23	43	81
6	60	−25	26	45	83
7	70	−27	22	44	79
8	80	−25	21	45	79
9	90	−27	17	44	75
10	100	−28	16	43	74
11	110	−28	15	43	73
12	120	−29	13	42	71

To relate the assumptions to the actual temperature loading of the chimney, the calculated gradients were compared with the temperatures appearing in the thermographic pictures of the existing chimney. The investigated real chimney is a 120 m high reinforced concrete structure serving a coke oven battery. The temperature of the combustion gas at the investigated chimney's inlet reaches 270 °C (maximally 340 °C) while the flue gas temperature at the outlet amounts to 220 °C (maximally 300 °C). The chimney is made of concrete C25/30 and reinforced with steel A-II. The chimney was divided into 9 segments, each about 15 m high. The chimney's insulation is made of semi-hard mineral wool boards and batts and its lining is made of fire brick. The thicknesses of the particular layers and their diameters are presented in Table 1.

The thermographic surveys revealed temperature differences on the chimney's surface, which indicates that the insulation was not uniform, there were gaps in it, and in places the insulation had slid down and was damp. The temperature differences can also be due to the degradation of the reinforced concrete chimney shaft. Since the measurements were carried out outside the chimney, the places with deteriorated thermal insulation are visible as dark red (the hottest places). The suspected places where the insulation is missing are visible in the thermograms as bands running at regular intervals around the circumference of the chimney. One can discern a certain pattern in insulation discontinuity: the bands are spaced at every 1 to 3 m. The insulation loss is most visible in the lower part of the chimney, and it is uniform there. In the thermographic picture, the surface temperature in this part reaches about 34 °C. In the segments situated higher one can see more distinct temperature differences on the outside surface, ranging from 22 to 34 °C.

Since the temperature of the flue gas in the existing chimney is comparable with the assumed emergency temperature in the chimney and also the diameters and thicknesses of the layers are similar, the temperature on the surface of the RC shaft, indicated by the thermographic surveys should be similar to the one yielded by the calculations. As indicated by the high background temperature, the surveys were carried out in the summer season. The ambient temperature was assumed to be 15 °C. The temperatures on the shaft surface calculated for the case with insulation and the case with insulation loss are comparable with the ones revealed by the thermographic surveys of the real chimney, as shown in the tables. For the assumed value of $\lambda_{iz} = 0.5$ W/mK, the temperature on the RC shell is comparable with the temperature indicated by the thermographic surveys.

Temperature loading in 2.5 m high bands for alternately full insulation and damaged insulation was assumed. The values were assumed according to Tables 2 and 3.

Figure 8. Thermographic picture of the lower part of chimney shaft.

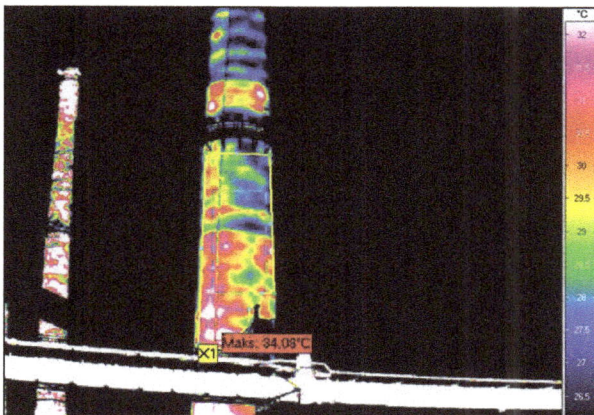

Figure 9. Thermographic picture of the middle part of chimney shaft.

Figure 10. Thermographic picture of the upper part of chimney shaft.

5.5. Surface Temperature Load

Since the chimney works at a temperature different than the one prevailing during its erection, a temperature load generating internal forces due to the chimney's limited freedom of deformation was assumed. Two load cases, the summer season load and the winter season load, were considered. In each of the cases, a load at full insulation and a load at damaged insulation were analysed.

The mean temperature in the wall can be calculated from the formula

$$\overline{T} = T_{b,wew} - \frac{\Delta T_b}{2}, \tag{5}$$

where $T_{b,wew}$—the temperature on the inside surface of the reinforced concrete shaft; ΔT_b—the temperature drop in the reinforced concrete shaft.

The temperature at which the chimney had been erected was assumed as $T_o = 10\,^{\circ}$C.

The difference between the chimney operating temperature and the chimney erection temperature was calculated from the formula

$$\Delta T = \overline{T} - T_o, \tag{6}$$

5.6. Increase in Bending Moments Due to the Increase in the Temperature Gradient

The difference in the temperature distribution across the thickness of the reinforced concrete wall between the case with damage insulation and the case with undamaged insulation amounts to

$T = 113 - 28 = 85°$

Such a big gradient causes additional bending moment and additional shear force.

The model of these calculations is presented in Figure A1.

Calculations of this influence are included in Appendix A

Additional thermally induced internal forces.

Let us consider the chimney as a cylindrical shell in which the unexpandable (well thermally insulated) band can be regarded as an element resembling a clamping tendon. Circumference L of the surrounding area with damaged thermal insulation is increased by L due to higher temperature T. The tendon with the introduced compressive force reduces the circumference to its size before the thermal expansion.

Using such a static load model one gets additional bending moment M and additional shear force Q (model is presented on Figure A2. The calculation of these additional factors is presented in Appendix B.

A few remarks

The presented computing models do not average their results. They assume quite a sharp transition from one temperature field to another without a transition area. In fact, the intermediate area can smooth calculations outcomes.

6. Conclusions Emerging from Investigations of Thermal Insulation of Chimneys

The method of thermographic surveys, used to monitor thermal insulation in housing, also finds an important application in the investigation of the durability and failure hazard of industrial structures in which the condensation of chemically aggressive gases is likely to occur. This particularly applies to chimneys where invasive tests are not preferred because of the consequences of taking samples from the structure.

Owing to thermographic surveys, one can monitor the hazards leading to the degradation of the chimney structure.

It is possible to indicate the area where flue gas condensate penetrates through cracks and leaks into the inner fire-brick lining into the insulation layer made of mineral wool. The condensate penetrating through cracks in the RC shaft to the structural reinforcement intensifies the corrosion of the steel and the concrete. As a result of the decrease in insulating power the temperature in the chimney RC core rises, whereby the bending moments and the tensile forces increase and the reinforced concrete shell cracks under the additional stresses which had not been taken into account in the design of the walls [2].

Thermograms can show increased shear and tensile stresses that can cause these large cracks in concrete.

Soaked by condensate or damaged mineral wool areas are visible on thermograms.

The calculations presented in the paper indicate the possibility of assessing the values of shear stresses from thermal interactions.

The authors took part in the design of chimney repairs with large diagonal cracks caused by high shear stresses.

In the case of chimneys operating in the uninterruptable process mode, special measures are required to improve the condition of the insulation. The repair of the reinforced concrete chimney consisting solely in injection filling the external cracks [10] stops the corrosion of the reinforcing steel, but it does not improve the insulation of the chimney.

Author Contributions: Conceptualisation, M.M. and A.U.; Methodology, M.M.; Software, H.H. and F.A.; Validation, M.M., A.U., H.H. and F.A.; Formal Analysis, M.M.; Investigation, M.M. and A.U.; Resources, M.M., A.U., H.H. and F.A.; Data Curation, M.M. and A.U.; Writing—Original Draft Preparation, M.M. and A.U.; Writing—Review and Editing, M.M., A.U., H.H. and F.A.; Visualisation, M.M., H.H. and F.A.; Supervision, A.U.

Funding: This research received no external funding

Acknowledgments: Developed as part of the research project "Industrialized construction process (Construction 4.0). Technological and methodological conditions of application of selected composite elements in civil engineering". This project is carried out by the Wroclaw University of Technology together with the Peoples' Friendship University of Russia in Moscow, Research Project PWr-RUDN 2017 no. 45WB/0001/17 Industrialized Construction Process.

Conflicts of Interest: The authors declare no conflict of interest.

Appendix A

$M_t = {} * T * E_c J_c / t = 0.00001 * 85 * 33,000,000 * 1 * 0.24^2 / 12 = 134.5$ kNm

The cracking moment for the concrete

$M_{cr} = f_{ctm} W_1$ (1)

where $E_c = 33$ GPa and $f_{ctm} = 2.9$ Mpa—the mean tensile strength of the concrete (C30/37),

A_c—the cross-sectional area $A = 1 * 0.24 = 0.24$ m^2, $W_1 = bt^2/6$, $b = 1$ m, $t = 0.24$ m, the bending section modulus for phase I.

$M_{cr} = 2.9 * 1 * 0.24^2 / 6 = 117$ kNm

Shear stress at damaged insulation/undamaged insulation boundary

The difference between the elongations of the adjacent rings depends on (for diameter $D = 6.68$ m):

$(\pi D) = \pi*D* * T = 3.14*6.68*0.00001*85 = 20.98*0.00001*85 = 0.018$ m

The annular strain is

$= *(\pi D)/(\pi*D) = 0.018/20.98 = 0.000858$.

The annular force generated due to strain locking is

$N = A_c * *E_c = 0.24*0.000858*33,000,000 = 6795.4$ kN/m

The force is blocked by horizontal surfaces with width t, whereby shear stress $_c$ arises

$_c = N/(2*t) = 14.2$ MPa

at the mean shear strength of the concrete

$f_{cm} = 0.5 * f_{ctm} = 0.5*2.9 = 1.45$ MPa.

The above bending moments due to gradient T, and shear forces are additional quantities, whereby the forces assumed in the design are exceeded.

Calculations were carried out for simplified static load diagrams. The values of the bending moments and shear stresses indicate high additional stresses in the concrete.

As the horizontal band in the concrete wall with damaged insulation heats up, the band is "pushed" outside more strongly, whereby a bending moment arises.

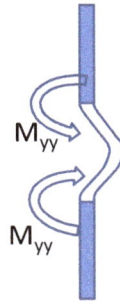

Figure A1. Change in length of cylindrical shell results in the meridional bending moment.

As this displacement is blocked, annular force $N = 6795.4$ kN/m is generated.

Appendix B

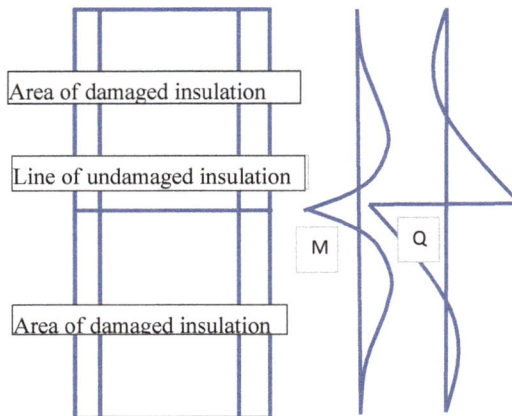

Figure A2. Model of influence damaged on the undamaged area of insulation.

In this case, calculations of the internal forces for design value T obtained from the thermograms are as follows ([16]):

E_c = 33 GPa

t – thickness = 0.24 m

$\ $ = 0.2

T_{design} = 15°.

$D_{diameter}$ = 6.68 m

r = D/2 = 3.5 m

L = 20.986 m

L = 0.003 m

N (force in shell, generated by T) = 1188 kN/m

$_p$ = $(12*(1 - {}^2)/(D^2h^2))^{(1/4)}*1.42$ ([9,16])

$/ (4\ _p)$ = 0.55 m

F_{des} = N/r = 339.43 kN/m

$M = F_{des}/(4*\ _p)$ = 59.70 kNm/m

$Q = F_{des}/2$ = 169.71 kN/m

The above values added to the stress resulting from the permanent loads and the variable loads may contribute to horizontal cracking.

References

1. Dörr, R.; Noakowski, P.; Breddermann, M.; Leszinski, H.; Potratz, S. Verstärkung eines Stahlbetonschornsteins. *Beton- und Stahlbetonbau* **2004**, *99*, 670–674.

2. Schnell, J.; Kautsch, R.; Noakowski, P.; Breddermann, M. Verhalten von hochbaudecken bei zugkräften aus zwang: Einfluß von kriechen, betonfestigkeit, temperaturdifferenz, plattendicke und spannweite—Auswirkung auf schnittgrössen, stahlspannung, rissbreite, druckzonenhöhe und durchbiegung. *Beton- und Stahlbetonbau* **2005**, *100*, 406–415. [CrossRef]

3. Jäger-Cañás, A.; Pasternak, H. Influence of closely spaced ring-stiffeners on the axial buckling behavior of cylindrical shells. *Ce/Papers* **2017**, *1*, 928–937. [CrossRef]

4. Bobkiewicz, J. Badania termograficzne budynków jako podstawa skutecznej termomodernizacji. *Pomiary Autom. Robotyka* **2003**, *7*, 23–25. (In Polish)

5. Jaworski, J. *Termografia Budynków. Wykorzystanie Obrazów Termalnych w Diagnostyce Budynków*; Dolnośląskie Wydawnictwo Edukacyjne: Wrocław, Poland, 2000.

6. Nowak, H. *Zastosowanie Badań Termowizyjnych w Budownictwie*; Oficyna Wydawnicza Politechniki Wrocławskiej: Wrocław, Poland, 2012.

7. Pichniarczyk, P.; Zduniewicz, T. Wykorzystanie w budownictwie metody termowizji w podczerwieni. *Izolacje* **2010**, *15*, 21–23. (In Polish)

8. Żurawski, J. Termowizja jako weryfikacja jakości prac izolacyjnych. *Izolacje* **2008**, *11–12*, 19–23.

9. Sendkowski, J.; Tkaczyk, A.; Tkaczyk, Ł. Termowizja i termografia w diagnostyce kominów przemysłowych. Przykłady, możliwości. *Prz. Bud.* **2013**, *84*, 21–25.

10. Kaminski, M.; Maj, M.; Ubysz, A. Chimney cracked reinforced concrete walls as a problem of durability exploitation. In Proceedings of the Fifth International Conference on Structural Engineering (SEMC 2013), Cape Town, South Africa, 2–4 September 2013.

11. Horváth, L.; Iványi, M.; Pasternak, H. Thermovision: An efficient tool for monitoring of steel members. *IABSE Symposium Rio De Janerio* **1999**, 142–143.

12. Świderski, W. Metody i techniki termografii w podczerwieni w badaniach nieniszczących materiałów kompozytowych (Methods and techniques of infrared thermography in non-destructive testing of composite materials). *Probl. Tech. Uzbroj. (Probl. Armament Technol.)* **2009**, *38*, 75–92.

13. Maj, M.; Ubysz, A. Some methods of repairing of chimney cracked reinforced concrete wall. *TISNOB Pozn.* **2015**, 213–224.

14. Oddziaływania na konstrukcje. Część 1-5: Oddziaływania ogólne. Oddziaływania termiczne. *PN-EN 1991-1-5 Eurokod 1.*

15. Brown, C.J.; Nielsen, J. *Silos Fundamentals of Theory Behavior and Design*; E & FN Spon: London, UK, 1998.
16. Timoshenko, S.; Woinowsky Krieger, S. *Theory of Plates and Shells*, 2nd ed.; McGraw-Hill Book Company: New York, NY, USA, 1959; pp. 466–532.

materials

MDPI

Article

Pore Structure Damages in Cement-Based Materials by Mercury Intrusion: A Non-Destructive Assessment by X-Ray Computed Tomography

Xiaohu Wang, Yu Peng, Jiyang Wang and Qiang Zeng *

College of Civil Engineering and Architecture, Zhejiang University, Hangzhou 310058, China
* Correspondence: cengq14@zju.edu.cn

Received: 18 June 2019; Accepted: 6 July 2019; Published: 10 July 2019

check for updates

Abstract: Mercury intrusion porosimetry (MIP) is questioned for possibly damaging the micro structure of cement-based materials (CBMs), but this theme still has a lack of quantitative evidence. By using X-ray computed tomography (XCT), this study reported an experimental investigation on probing the pore structure damages in paste and mortar samples after a standard MIP test. XCT scans were performed on the samples before and after mercury intrusion. Because of its very high mass attenuation coefficient, mercury can greatly enhance the contrast of XCT images, paving a path to probe the same pores with and without mercury fillings. The paste and mortar showed the different MIP pore size distributions but similar intrusion processes. A grey value inverse for the pores and material skeletons before and after MIP was found. With the features of excellent data reliability and robustness verified by a threshold analysis, the XCT results characterized the surface structure of voids, and diagnosed the pore structure damages in terms of pore volume and size of the paste and mortar samples. The findings of this study deepen the understandings in pore structure damages in CBMs by mercury intrusion, and provide methodological insights in the microstructure characterization of CBMs by XCT.

Keywords: non-destructive method; damage; mercury intrusion porosimetry; X-ray computed tomography

---

1. Introduction

Pore structure characteristics of cement-based materials (CBMs) importantly indicate their mechanical property and durability performance. Determining the pore structure of CBMs, however, still faces big challenges because (1) pore structure testing methods, more or less, have intrinsic shortages, and (2) the microstructure of cement hydrates is rather sensitive to environments [1]. Mercury intrusion porosimetry (MIP) is probably one of the most widely used techniques for characterizing the pore structure of CBMs due to its advantages of simple physic principle, broad pore rang (depending on the maximum pressure applied), and low costs in time and manpower (fast and easy operation and sample preparation). With those features, the pore structure characteristics by MIP may become a benchmark when assessing the pore structure of CBMs with different methods [2].

Because liquid mercury is hydrophobic to most solids, it cannot invade into pores spontaneously without sufficient external pressures. The surface forces of the mercury fronts in pores, inversely depending on the pore curvatures, will resist the forces applied. By recording the stepwise increased mercury volume with pressure, the volume–pressure data can be obtained. Those original data generally are not directly used without the specific relation between pressure and pore size. With the assumptions of cylindrical, size-graded and connected pores, the pore size-pressure relation gives: $D = -4\gamma \cos\theta / P$, with D: the pore diameter; P: the applied pressure; γ: the surface tension of mercury;

and θ: the contact angle between mercury and pore wall. This relation is known as the Washburn equation [3]. Through this simple equation, MIP provides various size-related pore parameters, such as accumulative and differential pore size distributions (PSDs), mean size, threshold size and fractal dimension [4–7].

Whilst MIP has been popularly used, its accuracy in pore structure characterization is always debatable [8–14]. Generally, the debates of MIP pore data focus on: (1) the microstructure damages by sample pretreatment (drying), (2) the oversimplifications of pore topology (geometry and connectivity), (3) the constants of MIP parameters, (4) the conformance effect of samples, and (5) the pore damages by high pressures during mercury intrusion. The first term on the microstructure damages by pretreatment is inevitable but may be mitigated by using the relatively mild drying methods (e.g., solvent exchange [1,4,15]). The second term on the oversimplifications of pore topology that are intrinsically related to the physical bases of MIP would yield the so-called "ink-bottle" effect [8]. To compensate for this shortage, a multi-cycled intrusion-extrusion test scheme was often applied [14,16,17]. The third term on the constant MIP parameters is argued with the significant influences of the contact angle between mercury fronts and pore walls [11,18]. The parameters of MIP may be greatly different when the tested pores are narrowed to nano sizes [18]. However, due to the lack of data at nano sizes, the improvements in size-associated MIP parameters for pore structure characterization are limited. The fourth term on the conformance effect of samples is rarely mentioned because the samples of CBMs generally have no big differences. Our recent tests, however, showed that the surface conformance control may greatly narrow the threshold pore size [13]. The last term on the pore structure damages of CBMs during mercury intrusion, albeit noticed by previous researcher [19–21], can hardly be quantitatively characterized.

Feldman [20] used a repeat-intrusion testing scheme to detect the possible pore structure alterations of blend cement pastes. Note that the repeat-intrusion testing scheme used by Feldman [20] is different from the multi-cycled intrusion-extrusion test that is often operated in a stepwise loading-unloading way without expelling the mercury entrapped in the pores. In his tests, the second mercury intrusion was operated after the entrapped mercury during the first intrusion was completely removed, so the differences in PSDs between the first and second MIP tests can reflect the sizes of the pores damaged. It was observed by Feldman that damages to the pore structure occurred at 70 MPa in the hydrated blends. Olson et al. [21] employed an environmental scanning electron microscopy (ESEM) to in situ observe the damages to pore structure of a hardened Portland cement paste by mercury intrusion. Analyses indicated that the connectivity of the pores between 1–10 μm was raised after the intrusion pressure reached the threshold value. However, the results of Feldman [20] may fail to capture the real damaged pore sizes due to the oversimplifications of pore topology by MIP as mentioned above. Despite of the direct and obvious evidence of pore structure damages by MIP documented by Olson et al. [21], the ESEM tests on the open samples without specific treatments would be unsafe to the operators because mercury is highly evaporable at room temperature and poison to humans. Therefore, seeking a non-destructive method to assess the microstructure damages of CBMs before and after MIP is urgently wanted. X-ray computed tomography (XCT) may be a preferable candidate because XCT not only is a non-destructive method, but also provides the component and spatial information of the object tested.

By delivering X-ray beams at different angles, numerous 2D radiographic projections of a scanned object can be gathered and treated in a digital geometry processing to construct the 3D digital structures of the object [22]. The continual methodological developments of XCT test make it extensively used for the phase characterization in CBMs for predicting their mechanical and transport performances [23–26]. Because mercury has much stronger X-ray absorptivity than any constituent in CBMs and other nonmetallic materials, the combination of XCT and MIP may provide an effective way to enhance the ability of XCT to detect the pores beyond the normal voxel resolution [27]. This further provides a routine to detect the pore damages of CBMs after MIP because mercury drops can be entrapped in the damaged and/or undamaged pores [13]. The method also generates significances for pore structure characterization because mercury entrapment may also, to some extent, reflect the connectivity of pores [28].

In the present study, XCT tests were operated on ordinary Portland cement (OPC) paste and mortar before and after mercury intrusion to evaluate their pore structure changes with deepened analyses and discussions. The findings of this study provide a new and effective routine to non-destructively characterize the pore structure damages of CBMs by MIP.

2. Materials and Experiments

2.1. Materials and Sample Preparation

A PI 42.5 OPC cement (corresponding to ASTM Type I) was used as the only binding phase to prepare the porous paste and mortar samples. The chemical component and physical properties of the cement are shown in Table 1. When preparing the samples, no agents were used to control the fluidity of the fresh paste and mortar slurries. However, to obtain the similar fluidity, the water-to-cement (w/c) ratios of 0.4 and 0.5 were adopted for the paste and mortar, respectively. Commercial standard quartz sands with the fineness modulus of 2.6 and the SiO_2 content above 95% (Xiamen ISO Standard Sand Co., Ltd., Xiamen, China) were used as the fine aggregates to prepare the mortar. The cement/sand ratio was controlled as 1/3. Following standard casting, moulding and demoulding procedures, macro paste and mortar specimens were prepared, and then cured in a chamber with temperature at $20 \pm 2\,°C$ and relative humidity above 95%. After 28 days, the well cured specimens were crushed into small pieces (around 1 mL in volume or 2 g in mass) for further experiments.

Table 1. Chemical component and physical properties of cement.

Oxides	Content (%)	Minerals	Content (%)	Physical Properties	Value
SiO_2	21.68	C_3S	57.34	Density (g/mL)	3.10
Al_2O_3	4.80	C_2S	18.09	Specific area (m^2/kg)	345
Fe_2O_3	3.70	C_3A	6.47	Mean size (μm)	11
CaO	64.90	C_4AF	11.25		
MgO	2.76	Others	6.04		
SO3	0.29				
$Na_2O(eq)$	0.56				
CaO(f)	0.93				

The crushed OPC paste and mortar pieces were then immersed into pure ethanol to cease the hydration of cement. After two days of immersion in ethanol, the samples were then removed into an oven at 105 °C to expel the water physically absorbed in the pores. After 24 h, the samples were then stored in a sealed desiccator to eliminate the possible influences of water and carbon oxide in the air on the microstructure, and were readily prepared for XCT and MIP tests. Note that the drying temperature used here may be too severe to preserve the microstructure of the samples because the rapid water loss in the pores may yield high capillary stresses to damage the material matrix [1] and to alter the status of calcium-silicate-hydrate (C-S-H) gels [29]. However, because the microstructure alterations by drying are stable and will not recover during the following MIP and XCT tests, those alterations can be treated as the intrinsic microstructure features, and thus would be not considered here.

2.2. MIP Tests

The pre-dried samples were then placed into a sample chamber for MIP test in a device of Autopore IV 9510 (Micromeritics, Norcross, GA, USA). After a pre-equilibrium step to fill the gaps between the sample and chamber wall at 0.5 psi (3.45 KPa), pressures on mercury were automatically and stepwise raised to 60,000 psi (413.69 MPa) and then unloaded to certain values. With the equilibrium time of 10 s, a complete MIP test lasted about 130 min.

Immediately after the MIP tests finished, the samples were carefully and rapidly removed into small plastic tubes. A quick-hardening epoxy resin was rapidly poured into the tubes to completely cover the samples, and those tubes were quickly and tightly lidded (Figure 1). Those steps were to cease the leakages of mercury from the mercury-filled samples for diminishing the possible dangers to the technicians when handling those CBM samples.

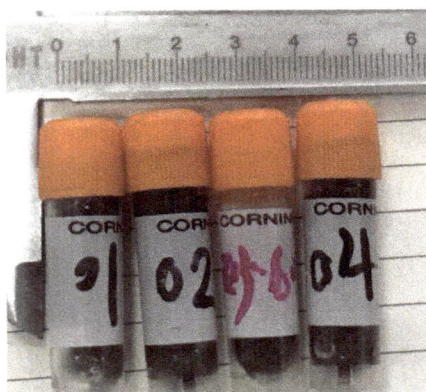

Figure 1. Paste and mortar samples encased in epoxy resin after mercury intrusion.

2.3. XCT Tests

Before and after the MIP tests, XCT scans were performed on the samples by an X-CT scanner of Nikon XTH 255/320 LC (Nikon, Tokyo, Japan). For the XCT tests before and after MIP processes, the voltages to deliver the X-ray beams were set as 100 KeV and 150 KeV, respectively (Table 2; see below for detailed explanations). The penetrated X-ray beams were detected by a high-sensitive detector (DRZplus Scintillator with the pixels of 2000(h) × 2000(v)) at the back of the objects synchronously. During testing, the samples rotated in the rate of 12 °/min. The collected data were then loaded into the software of VGStudio Max (version 3.1, Volume Graphics, Inc., Charlotte, NC, USA) for further analyses. Because the pre-MIP sample size was smaller than the post-MIP one (CBM sample plus tube), the pixel resolution of the former case was slightly higher (5.02 µm for the pre-MIP sample versus 5.60 µm for the post-MIP sample).

Table 2. XCT parameters used before and after MIP tests.

Condition	Voltage (KeV)	Pixel Resolution (µm)
Before MIP (Pre-MIP)	150	5.02
After MIP (Post-MIP)	100	5.60

The reason for using different delivering voltages for the pre-and post-MIP tests was because mercury has far higher X-ray mass attenuation coefficient (MAC) than the main solid phases in the CBM samples. Figure 2 shows the MAC curves of mercury, SiO_2, C-S-H and cement in the generally used photon energy interval (90–160 KeV). At 100 KeV (the photon energy used for the pre-MIP XCT tests), the MACs of SiO_2, C-S-H and cement are rather close (0.17–0.21 m^2/g). The slight MAC differences among those phases will even become less at a higher photon energy (see the inserted panel in Figure 2), so to obtain the high quality images of the pre-MIP samples, the photon energy of 100 KeV was selected. When mercury was intruded into the CBM samples, the MACs are greatly altered. As displayed in Figure 2, the MAC of mercury is 25–30 times higher than that of SiO_2, C-S-H and cement at 100 KeV. Such strong X-ray absorption by mercury would induce significant beam hardening artifacts [30]. Since the MAC gaps between mercury and the other phases will be narrowed with increasing photon energy (Figure 2), it is thus expected that a higher photon energy may bring less beam hardening artifacts. Our practices indeed indicated the images at the photon energy of 150 KeV would achieve high quality images for further analysis.

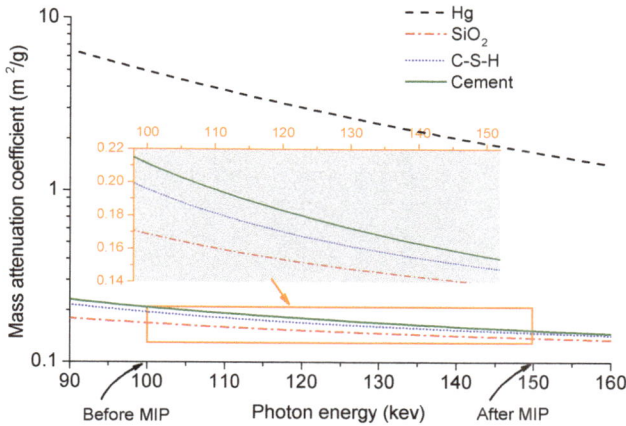

Figure 2. Spectra of mass attenuation coefficient of Hg, SiO$_2$, C-S-H and cement between 90 KeV and 160 KeV (Data from Ref. [31]).

Because the samples used have irregular shapes and rough surfaces, it is unrealistic to analyse the entire volume of the samples. Instead, some volumes of interest (VOI) inside the samples (or region of interest in 2D analysis) were selected for image analysis, which can prevent edge effects and increase data efficiency. Due to the heterogeneity in microstructure and the chaos in pore structure of CBMs [6], a random selection of VOI that would be preferred for obtaining representative and reproductive results was not adopted here. In order to easily and precisely identify the same VOI of the samples before and after mercury intrusion, the microstructure of the VOI must have distinguished characteristics. In this study, we intentionally selected the VOIs containing big voids. Figure 3 shows an example of the best fit registration of two cubic VOIs in the pre-MIP paste sample. Clearly, with these big voids (dark circles in the VOIs shown in Figure 3), the same VOIs can be easily identified for the same sample after mercury intrusion.

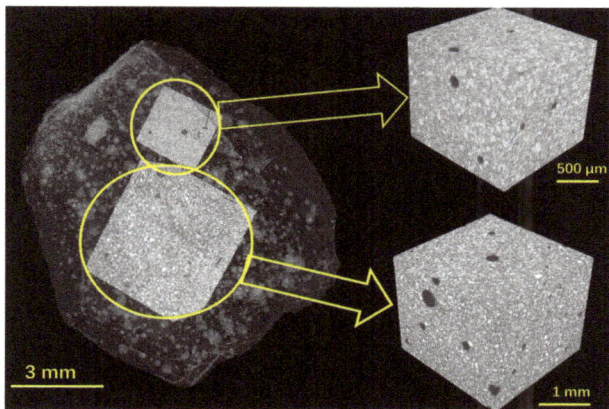

Figure 3. An example of VOI selections from a pre-MIP paste sample.

3. Results and Discussion

3.1. MIP Outcomes

For pore structure characterization, the classic Washburn equation was used to interpret the MIP data with the mercury surface tension of 485 mN/m and the contact angle between mercury and substrate of 130°. With those data, some characteristic pore parameters of the paste and mortar samples, i.e., total porosity, volume-median pore size, specific surface area and threshold pore size, can be evaluated (Table 3). Obviously, compared with mortar, paste showed the higher total porosity and specific surface area, the similar threshold pore size, but the lower volume-median pore size. The results are reasonably due to the fact that the impermeable sands in the mortar occupied more than 60% of the total volume. The looser compactness and more porous cement hydrates of the mortar induced by the higher w/c ratio, as well as the porous interfacial transition zones (ITZs) between cement matrix and aggregates, caused the higher volume-median pore size (independent of the absolute pore volumes), but remained unable to compensate for the reductions in porosity and specific surface area (Table 3). The similar threshold pore sizes between the paste and mortar suggested that the connected throats formed from the interparticle continuum had the similar widths.

Table 3. Characteristic pore parameters of paste and mortar form MIP.

Sample	Total Porosity (%)	Volume-Median Pore Size (nm)	Specific Surface Area (m²/g)	Threshold Pore Size (nm)
Paste	20.0	66.1	12.3	76.5
Mortar	15.1	85.1	9.3	76.9

Figure 4 shows the (top) accumulative and (bottom) differential PSDs of the paste and mortar. While the PSDs of the paste and mortar had the different shapes, they both displayed the similar five-stage characteristics: surface conformance, non-channel stage, capillaries by flaws and ITZs, capillaries by interparticle space, and gel pores.

1. Mercury first covered the open cracks, gaps, cavities and irregularities on the sample surfaces in relatively low pressures (termed as the surface conformance effect) [13,32]. Our previous study [13] suggested that the surface conformance effect might not be avoided because these cavities, cracks and flaws can be inevitably induced during sample pretreatments such as cutting and drying [1,15]. However, the volume increases at the very beginning stage of MIP by the surface conformance effect can be mitigated by controlling the exposed areas of the MIP samples [13]. In this study, the surface conformance effect was insignificant for both the paste and mortar samples (<0.005 mL/g).

2. Later, almost no mercury intrusion was recorded between 2 μm and 100 μm (Figure 4). This meant that no open channels (not the pores inside the materials) in such size interval can be recognized by MIP, which was termed as the non-channel stage.

3. As the size decreased further, the mercury increases of both the paste and mortar became obvious (Figure 4). Generally, for normally cured cement paste, these increases can be rarely observed [8]. In this study, the very severe drying scheme (105 °C) was used, so the microstructure flaws or damages by drying [1] would account for the abnormal mercury rises in this stage. For the mortar sample, the porous ITZs, together with the capillary flaws by drying, were responsible for the higher PSD data (see the shadowed areas shown in Figure 4).

4. After that, the intrusion volumes rose rapidly and significantly with obvious peaks around 70 nm (Figure 4). The peak size was identical to the threshold pore size form the percolated pore continuum [29,33,34]. Because of the 'ink-bottle' effect [8], the volumes at or below the threshold size could partially represent the capillaries of the interparticle space that remained unfilled by cement hydration. Compared with the mortar sample, the paste sample showed the faster raising rate and higher peak intensity because of the higher capillary pores.

5. Under the higher pressures, the mercury rising rates became slower and the differential PSDs were depressed (Figure 4) because only limited space (mainly gel pores) was available to accommodate the mercury after the capillaries were filled. Since the MIP parameters in nano scales remained debatable [11,18], those data would not shed much light on gel pore characterization.

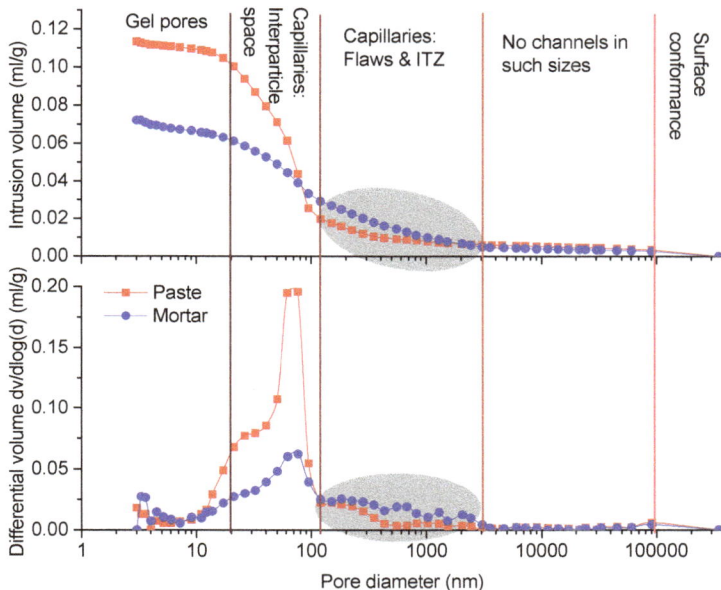

Figure 4. Accumulative (**top**) and differential (**bottom**) pore size distributions of paste and mortar samples (The specific contribution of ITZ in the mortar was singled out in the shadowed areas).

3.2. Threshold Analysis

Despite the fact that XCT is a powerful tool to non-destructively characterize the microstructure of various materials, the results, as cautioned elsewhere [22,27], are highly depending on the process of threshold segmentation. In this section, the effect of threshold segmentation on the reconstructed results of the pore phase was discussed, so the reliability and robustness of the damage diagnoses by XCT could be guaranteed.

Figure 5 shows the voxel-grey value distributions of a VOI in the paste sample before and after mercury intrusion. Generally, due to the lower X-ray absorption, the empty pores in a CBM sample were captured by the low grey value area, and the cement skeletons (including the hydration products and unhydrated cement clinkers) that can absorb more X-rays thus were represented by the high grey value area (Figure 5a). When mercury was intruded into the empty pores, the characteristic areas of grey value were exchanged. Specifically, the high grey value area represented the mercury-filled pores, while the low grey value area denoted the cement matrix (Figure 5b). This feature of grey-value inverse was recently used to determine the mercury drops entrapped in the pores of HCP samples with/without surface conformance control for pore structure characterization [13].

The voxel-grey value distributions displayed in Figure 5 clearly showed two peaks, so the threshold segmentation should be operated at the minimum between the two peaks. Here, to discuss the threshold sensitivity, three threshold points were selected, i.e., the middle, low (−5%) and high (+5%) threshold values shown in Figure 5a. In Figure 6, the 2D and 3D images of the pores segmented from the paste skeleton are comparatively plotted with the designed three threshold values. Apparently, no obvious differences can be seen from those images. To specifically compare the pore information in different scales, the volume-size plots of all the extracted pores in a VOI of the cement with three threshold values were illustrated in Figure 7. The results showed that both the number and size of those objects had no obvious differences (Figure 7). A similar threshold analysis was also performed on the post-MIP samples to obtain the appropriate threshold grey values with the reliability data.

Overall, the data of Figures 6 and 7 implied that the XCT analyses used in this study provided reliable and robust pore structure information for identifying the pore structure damages of CBMs by MIP. In the following contexts, all discussions were based on the XCT results with the middle threshold segmentation.

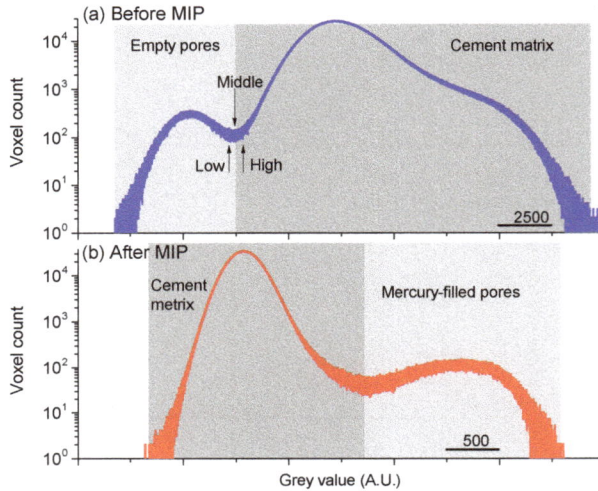

Figure 5. Voxel-grey value distributions of a VOI in the paste sample (**a**) before and (**b**) after mercury intrusion. Three threshold values (low, middle and high) were selected to test the influence of threshold process on pore segmentation.

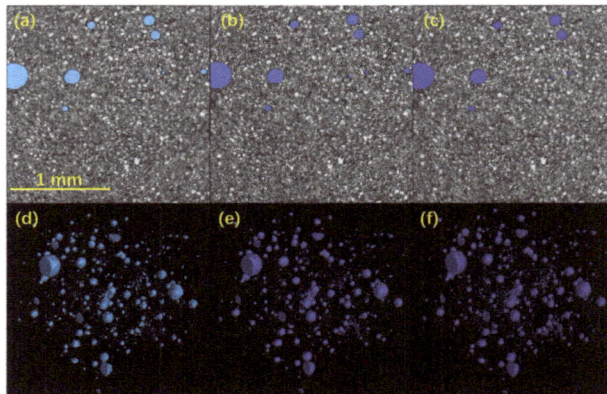

Figure 6. 2D (**a–c**) and 3D (**d–f**) images of pores segmented from paste skeleton at (**a,c**) low, (**b,d**) middle, and (**c,f**) high threshold values from Figure 5.

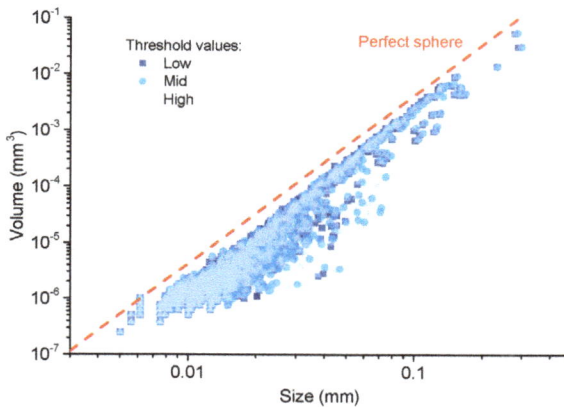

Figure 7. Volume-size distributions of the pores segmented from paste skeleton at low, middle, and high threshold values from Figure 5.

3.3. Characteristics of XCT Results

Figures 8 and 9 show the 2D and 3D representative images of a VOI and a localized big pore of the cement and mortar sample, respectively. In a much clearer way, the grey-value inverse of the pores before and after mercury intrusion can be displayed. For instance, the same pores were illustrated in the darkest color before MIP and the brightest color after MIP (Figure 9b,e).

Some features in microstructure characterization of CBMs by the combination of MIP and XCT can be pointed out from Figures 8 and 9. Firstly, all the visible air voids in the paste and mortar samples were fully filled with mercury. For ordinary CBMs, the air voids with the size range of 10–500 μm and the content less then 2.5% were reported [8]. Those air voids, however, can not be detected by MIP because it only measures the open channels rather than the pore chambers [13]. This finding again made evident that MIP fails to detect the 'ink-bottle' like pores [8]. Secondly, the air voids in the paste and mortar samples showed different surface structures. Specifically, the surfaces of the voids in the paste were rather rough with and without mercury intrusion (Figure 8c,f), while those in the mortar were much smoother (Figure 9c,f). Although the akin rough pore surfaces of CBMs were documented [13,35], the mechanisms for the surface structure differences between paste and mortar remained unclear and deserved further rigorous studies. Thirdly, the mercury intrusion process enhanced the contrast of the mortar images to figure out the aggregates (quartz). As shown in Figure 9d, the aggregates were illustrated as the darkest phase (due to the lowest X-ray absorption, Figure 2) embedded in the much brighter cement matrix and the brightest mercury-filled voids. Because of the closely valued MACs of quartz, C-S-H and cement clinkers (Figure 2), separating quartz from the other two phases would be difficult. The post-MIP XCT test used in this study may provide an effective way to obtain the packing pattern of the aggregates in mortar.

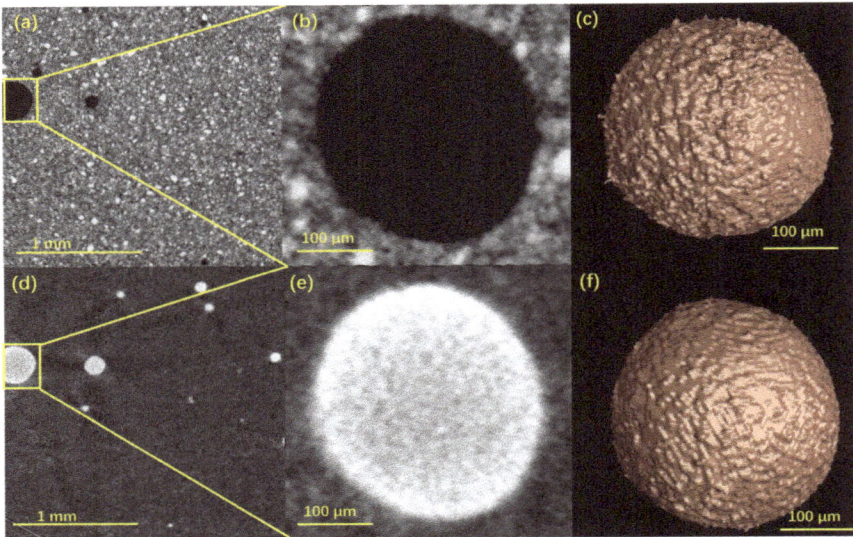

Figure 8. Representative images of a VOI in the cement sample (**a**–**c**) before and (**d**–**f**) after mercury intrusion: (**a**,**d**) the 2D sectional view with (**b**,**e**) the magnified pore and (**c**,**f**) its 3D structure.

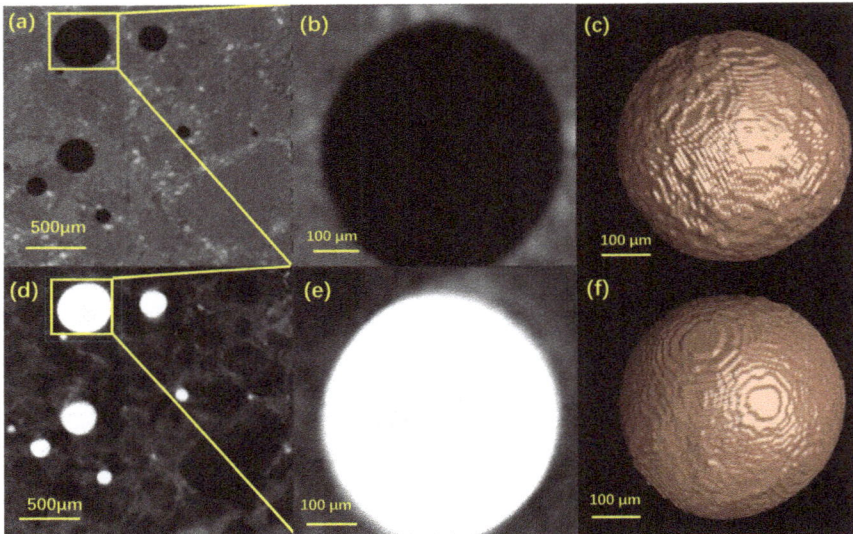

Figure 9. Representative images of a VOI in the mortar sample (**a**–**c**) before and (**d**–**f**) after mercury intrusion: (**a**,**d**) the 2D sectional view with (**b**,**e**) the magnified pore and (**c**,**f**) its 3D structure.

Last but not least, the mercury intrusion paths were identified from the XCT images of the post-MIP mortar. For instance, Figure 10 displays a 2D XCT image of a local area of the mortar sample after mercury intrusion, where the bright areas along the ITZs around the aggregates indicated the thoroughly penetrated path to the air void. However, these mercury penetration paths cannot be identified from the XCT images of the post-MIP paste sample (Figure 8d) because the penetration sizes in the paste (0.2 μm [8]) would be beyond the resolution of our XCT tests.

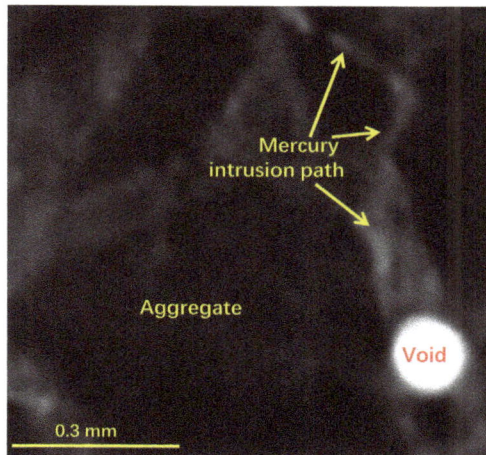

Figure 10. A 2D sectional image of a local area in the mortar showing a mercury intrusion path along the ITZs between aggregates and cement matrix to an air void.

3.4. Damage Diagnosis

We then used the XCT data of the pre-and post-MIP samples to diagnose whether or not the microstructure of those samples was damaged after MIP and in what sizes the damages occurred.

Figures 11 and 12 comparatively plot the PSDs and the statistics in pore volume of VOIs, respectively, in the paste and mortar samples before and after mercury intrusion. Note that the minimum diameters of the reconstructed pores (around 10 μm) shown in Figures 11 and 12 were higher than the minimum detectable pixel sizes (Table 2) because the voxel resolution for a 3D reconstructed object (depending on the geometry of the object) would be always lower than the 2D pixel resolution [22]. After the MIP tests finished, clearly, the total pore volume was increased from 0.29 mm^3 to 0.31 mm^3 by 6.7% for the paste sample (Figure 11a), and from 0.13 mm^3 to 0.14 mm^3 by 7.6% for the mortar sample (Figure 12a). In a statistic manner, the mean pore volume was largely augmented from 5×10^{-6} mm^3 to 8×10^{-5} mm^3 by around 16 times for the paste sample (Figure 11b), and slightly from 5×10^{-5} mm^3 to 6×10^{-5} mm^3 by 20% for the mortar sample (Figure 12b). However, the heavy increases in mean pore volume shown in Figure 11b may be misleading because the numbers of the pores recognized (especially the thin-sized pores) were largely decreased. The increases in pore volume and decreases in pore number for CBMs were in line with the results reported by Olson et al. [21] through ESEM observations. From Figures 11 and 12, one could further read the sizes of the pores damaged directly. Two obvious pore volume increases shown in Figure 11a indicated that the damages mainly occurred to the pores of 100–200 μm and 300–500 μm for the paste sample. For the mortar sample, the damages concentrated at the size interval of 100–400 μm (Figure 12b).

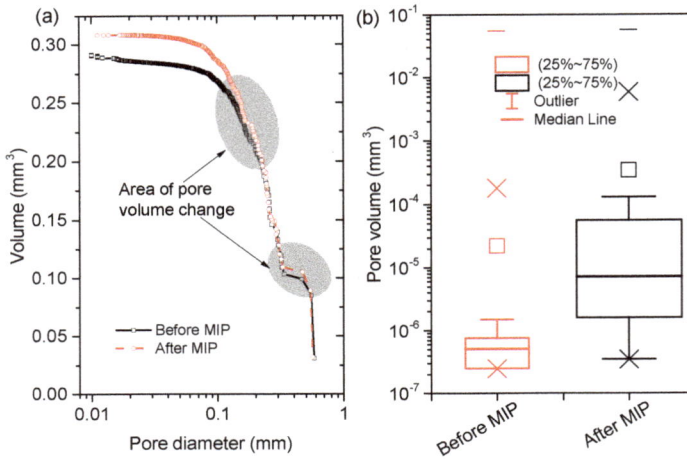

Figure 11. (**a**) comparative plots of pore size distribution of a VOI in the paste sample before and after mercury intrusion, and (**b**) the statistic results of the pore phase.

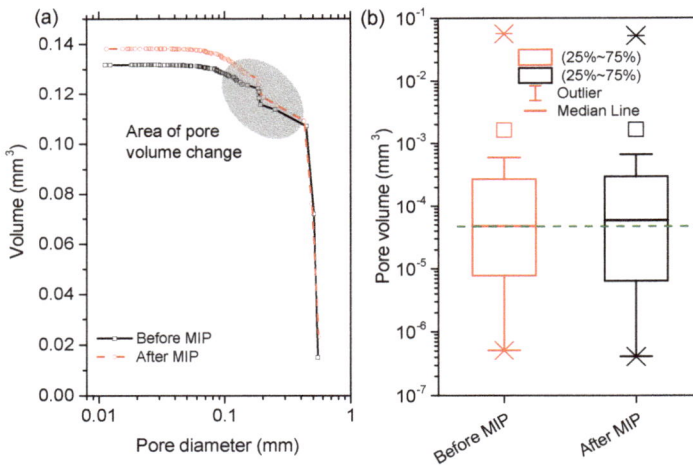

Figure 12. (**a**) comparative plots of pore size distribution of a VOI in the mortar sample before and after mercury intrusion, and (**b**) the statistic results of the pore phase.

3.5. Further Discussion

When mercury is enforced to invade into a porous CBM sample, the capillaries among the compacted particles and the porous cement hydrates deform to sustain the applied pressures. If the pressures are (even locally) higher than the strength of the phases in contact with the stressed mercury, damages take place. Those mechanically reasonable damages to CBMs by mercury intrusion can be schematically illustrated in Figure 13. Before MIP, the voids in a CBM sample (generally in 10–500 μm [8]) may be isolated by the material matrix consisting of the closely compacted cement particles and their hydration products (Figure 13a). The capillaries (channels) connecting those voids are generally too thin to be diagnosed by normal XCT. However, after mercury intrusion under sufficient pressures, the voids as well as these throats will be filled with mercury. Since mercury can strongly absorb the X-ray penetrated, the signals of the mercury-entrapped channels (albeit below the resolution of the XCT) may become detectable (Figure 13b). This is the regime applied in this study to diagnose the pore damages of CBM samples after mercury intrusion.

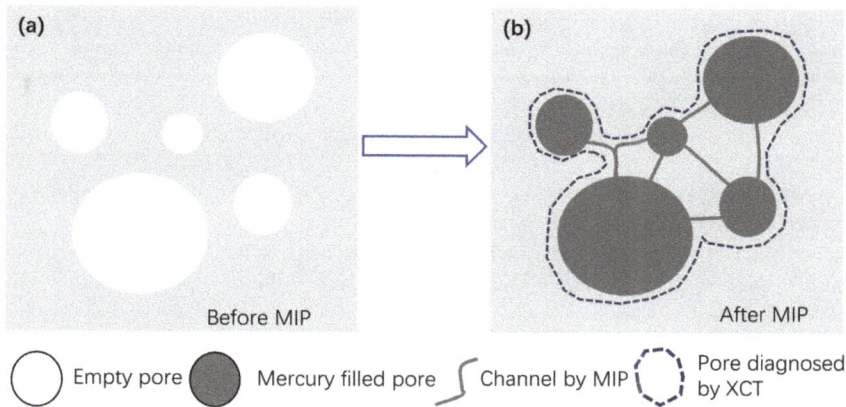

Figure 13. Mechanisms of pore damages induced by mercury intrusion under high pressures with the possibility of an overestimation on the pore size of a CBM sample after MIP (right) than that before MIP (left) by XCT.

In our tests, the damaged pore sizes measured by XCT were much higher than the data obtained by Olson et al. [21], who found that the connectivity of the pores in the 1–10 µm size range was greatly increased, and the average size was enlarged from 1.60 µm to 2.36 µm after MIP. A much lower size of the damaged pores was reported by Feldman [20] (around 18 nm corresponding to the applied pressure of 70 MPa). The size differences between our data and those reported in the literature [20,21] were mainly due to the different methods used and different objects concerned. In Feldman's tests [20], the pore damages were assessed by the PSD differences of cement blends before and after mercury intrusion, so the obtained sizes were always underestimated due to the intrinsic biases in the pore sizes of MIP, e.g., the 'ink-bottle' effect [8]. In the tests by Olson et al. [21], 2D images from limited local areas were obtained from ESEM. In our tests because of the limited resolution of the XCT used, the pores below 10 µm, mainly the throats to connect the voids [8], can not be detected. Instead, the damaged voids under high pressures were diagnosed.

While the present study reported obvious damages to the pore structure of the paste and mortar samples, several themes remained to be discussed further. Firstly, one must understand that the obvious enhancement in image contrast by mercury may induce some biases in pore structure characterization due to the beam hardening artifacts [27,36]. For example, if the voids were neighbours, the signals of X-ray beams may overlap so the individual pores as well as the connecting channels may be diagnosed as a big pore. This would significantly decrease the detected pore numbers but increase the pore volumes (Figure 13b). This regime may also partially account for the results shown in Figures 11 and 12. Secondly, the pore structure alterations by the severe drying process (105 °C) may induce additional variances when assessing the pore damages by mercury intrusion. It has been recognized that severe drying can greatly impact the packing patterns of C-S-H gels and the connectivity of pores [15,37]. Those may increase the difficulties in diagnosing the pore damages to CBMs after mercury intrusion. Future tests on the CBM samples with the milder drying schemes are preferred to mitigate this effect.

4. Conclusions

- XCT is a powerful technique to non-destructively characterize the microstructure of CBMs. The significant differences in X-ray MACs between mercury and the phases in CBMs can greatly enhance the contrast gradients in XCT images and facilitate the reconstruction of 3D microstructure.

- MIP tests indicated that, compared with the mortar sample, the paste sample had the higher porosity and specific surface area, similar threshold pore size, but lower median pore size. The MIP PSDs of the paste and mortar samples showed the similarly five-stage intrusion curves but the different specific spectra. The drying at 105 °C brought additional flaws just before the threshold stage to the paste and mortar samples.
- The grey values for the pores and material skeletons in the CBM samples were inversely distributed due to the shifts in X-ray absorptivity when the pores were filled with mercury.
- A threshold analysis indicated that the obtained XCT results showed good reliability and robustness in pore phase segmentation.
- The surfaces of the voids in the paste were rough, while those in the paste were smooth. Mercury intrusion paths along the ITZs around aggregates in the mortar sample were visible in the post-MIP XCT images.
- Mercury intrusion in the paste and mortar samples caused the increases in pore volume and the decreases in pore number as determined by XCT. The results were consistent with those reported in the literature.

Overall, the damages to the pore structure of CBMs after mercury intrusion can be non-destructively diagnosed by XCT with quantitative parameters. Going beyond this, the combination of MIP and XCT may provide a powerful tool to probe the pore structure alterations in CBMs under different environments.

Author Contributions: X.W. and Y.P. conducted the experiments and analysed the data, Q.Z., Y.P. and J.W. designed this work, and Q.Z. wrote this paper.

Funding: The research was funded by the National Natural Science Foundation of China (No. 51878602).

Conflicts of Interest: The authors declare no conflict of interest.

References

1. Zhang, Z.; Scherer, G.W. Evaluation of drying methods by nitrogen adsorption. *Cem. Concr. Res.* **2019**, *120*, 13–26. [CrossRef]
2. Zuo, Y.; Ye, G. Pore structure characterization of sodium hydroxide activated slag using mercury intrusion porosimetry, nitrogen adsorption, and image analysis. *Materials* **2018**, *11*, 1035. [CrossRef]
3. Washburn, E.W. Note on a method of determining the distribution of pore sizes in a porous material. *Proc. Natl. Acad. Sci. USA* **1921**, *7*, 115–116. [CrossRef] [PubMed]
4. Zeng, Q.; Li, K.; Fen-Chong, T.; Dangla, P. Pore structure characterization of cement pastes blended with high-volume fly-ash. *Cem. Concr. Res.* **2012**, *42*, 194–204. [CrossRef]
5. Zeng, Q.; Li, K.; Fen-Chong, T.; Dangla, P. Surface fractal analysis of pore structure of high-volume fly-ash cement pastes. *Appl. Surf. Sci.* **2010**, *257*, 762–768. [CrossRef]
6. Zeng, Q.; Luo, M.; Pang, X.; Li, L.; Li, K. Surface fractal dimension: An indicator to characterize the microstructure of cement-based porous materials. *Appl. Surf. Sci.* **2013**, *282*, 302–307. [CrossRef]
7. Leóny León, C.A. New perspectives in mercury porosimetry. *Adv. Colloid Interface Sci.* **1998**, *76*, 341–372. [CrossRef]
8. Diamond, S. Mercury porosimetry: An inappropriate method for the measurement of pore size distributions in cement-based materials. *Cem. Concr. Res.* **2000**, *30*, 1517–1525. [CrossRef]
9. Moro, F.; Boehni, H. Ink-Bottle effect in mercury intrusion porosimetry of cement-based materials. *J. Colloid Interface Sci.* **2002**, *246*, 135–149. [CrossRef]
10. Ma, H. Mercury intrusion porosimetry in concrete technology: Tips in measurement, pore structure parameter acquisition and application. *J. Porous Mater.* **2014**, *21*, 207–215. [CrossRef]
11. Muller, A.C.A.; Scrivener, K.L. A reassessment of mercury intrusion porosimetry by comparison with 1H nmR relaxometry. *Cem. Concr. Res.* **2017**, *100*, 350–360. [CrossRef]

12. Dong, H.; Zhang, H.; Zuo, Y.; Gao, P.; Ye, G. Relationship between the Size of the Samples and the Interpretation of the Mercury Intrusion Results of an Artificial Sandstone. *Materials* **2018**, *11*, 201. [CrossRef] [PubMed]

13. Zeng, Q.; Wang, X.; Yang, P.; Wang, J.; Zhou, C. Tracing mercury entrapment in porous cement paste after mercury intrusion test by X-ray computed tomography and implications for pore structure characterization. *Mater. Charact.* **2019**, *151*, 203-215. [CrossRef]

14. Zhang, Y.; Yang, B.; Yang, Z.; Ye, G. Ink-bottle effect and pore size distribution of cementitious materials identified by pressurization–depressurization cycling mercury intrusion porosimetry. *Materials* **2019**, *12*, 1454. [CrossRef] [PubMed]

15. Galle, C. Effect of drying on cement-based materials pore structure as identified by mercury intrusion porosimetry: A comparative study between oven-, vacuum-, and freeze-drying. *Cem. Concr. Res.* **2001**, *31*, 1467–1477. [CrossRef]

16. Zhou, J.; Ye, G.; Breugel, K.V. Characterization of pore structure in cement-based materials using pressurization depressurization cycling mercury intrusion porosimetry (PDC-MIP). *Cem. Concr. Res.* **2010**, *40*, 1120–1128. [CrossRef]

17. Gao, Z.; Hu, Q.; Hamamoto, S. Using multicycle mercury intrusion porosimetry to investigate hysteresis of different porous media. *J. Porous Med.* **2018**, *21*, 607–622. [CrossRef]

18. Wang, S.; Javadpour, F.; Feng, Q. Confinement correction to mercury intrusion capillary pressure of shale nanopores. *Sci. Rep.* **2016**, *6*, 20160. [CrossRef]

19. Shi, D.; Winslow, D.N. Contact angle and damage during mercury intrusion into cement paste. *Cem. Concr. Res.* **1985**, *15* , 645–654. [CrossRef]

20. Feldman, R.F. Pore structure damage in blended cements caused by mercury intrusion. *J. Am. Ceram. Soc.* **1984**, *67*, 30–33. [CrossRef]

21. Olson, R.A.; Neubauer, C.M.; Jennings, H.M. Damage to the pore structure of hardened Portland cement paste by mercury intrusion. *J. Am. Ceram. Soc.* **1997**, *80*, 2454–2458. [CrossRef]

22. Cnudde, V.; Boone, M.N. High-resolution X-ray computed tomography in geosciences: A review of the current technology and applications. *Earth-Sci. Rev.* **2013**, *123*, 1–17. [CrossRef]

23. Promentilla, M.; Cortez, S.; Papel, R.; Tablada, B.; Sugiyama, T. Evaluation of microstructure and transport properties of deteriorated cementitious materials from their X-ray computed tomography (CT) images. *Materials* **2016**, *9*, 388. [CrossRef] [PubMed]

24. Erdem, S.; Gürbüz, E.; Uysal, M. Micro-mechanical analysis and X-ray computed tomography quantification of damage in concrete with industrial by-products and construction waste. *J. Clean. Prod.* **2018**, *189*, 933–940. [CrossRef]

25. Yang, S.; Cui, H.; Poon, C.S. Assessment of in-situ alkali-silica reaction (ASR) development of glass aggregate concrete prepared with dry-mix and conventional wet-mix methods by X-ray computed micro-tomography. *Cem. Concr. Compos.* **2018**, *90*, 266–276. [CrossRef]

26. Buljak, V.; Oesch, T.; Bruno, G. Simulating fiber-reinforced concrete mechanical performance using CT-based fiber orientation data. *Materials* **2019**, *12*, 717. [CrossRef] [PubMed]

27. Fusi, N.; Martinez-Martinez, J. Mercury porosimetry as a tool for improving quality of micro-CT images in low porosity carbonate rocks. *Eng. Geol.* **2013**, *166*, 272–282. [CrossRef]

28. Zeng, Q.; Li, K.; Fen-Chong, T.; Dangla, P. Analysis of pore structure, contact angle and pore entrapment of blended cement pastes from mercury porosimetry data. *Cem. Concr. Compos.* **2012**, *34*, 1053–1061. [CrossRef]

29. Zhou, C.; Ren, F.; Zeng, Q.; Xiao, L.; Wang, W. Pore-size resolved water vapor adsorption kinetics of white cement mortars as viewed from proton nmR relaxation. *Cem. Concr. Res.* **2018**, *105*, 31–43. [CrossRef]

30. Katsura, M.; Sato, J.; Akahane, M.; Kunimatsu, A.; Abe, O. Current and novel techniques for metal artifact reduction at CT: Practical guide for radiologists. *Radiographics* **2018**, *38*, 450–461. [CrossRef]

31. National Institute of Standards and Technology. Avaliable on line: https://physics.nist.gov/PhysRefData/FFast/html/form.html (accessed on 11 May 2019).

32. Peng, S.; Zhang, T.; Loucks, R.G.; Shultz, J. Application of mercury injection capillary pressure to mudrocks: Conformance and compression corrections. *Mar. Pet. Geol.* **2018**, *88*, 30–40. [CrossRef]

33. Katz, A.J.; Thompson, A.H. Quantitative prediction of permeability in porous rock. *Phys. Rev. B* **1986**, *34*, 8179. [CrossRef] [PubMed]

34. Zhou, C.; Ren, F.; Wang, Z.; Chen, W.; Wang, W. Why permeability to water is anomalously lower than that to many other fluids for cement-based material? *Cem. Concr. Res.* **2017**, *100*, 373–384. [CrossRef]

35. Wang, Z.; Zeng, Q.; Wang, L; Li, X.; Xu, S.; Yao, Y. Characterizing frost damages of concrete with flatbed scanner. *Constr. Build. Mater.* **2016**, *102*, 872–883. [CrossRef]

36. Hiller, J.; Hornberger, P. Measurement accuracy in X-ray computed tomography metrology: Toward a systematic analysis of interference effects in tomographic imaging. *Precis. Eng.* **2016**, *45*, 18–32. [CrossRef]

37. Gajewicz, A.M.; Gartner, E.; Kang, K.; McDonald, P.J.; Yermakou, V. A 1H nmR relaxometry investigation of gel-pore drying shrinkage in cement pastes. *Cem. Concr. Res.* **2016**, *86*, 12–19. [CrossRef]

![materials logo] **materials**

MDPI

Article

Residual Magnetic Field Non-Destructive Testing of Gantry Cranes

Janusz Juraszek

Faculty of Materials, Civil and Environmental Engineering, University of Bielsko-Biala; 43-309 Bielsko-Biala, Poland; jjuraszek@ath.bielsko.pl; Tel.: +48-33-8279191

Received: 29 December 2018; Accepted: 11 February 2019; Published: 14 February 2019

check for updates

Abstract: Non-destructive tests of gantry cranes by means of the residual magnetic field (RMF) method were carried out for a duration of 7 years. Distributions of the residual magnetic field tangential and the normal components of their gradients were determined. A database of magnetograms was created. The results show that the gradients of tangential components can be used to identify and localize stress concentration zones in gantry crane beams. Special attention was given to the unsymmetrical distribution of the tangential component gradient on the surface of the crane beam No. 5 (which was the most loaded one). The anomaly was the effect of a slight torsional deflection of the beam as it was loaded. Numerical simulations with the finite element method (FEM) were used to explain this phenomenon. The displacement boundary conditions introduced into the simulations were established experimentally. Validation was carried out using the X-ray diffraction method, which confirmed the location of strain concentration zones (SCZs) identified by means of RMF testing.

Keywords: gantry crane; RMF technique; civil engineering

1. Introduction

The analysis of classical standard-related crane tests makes it possible to offer a general assessment of the gantry crane technical condition. This is usually a *post-factum* activity, i.e., an activity performed after the occurrence of a crack or damage. In the case of overhead cranes operated for over 30 years, a different, more thorough assessment is necessary. Standard tests are unable to determine whether stress concentration zones that cause cracks have appeared in the crane structure. This paper presents a new approach that enables a priori diagnostics, allowing the identification of potentially hazardous areas in advance.

The residual magnetic field (RMF) has been used as an inspection tool in transportation, power engineering and in the metal industry [1]. For a ferromagnetic structure, such as a crane, the RMF technique is expected to be one of the solutions that may enable early damage evaluation. It is proved in [2] that the size of the sample does not change the magnetization curve profile but it affects the RMF value. Four sets of Q345 steel samples with different widths and thicknesses were tested in the laboratory. The experimental results are explained by the theory of the interaction between dislocation and the domain wall, as well as the theory of the demagnetizing field. It is proved in [3] that the size of the sample does not change the magnetization curve profile but it affects the RMF value. Four sets of Q345 steel samples with different widths and thicknesses were tested in the laboratory. The experimental results are explained by the theory of the interaction between dislocation and the domain wall, as well as the theory of the demagnetizing field.

Significant static and fatigue experiments were performed with the use of the RMF technique [4–14]. There are numerous publications on the analysis of damage identified in flat specimens using the RMF technique. For example, damage identification in flat 12-mm-thick specimens subjected to tensile stresses was analyzed in [15] by Shui et al. The places in which the normal component Hn is zero indicate locations of damage to the specimen. The RMF also enables the identification of stress concentration zones. An interesting work on this subject is [16], which discussed the impact of the occurrence of stress concentration zones in three-point and four-point bending tests on the distribution of the Hp and Hn components, as well as the reduced stress according to the Huber–Misses–Hencky hypothesis. The RMF technique is used in the diagnostics of ferromagnetic structural elements. The RMF enables the detection of cracks, micro-cracks and closed cracks, which are difficult to detect using traditional methods. Knowing the previous magnetic image of the examined component, it is possible to detect stress concentration zones before actual cracks or micro-cracks appear. It is one of few methods that indicates, a priori, hazardous areas in the structure [17].

2. Materials and Methods

The analyzed gantry crane works in an open area and is used to transport coal from a storage site to a furnace. It was built in 1980. The structure consists of a truss bridge rolling on rails installed on two parallel 13-span assemblies consisting of crane beams. One span is less than 12 m long. The width of the crane is 32.01 m, and the trestle bridge reaches a height of 8.32 m (Figure 1). The cross-section of the crane beam is designed as an I-bar with a height of 824 mm. The web was made of sheet metal with a width of 800 mm and a thickness of 10 mm, while the feet were made of sheet metal with a width of 250 mm and a sheet thickness of 12 mm.

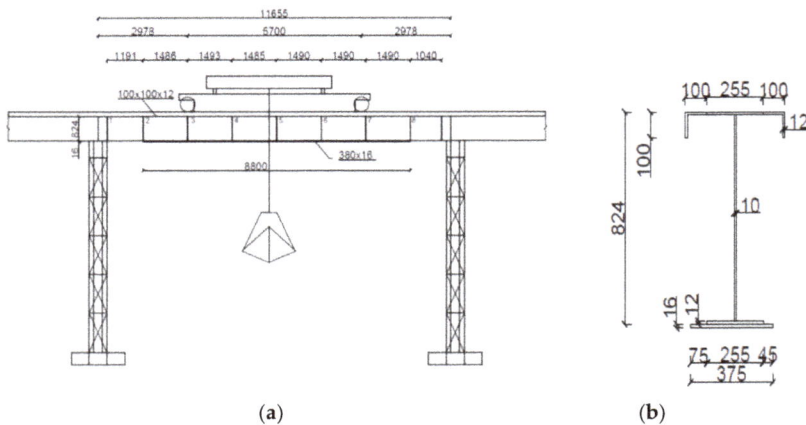

(a)   (b)

Figure 1. Diagram: (**a**) crane beam, (**b**) cross-section of the beam.

The approximate weight of the crane carried by the beam and the maximum load weight that can be transported by the structure were taken into account. The load was divided into two spot forces acting in the place where the gantry crane wheel rests on the beam. The total operating load was Fo = 40 kN, while the crane structure weight load was 150 kN distributed over two road wheels, 75 kN each. At this load, the highest stress total was 40.5 MPa.

RMF Methods

In general, local irregularities in the material homogeneity resulting from stress, plastic strain or fatigue-induced changes in the structure generate corresponding, as well as local, changes in the degree of the material magnetization, and thus also in induction. In this situation, an induction component perpendicular to the surface of element B_p appears, which results in the creation of a vertical magnetic field component H_p outside the element, the distributions of which are recorded by the probes used in the RMF method. According to Equation (1):

$$H_p = \frac{1}{\mu_o} B_p \tag{1}$$

where:

B_p is the induction component perpendicular to the metal surface;

$\mu_0 = 2\pi * 10^{-7}$ is the air magnetic permeability.

Since B_p and thus H_p are proportional to the rate of changes in induction B along the tested element, it can be assumed that:

$$H_p \sim \frac{1}{\mu_o} \frac{\Delta B}{\Delta z} \tag{2}$$

where: Δz is the length at which induction B changes.

Juraszek [7] claims that when entering the material area with a permeability μ lower than in the rest of the material, e.g., in the area of a strong plastic strain, the induction flux is scattered and the RMF vertical component above the surface has a positive sign. Moving to an area of higher magnetic permeability (e.g., in the area of increased elastic stresses), the lines of the magnetic field inside the material are focused and the vertical component H_p has a negative sign. Away from areas of different magnetic properties, when the material is homogeneous, the residual magnetic field disappears. In this method, the RMF component H_p is measured by determining the following components:

- tangential component $H_{p(x)}$;
- normal component $H_{p(y)}$.

Based on [11,14], Equation (3) and Equation (4) respectively present simplified formulae describing the values of components $H_{p(x)}$ and $H_{p(y)}$. The values have been confirmed experimentally.

$$H_x = \int_0^b dH_x = \frac{1}{2\pi\mu_0} \int_0^b \rho(l) * \frac{(x-l)}{||r||^2} dl \tag{3}$$

$$H_y = \int_0^b dH_y = \frac{1}{2\pi\mu_0} \int_0^b \rho(l) * \frac{(y)}{||r||^2} dl \tag{4}$$

where:

$r = [x - l, y]$ is a vector, a variable related to the element surface;

ρ is a function defining the boundary condition presenting the distribution of the magnetic charge density on the specimen surface, which depends directly on the dislocation density and indirectly on the stress;

x, y are the coordinates of the point on the tested element surface;

l is the dislocation length;

b is the dislocation width.

In the RMF method, the so-called magnetic stress intensity factor, or the gradient of the magnetic field normal component, was adopted as a measure of the quantitative assessment of the stress concentration level. The gradient of the residual magnetic field is defined by Equation (5):

$$K_{in} = \frac{|\Delta H_p|}{2l_k} \tag{5}$$

where:

K_{in} is the RMF gradient;

$|\Delta H_p|$ is the absolute value of the difference in H_p between two control points located at equal distances l_k on both sides of the line $H_p = 0$.

Due to the magnetoelastic effect, the defective area is very easily detectable under the influence of the load. It is characterized by a different modulus of elasticity, magnetic susceptibility or magnetoelastic properties. Therefore, under the influence of the load, its deformation value as well as the magnetoelastic increase in magnetization will be different from that of the matrix.

The difference in magnetization between the defective area and the matrix can be described using the following formula:

$$\Delta M = \Delta M(H_0) + \Delta M^{\sigma}[H_0; \sigma(x, \alpha)] \tag{6}$$

where:

$\Delta M(H_0)$ is the rise in magnetization under the influence of a weak magnetic field;

$\Delta M^{\sigma}[H_0; \sigma(x, \alpha)]$ is the rise in magnetization under the influence of elastic stresses in the presence of a weak magnetic field. %endenumerate

The result is a clear increase in the residual magnetic field within the defect area, since its value is directly proportional to ΔM.

3. Results of the RMF Scanning

Span No. 5 from the set of 13 spans was selected for analysis. The span is located directly above the coal chute to the conveyor that transports fuel to the turbine boiler (cf. Figure 1, showing a diagram of a single span). Each coal load is moved to span No. 5 using the crane. The crane with the load must stop on span No. 5, causing additional dynamic loads. Then, coal is put into the chute, which in turn causes vibrations of the crane beam. Due to this, the load of span No. 5 is the highest in the entire crane. The measurements involved scanning the bottom surface of the crane beam and determining the distributions of the normal and tangential components of the residual magnetic field. Gradients of both RMF components were also determined.

Next, based on the analysis of the gradient distribution, the highest gradient value of 10 A/m/mm was adopted as the limit criterion based on numerous previous works by the author [7,11]. Areas where the limit gradient value was exceeded were selected. They were then subjected to a further detailed analysis. The beam measurements were carried out in two stages.

The first stage of the analysis consisted in scanning the entire underside of span No. 5 along the previously determined three uniformly distributed measuring lines L1, L2, L3, as shown in Figure 2. It should be noted that line 3 is placed on the chute side.

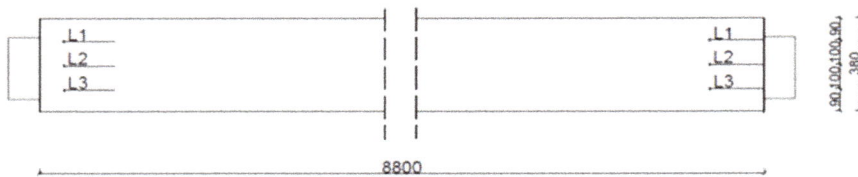

Figure 2. Diagram of measuring lines L1, L2, L3.

Example results of the beam surface scanning in the form of distributions of the tangential and the normal component (Hpx and Hpy, respectively) are shown in Figures 3 and 4 below. The value of the tangential component along the length of measuring line 3 ranges from −100 A/m to 69 A/m. The value of the orthogonal component is in the range of 340 A/m to −80 A/m. The gradient of the tangential and normal components is many times higher than the limit value of 10 A/m/mm.

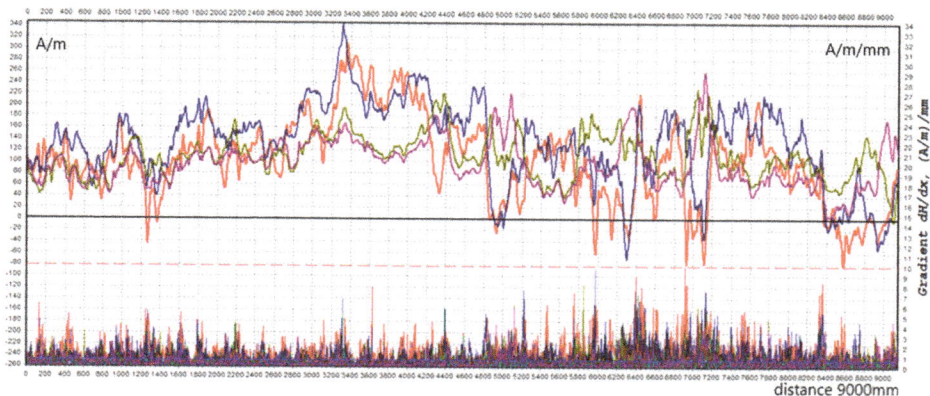

Figure 3. Distribution of tangential component Hpx along L3.

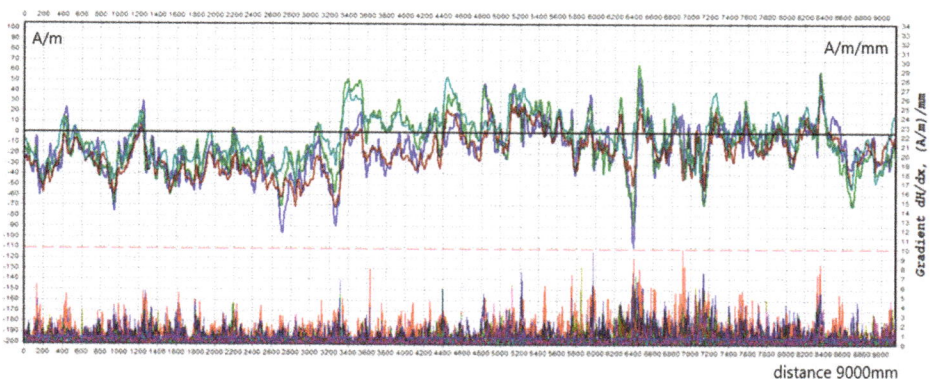

Figure 4. Distribution of normal component Hpy along L3.

This section provides a concise and precise description of the experimental results, their interpretation as well as the conclusions that can be drawn from the experiments. In the second stage, the area with the highest gradient value was selected. For this purpose, the measurements were first carried out with no coal load on the crane, and then the operation was repeated after the crane bucket was filled with coal. In both cases, based on the analysis of the obtained magnetograms and comparing the gradient values of the RMF component Hp with the limit value of 10 A/m/mm, the measurements demonstrated the occurrence of stress concentration zones between 5655 and 7145 mm from the crane left support. Then, the most loaded part of the beam, indicated in red in Figure 5, was tested again. An additional 22 measuring lines with numbers from L27 to L49 were introduced in this area.

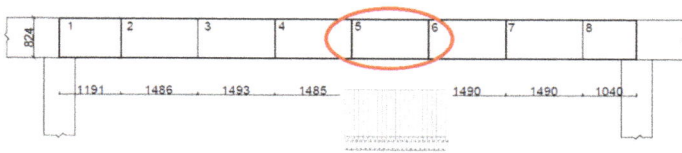

Figure 5. The beam's most loaded part and the additional 22 measuring lines.

The distribution of the changes in the residual magnetic field tangential component gradient for measuring line 3, for the selected sub-area, is presented in Figure 6.

A/m/mm

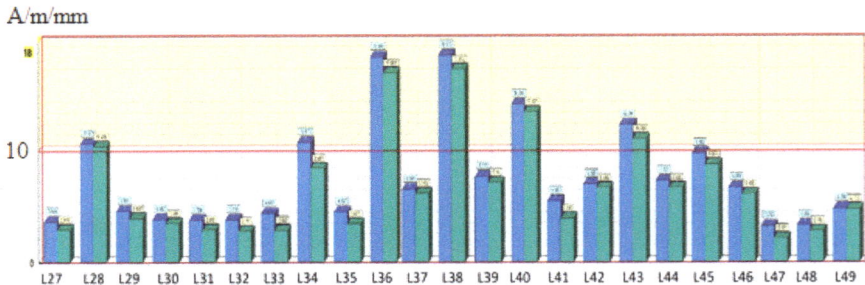

Figure 6. Changes in the tangential component gradient for the selected sub-area under analysis.

The adopted procedure was a gradual refinement of the observation area. Based on the analysis of the tangential component gradient distribution, stress concentration zones were found to occur in places where the limit of 10 A/m/mm was exceeded, as indicated in earlier studies by the author [7]. These are areas covering measuring lines 36–38, while the second area covered lines 40–45. In the next stage of the analysis, 2-D maps of the distribution of the magnetic field component gradients were prepared. The data for the residual magnetic field intensity maps were imported from J&J System software, while the gradients were calculated as a derivative of the RMF component H_p over the measurement line length using the finite difference method according to Equation (7):

$$G = H_p(x)' \approx \frac{H_p(x+a) - Hp(x)}{a} \qquad (7)$$

where:

$Hp(x)$ is the RMF intensity value in the point where the gradient was determined;

$H_p(x+a)$ is the RFM intensity value for the next measuring point;

a is the distance between consecutive points on the measuring line.

It was decided that for the magnetic field intensities, the colors would change every 10 units. The intensities of 0–10 A/m are presented as white. The values of positive intensities depending on the value are represented by warm colors from yellow to brown, while negative values are represented by cold colors from green to purple. For the gradient maps, colors changed every 1 unit. Because the gradient is non-negative, the colors range from white to dark brown. The distribution of the RFM tangential component on the 2-D map for the selected area is shown in Figure 7. It can be noticed that the highest values of 130 A/m were in the range of 607 to 1290 mm (x) and 152 to 380 mm (y). A relatively small area was found for the coordinates of 480 (x) and 16 mm (y).

Further analyses in the selected area concern the distribution of the gradient of the tangential and normal components. Figure 8 presents the distribution of the tangential component gradient for the full load on the crane beam.

A/m/mm

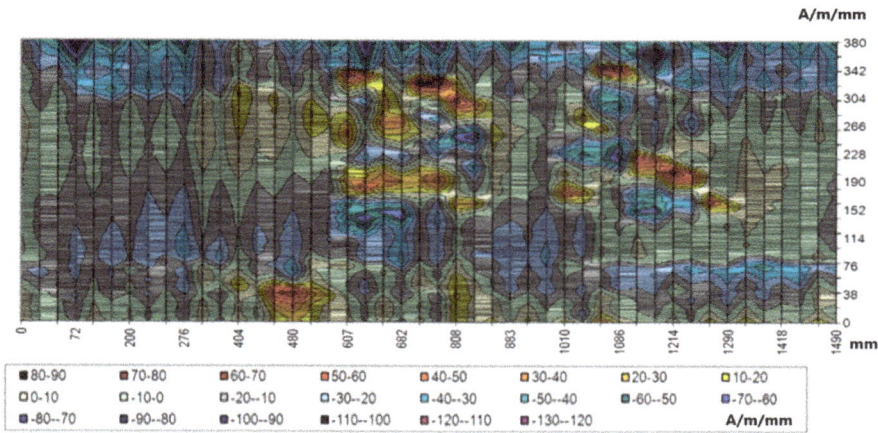

Figure 7. Distribution of tangential component Hpy for the selected area.

A/m/mm

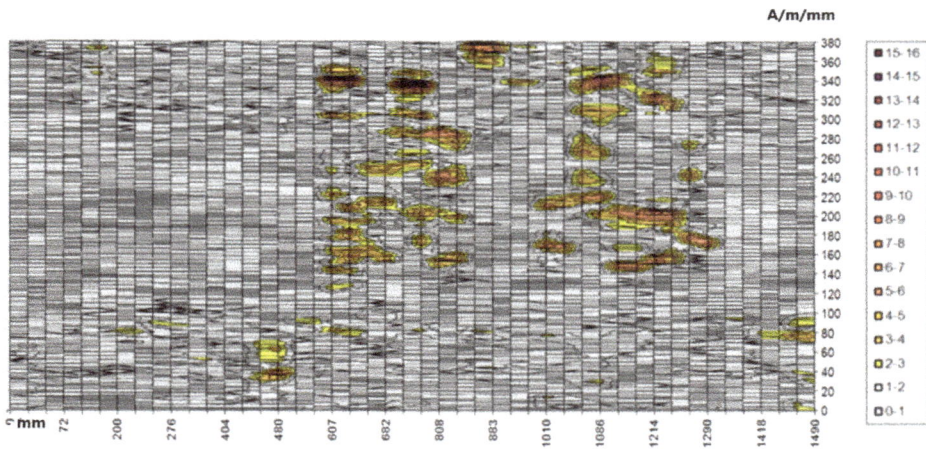

Figure 8. Distribution of the gradient of the tangential component dHpx/dx for the selected area.

If the crane is loaded with a truck only, the gradient values are lower by approximately 2 A/m/mm. Areas with values exceeding 10 A/m/mm also coincide with the areas of the highest values of the tangential component. They are located in the upper left quadrant of the map in the range of 607–1290 mm (x) and 140–380 mm (y). This area is located on the coal chute side—the analyzed part of the beam is located exactly above the coal chute. It is also the most heavily loaded beam, because during each cycle coal is transported at various points of the coal storage, always above the chute, where the load of a full skip hoist is applied first, and then rapid unloading takes place by opening the skip hoist and pouring out the coal. The estimated number of cycles in the heating season is about 60,000. A similar distribution of the normal component gradient confirmed the occurrence of the identified stress concentration zones. The highest values of the gradient of the RMF normal component reached 14 A/m/mm. Figure 9 presents the normal component gradient distribution. The area with the highest gradient value appears in the top left quarter of the 2-D map.

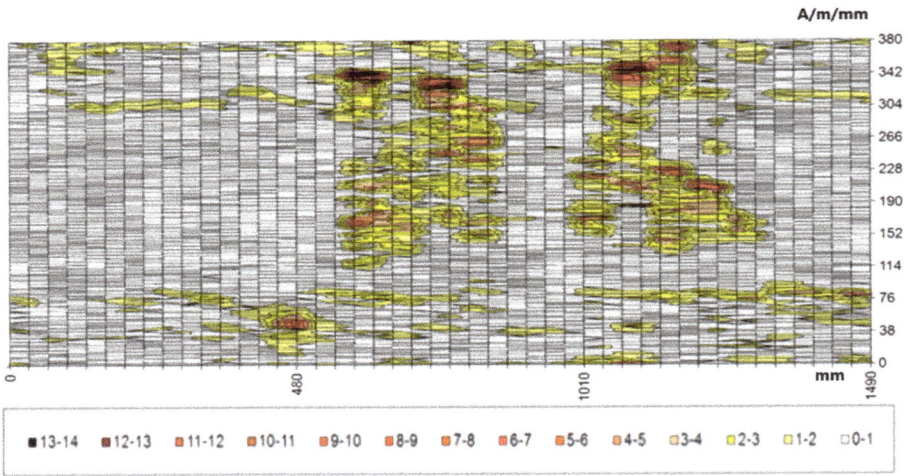

Figure 9. Distribution of the gradient of the normal component for the selected area.

4. Validation and Numerical Simulation

During the crane beam overhaul, samples were taken to determine the distribution of stress components. The samples were collected from places indicated as SCZs. The tests were carried out by means of the X-ray angle diffraction using an AvanceD-8 diffraction instrument. The testing results confirmed stress concentration zones determined using the RMF method. Stresses in the SCZs reached a value of 186 MPa, and outside the zone, of about 60 MPa. The diffraction pattern is shown in Figure 10.

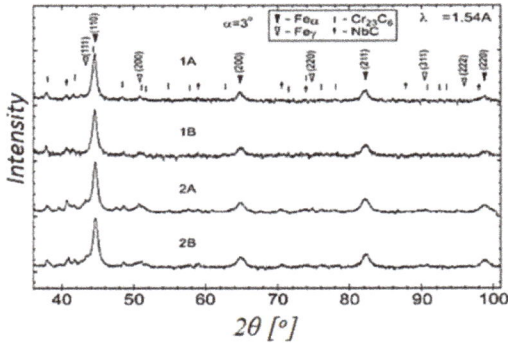

Figure 10. Diffraction pattern of a sample taken from the crane beam.

The detailed and comprehensive experimental and numerical verification consisted in the determination of deflection values halfway through the span length. The values were determined in three locations (points) marked in the figure presenting the crane beam cross-section (cf. Figure 11).

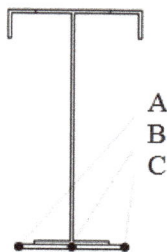

Figure 11. Arrangement of measuring points in the cross-section halfway through the span length.

Figure 12 presents influence lines for span No. 5 determined experimentally (values measured during the gantry crane passage). The measuring points located on the side of the coal chute (inner points) were characterized by the biggest value of deflection of 7.9 mm, whereas the smallest deflection value of 5.2 mm was observed for the opposite (outer) points. The deflection value of the measuring point above the I-beam web (the central point) was included between the values found for the outermost points. It totaled 6.35 mm. The deflection values determined in the crane beam measuring points gave evidence of beam warp. It turns out that this is the effect of the eccentric load of the analyzed crane beam I-section (which does not comply with the gantry crane technical documentation).

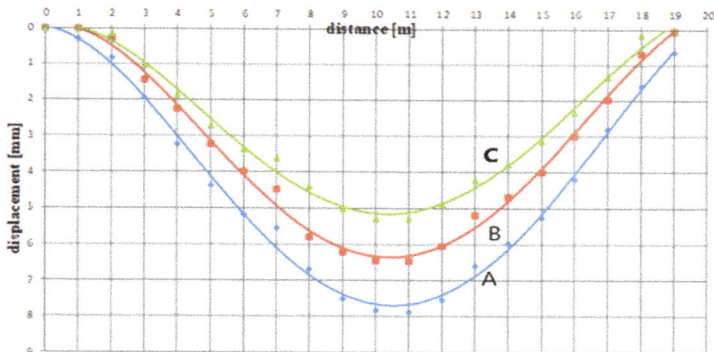

Figure 12. Crane beam deflection during the crane crab passage.

The experimentally determined values were entered into the FEM numerical model. The element side characteristic length was 2–10 mm. Shell 63 element was used. As the aim of the analysis was to determine the deformation of the crane beam and establish the cause of the unsymmetrical distribution of the tangential component gradients at set boundary conditions, a linear elastic material model was used. The constitutive model was linear. The total number of elements was 1.4 million. The numerical testing results confirm the twisting of the beam, which causes a higher level of stress. The experimental results were introduced into the FEM numerical model in the form of boundary conditions for displacement. A 3-D numerical model of the analyzed crane girder was built based on the finite element method ANSYS code, taking into account the upgrades and the span axis displacement detected in relation to the half-distance between the truss poles. The results of the numerical simulations (Figure 13) confirmed the previously indicated two areas, i.e., the area in the middle of the span length and in the places of welding the reinforcing cover. The stress levels in the analyzed areas were also determined. In the middle of the span length, the stresses totaled ~40 MPa.

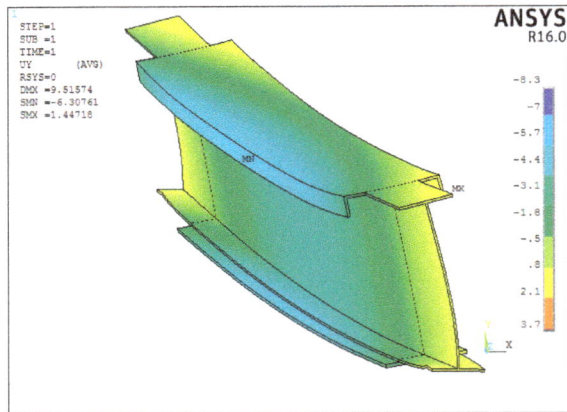

Figure 13. Distribution of the UY displacement.

The reason for this is the slight displacement of the crane travelling wheel towards the coal chute. The determined numerical distribution of the crane beam displacement is shown in Figure 11.

5. Conclusions

1. The RMF technique enabled the effective location of stress concentration zones in the analyzed crane beam by determining the gradient of the RMF tangential component.
2. The experimental verification using the X-ray diffraction method confirmed the stress concentration zones in the beam.
3. Creating a relational database of magnetograms of the ferromagnetic structure from the beginning of its operation seems to be an interesting addition to standard magnetic diagnostic methods.
4. Combining different methods and measurement techniques, for example a fiber optic beam deflection measurement system using the FEM method which implements boundary conditions from optical fiber measurements, with residual magnetic field techniques can significantly contribute to the identification of stress concentration zones.

Funding: This research received no external funding.

Conflicts of Interest: The author declares no conflict of interest.

References

1. Shannon, R.W.F.; Braithwaite, J.C.; Morgan, L.L. Flux-leakage vehicles pass tests for pipeline inspection. *Oil Gas J.* **1988**, *32*. [CrossRef]
2. Fu, M.L.; Bao, S.; Zhao, Z.Y.; Gu, Y.B. Effect of sample size on the residual magnetic field of ferromagnetic steel subjected to tensile stress. *Non-Destr. Test. Cond. Monit.* **2018**, *60*, 90–94. [CrossRef]
3. Bao, S.; Fu, M.L.; Lou, H.J.; Bai, S.Z. Defect identification in ferromagnetic steel based on residual magnetic field measurements. *J. Magn. Magn. Mater.* **2017**, *441*, 590–597. [CrossRef]
4. Yao, L.K.; Wang, Z.D.; Deng, B. Experimental research on metal magnetic memory method. *Exp. Mech.* **2012**, *3*, 305–314. [CrossRef]
5. Wang, Z.D.; Yao, K.; Ding, K.Q. Quantitative study of metal magnetic memory signal versus local stress concentration. *NDT E Int.* **2010**, *6*, 513–518. [CrossRef]
6. Wang, Z.D.; Yao, K.; Ding, K.Q. Theoretical studies of metal magnetic memory technique on magnetic flux leakage signals. *NDT E Int.* **2010**, *43*, 354–359. [CrossRef]
7. Juraszek, J. *Innovative Non–destructive Testing Methods Monography*; ATH, Univ. of Bielsko-Biala: Bielsko-Biała, Poland, 2013.

8.   Ren, S.K.; Song, K.; Ren, J.L. Influence of environmental magnetic field on stress magnetism effect for 20 steel ferromagnetic specimen. *Insight* **2009**, *51*, 672–675. [CrossRef]

9.   Wilson, J.W.; Tian, G.Y.; Barrans, S. Residual magnetic field sensing for stress measurement. *Sens. Actuators A* **2007**, *135*, 381–387. [CrossRef]

10. Yao, K.; Deng, B.; Wang, Z.D. Numerical studies to signal characteristics with the metal magnetic memory-effect in plastically deformed samples. *NDT E Int.* **2012**, *47*, 7–17. [CrossRef]

11. Juraszek, J. Hoisting machine brake linkage strain analysis. *Arch. Min. Sci.* **2018**, *63*, 583–597.

12. Shi, C.L.; Dong, S.Y.; Xu, B.S.; Peng, H. Stress concentration degree affects spontaneous magnetic signals of ferromagnetic steel under dynamic tension load. *NDT E Int.* **2010**, *43*, 8–12.

13. Dong, L.H.; Xu, B.S.; Dong, S.Y.; Chen, Q.Z.; Wang, D. Stress dependence of the spontaneous stray field signals of ferromagnetic steel. *NDT E Int.* **2009**, *42*, 323–327. [CrossRef]

14. Guo, P.J.; Chen, X.D.; Guan, W.H.; Cheng, H.Y.; Jiang, H. Effect of tensile stress on the variation of magnetic field of low-alloy steel. *J. Magn. Magn. Mater.* **2011**, *23*, 2474–2477.

15. Shui, G.S.; Li, C.W.; Yao, K. Non-destructive evaluation of the damage of ferromagnetic steel using metal magnetic memory and nonlinear ultrasonic method. *Int. J. Appl. Electromagn. Mech.* **2015**, *47*, 1023–1038. [CrossRef]

16. Yi, S.C.; Wei, W.; Su, S.Q. Bending experimental study on metal magnetic memory signal based on von Mises yield criterion. *Int. J. Appl. Electromagn. Mech.* **2015**, *49*, 547–556. [CrossRef]

17. Chen, H.L.; Wang, C.L.; Zuo, X.Z. Research on methods of defect classification based on metal magnetic memory. *NDT E Int.* **2017**, *92*, 82–87. [CrossRef]

materials

MDPI

Article

The Use of the Acoustic Emission Method to Identify Crack Growth in 40CrMo Steel

Aleksandra Krampikowska [1], Robert Pała [2], Ihor Dzioba [2] and Grzegorz Świt [1,*]

[1]   Department of Strength of Materials, Concrete and Bridge Structures, Kielce University of Technology,
       Al. 1000-lecia PP 7, 25-314 Kielce, Poland
[2]   Faculty of Mechatronics and Mechanical Engineering, Department of Machine Design,
       Kielce University of Technology, Al. 1000-lecia PP 7, 25-314 Kielce, Poland
*   Correspondence: gswit@tu.kielce.pl

Received: 30 May 2019; Accepted: 1 July 2019; Published: 3 July 2019

check for
updates

Abstract: The article presents the application of the acoustic emission (AE) technique for detecting crack initiation and examining the crack growth process in steel used in engineering structures. The tests were carried out on 40CrMo steel specimens with a single edge notch in bending (SENB). In the tests crack opening displacement, force parameter, and potential drop signal were measured. The fracture mechanism under loading was classified as brittle. Accurate AE investigations of the cracking process and SEM observations of the fracture surfaces helped to determine that the cracking process is a more complex phenomenon than the commonly understood brittle fracture. The AE signals showed that the frequency range in the initial stage of crack development and in the further crack growth stages vary. Based on the analysis of parameters and frequencies of AE signals, it was found that the process of apparently brittle fracture begins and ends according to the mechanisms characteristic of ductile crack growth. The work focuses on the comparison of selected parameters of AE signals recorded in the pre-initiation phase and during the growth of brittle fracture cracking.

Keywords: pattern recognition; acoustic emission; Structural Health Monitoring; brittle fracture; diagnostics

1. Introduction

Diagnosing and monitoring the condition of structures is an important and very topical area in construction. Aging infrastructure and increasing service load of engineering structures are the main drivers of fast-progressing research in the new interdisciplinary field of knowledge called structural health monitoring (SHM), which is closely related to the durability and safe operation over the life of the structural elements.

The load is a typical random load that is difficult to model with currently known fatigue calculation procedures, and can be modelled only for steady loads with low-level amplitudes. As a result, the fatigue life calculation models for steel structures fail to provide accurate information about the risk of fatigue failure. Where calculation methods provide only limited information, the existing structures are subjected to tests. Fatigue damage testing is carried out during inspection. Cracks, which occur for a number of reasons, are detected and, in the case of fatigue cracks, the rate of their growth is assessed.

Depending on the operating conditions, the loading and materials used, the failure takes place according to a ductile or brittle fracture mechanism.

Ductile cracking usually occurs in steels and is preceded by significant plastic deformation that manifests itself long before the failure of the element, thus allowing preventive measures to be undertaken. The brittle fracture mechanism is much more dangerous.

Brittle cracking proceeds without visible deformation of the element and occurs almost immediately. The occurrence of brittle fracture can be expected mainly in high-strength structural steels or in low-temperature operation and overloading of the element with simultaneous increase in the load speed. Welded joints are also exposed to brittle fracture due to their inhomogeneous microstructure, welding defects and residual stresses. The presence of these factors together with the impact of fatigue loads and corrosive environment leads to the degradation of the material and development of micro-cracks, resulting in the brittle failure of steel.

Leading research centers worldwide have devoted a great deal of attention to the development of methods for assessing the condition of structural elements and preventing their failure. As a result, many methods for evaluation of component durability [1–5] are currently in wide use. These methods are based on aversion to change assuming prior knowledge of the element microstructure, load history, and in-service conditions. However, more accurate results will be obtained when as much information as possible can be collected on in-service parameters and material.

Hence, the modelling of bearing capacity and durability of a structure under real operating conditions requires that:

- real operational loads are defined;
- the material model is defined, in particular welded joints material that changes over time under the influence of operating conditions;
- the load around the defect after its initiation and during development (redistribution of stresses) is determined; and
- the interaction of various damage types is identified.

As determining the factors above is very difficult, if not impossible, most analyses are characterized a high degree of conservatism, which significantly reduces assessment accuracy. Appropriate solution should include developing an NDT-based monitoring system able to signal the structural safety risk.

Finding and determining the "destructive characterization" of all hot spots in large-sized structures using the NDT methods is nearly impossible. In steel structures, the volume of "hot spot" is of the order of cubic centimeters, with dimensions of the structure reaching tens and more meters. In our opinion, the solution to the problems of diagnosing engineering structures is the use of continuous, long-term monitoring using the passive NDT methods. One of them is the acoustic emission (AE) method [4–14].

The paper presents a proposal for the use of acoustic emission for the diagnosis of steel structures vulnerable to brittle fracture.

The choice of acoustic emission as the research method was determined by its advantages in relation to other non-destructive methods. These are:

- locating the faults that are undetectable with conventional methods;
- recording only active damage, i.e., the defect growth as it occurs;
- continuous monitoring of structures while in service or during load tests, with continuous data recording;
- detecting all types of damage, whereas most other methods focus on particular defects;
- characterising the rate of damage development during the operation of the structure; and
- enabling characterization of AE signal sources.

The main task of the measurement system consisting of an AE processor is to detect, record, filter and analyze the signals generated by AE sources.

Each destructive process is a source of acoustic emission. The source is described by the parameters of the recorded AE signal. The values of the parameters are used to classify the signals (and, thus, damage processes). The similarity of signals is used to attribute particular signals to the defects caused by specific damage processes, and then by applying statistical grouping methods to identify the existing defects.

For the statistical methods used in identification, an important issue is the optimal selection of recorded 13 AE parameters (counts, counts to peak, amplitude, RMS—root means square voltage, ASL—average signal level, energy, absolute energy, signal strength, rise time, duration, initiation frequency, average frequency, reverberation frequency). The parameters must be characterized by low mutual correlation. A set of diagnostic variables must describe the most important aspects of the studied phenomenon [15–18]. Techniques such as hierarchical and non-hierarchical clustering methods and Kohonen's neural networks are used to build the reference signals data base for identification of destructive processes in steel structures (IPDKS) [4,5].

This paper uses unsupervised pattern recognition methods to characterize different AE activities corresponding to different fracture mechanisms. A sequential feature selection method based on a k-means clustering algorithm is used to achieve high classification accuracy. Fatigue damage propagation represents the main failure factor. To study the contributions of different types of damage at progressive fatigue stages, a tool with the ability to detect the damage initiation and to monitor failure progress online is needed. Acoustic emission (AE) testing has become a recognized suitable and effective non-destructive technique to investigate and evaluate failure processes in different structural components. The main advantage of AE over other condition monitoring techniques is that detected AE signals can be used to characterize the different damage mechanisms. The approach of using parameter distribution has the advantage of real-time damage detection, but can also lead to false conclusions due to noise effects in the AE signals [18]. This is especially true for materials that are working under the fatigue conditions which are usually present in damage mechanism interactions. Therefore, AE techniques are needed to account for more intricate wave propagation features caused by the anisotropic nature of materials and to enable the identification of a large variety of failure modes. It is now possible to detect and capture very large numbers of AE signals, driving a trend for seeking computationally complex algorithms, such as pattern recognition, to determine the onset of significant AE. For a real structure it is not possible to provide a set of training patterns belonging to multiple damage mechanisms, which thus makes the use of unsupervised pattern recognition techniques more appropriate for these studies [18].

The first step of developing an unsupervised pattern recognition process is to classify signals into groups based on similarities. This process involves statistical effects, and the key point of successful feature selection to construct fine classification accuracy. A paper by Doan et al. [16] presented a feature selection method that introduced a sequential method based on the Gustafson–Kessel clustering algorithm. In these the method, the subset of features is selected by minimizing the Davies-Bouldin (DB) index, which is a metric for the evaluation of classification algorithms. The signals are classified into four groups by comparing their features and deciding upon their similarity.

The assignment of the clustering results to the fracture mechanisms is achieved by a detailed analysis of the physical meaning of the data [19–22]. This is the first time that the pattern recognition technique has been applied to a database acquired from such a complex structure in a fatigue testing environment [16]. The applied feature selection algorithm proves to be a powerful tool providing relevant clustering when used together with a k-means algorithm. When a crack occurs in the material, it results in a rapid release of energy, transmitting in the form of an elastic wave, namely acoustic emission (AE). The AE-based detection method has been intensively used also in non-destructive assessments of cracks in papers [23–25]. Additionally, Rabiei and Modarres revealed a log-linear relationship between the AE features and crack growth rate, and presented an end-to-end approach for structural health management [10]. Qu et al. presented a comparative study of the damage level diagnostics of gearbox tooth using AE and vibration measurements; the results indicated that vibration signals were easily affected by mechanical resonance, while the AE signals showed a more stable performance [26]. Zhang et al. have studied defect detection of rails using AE and wavelet transform at a high speed [27]. Li and al. created a template library of cracking sounds and designed a detection device using voiceprint recognition with an accuracy of 77% [28]. Hase et al. combined the Hurst exponent and the neural network to develop a crack detection algorithm of carbide anvils [27]. In the above methods, the AE sensors are usually attached to the surface of the monitoring object, what makes measurement difficult.

A more practical crack identification and detection method is still lacking. Aiming to improve recognition accuracy and generalization, in papers [29–31] a novel crack identification method based on acoustic emission and pattern recognition is proposed. In these methods, the sound pulses from cracks are firstly separated from the original signal by pre-processing. The high-dimensional features are reduced adaptively by using principal component analysis (PCA). The algorithm combines a k-nearest neighbor (kNN) classifier with a support vector machine (SVM) to refine the classification outcome. While debris monitoring does not require any electronics, it is simple to interpret and has excellent sensitivity to wear-related failure; this method is insufficient to non-benign cracks as no debris is produced. Acoustic emission (AE) and vibration signals have more quantitative results to detect the earliest stage of damage in rotating machinery. AE and vibration methods are based on recording transient signals in two different frequency spectrums. While vibration method is based on features that are extracted from time and frequency domain signals recorded by low frequency accelerometers in order to assess the changes in vibrational properties as related to the damage [29–31], the AE method is based on detecting propagating elastic waves released from active flaws. Once transient signals are collected, signal processing methods, such as wavelet decomposition [11], empirical mode decomposition [11], and multivariate pattern recognition [20], are applied. Typical parameters extracted from the transient signals are root mean square value, frequency domain characteristics, energy, spectral kurtosis, and peak-to-peak vibration level. Due to the difference in the frequency bandwidth, the AE method is more sensitive to microcracks as compared to the vibration method. Typical AE data acquisition approach is based on threshold: an AE signal is detected when the signal level is above threshold. As crack growth is a stochastic process, it is considered that while some data will be lost due to the idle time of data acquisition system between waveform recording intervals, crack information will be stochastically detected. However, it is important to identify how to analyze long-duration signals in order to reduce the influence of background noise from the extracted features. In studies, the AE signals, which accompanying of the fatigue crack growth is important obtained from the scaled laboratory experiments. Acoustic emission (AE) is a health monitoring approach which acts as a passive receiver to record internal activities in structures. This method is capable of continuous monitoring, which is not the case with most traditional methods. In addition, AE sensors are very sensitive and can capture signals due to micro-scale defect formations coming from the internal regions of structures rather than only those at the surface [18,23]. An unsupervised and supervised pattern recognition algorithm was employed to classify the AE signals. Different damage mechanisms for specimens during cracking were identified using reference database AE signals created from signals registered in the tests described below.

2. Materials and Methods

The tests discussed below were carried out using Zwick-100 testing machine (ZwickRoel Ulm, Germany) on SENB (Single Edge Notch in Bending) specimens (Figure 1) made of 40CrMo steel. The specimens with dimensions 12.5 mm × 25 mm × 110 mm were subjected to heating at 850 °C (15 min), quenching in oil, tempering at 250 °C (3 h) and cooling in oil. As a result, a material with a tempered martensite microstructure was obtained (Figure 2a). The signals of loading, specimen deflection and acoustic emission (AE) were recorded during testing.

Figure 1. The view of the SENB specimen with AE sensors (Zwick-100, ZwickRoel Ulm, Germany) during test: 1—a sample, 2—support, 3—load cylinder, 4—COD extensometer, 5—AE sensors—100–1200 kHz, 6—AE sensors—30–80 kHz.

SENB type specimens with a total fracture length, which includes a notch + previously fatigue crack, equal to about $0.5 \cdot a/W$ are recommended by the ASTM and PN-EN standards [32–35] and commonly used to determine fracture toughness characteristics—critical values of the stress intensity factor (SIF) K_{IC} or the critical *J-integral*, J_{IC}. If during the entire process of loading the SENB specimen in the net-section area before the crack tip there is a definite predominance of the linear-elastic nature of the stress and strain fields before the crack tip are described by SIF and the fracture toughness is characterized by the critical value of K_{IC}. The condition allowing the use of SIF is to limit the size of the yielding before the crack tip: $r_p \leq 0.01 \cdot a$, where for a plane strain $r_p = (1/6\pi) \cdot (K_I/\sigma_y)^2$. If the plastic zone is larger, it means that the material in front of the fracture tip is elastic-plastic and for the description of mechanical fields an integral J must be used, and fracture toughness is represented by the critical integral $J - J_{IC}$ [36,37].

For determining material strength characteristics, cylindrical specimens 10 mm in diameter were prepared with the measuring section of 50 mm. For determining the critical value of fracture toughness, K_{IC}, the specimens with dimensions 12.5 mm × 25 mm × 110 mm with one-sided notch and crack fatigue (SENB) with a total length of about 12.5 mm were made. Figure 2b shows an example plot of loading a cylindrical specimen. On the basis of the graphs, strength characteristics of 40CrMo steel were determined: $R_e = \sigma_y = 1475$ MPa; $R_m = \sigma_{uts} = 1800$ MPa; $E = 205$ GPa. During the loading of the SENB specimen, it was determined that the crack growth process takes place according to the brittle fracture mechanism. All standard conditions were met. The obtained critical value of fracture toughness is a material characteristic and is equal to $K_{IC} = 39$ MPa·m$^{1/2}$.

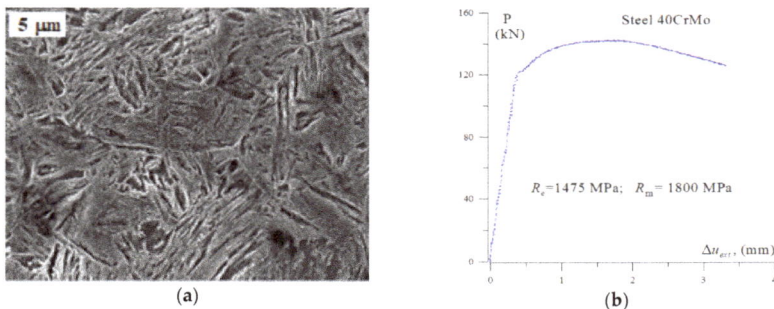

Figure 2. (a) Microstructure of tempered martensite of 40CrMo steel; and (b) loading plot of a cylindrical specimen in uniaxial tensile test.

The reference signal database in the IPDKS method for steel structures was developed using the Fuzzy k-means algorithm. The k-means algorithm belongs to a group of non-hierarchical clustering approaches. Its essence lies in the random choice of initial centres. In each iteration further approximations of the patterns are searched for and enumerated using the given methods. Depending on the assumptions made, the reference element may be one of the elements of the X population or belong to a certain universe $U \supseteq X$.

In metric spaces, a reference element can be calculated as an arithmetic mean to represent the centre of gravity of the cluster.

In general, the *k-clustering* algorithm input is a set of X objects and the expected number of clusters k, and the output is the division into subsets $\{C_1, C_2, \ldots, C_k\}$. Frequently, **k-clustering** algorithms belong to the category of optimization algorithms. Optimization algorithms assume that there is a loss function $k: \{x \mid X \subseteq S\} \rightarrow R +$ specified for each S subset. The aim is to find a branch due to minimizing the sum of losses described by Equation (1):

$$E_q = \sum_{i=1}^{k} k(C_i) \tag{1}$$

In order to apply the iterative algorithm, the distance measure used should be determined in the grouping process. Instead of the median, you can use a point that is the resultant point for a given cluster (representation element) calculated, e.g., as a geometric or arithmetic mean. Then we deal with k-means (k-means) algorithms described by Equation (2):

$$k(C_i) = \sum_{r=1}^{|C_i|} d\left(\overline{x}^i, x_r^i\right) \tag{2}$$

where: $\overline{x}^i$—the arithmetic mean (or geometric) of the cluster.

Doing so means that the centroids search for their correct positions using Equation (3) [4,5]:

$$\underline{\mu}_j = \frac{\sum\limits_{j=1}^{n} P\left(\omega_i \middle| \underline{x}_j\right)^b \underline{x}_j}{\sum\limits_{j=1}^{n} P\left(\omega_i \middle| \underline{x}_j\right)^b}, \tag{3}$$

where $P\left(\omega_i \middle| \underline{x}_j\right)$ is the conditional probability of belonging j-th element to the i-th group, b is the parameter that has to take values other than 1, x_{-j} is the j-th element.

The probability function is normalized according to Equation (4):

$$\sum P\left(\omega_i \middle| x_{-j}\right) = 1, \; where \; j = 1, \ldots, n, \tag{4}$$

The probability of the membership of the element in each of the clusters $P\left(\omega_i \middle| \underline{x}_j\right)$ is calculated from Equation (5):

$$P(\omega_i \middle| \underline{x}_j) = \frac{\left(\frac{1}{d_{ij}}\right)^{\frac{1}{b-1}}}{\sum\limits_{r=1}^{c} \left(\frac{1}{d_{rj}}\right)^{\frac{1}{b-1}}}, \tag{5}$$

where $d_{ij}^2 = \|\underline{x}_j - \underline{\mu}_i\|^2$ is the distance of a data point $\underline{x}_j$ from the center of the group $\underline{\mu}_i$.

The *k-means* algorithm consists of the following steps:

1. Randomly select initial centroids.
2. Compute the distances between the data points and the cluster centroids.

3. Compute the membership function value of all elements $P\left(\omega_i|\underline{x}_j\right)$.

4. Compute cluster centroids $\underline{\mu}_i$.

5. If:

   - there are no changes in $\underline{\mu}_i$ and $P\left(\omega_i|\underline{x}_j\right)$ return $\mu_{-1} \cdots \mu_{-c}$,
   - otherwise go back to Step 2.

When this algorithm is used, the number of clusters is pre-determined. However, the speed of computation compensates for this inconvenience. The reference signals database is created with 13 correlated AE parameters (rise time, durability, counts, counts to peak, energy, RMS, ASL, amplitude, average frequency, initiation frequency, reverberation frequency, absolute energy, signal strength). Reference signals obtained in this way make it possible to identify individual destructive processes in all tested structures. The database can be supplemented as per the diagram above whenever new measurement data are obtained.

The use of the grouping method is particularly important for the identification of fracture mechanisms as it allows for timely response to the possible averse effects.

The study used a 24-channel "μSamos" acoustic emission processor, 40 dB preamplifiers and two resonance sensors with flat characteristics in the 30–80 kHz range and two broadband sensor in the 100–1200 kHz range. The use of four sensors with flat characteristics was aimed at determining the lower and upper discrimination thresholds of the AE signal, while the broadband sensor had the task of controlling whether during the measurement there are no signals generated in other frequency bands.

3. Results and Discussion

During the loading of the SENB specimen, the crack growth process takes place according to the brittle fracture mechanism and critical value of fracture toughness is a material characteristic equal to $K_{IC} = 39$ MPa·m$^{1/2}$. At the critical moment, the length of the plastic zone according to the Irwin model for plane strain equals: $2r_p = 2(1/6\pi)\cdot(K_{IC}/\sigma_y)^2 = 0.074$ mm. According to Williams' formulas in front of the plastic zone in the crack plane direction, the level of stress tensor components is $\sigma_{xx} = \sigma_{yy} = 1808$ MPa, $\sigma_{zz} = \nu\cdot(\sigma_{xx} +\sigma_{yy}) = 1193$ MPa. Directly in the plastic zone the stress components are significantly higher and the stress opening component, σ_{yy}, can obtain values higher than $3\sigma_y$, i.e., over 4500 MPa [38]. The σ_{zz} and σ_{xx} components are lower, but also exceed the level of $2\sigma_y$.

High stress levels in front of the crack tip cause material damage processes, cracking development through various mechanisms: brittle fracture on the cleavage planes, brittle intergranular cracking, ductile fracture through nucleation, growth and coalescence of voids. The predominance of any fracture mechanism depends on the stress-strain state in the tested element, on the microstructure type, loading method, the temperature of the environment, and on other effects. Here, the situation was seemingly simple—SENB specimens crack according to the brittle mechanism. The AE method was used to monitor the processes during loading of steel specimens.

Five specimens were tested. The load plots and the character of AE signals distribution were similar in all tested samples. Figure 3b shows the characteristic loading curve and numerous points of the absolute energy of AE signal and the graph illustrating the cumulative absolute energy vs. load on time from 3.5 to 6.2 s. The graphs show AE parameters without using the clustering method (identification) of destructive processes.

Figure 3a shows the change in the absolute energy parameter (aJ), and the specimen loading as a function of time and COD (crack opening displacement). Acoustic emission signals generated during the study showed an increase in their numbers and increase in the value of absolute energy parameter at COD of 0.045 (mm) and 4.2 kN of load. It can be noticed that the increase in the value of the absolute energy parameter occurs when the cracking process begins (initiation, propagation).

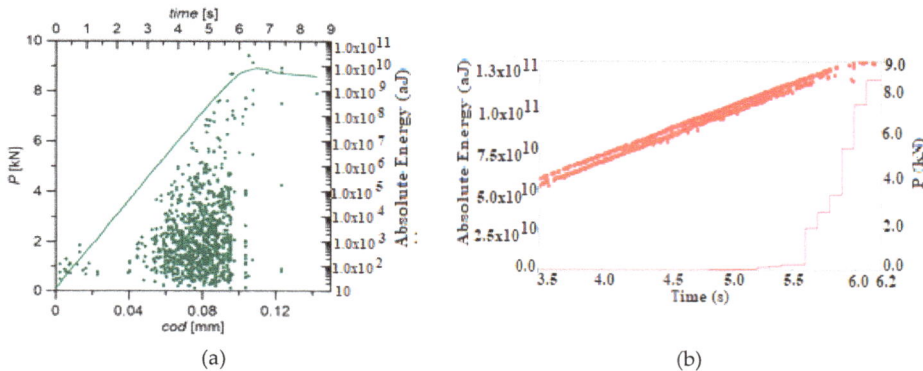

(a)   (b)

Figure 3. The graph of specimen loading (**a**) with absolute energy of AE signals (aJ) and (**b**) graph cumulative absolute energy vs. load on time from 3.5 to 6.2 s. The graphs show AE parameters without using the clustering method (identification) of destructive processes.

During the observations of the fracture surface of the specimens examined using the scanning microscope, three basic mechanisms in the crack development process were identified. The images shown in Figure 4 are presents the same fragment the fracture surface, but at different magnitude. In Figure 4a, at 500×, we can see the general character of brittle fracture surface. In Figure 4b, at 1500×, the details of fracture mechanisms are shown. At the fracture surface, two types of brittle fracture were present: *intercrystallite* cracking and *transcrystallite cleavage*. (Figure 4a,b). There were also areas where cracking developed according to the ductile mechanism, that is, where voids nucleated, grew, and coalesced around the small particles of precipitations (Figure 4b).

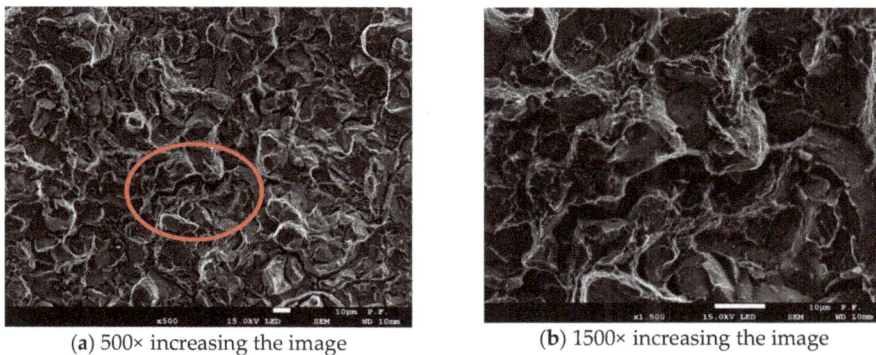

(**a**) 500× increasing the image   (**b**) 1500× increasing the image

Figure 4. Morphology of the fracture surface the SENB specimen breakthrough: (**a**) general character of fracture surface; and (**b**) details of fracture mechanisms.

Observations of the process of cracking over time shows that at first cracking proceeds according to brittle mechanisms, while cracking due to sudden immediate coalescence of voids is the last stage of crack growth. These observations were confirmed by the analysis of AE signals generated in the tested SENB specimen. However, in order to be able to detect the differences in the processes accompanying the cracking of steel specimens, the signal grouping method must be used. Figure 4a, b shows the selected AE parameters (absolute energy) not analyzed using the clustering methods. It can be noticed that it is difficult to interpret the changes in the work of the tested element. It is impossible to assess the moment of initiation of the crack or its subsequent propagation. It can be said that individual parameters not analyzed by clustering methods do not help in the interpretation of changes in the tested samples. That is why it is so important to use Big Data analysis to create and identify destructive processes on the basis of the pattern recognition methods.

Using one of the grouping methods, namely the fuzzy k-means algorithm, we developed the reference signal data base describing the processes of brittle fracture during destructive tests of steel structures. The AE signals generated and recorded in the crack initiation and development area were subjected to a grouping analysis using the NOESIS 5.8 program, which allowed to separate four basic classes. Figure 5a–c present the effects of using method of grouping signals—a non-hierarchical k-means method to identify mechanisms accompanying the brittle fracture.

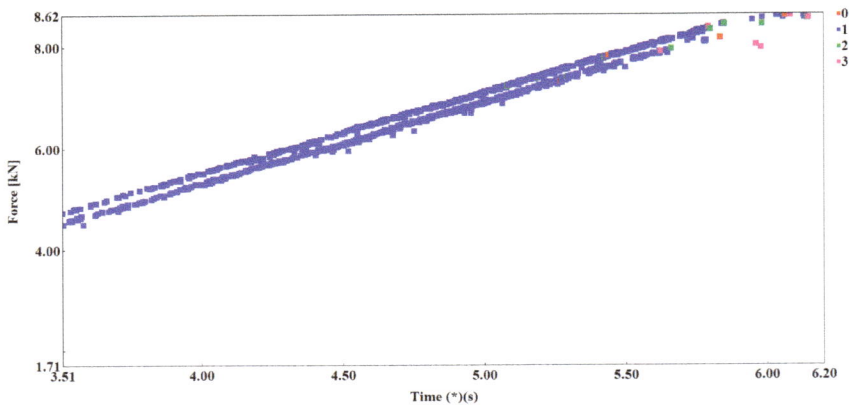

(a)

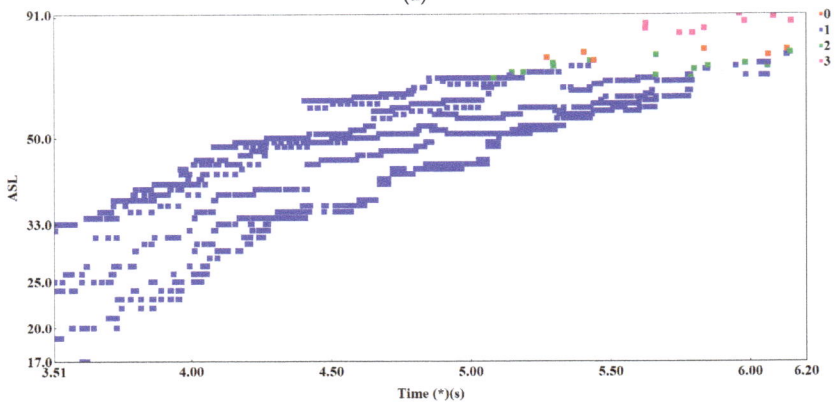

(b)

Figure 5. *Cont.*

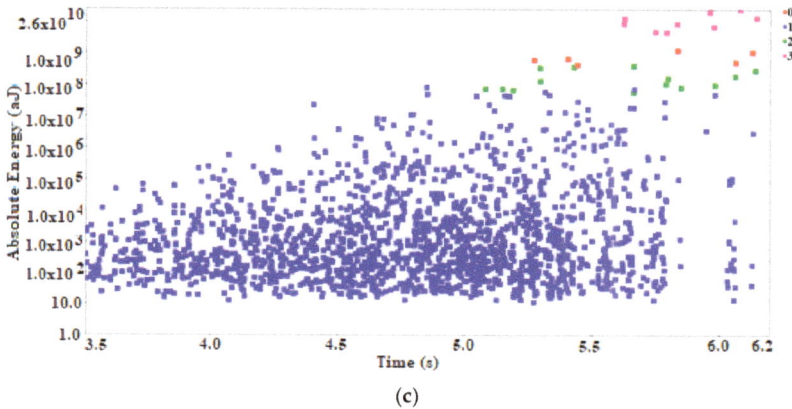

(c)

Figure 5. Scatterplots of acoustic emission signals: (a) Force (kN) vs. time (s); (b) ASL (dB) vs. time (s); and (c) absolute energy (a) vs. time (s).

Analyzing the division of AE signals into the classes marked in Figure 5 with colors and figures and using photographs (Figure 4) from the scanning microscope, it can be noticed that cracking due to sudden immediate combination of the voids, breaking bridges between them are pink signals marked with the number 3. The same type of signals recorded in the last stage of cracking the sample tensile steel S355JR, when there was no cracking at the brittle mechanisms (Figure 6). Signals marked in blue and number 1 come mainly from elastic and plastic deformations, as well as from individual noise generated by the strength machine and the loading system. These signals, therefore, probably can be equated with the complex, rapid dislocation motion process in ferrite and the act of fracture.

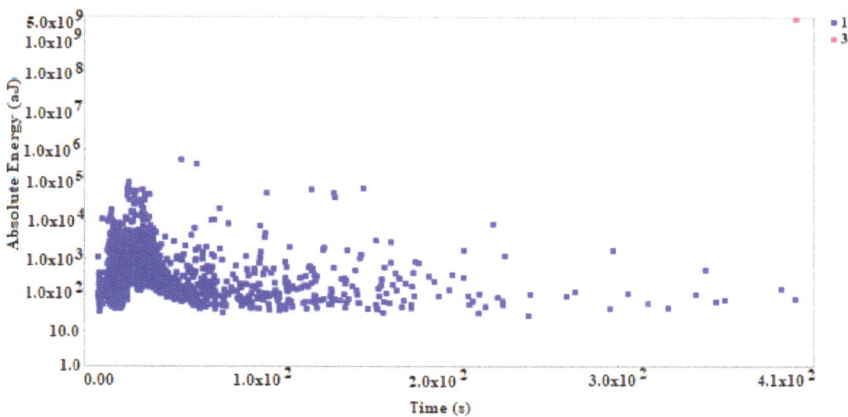

Figure 6. Scatterplot: Absolute energy (aJ) vs. time (s) for uniaxial tested specimen.

Figure 7a,b presents the parameters of the acoustic emission signal: initiation frequency (IF) (kHz) and reverberation frequency (RF) (kHz) vs. time and signal classes. Analysing these graphs, it can be seen that green signals have a similar frequency range in the initial stage (IF) parameter as well as in the later stage (RF) parameter during signal duration. This range falls within the frequency range of pink signals.

We can, therefore, assume that the signals of green color characterize the process of dislocation motion, and cracking on the cleavage surfaces in ferrite, i.e., brittle cleavage fracture. The red signals show different frequencies at different stages of the load frequency band. Parameter (IF) signals occurring in the earlier stage of the load have much higher than others, which allows to assume that these signals are generated in a different environment than ferrite. These may be signals from the cracking of precipitates segregated at the ferrite grain boundaries or connected with the process of de-bonding from the ferrite matrix. These signals can be imitated, in ferrite, and in precipitations. That is why there is a different frequency in parameter (IF).

The signal imitated in the precipitations should be short. However, the signal imitated in ferrite is longer and therefore parameter (RF) is similar. In the next stage of loading, red signals characterize the development of intercrystallite fracture growth and are imitated in ferrite.

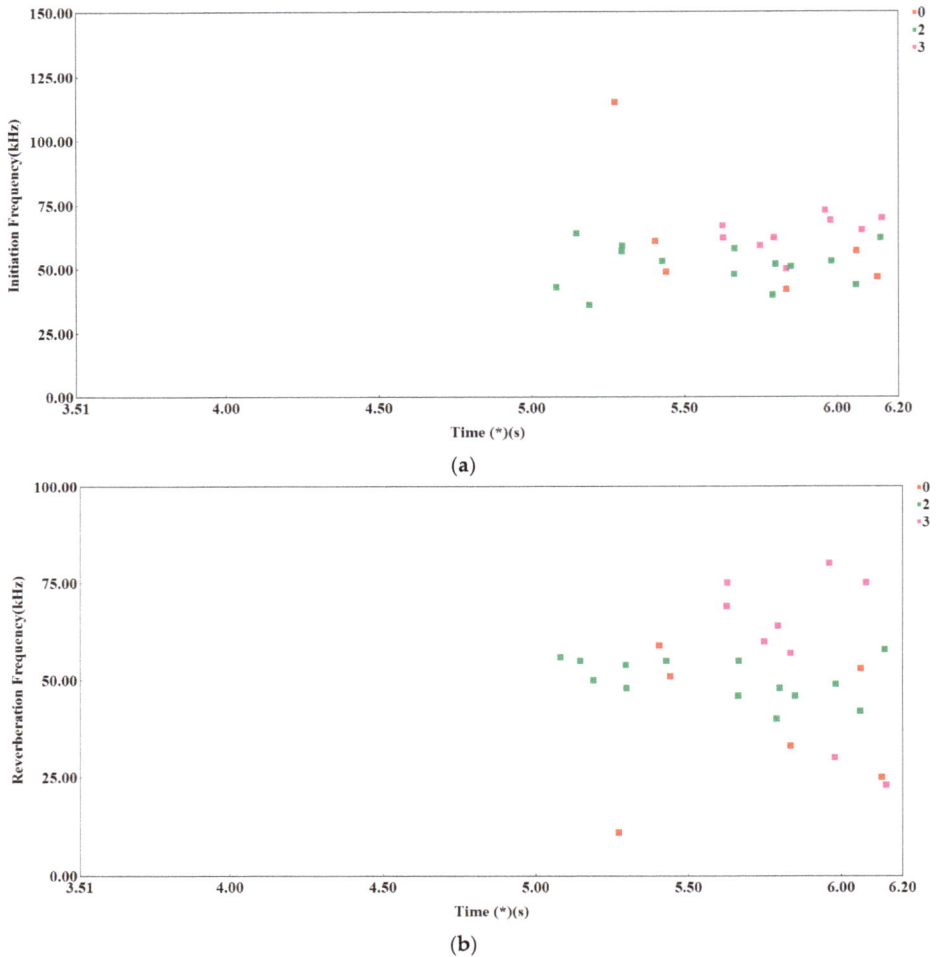

Figure 7. Scatterplot: (**a**) Initiation frequency (IF) (kHz) vs. time (s); and (**b**) reverberation frequency (RF) (kHz) vs. time (s).

4. Conclusions

4.1. General Conclusions

To raise AE technology to the next level, we must improve it from several fronts. Only then can AE become a serious player in structural integrity assessment. We need to better characterize AE sources; to extract more from AE signals reaching sensors; to devise effective AE parameters; to integrate new analysis methods, such as wave transformation analysis(WT), moment tensor analysis (MTA), and pattern recognition analysis (PRA); to include simulation tools in analyses; to accumulate basic data on structures with standardized procedures; and to devise combinatorial approach between localized damage evaluation and long-range detection and to develop regional or global database under international cooperation. Using systematic approach along with NDT, fracture mechanics, etc., improved AE can offer more value to industrial users and to contribute substantially to structural integrity assessment. Development of the proposed AE monitoring technique reported in this paper facilitates for the prognostics and life predictions of the structure. The developed methodology can be utilized for continuous in-service monitoring of structures and has proven to be promising for use in practice.

4.2. Specific Applications

1. Application of the k-means grouping method based on the analysis in the 13 parameter space of AE signals provides positive results in the classification of destructive processes in fracture toughness K_{Ic} tests.
2. Application of the k-means grouping method allows differentiating between brittle fractures mechanisms.
3. The process of brittle fracture was shown to be a sequential process composed of inter-grain cracking, brittle-cleavage and ductile fracture.
4. Acoustic emission technology was found to be able to successfully monitor and detect real-time crack initiation and also separate and identify AE signals that correspond to different fracture mechanisms
5. The grouping analysis shows that some AE parameters noticeably change after the initiation of the crack (absolute energy, initial frequency, reverberation frequency), which indicates their high independence and suitability for monitoring the cracking process.
6. The developed base of reference signals forms the theoretical basis for using pattern-based AE technology to detect and monitor crack propagation in structures that do not have a "load history" to assess the safety of steel structures using the principles of fracture mechanics.

Author Contributions: Conceptualization: G.Ś. and I.D.; methodology: G.Ś., A.K.; software: R.P.; validation: G.Ś., I.D.; formal analysis: A.K.; investigation: A.K., R.P.; resources: G.S, R.P.; data curation: A.K., R.P.; writing—original draft preparation: G.Ś., I.D.; writing—review and editing: G.Ś., I.D.; visualization: R.P., A.K.; supervision: G.Ś.; project administration: G.Ś.; funding acquisition: G.Ś., R.P.

Funding: The research was financed in equal parts and the project is supported by the program of the Minister of Science and Higher Education under the name: "Regional Initiative of Excellence" for 2019–2022 (project number: 025/RID/2018/19; financing amount: PLN 12,000,000) and the National Science Centre, Poland (no. 2017/25/N/ST8/00179).

Conflicts of Interest: The authors declare no conflict of interest.

References

1. FITNET: Fitness-for-Service. *Fracture-Fatigue-Creep-Corrosion*; Koçak, M., Webster, S., Janosch, J.J., Ainsworth, R.A., Koerc, R., Eds.; GKSS Research Centre Geesthacht GmbH: Stuttgart, Germany, 2008.
2. Ono, K. Application of Acoustic Emission for Structure Diagnosis. *Diagn.-Diagn. Struct. Health Monit.* **2011**, *2*, 3–18.

3. Adamczak, A.; Świt, G.; Krampikowska, A. Application of the Acoustic Emission Method in the Assessment of the Technical Condition of Steel Structures. *IOP Conf. Ser. Mat. Sci. Eng.* **2019**, *471*, 032041. [CrossRef]

4. Świt, G.; Krampikowska, A. Influence of the number of acoustic emission descriptors on the accuracy of destructive process identification in concrete structures. In Proceedings of the 7th IEEE Prognostics and System Health Management Conference (PHM-Chengdu), Chengdu, China, 19–21 October 2016; Article no. 7819756.

5. Świt, G. Acoustic emission method for locating and identifying active destructive processes in operating facilities. *Appl. Sci.* **2018**, *8*, 1295. [CrossRef]

6. ASTM E647. *Standard Test Method for Measurement of Fatigue Crack Growth Rates*; ASTM International: West Conshohocken, PA, USA, 2010.

7. Biancolini, M.E.; Brutti, C.; Paparo, G.; Zanini, A. Fatigue cracks nucleation on steel, Acoustic emission and fractal analysis. *Int. J. Fatigue* **2006**, *28*, 1820–1825. [CrossRef]

8. Caesarendra, W.; Kosasih, B.; Tieu, A.K.; Zhu, H.; Moodie, C.A.S.; Zhu, Q. Acoustic emission-based condition monitoring methods: Review and application for low speed slew bearing. *Mech. Syst. Signal Process.* **2016**, *72–73*, 134–159. [CrossRef]

9. Keshtgar, A.; Modarres, M. Detecting Crack Initiation Based on Acoustic Emission. *Chem. Eng. Trans.* **2013**, *33*, 547–552.

10. Rabiei, M.; Modarres, M. Quantitative methods for structural health management using in situ acoustic emission monitoring. *Int. J. Fatigue* **2013**, *49*, 81–89. [CrossRef]

11. Zhang, X.; Feng, N.; Wang, Y.; Shen, Y. Acoustic emission detection of rail defect based on wavelet transform and Shannon entropy. *J. Sound Vib.* **2015**, *339*, 419–432. [CrossRef]

12. Keshtgar, A.; Modarres, M. Acoustic Emission-Based Fatigue Crack Growth Prediction. In Proceedings of the Reliability and Maintainability Symposium (RAMS), Orlando, FL, USA, 28–31 January 2013.

13. Yan, Z.; Chen, B.; Tian, H.; Cheng, X.; Yang, J. Acoustic detection of cracks in the anvil of a large-volume cubic high-pressure apparatus. *Rev. Sci. Instrum.* **2015**, *86*, 124904. [CrossRef]

14. Gao, L.X.; Zai, F.L.; Su, S.B.; Wang, H.Q.; Chen, P.; Liu, L.M. Study and application of acoustic emission testing in fault diagnosis of low-speed heavy-duty gears. *Sensors* **2011**, *11*, 599–611. [CrossRef]

15. Qu, Y.Z.; He, D.; Yoon, J.; Van Hecke, B.; Bechhoefer, E.; Zhu, J.D. Gearbox tooth cut fault diagnostics using acoustic emission and vibration sensors—A comparative study. *Sensors* **2014**, *14*, 1372–1393. [CrossRef] [PubMed]

16. Doan, D.D.; Ramasso, E.; Placet, V.; Zhang, S.; Boubakar, L.; Zerhouni, N. An unsupervised pattern recognition approach for AE data originating from fatigue tests on polymer-composite materials. *Mech. Syst. Signal Process.* **2015**, *64*, 465–478. [CrossRef]

17. Ting-Hua, Y.; Stathis, C.; Stiros, X.-W.Y.; Jun, L. *Structural Health Monitoring—Oriented Data Mining, Feature Extraction, and Condition Assessment*; Hindawi Publishing Corporation: London, UK, 2014.

18. Tang, J.; Soua, S.; Mares, C.; Gan, T.-H. An experimental study of acoustic emission methodology for in service condition monitoring of wind turbine blades. *Renew. Energy* **2016**, *99*, 170–179. [CrossRef]

19. The MathWorks, Inc. *Statistics and Machine Learning Toolbox*; R2017a; The MathWorks, Inc.: Natick, MA, USA, 2017.

20. Li, L.; Lomov, S.V.; Yan, X.; Carvelli, V. Cluster analysis of acoustic emission signals for 2D and 3D woven glass/epoxy composites. *Compos. Struct.* **2014**, *116*, 286–299. [CrossRef]

21. Crivelli, D.; Guagliano, M.; Monici, A. Development of an artificial neural network processing technique for the analysis of damage evolution in pultruded composites with acoustic emission. *Compos. Part B* **2014**, *56*, 948–959. [CrossRef]

22. Desgraupes, B. *Clustering Indices*; University of Paris Ouest-Lab Modal'X: Paris, France, 2013.

23. Eftekharnejad, B.; Mba, D. Monitoring natural pitting progress on helical gear mesh using acoustic emission and vibration. *Strain* **2011**, *47*, 299–310. [CrossRef]

24. Zhu, X.; Zhong, C.; Zhe, J. A high sensitivity wear debris sensor using ferrite cores for online oil condition monitoring. *Meas. Sci. Technol.* **2017**, *28*, 75102. [CrossRef]

25. Li, R.; He, D. Rotational machine health monitoring and fault detection using EMD—Based acoustic emission feature quantification. *IEEE Trans. Instrum. Meas.* **2012**, *61*, 990–1001. [CrossRef]

26. Loutas, T.H.; Roulias, D.; Pauly, E.; Kostopoulos, V. The combined use of vibration, acoustic emission and oil debris on-line monitoring towards a more effective condition monitoring of rotating machinery. *Mech. Syst. Signal Process.* **2011**, *25*, 1339–1352. [CrossRef]

27. Hase, A.; Mishina, H.; Wada, M. Correlation between features of acoustic emission signals and mechanical wear mechanisms. *Wear* **2012**, *292*, 144–150. [CrossRef]

28. Li, R.; Seçkiner, S.U.; He, D.; Bechhoefer, E.; Menon, P. Gear fault location detection for split torque gearbox using AE sensors. *IEEE Trans. Syst. Man Cybern. Part C Appl. Rev.* **2012**, *42*, 1308–1317. [CrossRef]

29. Gu, D.; Kim, J.; An, Y.; Choi, B. Detection of faults in gearboxes using acoustic emission signal. *J. Mech. Sci. Technol.* **2011**, *25*, 1279–1286. [CrossRef]

30. Zhang, L.; Yalcinkaya, H.; Ozevin, D. Numerical Approach to Absolute Calibration of Piezoelectric Acoustic Emission Sensors using Multiphysics Simulations. *Sens. Actuators A Phys.* **2017**, *256*, 12–23. [CrossRef]

31. Zhang, L.; Ozevin, D.; Hardman, W.; Timmons, A. Acoustic emission signatures of fatigue damage in idealized bevel gear spline for localized sensing. *Metals* **2017**, *7*, 242. [CrossRef]

32. ASTM E8. *Standard Test Method for Tension Testing of Metallic Materials*; ASTM International: West Conshohochen, PA, USA, 2003.

33. ASTM E1820-09. *Standard Test. Method for Measurement of Fracture Toughness*; ASTM International: West Conshohochen, PA, USA, 2011.

34. ISO 12135:2002. *Metallic Materials—Unified Method of Test for the Determination of Quasistatic Fracture Toughness*; International Organization for Standardization: Geneva, Switzerland, 2002.

35. Schwalbe, K.H.; Landes, J.D.; Heerens, J. *Classical Fracture Mechanics Methods*; GKSS 2007/14.2007; Elsevier: Amsterdam, the Netherlands, 2007.

36. Anderson, T.L. *Fracture Mechanics: Fundamentals and Applications*; CRC Press: Boca Raton, FL, USA, 2008.

37. Neimitz, A. *Mechanika Pękania*; Powszechne Wydawnictwo Naukowe (PWN): Warsaw, Poland, 1999. (In Polish)

38. Pała, R.; Dzioba, I. Influence of delamination on the parameters of triaxial state of stress before the front of the main crack. *AIP Conf. Proc.* **2018**, *2029*, 020052. [CrossRef]

materials

MDPI

Article

Validation of Selected Non-Destructive Methods for Determining the Compressive Strength of Masonry Units Made of Autoclaved Aerated Concrete

Radosław Jasiński *[ID], Łukasz Drobiec[ID] and Wojciech Mazur[ID]

Department of Building Structures, Silesian University of Technology; ul. Akademicka 5, 44-100 Gliwice, Poland; lukasz.drobiec@polsl.pl (Ł.D.); wojciech.mazur@polsl.pl (W.M.)
* Correspondence: radoslaw.jasinski@polsl.pl; Tel.: +48-32-237-1127

Received: 23 December 2018; Accepted: 21 January 2019; Published: 26 January 2019

check for
updates

Abstract: Minor-destructive (MDT) and non-destructive (NDT) techniques are not commonly used for masonry as they are complex and difficult to perform. This paper describes validation of the following methods: semi-non-destructive, non-destructive, and ultrasonic technique for autoclaved aerated concrete (AAC). The subject of this study covers the compressive strength of AAC test elements with declared various density classes of: 400, 500, 600, and 700 (kg/m^3), at various moisture levels. Empirical data including the shape and size of specimens, were established from tests on 494 cylindrical and cuboid specimens, and standard cube specimens 100 mm × 100 mm × 100 mm using the general relationship for ordinary concrete (Neville's curve). The effect of moisture on AAC was taken into account while determining the strength f_{Bw} for 127 standard specimens tested at different levels of water content (w = 100%, 67%, 33%, 23%, and 10%). Defined empirical relations were suitable to correct the compressive strength of dry specimens. For 91 specimens 100 mm × 100 mm × 100 mm, the P-wave velocity c_p was tested with the transmission method using the ultrasonic pulse velocity method with exponential transducers. The curve (f_{Bw}–c_p) for determining the compressive strength of AAC elements with any moisture level (f_{Bw}) was established. The developed methods turned out to be statistically significant and can be successfully applied during in-situ tests. Semi-non-destructive testing can be used independently, whereas the non-destructive technique can be only applied when the developed curve f_{bw}–c_p is scaled.

Keywords: autoclaved aerated concrete (AAC); compressive strength; shape and size of specimen; moisture of AAC; ultrasonic testing

1. Introduction

Significant variations in materials, technology, and performance cause that masonry structures are much more difficult to be diagnosed than concrete or reinforced concrete, for which the standard EN 13791:2011 [1] specifies both the methodology of tests and conclusions. Regarding masonry structures, there are no standards that classify testing methods. Methods which directly determine compressive or shear strength of a wall, are commonly assumed as destructive testing (DT). Those studies consist in testing fragments of a masonry wall [2,3]. Destructive (direct) techniques use fragments of walls or flat jacks in bed joints, and deliver test results in the form of compressive strength of the wall f_k. These methods cause quite a significant damage to the wall. Consequently, the number of tests to be performed becomes sharply limited.

Non-destructive testing (NDT) conducted on masonry walls, which is per analogiam to reinforced concrete structure, include the following methods: sclerometric method, ultrasound method, and pull-out method, which are not commonly used and have not been normalized so far [4–7]. There are some recommendations [8] and general guidelines, but they are not regarded as the European document. Tests can be performed to evaluate compressive strength providing that the appropriate standard curve will be scaled taking into account destructive tests conducted on cores from the structure (or on masonry units, or the mortar). The number of tests conducted with these techniques is significantly high, the damage of masonry structures is not severe and can be easily repaired. Unfortunately, there are no standard curves for adequate scaling except for original solutions.

Minor-destructive testing (MDT) is most commonly applied for masonry structures. This technique consists of taking small cores from such a structure or applying flat-jack. As it is in the case of NDT, there are no uniform regulations at the European level. The practical application of flat-jack technique involves American standards [9–12]. For small cores from the wall structure, some recommendations [13], which specify conversion factors for solid brick wall to determine its compressive strength. Besides, compressive strength can be determined on the basis of tests conducted on wall components (masonry units and the mortar). This technique consists of converting compressive strength of small specimens into the strength of standard specimens (f_b into f_m), and using standard equations in their exponential form $f_k = Kf_b{}^\alpha f_m{}^\beta$ (K—coefficient specified in EC-6). There are not many tests in this field, and the performed ones are rather single cases [14–16] and usually refer to solid brick and traditional mortar. Non-destructive and semi-non-destructive tests are indirect techniques because they do not determine compressive strength of the wall, but the strength of its component (masonry unit or mortar). Neither NDT nor MDT techniques can be used to determine compressive strength of the wall without performing destructive tests to scale the suitable correlation curve to convert obtained strength values into the requested value f_k [17].

Determining the compressive strength of modern masonry walls with thin joints, where mortar levels any irregularities of support areas and head joints are unfilled, requires only the properly determined compressive strength of the masonry unit f_B and calculated (with empirical factors η_w and δ expressing the specimen moisture and shape) an average normalized compressive strength. This procedure involves the relationship according to Eurocode 6 [18], and is used to calculate the specific compressive strength of the masonry wall:

$$f_k = Kf_b^{0.85} = K(\eta_w \delta f_B)^{0.85} \rightarrow K(f_{Bw})^{0.85} \tag{1}$$

where $K = 0.75$ or 0.8, f_b—average normalized compressive strength of masonry unit determined for specimens 100 mm × 100 mm × 100 mm, f_B—average compressive strength of the whole masonry unit or a specimen with moisture content $w = 0$, f_{Bw}—compressive strength of the specimen from the masonry with real moisture content.

If tests are performed on specimens having different dimensions than a cube with a 10 mm side, the normalized strength is determined using δ factors specified in the standard PN-EN 772-1 [19]. However, the standard does not specify conversion factors for non-standard specimens, such as cores or micro-cores. Consequently, the conversion of results is burdened with a default error that is difficult to be estimated. The literature [20–22] describes conversion factors obtained from tests on other materials, such as concrete, ceramics, or masonry units [23]. No relations to AAC have been presented so far.

Autoclaved aerated concrete (AAC) contains cement, calcium, and lime as binding material, sand used as a filler and tiny quantities of aluminium powder (or paste), which is used as a blowing agent. Density of this type of concrete ranges from 300 to 1000 kg/m³, and its compressive strength varies from 1.5 to 10 N/mm². Taking into account all construction materials, AAC is characterized by the highest thermal insulation power (thermal conductivity coefficient λ is 8–10 times lower compared to brick or reinforced concrete). AAC has been commonly used since the middle of the 1950s. This material (>40% of the construction segment in Europe) is used for masonry structures, precast wall or floor elements, and lintels [24]. The open-pore structure explains why AAC is sensitive to direct exposure to moisture, which results in worse insulating and strength properties. The available articles, apart from general relations specified in standards, do not contain detailed references expressed as empirical relations.

There are no procedures for determining specific compressive strength of the existing masonry wall with the actual density and moisture content. However, in some situations, drilling micro-cores, and even performing sclerometic tests is impossible. Therefore, only ultrasonic non-destructive techniques can be used to determine compressive strength.

The main aim of this article was to present the complex analysis of strength issues, which included developing empirical relations to determine compressive strength f_b on the basis of tests performed on specimens of any shape and real moisture content, and to develop the universal curve representing ultrasound velocity c_p and compressive strength, taking into account moisture content.

This article describes an attempt to establish the empirical curve for determining the normalized compressive strength of the AAC masonry unit, with unspecified density and moisture content f_{Bw} using semi-non-destructive techniques. Neville's curve [20], in the commonly known form from diagnosing ordinary concrete, was used and calibrated to nominal density classes of AAC (400, 500, 600, and 700). Knowing that, apart from the effect of rising and hardening [25,26], also moisture content in AAC influences the compressive strength, tests were performed and additional empirical relations were defined. The analysis included test results [27] from 494 + 127 cylindrical and cuboid specimens used to develop empirical curves. Results obtained from destructive tests on standard cube specimens 100 mm × 100 mm × 100 mm at different moisture content were correlated with results from testing velocity of P-wave generated by point transducers with the transmission method. Developed curves and the test procedure can be employed in a widely understood diagnostic of masonry structures to evaluate the safety of AAC structures.

2. Minor-Destructive Testing

2.1. Specimens, Technique of Tests, and Analysis

Tests included four series of masonry units with thickness within the range of 180–240 mm and different classes of density: 400, 500, 600, and 700, from each 20 masonry units were randomly selected. Six series of cores with varying diameters were taken from each masonry unit. Six series of square specimens having different side length and height were drilled from masonry units using a diamond saw. Cuboid specimens included blocks with dimensions of 100 mm × 100 mm × 100 mm, which were used as basic specimens for determining the strength f_B (in accordance with Appendix B to the standard EN 771-4 [16]). Drilled core and cube specimens are illustrated in Figure 1. All specimens drilled from blocks were dried until constant weight at a temperature of 105 °C ± 5 °C (for at least 36 h).

Depending on the specimen size, loading rate was 2400 N/s or 100 N/s. Due to the size of specimens, two types of machines having an operating range of 100 kN and 3000 kN, and the class of accuracy of 0.5, were used (according to EN 772-1 [28]; Figure 2). Compressive strength f_B was determined for cube specimens 100 mm × 100 mm × 100 mm (dried until constant weight). The summary of our test results for core and cube specimens is shown in Tables 1 and 2. Tables show dimensions and strength of each tested specimen, average strength and coefficient of variation for each tested series. Arrows indicate the direction of AAC growth. When dried until constant weight, each cuboid specimen was weighed and its apparent density was calculated (Table 3).

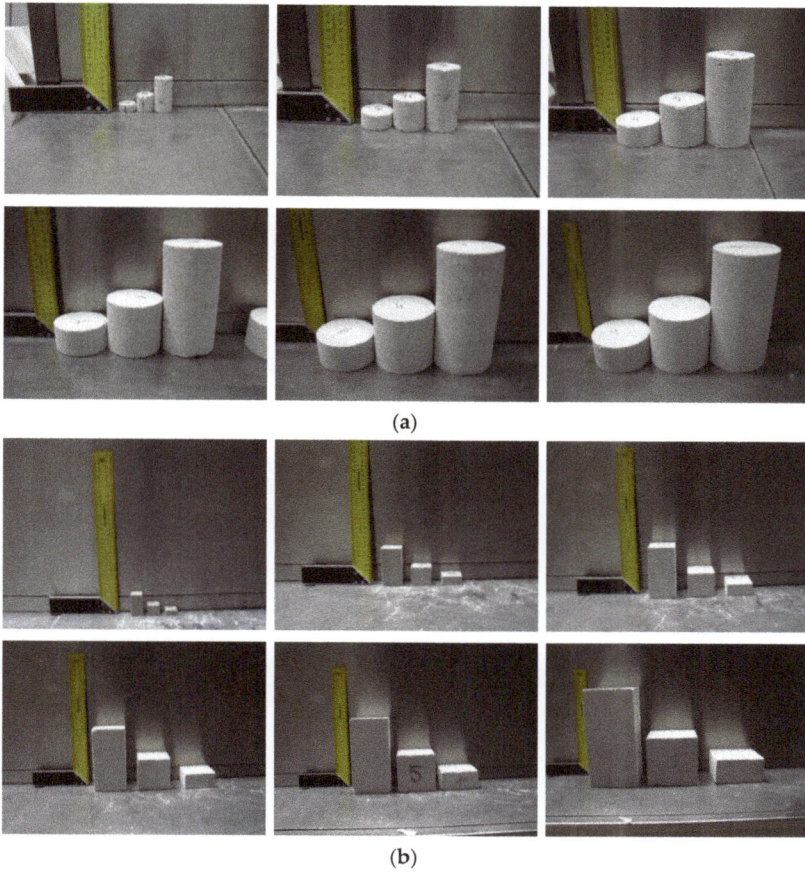

Figure 1. Specimens before tests [27]: (**a**) core specimens, and (**b**) cube specimens.

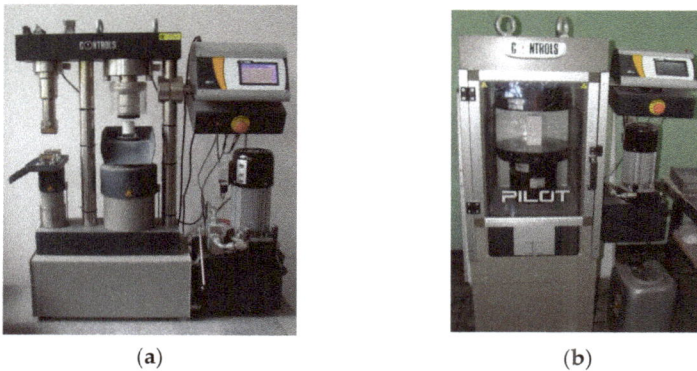

Figure 2. Testing compressive strength of AAC specimens [27]: (**a**) tests on cores using a strength testing machine with an operating range of 100 kN, and (**b**) tests on cuboid specimens using a strength testing machine with an operating range of 3000 kN.

Table 1. Results from compressive tests performed on core (cylindrical) specimens.

No.	Class of Density kg/m³	Specimen Type	Dimensions, mm		No. of Specimens n	Average Compressive Strength f_{ci}, N/mm²	Standard Deviation $s=\sqrt{\frac{(f_i-f_{ci})^2}{n-1}}$, N/mm²	C.O.V $\frac{s}{f_{ci}}$, %
			Diameter, ø	Height, h				
1	2	3	4	5	6	7	8	9
1				150	6	2.84	0.40	14
2			150	301	3	2.33	0.13	6
3				76	3	2.70	0.06	2
4				97.8	4	2.61	0.29	11
5			97.6	195	3	2.16	0.21	10
6				49	4	2.81	0.39	14
7				79.2	3	2.53	0.25	10
8	400		79.4	159	3	2.26	0.26	12
9				40.6	3	2.85	0.17	6
10				61	4	2.77	0.17	6
11			61	121.8	3	2.65	0.12	5
12				31.8	5	2.51	0.39	15
13				40	5	2.82	0.42	15
14			39.5	79	4	2.28	0.34	15
15				20.5	4	2.60	0.35	14
16				24.4	3	2.33	0.30	13
17			25	49.2	3	2.69	0.42	16
18				12.5	3	3.56	0.33	9
1				150	6	2.94	0.25	9
2			150	301	3	3.28	0.18	6
3				76	3	3.01	0.12	4
4				97.8	4	2.88	0.16	6
5			97.6	195	3	3.09	0.06	2
6				49	4	3.15	0.29	9
7				79.2	3	3.30	0.12	4
8	500		79.4	159	3	2.90	0.21	7
9				40.6	4	3.27	0.45	14
10				61	4	3.21	0.23	7
11			61	121.8	3	3.17	0.26	8
12				31.8	5	3.19	0.18	6
13				40	4	2.94	0.30	10
14			39.5	79	4	2.89	0.32	11
15				20.5	4	3.63	0.36	10
16				24.4	4	2.91	0.27	9
17			25	49.2	3	3.16	0.18	6
18				12.5	4	4.06	0.25	6
1				150	5	5.06	0.36	7
2			150	301	3	4.23	0.21	5
3				76	3	5.11	0.68	13
4				97.8	4	4.49	0.22	5
5			97.6	195	3	4.26	0.18	4
6				49	4	5.01	0.61	12
7				79.2	4	4.43	0.09	2
8			79.4	159	3	4.73	0.25	5
9	600			40.6	4	5.14	0.52	10
10				61	4	4.65	0.47	10
11			61	121.8	3	4.54	0.16	3
12				31.8	5	5.19	0.66	13
13				40	4	4.87	0.53	11
14			39.5	79	3	4.18	0.31	8
15				20.5	4	6.00	0.81	14
16				24.4	4	5.17	0.27	5
17			25	49.2	3	4.79	0.64	13
18				12.5	4	6.88	0.76	11

Table 1. *Cont.*

No.	Class of Density kg/m³	Specimen Type	Dimensions, mm		No. of Specimens n	Average Compressive Strength f_{ci}, N/mm²	Standard Deviation $s=\sqrt{\frac{(f_i-f_q)^2}{n-1}}$, N/mm²	C.O.V $\frac{s}{f_q}$, %
			Diameter, ø	Height, h				
1	2	3	4	5	6	7	8	9
1				150	5	7.12	0.96	14
2			150	301	3	7.25	0.56	8
3				76	4	7.69	0.63	8
4				97.8	4	7.37	0.76	10
5			97.6	195	3	7.22	0.42	6
6				49	4	7.93	0.28	4
7				79.2	3	6.77	0.35	5
8	700		79.4	159	3	7.25	0.57	8
9				40.6	4	8.87	0.36	4
10				61	4	7.25	1.04	14
11			61	121.8	3	7.05	0.51	7
12				31.8	5	8.57	0.35	4
13				40	3	7.55	0.32	4
14			39.5	79	3	7.21	1.08	15
15				20.5	4	9.18	0.77	8
16				24.4	3	7.66	0.77	10
17			25	49.2	3	7.73	0.40	5
18				12.5	4	13.42	0.95	7

Table 2. Results from compressive tests performed on cuboid specimens.

No.	Class of Density kg/m³	Specimen Type	Dimensions, mm			No. of Specimens n	Average Compressive Strength f_{ci}, N/mm²	Standard Deviation $s=\sqrt{\frac{(f_i-f_q)^2}{n-1}}$, N/mm²	C.O.V $\frac{s}{f_q}$, %
			Width, d	Thickness, b	Height, h				
1	2	3	4	5	6	7	8	9	10
1					143	3	2.80	0.18	6
2			143	143	72	3	2.91	0.14	5
3					285	3	2.47	0.05	2
4					100 *	6	2.88	0.36	12
5			100	100	50	3	2.59	0.24	9
6					200	3	3.16	0.13	4
7					80	3	3.12	0.23	7
8	400		80	80	39	3	3.60	0.44	12
9					158	3	2.71	0.15	5
10					59	3	2.99	0.11	4
11			59	59	30	3	3.16	0.17	5
12					121	3	2.98	0.08	3
13					40	3	2.85	0.07	3
14			40	40	19.6	3	3.02	0.07	2
15					78.5	3	2.77	0.41	15
16					24	3	2.80	0.20	7
17			24	24	12.5	3	3.23	0.56	17
18					49	3	2.43	0.26	11
1					143	3	2.33	0.28	12
2			143	143	72	3	3.74	0.06	2
3					285	3	2.16	0.11	5
4					100 *	6	3.59	0.13	4
5			100	100	50	3	3.29	0.13	4
6					200	3	3.40	0.06	2
7					80	3	3.31	0.15	5
8	500		80	80	39	3	3.67	0.11	3
9					158	3	2.48	0.22	9
10					59	3	2.83	0.09	3
11			59	59	30	3	3.20	0.55	17
12					121	3	2.94	0.17	6
13					40	3	2.90	0.04	1
14			40	40	19.6	3	3.28	0.21	6
15					78.5	3	2.77	0.44	16
16					24	3	4.78	0.39	8
17			24	24	12.5	3	4.92	0.90	18
18					49	3	1.79	0.10	6

Table 2. *Cont.*

No.	Class of Density kg/m³	Specimen Type	Width, d	Thickness, b	Height, h	No. of Specimens n	Average Compressive Strength f_{ci}, N/mm²	Standard Deviation $s=\sqrt{\frac{\sum(f_i-f_{ci})^2}{n-1}}$, N/mm²	C.O.V $\frac{s}{f_{ci}}$, %
1	2	3	4	5	6	7	8	9	10
1					143	3	3.97	0.10	2
2			143	143	72	3	5.69	0.10	2
3					285	3	3.58	0.25	7
4					100 *	6	4.95	0.35	7
5			100	100	50	3	5.80	0.35	6
6					200	3	5.34	0.61	11
7					80	3	6.01	0.75	12%
8	600		80	80	39	3	6.60	0.12	2
9					158	3	4.45	0.19	4
10					59	3	4.58	0.08	2
11			59	59	30	3	5.85	0.04	1
12					121	3	4.84	0.09	2
13					40	3	5.81	0.41	7
14			40	40	19.6	3	5.06	0.17	3
15					78.5	3	5.65	0.20	4
16					24	3	6.02	0.74	12
17			24	24	12.5	3	6.30	0.19	3
18					49	3	4.19	0.91	22
1					143	3	4.88	1.03	21
2			143	143	72	3	7.21	0.27	4
3					285	3	5.28	0.44	8
4					100 *	6	8.11	0.58	7
5			100	100	50	3	7.02	1.07	15
6					200	3	7.56	0.25	3
7					80	3	6.31	0.27	4
8	700		80	80	39	3	8.79	0.89	10
9					158	3	6.50	0.98	15
10					59	3	5.76	0.34	6
11			59	59	30	3	6.31	1.10	17
12					121	3	4.65	0.95	20
13					40	3	5.48	0.52	10
14			40	40	19.6	3	7.00	0.22	3
15					78.5	3	6.71	0.29	4
16					24	3	5.38	1.98	37
17			24	24	12.5	3	9.37	1.96	21
18					49	3	4.89	1.66	34

* cube specimens according to PN-EN 771-4:2012 [28] used to determine compressive strength f_B.

Table 3. Test results for AAC density.

No.	Nominal Class of Density, kg/m³	No. of Cuboid Specimens (see Table 2)	Average Density, kg/m³	Standard Deviation s, kg/m³	C.O.V., %
1	400	57	397	22.01	6
2	500	57	492	15.86	3
3	600	57	599	13.39	2
4	700	57	674	19.83	3

Development of cracks in cuboid specimens of different dimensions was recorded with an optical measuring system (Figure 3). In dense specimens with slenderness ratio $h/b = 1$, diagonal cracks developed at the upper edges, and they formed two truncated pyramids at failure (Figure 3a). In specimens with slenderness ratio $h/d = 2$, a vertical crack in the mid-length of the base appeared first, and then secondary diagonal cracks formed near corners of specimens (Figure 3b). The arrangement of cracks in specimens of bigger volume at failure was similar to dense specimens (Figure 3c).

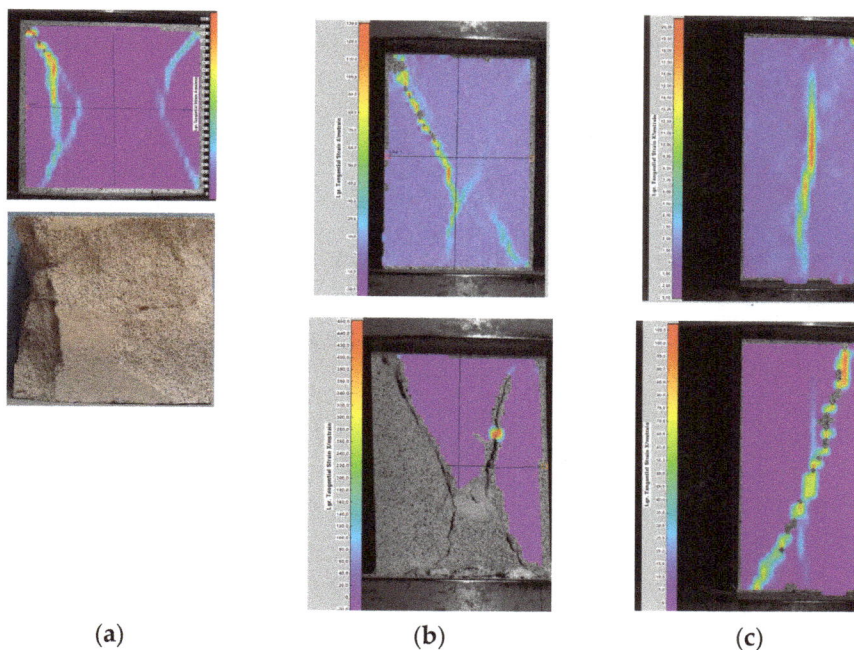

(a)   **(b)**   **(c)**

Figure 3. Destruction of specimens with varying slenderness ratio [27]: (**a**) specimen 143 mm × 143 mm × 143 mm, (**b**) specimen 100 mm × 100 mm × 200 mm, and (**c**) specimen 80 mm × 80 mm × 158 mm.

2.2. Determining an Empirical Curve in Air-Dry Conditions

If strength of the material depends on its defects, such as pores or voids, then individual specimens of different shapes can have significantly different values. These aspects are covered by Weibull's statistical theory of material strength [29,30], which states that strength of the material is reversely proportional to the volume of the tested specimen at the same probability of failure:

$$\frac{\sigma_1}{\sigma_2} = \left(\frac{V_2}{V_1}\right)^{1/m} \tag{2}$$

where σ_1 and σ_2 are failure stresses for specimens with volume V_1 and V_2, respectively; m is constant.

The exponential type of Equation (2) is similar to hyperbole and is used during tests on compressive and tensile strength of dense specimens. Neville [20] developed a similar hyperbolic relation with regard to its course, while testing specimens of different slenderness. This relation is used to determine compressive strength of concrete in specimens with shape and dimensions different from those of standard specimens (blocks 150 mm × 150 mm × 150 mm). The empirical curve for ordinary concrete is expressed as:

$$\frac{f_c}{f_{c,cube\ 150}} = 0.56 + \frac{0.697}{\frac{V}{152hd} + \frac{h}{d}} \tag{3}$$

where V is specimen volume, h is specimen height, and d is the smallest side dimension of the specimen.

Replacing strength $f_{c,cube150}$ obtained from standard specimens 150 mm × 150 mm × 150 mm with strength f_B for specimens 100 mm × 100 mm × 100 mm drilled from masonry units, and the ratio 152hd with volume of the standard specimen 100hd, the relationship (3) can be expressed as:

$$\frac{f_c}{f_B} = b + \frac{a}{\frac{V}{100hd} + \frac{h}{d}} \rightarrow y = b + \frac{a}{x} \tag{4}$$

where f_B is the compressive strength of normalised specimen 100 mm × 100 mm × 100 mm with moisture content $w = 0$, f_c is the compressive strength of a specimen with any shape and dimensions, and moisture content $w = 0$, a and b are constant coefficients for the curve, $y = f_c/f_B$ is the ratio of compressive strength, and $x = V/(100hd) + h/d$ is the dimensionless coefficient representing the effect of specimen volume and slenderness.

Requested parameters of the curve (4) were determined by searching a local minimum sum of squares difference:

$$S(a,b) = \sum_{i=1}^{n} [y_i - y(x_i)]^2 = \sum_{i=1}^{n} \left[y_i - \left(\frac{a}{x_i} + b \right) \right]^2, \tag{5}$$

using the following relationships:

$$\frac{\partial S(a,b)}{\partial a} = 0, \tag{6}$$

$$\frac{\partial S(a,b)}{\partial b} = 0. \tag{7}$$

When the system of linear equations was differentiated and solved, the following relations were obtained expressed in the form facilitating the construction of a correlation table:

$$a = \frac{\sum_{i=1}^{n} \frac{y_i}{x_i} - \frac{1}{n} \sum_{i=1}^{n} y_i \sum_{i=1}^{n} \frac{1}{x_i}}{\left(\sum_{i=1}^{n} \frac{1}{x_i^2} - \frac{1}{n} \sum_{i=1}^{n} \frac{1}{x_i} \sum_{i=1}^{n} \frac{1}{x_i} \right)}, \tag{8}$$

$$b = \frac{1}{n} \sum_{i=1}^{n} y_i - \frac{1}{n} \left(\frac{\sum_{i=1}^{n} \frac{y_i}{x_i} - \frac{1}{n} \sum_{i=1}^{n} y_i \sum_{i=1}^{n} \frac{1}{x_i}}{\left(\sum_{i=1}^{n} \frac{1}{x_i^2} - \frac{1}{n} \sum_{i=1}^{n} \frac{1}{x_i} \sum_{i=1}^{n} \frac{1}{x_i} \right)} \right) \sum_{i=1}^{n} \frac{1}{x_i}. \tag{9}$$

For defining compliance of the curve, some uncertainty was assumed to be neglected during measurements x (the specimen geometry). Additionally, uncertainties of all y values were the same (the same significance of measurements resulting from identical measuring techniques). To estimate the coefficient of correlation, the following values were calculated:

-error of estimate

$$S_{tN} = \frac{1}{n} \sum_{i=1}^{n} (y_i - y_m)^2, \tag{10}$$

where: $y_m = \frac{1}{n} \sum_{i=1}^{n} y_i$,

-sum of errors:

$$S_{rN} = \frac{1}{n} \sum_{i=1}^{n} \left(y_i - \left(\frac{a}{x_i} + b \right) \right)^2, \tag{11}$$

and then coefficient of correlation:

$$R = \sqrt{\frac{S_{tN} - S_{rN}}{S_{tN}}}. \tag{12}$$

The paper [27] compares curve correlations developed for cuboid and cylindrical specimens. Obtained values of curve coefficients a and b are compared in Table 4. Comparison of test results and the common curve is shown in Figure 4.

Table 4. Comparison of coefficients and equations of empirical curves.

Density Range of AAC, Average Density ρ, (Nominal Class of Density) kg/m³	Coefficient for Curve		R	Additive Correction Factor Δb	Corrected Coefficient for Curve b_{kor}	Curve Equation	n	F_{exp} F_{α,f_1,f_2}
	a	b						
from 375 to 446 397, (400)	0.159	0.857	0.324	0.06	0.921	$\frac{f_c}{f_B} = 0.921 + \frac{0.159}{\frac{V}{100hd}+\frac{h}{d}}$	123	14.19 3.919
from 462 to 532 492, (500)	0.312	0.682	0.533	0.16	0.844	$\frac{f_c}{f_B} = 0.844 + \frac{0.312}{\frac{V}{100hd}+\frac{h}{d}}$	125	48.81 3.918
from 562 to 619 599, (600)	0.349	0.779	0.612	0.05	0.826	$\frac{f_c}{f_B} = 0.826 + \frac{0.349}{\frac{V}{100hd}+\frac{h}{d}}$	124	73.06 3.919
from 655 to 725 674, (700)	0.454	0.608	0.614	0.16	0.773	$\frac{f_c}{f_B} = 0.773 + \frac{0.454}{\frac{V}{100hd}+\frac{h}{d}}$	122	72.62 3.920
common curve	$a_w = 0.321$	$b_w =$ 0.730	0.512	0.11	0.840	$\frac{f_c}{f_B} = 0.840 + \frac{0.321}{\frac{V}{100hd}+\frac{h}{d}}$	494	174.8 3.860

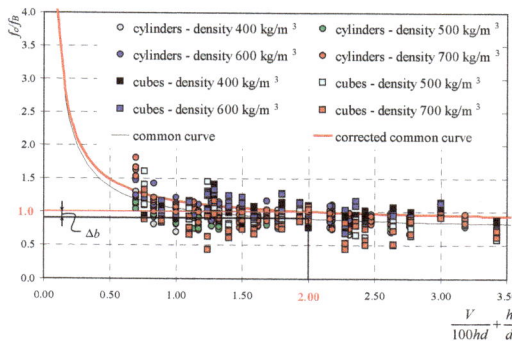

Figure 4. Test results for all core and cube specimens and determined curve of correlation.

When specimens 100 mm × 100 mm × 100 mm were used, the value of curve dominator was $V/100hd + h/d = 2$, and strength ratios calculated according to equations from Table 4 were $f_c/f_B \neq 1$. To obtain the ratio $f_c/f_B = 1$ from normalized specimens, curves needed to be translated in parallel to the intercept axis using the additive correction factor Δb for the common curve:

$$\frac{f_c}{f_B} = b + \Delta b + \frac{a}{\frac{V}{100hd} + \frac{h}{d}} \rightarrow \Delta b = 1 - b - \frac{a}{2}. \tag{13}$$

To demonstrate the correct scaling of curves, the approximate variance correlation test was applied. This approach is also adequate for linear and non-linear correlations [31]. Statistical values were calculated for each curve from Table 4 using the following formula: $F_{exp} = \frac{R^2}{(1-R^2)} \cdot \frac{f_2}{f_1}$ where degrees of freedom were $f_2 = n - k - 1$ and $f_1 = k$ ($k = 1$), and the assumed statistical significance $\alpha = 5\%$. The obtained statistical values were compared to critical values from the Fisher–Snedecor tables ($F_{\alpha,f1,f2}$). Statistical results are presented in Table 4. Analyses demonstrated that correlations were significant at the assumed statistical significance equal to 5%, thus the proposed model based on the general Neville relation was statistically significant. Besides, descriptive statistics based on the Guillford scale [32] was applied. It describes the correlation degree of individual curves. For concrete with the lowest density, obtained values R were sufficient for evaluating the relationship as poor, and for other classes of density $R > 0.5$, correlations could be regarded as moderate and the value of correlation factor as real. For the common curve, the obtained coefficient was $R = 0.512$. Thus, the relationship was moderate and real.

2.3. Calibrating a Curve in Air-Dry Conditions

Many curves developed for specific density of AAC were replaced with a curve that was more favourable for diagnostic purposes and could be used to determine the strength of AAC with any density and moisture content. Coefficients a and b determined for concrete with specific density within the defined ranges and presented in Table 4, as well as coefficients a_w and b_w of the common curve were used to develop correlations illustrated in Figure 5.

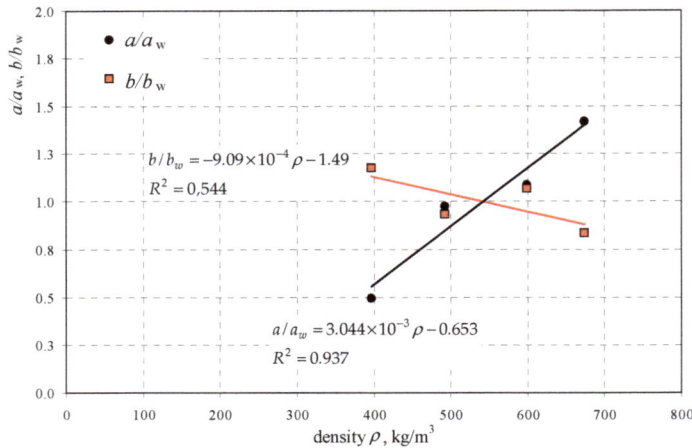

Figure 5. Relative coefficients of curves.

The following relationships describing curve coefficients as a function of AAC densities were developed on the basis of results shown in Figure 5, using the method of least squares:

$$a = a_w \times \left(3.044 \times 10^{-3}\rho - 0.653\right) = 0.321 \times \left(3.044 \times 10^{-3}\rho - 0.653\right), \tag{14}$$

$$b = b_w \times \left(9.09 \times 10^{-4}\rho - 1.49\right) = -0.730 \times \left(9.09 \times 10^{-4}\rho - 1.49\right). \tag{15}$$

where: $a_w = 0.321$, and $b_w = 0.730$ (see Table 4).

The formation of AAC curve with any density, when a and b values have been determined, requires a correction for the coefficient b which results in the strength ratio obtained from the curve (13) at $V/(100hd) + h/d = 2$.

2.4. Calibrating an Empirical Curve in Moisture Conditions

Properties of AAC and ordinary concrete depend on moisture contents [25,33,34], which cause a clear reduction in compressive and tensile strengths, and degradation of insulating parameters. Thus, other tests also focused on the effect of moisture content in AAC, which was a ratio of absorbed water to the mass of dry material:

$$w = \frac{m_w - m_s}{m_s} \cdot 100\%, \tag{16}$$

where m_w is the mass of wet specimen, and m_s is the mass of specimen dried until constant weight.

The maximum moisture content (absorbality) w_{max} in AAC corresponded to the level of water, at which no further increase in mass m_w was observed as the effect of passage of (capillary) water. Relative moisture was calculated as the ratio of current and maximum moisture w/w_{max}.

The total number of 127 specimens 100 m × 100 m × 100 m, divided into five six-element series, was prepared from AAC blocks with varying densities. Each specimen was put into containers filled with water to saturate it with water as the effect of passage of (capillary) water. Specimens were weighed every 6 h and moisture content w was calculated each time. Maximum moisture content in each type of AAC was assumed to be determined at first, and then specimens were dried until the required moisture content. Strength tests were expected to be performed at the following levels of relative moisture: w/w_{max} = 100%, 67%, 33%, 23%, 10%, and 0%. Average test results for individual series of specimens are shown in Table 5.

Maximum moisture content in AAC depended on nominal density. At the density increase in the range from ρ = 397 kg/m^3 to 674 kg/m^3, the maximum moisture content was varying within w_{max} = 53.3–89.9%, which made it possible to determine a straight line of the least square in the following form:

$$w_{max} = -1.23\frac{\rho}{1000} + 1.34 \text{ when } 397 \text{ kg/m}^3 \le \rho \le 674 \text{ kg/m}^3. \tag{17}$$

At each moisture level, destructive tests were performed to determine the strength of wet concrete f_{Bw}, and the results are illustrated in Figure 6a as a function of moisture w. Figure 6b presents the obtained strength values with respect to the strength f_B of dry (w = 0) AAC as a function of relative moisture w/w_{max}.

Two empirical lines were drawn on the basis of obtained results and used to determine the relative strength of AAC as a function of relative moisture in the following form:

$$\frac{f_{Bw}}{f_B} = -0.96\frac{w}{w_{max}} + 1 \to f_{Bw} = f_B\left(-0.97\frac{w}{w_{max}} + 1\right) \text{ when } 0 \le \frac{w}{w_{max}} \le 0.31. \tag{18}$$

$$\frac{f_{Bw}}{f_B} = -0.15\frac{w}{w_{max}} + 0.74 \to f_{Bw} = f_B\left(-0.15\frac{w}{w_{max}} + 0.74\right) \text{ when } 0.31 < \frac{w}{w_{max}} \le 1.0. \tag{19}$$

Strength f_{Bw} calculated from Equations (18) and (19) included the moisture effect, so it did not require conversion to average normalized compressive strength f_b.

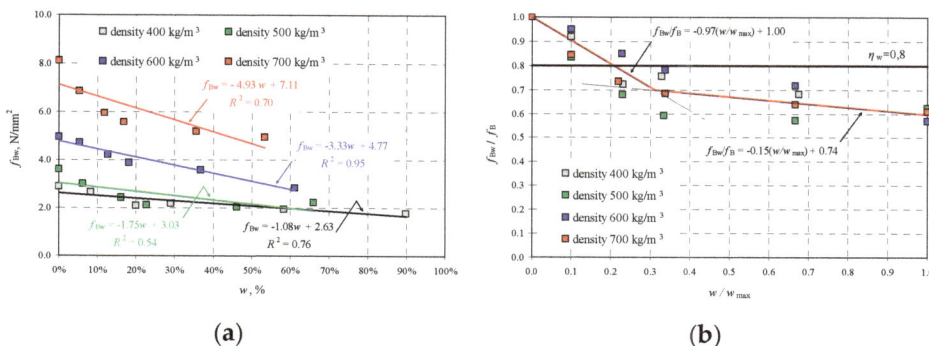

Figure 6. Test results for AAC strength, taking into account moisture level: (**a**) strength f_{Bw} as a function of moisture w, and (**b**) relative strength of AAC f_{Bw}/f_B as a function w/w_{max}.

Figure 6b also shows the value of factor η_w = 0.8 recommended by the standard EN 772-1 [19], and used to take into consideration the effect of moisture level. The standard recommendation provides the safe reduction of compressive strength only for the moisture level w/w_{max} = 0.2. Tests on walls with higher moisture content showed that compressive strength could be even reduced by 40%, that is, over twice more than the provisions recommend.

Table 5. Test results for AAC with varying moisture content.

No.	Density Range of AAC, Average Density ρ, (nominal class of density) kg/m^3	Average Moisture Content w, %	Average Relative Moisture w/w_{max}	Average Compressive Strength f_{Bw}, N/mm^2	Standard Deviation, s, N/mm^2	COV, %	Average Relative Compressive Strength f_{Bw}/f_B
1		0	0	2.88 *	0.36	12	1.0
2		8.3	0.10	2.64	0.21	8	0.92
3	from 375 to 446 397, (400)	20.1	0.23	2.09	0.11	5	0.72
4		29.1	0.33	2.18	0.16	8	0.76
5		58.3	0.67	1.96	0.14	7	0.68
6		89.9	1.00	1.78	0.13	7	0.62
7		0	0	3.59 *	0.13	4	1.0
8		6.2	0.10	3.00	0.22	7	0.84
9	from 462 to 532 492, (500)	16.2	0.23	2.44	0.49	20	0.68
10		22.8	0.33	2.12	0.21	10	0.59
11		46.1	0.67	2.06	0.29	14	0.57
12		66.0	1.00	2.24	0.23	10	0.62
13		0	0	4.95 *	0.35	7	1.0
14		5.40	0.10	4.71	0.49	10	0.95
15	from 562 to 619 599, (600)	12.6	0.23	4.21	0.38	9	0.85
16		18.2	0.34	3.88	0.52	13	0.78
17		58.3	0.67	1.96	0.33	9	0.68
18		61.1	1.00	2.82	0.28	10	0.57
19		0	0	8.11 *	0.58	7	1.0
20		5.30	0.10	6.86	0.63	9	0.85
21	from 655 to 725 674, (700)	11.7	0.22	5.96	0.71	12	0.74
22		16.8	0.34	5.56	0.58	10	0.69
23		46.1	0.67	2.06	0.70	13	0.57
24		53.3	1.00	4.95	0.41	8	0.61

$^*f_B$—compressive strength of dry AAC, when $w = 0$.

Test results for wet AAC were not thoroughly analyzed with reference to microstructure. It can be assumed that AAC structure will expand the most at moisture content in a range of 30%, and consequently compressive strength will be reduced. To sum it up, determination of compressive strength of the wall f_k required at first, taking into account varying shape and moisture, in-situ estimation of moisture content, specimen drilling, estimation of density, and compressive strength, and then the conversion relevant to moisture. Compressive strength calculated from the Equations (18) or (19) could be substituted to the Equation (1).

3. Ultrasonic Non-Destructive Method

The application of traditional cylindrical transducers may be difficult, as it requires the agent coupling with the tested surface. Tests on very porous and coarse materials, such as AAC, with cylindrical transducers may be also problematic. Measuring the distance of the wave is also difficult, especially if tests are performed only at one side [35–37]. Measurements are simpler and easier to perform when transducers having local contact with concrete are applied. Waveguides for this type of transducers are in cone shape or can be formed according to the exponential curve. As energy produced by ultrasound is lower than in cylindrical transducers with a larger contact surface, the spacing of transducers at one-side access to ordinary concrete with density of ca. 2500 kg/m^3 should not exceed 25 cm, and at both-side access—15 cm [37].

3.1. Testing Technique of Specimens

Non-destructive tests on AAC were performed using the ultrasonic testing, commonly applied for testing strength of concrete [38,39], and testing masonry walls [4–6]. Ultrasonic testing was conducted on block specimens 100 mm × 100 mm × 100 mm drilled from masonry units (Figure 7). Wet specimens with relative moisture w/w_{max} = 100%, 67%, 33%, 23%, and 10%, and specimens dried until constant weight w/w_{max} = 0% were used in tests. Each series of elements included at least > 20 specimens, and 91 specimens in total were tested.

The PUNDIT LAB instrument (Proceq SA, Schwerzenbach, Switzerland) was utilized for measurements of the ultrasonic pulse velocity. Commercial exponential transducers with the waveguide length L = 50 mm, diameters $\emptyset_1$ = 4.2 mm and $\emptyset_2$ = 50 mm, and frequency 54 kHz were employed. The applied research methodology and equipment was also used for testing also for ultrasonic tomography for concrete [40,41] or masonry [42,43].

Each specimen was put on a pad insulating from shock and outdoor noise, and then transducers were applied to walls and the measurement was made with the transmission method. Transducers were in contact with specimens at an angle of 90° within distance between transducers measured every time with an accuracy up to ±1 mm. Time was measured with an accuracy up to ±0.1 μs. The measurement results are presented in Table 6.

In AAC specimens dried until constant weight, the velocity of ultrasounds was varying from 1847 m/s in concrete of class 400 kg/m^3 to 2379 m/s in concrete of class 700 kg/m^3. An increase in P-wave velocity c_p was also proportional to density increase in wet specimens.

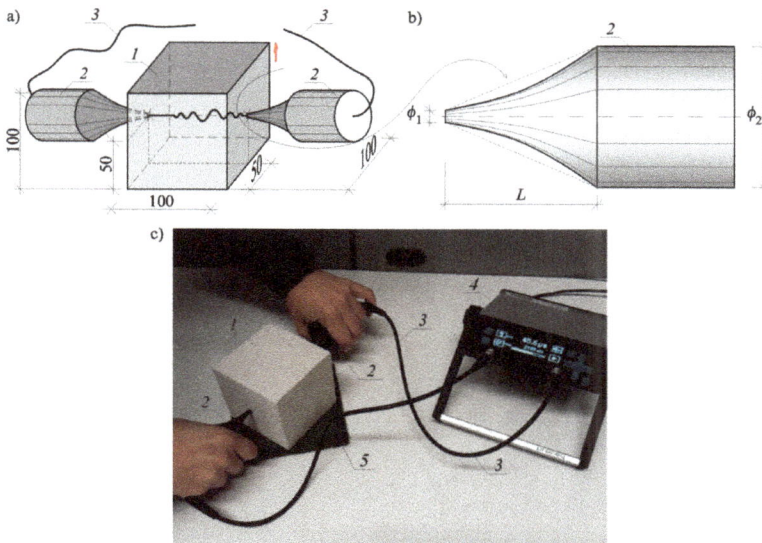

Figure 7. A test stand for measuring ultrasound velocity: (**a**) specimen geometry and elements of the stand (given in millimeters), (**b**) geometry of exponential transducer, and (**c**) a test stand; 1, tested AAC specimen 100 mm × 100 mm × 100 mm; 2, exponential transducers; 3, cables connecting transducers with recording equipment; 4, recording equipment; and 5, an insulating pad.

Table 6. Test results for ultrasound velocity in AAC with varying moisture content.

No.	Density Range of AAC, Average Density ρ, (nominal class of density) kg/m^3	w/w_{max}	Average Path Length L, mm	Average Passing Time of Wave t, µs	Average P-Wave Velocity $c_p = L/t$, m/s	N	Standard Deviation, $s=\sqrt{\frac{(c_{pi}-c_p)^2}{n-1}}$ s, m/s	C.O.V., $\frac{s}{f_{cl}}$ %
1		0		54.3	1847	21	35.9	1.9
2		0.10		57.4	1746	21	24.0	1.4
3	from 375 to 446 397, (400),	0.23	100.2	67.0	1501	21	37.7	2.5
4		0.33		67.6	1483	21	32.8	2.2
5		0.67		76.5	1315	21	25.6	1.9
6		1.00		72.7	1384	21	44.5	3.2
7		0		52.4	1917	23	51.4	2.7
8		0.10		56.3	1671	23	28.3	1.7
9	from 462 to 532 492, (500),	0.23	100.4	62.3	1614	23	33.6	2.1
10		0.33		63.0	1595	23	34.7	2.2
11		0.67		64.4	1562	23	70.2	4.5
12		1.00		62.0	1520	23	43.9	2.9
13		0		47.7	2101	24	49.7	2.4
14		0.10		50.5	1985	24	41.7	2.1
15	from 562 to 619 599, (600),	0.23	100.2	52.5	1910	24	59.6	3.1
16		0.34		54.7	1832	24	52.7	2.9
17		0.67		58.0	1738	24	69.1	4.0
18		1.00		55.6	1812	24	58.3	3.2
19		0		42.2	2379	23	46.2	1.9
20		0.10		44.3	2269	23	43.1	1.9
21	from 655 to 725 674, (700),	0.22	100.5	47.0	2139	23	52.4	2.4
22		0.34		47.6	2111	23	51.5	2.4
23		0.67		48.4	2085	23	56.1	2.7
24		1.00		48.2	2094	23	28.3	1.4

3.2. Calibrating a Curve in Air-Dry Conditions

Performed tests showed that density and relative moisture affected the velocity of P-waves in AAC. By performing steps described in Section 2.2, at first the correlation curve was determined which presented ultrasound velocity in AAC specimens in air-dry conditions as a function of compressive strength f_B. At the beginning, the curve representing the relationship between the average measured ultrasound velocity as a function of compressive strength f_{Bw} of wet AAC, grouping results by AAC density (Figure 8a). Higher sound velocity was found in concrete with greater density and compressive strength. Linear dependence, equations of which are illustrated in Figure 8a, are adequately precise approximations. Figure 8b illustrates results for compressive strength and corresponding ultrasound velocity of dry AAC (w/w_{max} = 0%), selected from each density class of AAC. Then, the relationship c_p-f_B was calculated with the least square method. For example, Figure 8b also shows the relationship of concrete with maximum moisture content (w/w_{max} = 100%), obtained similarly.

For concrete with moisture content w/w_{max} = 0%, the following empirical relationship was obtained:

$$f_B = a(c_p)^2 + bc_p + c = 5.73 \times 10^{-6}(c_p)^2 - 1.46 \times 10^{-2}c_p + 10.3, \tag{20}$$
$$\text{when } 1847 \text{ m/s} < c_p \leq 2379 \text{ m/s}.$$

Curve (20) covers results from testing all densities of AAC, where the obtained coefficient of correlation is R^2 = 0.98.

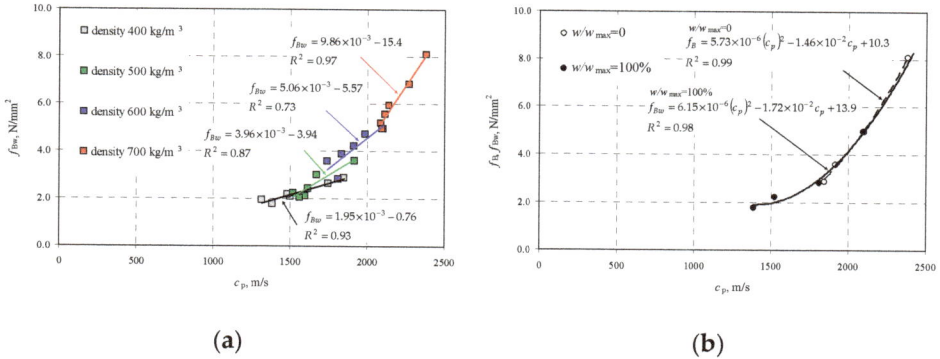

Figure 8. Results from P-waves velocity testing: (**a**) compressive strength of AAC including density classes, and (**b**) AAC strength in wet concrete (f_{Bw}) and totally dry concrete (f_B).

3.3. Calibrating a Curve in Moisture Conditions

The practical use of obtained test results required the common curve covering both the varying density of AAC and the moisture impact. For this purpose, the common curve including all moisture levels w/w_{max} and densities, was found with the least square method (Figure 9a). The equation of the common curve was:

$$f_{Bw} = a_w (c_p)^2 + b_w c_p + c_w \rightarrow f_{Bw} = 5.33 \cdot 10^{-6} (c_p)^2 - 1.39 \cdot 10^{-2} c_p + 10.9$$
$$\text{when } 1315 \text{ m/s} < c_p \leq 2379 \text{ m/s}. \tag{21}$$

Then, equations for individual curves were developed with reference to AAC density. Test results are presented in Table 7. The obtained coefficient values were compared to coefficients a_w, b_w, and c_w for the common curve, and then plotted to the graph Figure 9b.

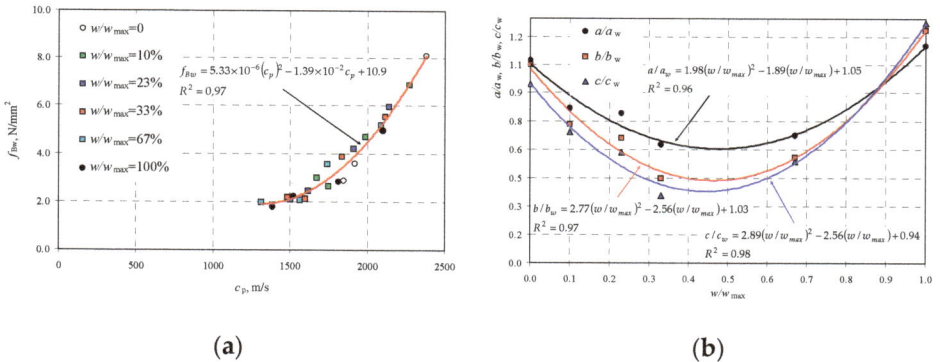

Figure 9. Results from ultrasound velocity testing: (**a**) common curve f_{Bw}–c_p for all AAC densities and moisture levels, and (**b**) equations for curve coefficients at varying moisture content in AAC f_{Bw}.

A parabolic relation of the compressive strength of wet AAC f_{bw}–c_p illustrated in Figure 9a was the same as for ordinary concrete [1]. Linear and parabolic relations were obtained for solid brick [21,44]. Generally, the result was similar to predictions. Taking into account that ultrasound velocity depends on the modulus of elasticity E, Poisson's ratio ν, and density ρ, and connected with the relationship $c_p = \sqrt{E(1-\nu)/\rho(1+\nu)(1-2\nu)}$, it was easily demonstrated that greater density caused by moisture content resulted in an increase in the modulus of elasticity. Obtained curves shown in Figure 9b are statistic. High R^2 values represent non-linear correlations. All curves had the minimum at moisture content in the range of $w/w_{max} = 0.4$–0.5. As a consequence, a difference in results with reference to the common curve will be the biggest. The curve obtained at this moisture content f_{bw}–c_p was likely to be shifted downwards. Further studies require additional tests at moisture content in the range of $w/w_{max} = 0.4$–0.5.

Table 7. Comparison of coefficients and equations of empirical curves.

w/w_{max}	Curve Coefficient			R^2	Curve Equation
	a	b	c		
0	5.73×10^{-6}	-1.46×10^{-2}	10.30	0.99	$f_{Bw} = 5.73 \times 10^{-6}(c_p)^2 - 1.46 \times 10^{-2}c_p + 10.3$
0.1	4.37×10^{-6}	-1.02×10^{-2}	7.56	0.97	$f_{Bw} = 4.37 \times 10^{-6}(c_p)^2 - 1.02 \times 10^{-2}c_p + 7.56$
0.23	4.22×10^{-6}	-9.19×10^{-3}	6.35	0.99	$f_{Bw} = 4.22 \times 10^{-6}(c_p)^2 - 9.19 \times 10^{-3}c_p + 6.35$
0.33	3.33×10^{-6}	-6.21×10^{-3}	3.88	0.98	$f_{Bw} = 3.33 \times 10^{-6}(c_p)^2 - 6.21 \times 10^{-3}c_p + 3.88$
0.67	3.59×10^{-6}	-7.75×10^{-3}	5.84	0.95	$f_{Bw} = 3.59 \times 10^{-6}(c_p)^2 - 7.75 \times 10^{-3}c_p + 5.84$
1	6.15×10^{-6}	-1.72×10^{-2}	13.90	0.98	$f_{Bw} = 6.15 \times 10^{-6}(c_p)^2 - 1.72 \times 10^{-3}c_p + 13.90$
common curve	$a_w = 5.33 \times 10^{-6}$	$b_w = -1.39 \times 10^{-2}$	$c_w = 10.90$	0.97	$f_{Bw} = 5.33 \times 10^{-6}(c_p)^2 - 1.39 \times 10^{-2}c_p + 10.9$

The method of least squares gave the following forms of empirical curves used to determine coefficients of the relationship f_{Bw}–c_p for AAC with any moisture level and density:

$$\frac{a}{a_w} = 1.99\left(\frac{w}{w_{max}}\right)^2 - 1.89\frac{w}{w_{max}} + 1.05, \ R^2 = 0.96, \tag{22}$$

$$\frac{b}{b_w} = 2.77\left(\frac{w}{w_{max}}\right)^2 - 2.56\frac{w}{w_{max}} + 1.03, \ R^2 = 0.97, \tag{23}$$

$$\frac{c}{c_w} = 2.89\left(\frac{w}{w_{max}}\right)^2 - 2.56\frac{w}{w_{max}} + 0.94, \ R^2 = 0.98. \tag{24}$$

Calculated coefficients a, b, and c should be put into the equation:

$$f_{Bw} = a(c_p)^2 + bc_p + c, \text{ when } 1315 \text{ m/s} < c_p \leq 2379 \text{ m/s}. \tag{25}$$

which gives the general form of the basic curve for AAC. In practice, ultrasonic testing should be associated with destructive tests for graduation. In this case, further steps can follow rules specified in the European standard EN 13791 [1] for ordinary concrete.

4. Procedure Algorithm for Determining Characteristic Compressive Strength of Masonry

Proposed empirical procedure for determining characteristic compressive strength of masonry with semi-non-destructive and non-destructive techniques can be described with the following steps shown in Table 8.

Table 8. Procedure algorithm for determining characteristic compressive strength of masonry with semi-NDT and NDT techniques.

Step	Description			
	Semi-Non-Destructive Testing	**Reference**	**Non-Destructive (ultrasonic) Testing**	**Reference**
1	Determining moisture content by weight w in AAC at the tested (in-situ) point	Equation (16)	Determining moisture content by weight w in AAC at the tested point	Equation (16)
2	Calculating maximum moisture content w_{max} in AAC	Equation (17)	Calculating maximum moisture content w_{max} in AAC	Equation (17)
3	Drilling specimens from AAC, drying them until constant weight and calculating density ρ	-	Determining P-waves velocity (c_p = L/t,) using the transmission method after measuring the path length L and time t.	-
4	Calculating coefficients a and b of the empirical curve	Equations (14) and (15)	Calculating coefficients a and b of the empirical curve	Equations (22)–(24)
5	Calculating the correction factor Δb	Equation (13)	Calculating compressive strength of AAC f_{Bw} acc. to the curve	Equation (25)
6	Performing destructive tests and determining compressive strength of dry AAC f_c	-	Graduating the curve according to the standard EN 13791:2008	-
7	Calculating compressive strength f_B acc. to the corrected curve	Equation (13)	Calculating compressive strength of AAC f_{Bw} acc. to the graduated curve	-
8	Calculating compressive strength f_{Bw} depending on moisture content in AAC	Equations (18) and (19)	Calculating characteristic compressive strength of AAC masonry	Equation (1)
9	Calculating characteristic compressive strength f_k of AAC masonry	Equation (1)		

5. Conclusions

The preformed tests confirmed the effect of specimen shape on compressive strength in the analyzed type of autoclaved aerated concrete and on the method of specimen failure. Regardless of AAC density, compressive strength determined at specific volumes and slenderness was found to be similar to the strength of standard specimens. The greatest strength was found in specimens with the smallest volume. Compressive strength of specimens with the greatest volume was much lower than in case of standard cube specimens with dimensions of 100 mm × 100 mm × 100 mm.

Maximum moisture content was increasingly reversely proportional to AAC density, and moisture significantly reduced strength with reference to the strength of AAC tested in air-dry conditions. The greatest 30% reduction in compressive strength was observed at moisture content $w = 0$–30%. This observation was particularly important because moisture content of masonry is ca. 10–15%. Higher moisture levels caused a drop in strength by 10%. AAC moisture coefficient $\eta_w = 0.8$ recommended by the standard PN-EN 772-1 may give dangerously overestimated strength of masonry with moisture content $w > 20\%$.

The non-destructive ultrasonic testing demonstrated the profound effect of density and moisture. An increase in P-waves velocity was proportional to density of AAC (maximum velocity was 2379 m/s in concrete with density of 700 kg/m^3, minimum velocity was 1847 m/s in concrete with density of 400 kg/m^3). Increasing density of AAC caused a significant reduction of the velocity.

Two complementary techniques were used. The first semi-non-destructive test can determine compressive strength of masonry units based on testing specimens of any shape. This technique can be used independently if at least 18 specimens can be prepared (cf. EN 13791:2009 [1])—like for concrete. The second is the non-destructive ultrasound method, which cannot be generally used without scaling the obtained curve. However, the great advantage of this solution is the reduced number of specimens to be prepared and scaled. The number of six drilled cores or cuboid specimens can be assumed as minimum. After scaling the curve, measurements can be made at any number of points and AAC strength can be determined. Performed tests indicated the impact of AAC density and moisture on both compressive strength and ultrasound velocity. These methods have some material limitations with reference to density $397 \text{ kg/m}^3 \leq \rho \leq 674 \text{ kg/m}^3$ and ultrasound velocity $1847 \text{ m/s} < c_p \leq 2379 \text{ m/s}$.

Author Contributions: Conceptualization, R.J. and Ł.D.; methodology, R.J. and W.M.; validation, R.J.; formal analysis, R.J. and W.M.; investigation, R.J., Ł.D., and W.M.; writing—original draft preparation, R.J.; writing—review and editing, R.J. and Ł.D.; visualization, R.J. and W.M.; supervision, R.J. and Ł.D.

Acknowledgments: Authors would like to express particular thanks to Solbet Sp. z o.o. Company for its technical support and supply of materials used during the research works.

Conflicts of Interest: The authors declare no conflicts of interest.

References

1. *EN 13791:2008 Assessment of In-Situ Compressive Strength in Structures and Pre-Cast Concrete Components*; European Committee for Standardization (CEN): Brussels, Belgium, 2008.
2. Łątka, D.; Matysek, P. The estimation of compressive stress level in brick masonry using the flat-jack method. *Procedia Eng.* **2017**, *193*, 266–272. [CrossRef]
3. Corradi, M.; Borri, A.; Vignoli, A. Experimental study on the determination of strength of masonry walls. *Constr. Build. Mater.* **2003**, *11*, 325–337. [CrossRef]
4. Suprenant, B.A.; Schuller, M.P. *Nondestructive Evaluation & Testing of Masonry Structures*; Hanley Wood Inc.: Washington, DC, USA, 1994; ISBN 978-0924659577.
5. Noland, J.; Atkinson, R.; Baur, J. *An Investigation into Methods of Nondestructive Evaluation of Masonry Structures*; National Technical Information Service Report No. PB 82218074; Report to the National Science Fundation: Springfield, VA, USA, 1982.
6. Schuller, M.P. Nondestructive testing and damage assessment of masonry structures. In Proceedings of the 2006 NSF/FILEM Workshop, In-Situ Evaluation of Historic Wood and Masonry Structures, Prague, Czech Republic, 10–16 July 2006; pp. 67–86.
7. McCann, D.M.; Forde, M.C. Review of NDT methods in the assessment of concrete and masonry structures. *NDT E Int.* **2001**, *34*, 71–84. [CrossRef]
8. Binda, L.; Vekey, R.D.; Acharhabi, A.; Baronio, G.; Bekker, P.; Borchel, G.; Bright, N.; Emrich, F.; Forde, M.; Forde, M.; et al. RILEM TC 127-MS: Non-destructive tests for masonry materials and structures. *Mater. Struct.* **2001**, *34*, 134–143.
9. *ASTM Standard C1196-91 In-Situ Compressive Stress within Solid Unit Masonry Estimated Using Flat-Jack Measurements*; ASTM International: West Conshohocken, PA, USA, 1991.
10. *ASTM Standard C1196-14a Standard Test Method for In Situ Compressive Stress within Solid Unit Masonry Estimated Using Flat-jack Measurements*; ASTM International: West Conshohocken, PA, USA, 2014.
11. RILEM Recommendation MDT.D.4. In-Situ Stress Tests Based on the Flat-Jack. In *Materials and Structures/Matériaux et Constructions*; RILEM Publications SARL: Paris, France, 2004; Volume 37, pp. 491–496.
12. RILEM Recommendation MDT.D.5. In Situ Stress-Strain Behaviour Tests Based on the Flat-Jack. In *Materials and Structures/Matériaux et Constructions*; RILEM Publications SARL: Paris, France, 2004; Volume 37, pp. 497–501.
13. *UIC Code Recommendations for the Inspection, Assessment and Maintenance Arch Bridges*; Final Draft; International Union of Railways, Railway technical publications: Paris, France, 2008; ISBN 978-2-7461-2525-4.
14. Schubert, P. Beurteilung der Druckfestigkeit von ausgefürtem Mauerwerk aus künstlichen Steinen und natur Steinen. In *Mauerwerk-Kalender*; Ernst und Sohn: Berlin, Germany, 1995; pp. 687–700. (In German)
15. Schubert, P. Zur Festigkeit des Mörtels im Mauerwerk, Prüfung, Beurteilung. In *Mauerwerk-Kalender*; Ernst und Sohn: Berlin, Germany, 1988; pp. 459–471. (In German)
16. Schrank, R. Materialeigenschaften historischen Ziegelmauerwerks im Hinblick auf Tragfähigkeitsberechnungen am Beispiel der Leipziger Bundwand. *Mauerwerk* **2002**, *6*, 201–207. (In German) [CrossRef]
17. Matysek, P. Compressive strength of brick masonry in existing buildings—Research on samples cut from the structures. In *Brick and Block Masonry—Trends, Innovations and Challenges*, 3rd ed.; Modena, C., da Porto, F., Valluzzi, M.R., Eds.; Taylor & Francis Group: London, UK, 2016; pp. 1741–1747. ISBN 978-1-138-02999-6.

18. *PN-EN 1996-1-1:2010+A1:2013-05P, Eurocode 6: Design of Masonry Structures. Part 1-1: General Rules for Reinforced and Unreinforced Masonry Structures*; Polish Committee for Standardization (PKN): Warsaw, Poland, 2010. (In Polish)

19. EN 772-1:2011 Methods of test for masonry units. In *Determination of Compressive Strength*; European Committee for Standardization (CEN): Brussels, Belgium, 2011.

20. Neville, A.M. *Properties of Concrete*, 5th ed.; Pearson Education Limited: Essex, UK, 2011.

21. Kadir, K.; Celik, A.O.; Tuncan, M.; Tuncan, A. The Effect of Diameter and Length-to-Diameter Ratio on the Compressive Strength of Concrete Cores. In Proceedings of the International Scientific Conference People, Buildings and Environment, Lednice, Czech Republic, 7–9 November 2012; pp. 219–229.

22. Bartlett, F.M.; Macgregor, J.G. Effect of Core Diameter on Concrete Core Strengths. *Mater. J.* **1994**, *91*, 460–470.

23. Beer, I.; Schubert, P. Zum Einfluss von Steinformate auf die Mauerdruckfestigkeit–Formfaktoren für Mauersteine. In *Mauerwerk Kalender*; Ernst und Sohn: Berlin, Germany, 2005; pp. 89–126. (In German)

24. Homann, M. *Porenbeton Handbuch. Planen und Bauen Mit System*; 6. Auflage Hannover, June 2008; Bauverlag: Gütersloh, Germany, 1991; ISBN 13 978-3-7625-3626-0.

25. Gębarowski, P.; Łaskawiec, K. Correlations between physicochemical properties and AAC porosity structure. *Mater. Bud.* **2015**, *11*, 214–216. (In Polish)

26. Zapotoczna-Sytek, G.; Balkovic, S. *Autoclaved Areated Concrete. Technology, Properties, Application*; Wydawnictwo Naukowe PWN: Warszawa, Poland, 2013. (In Polish)

27. Mazur, W.; Drobiec, Ł.; Jasiński, R. Effects of specimen dimensions and shape on compressive strength of specific autoclaved aerated concrete. *Ce/Pepers* **2018**, *2*, 541–556. [CrossRef]

28. *EN 771-4:2011 Specification for Masonry Units—Part 4: Autoclaved Aerated Concrete Masonry Units*; European Committee for Standardization (CEN): Brussels, Belgium, 2011.

29. Weibull, W. *A Statistical Theory of Strength of Materials*; Generalstabens Litografiska Anstalts Förlag: Stockholm, Sweden, 1939.

30. Weibull, W. A statistical distribution function of wide applicability. *J. Appl. Mech.* **1951**, *18*, 290–293.

31. Volk, W. *Applied Statistics for Engineers*; Literary Licensing, LLC: Whitefish, MT, USA, 2013.

32. Guilford, J.P. *Fundamental Statistics in Psychology and Education*; McGraw-HillBook, Inc.: New York, NY, USA, 1942.

33. Bartlett, F.M.; Macgregor, J.G. Effect of Moisture Condition on Concrete Core Strengths. *Mater. J.* **1993**, *91*, 227–236.

34. Jasiński, R. Determination of AAC masonry compressive strength by semi destructive method. *Nondestruct. Test. Diagn.* **2018**, *3*, 81–85. [CrossRef]

35. Matauschek, J. *Einführun in Die Ultraschalltechnik*; Verlag Technik: Berlin, Germany, 1957.

36. Stawiski, B.; Kania, T. Determination of the Influence of Cylindrical Samples Dimensions on the Evaluation of Concrete and Wall Mortar Strength Using Ultrasound Method. *Procedia Eng.* **2013**, *57*, 1078–1085. [CrossRef]

37. Noland, J.; Atkinson, R.; Kingsley, G.; Schuller, M. *Nondestructive Evaluation of Masonry Structure*; Project No. ECE-8315924; National Science Fundation: Springfield, VA, USA, 1990.

38. Stawski, B. *Ultarsonic Testing of Concrete and Mortar Using Point Probes*; Scientific Papers, Monographs. No. 39; Institute of Building Engineering of the Wrocław University of Technology: Wrocław, Poland, 2009. (In Polish)

39. Drobiec, Ł.; Jasiński, R.; Piekarczyk, A. Diagnostic testing of reinforced concrete structures. In *Methodology, Field Tests, Laboratory Tests of Concrete and Steel*; Wydawnictwo Naukowe PWN: Warszawa, Poland, 2013. (In Polish)

40. Haach, V.G.; Ramirez, F.C. Qualitative assessment of concrete by ultrasound tomography. *Constr. Build. Mater.* **2016**, *119*, 61–70. [CrossRef]

41. Schabowicz, K. Ultrasonic tomography—The latest nondestructive technique for testing concrete members—Description, test methodology, application example. *Arch. Civ. Mech. Eng.* **2014**, *14*, 295–303. [CrossRef]

42. Rucka, M.; Lachowicz, J.; Zielińska, M. GPR investigation of the strengthening system of a historic masonry tower. *J. Appl. Geophys.* **2016**, *131*, 94–102. [CrossRef]

43. Zielińska, M.; Rucka, M. Non-Destructive Assessment of Masonry Pillars using Ultrasonic Tomography. *Materials* **2018**, *11*, 2543. [CrossRef] [PubMed]

44. Binda, L. Learning from failure—Long-term behaviour of heavy masonry structures. In *Structural Analysis of Historic Construction*, 2nd ed.; D'Ayala, D., Fodde, E., Eds.; Taylor & Francis Group: London, UK, 2008; pp. 1345–1355, ISBN 978-0-415-46872-5.

materials

MDPI

Article

Variation in Compressive Strength of Concrete aross Thickness of Placed Layer

Jarosław Michałek

Faculty of Civil Engineering, Wrocław University of Science and Technology, 50-370 Wrocław, Poland;
jaroslaw.michalek@pwr.edu.pl; Tel.: +48-71-320-2264

Received: 18 June 2019; Accepted: 4 July 2019; Published: 5 July 2019

check for updates

Abstract: Is the variation in the compressive strength of concrete across the thickness of horizontally cast elements negligibly small or rather needs to be taken into account at the design stage? There are conflicting answers to this question. In order to determine if the compressive strength of concrete varies across the thickness of horizontally cast elements, ultrasonic tests and destructive tests were carried out on core samples taken from a 350 mm thick slab made of class C25/30 concrete. Special point-contact probes were used to measure the time taken for the longitudinal ultrasonic wave to pass through the tested sample. The correlation between the velocity of the longitudinal ultrasonic wave and the compressive strength of the concrete in the slab was determined. The structure of the concrete across the thickness of the slab was evaluated using GIMP 2.10.4. It was found that the destructively determined compressive strength varied only slightly (by 3%) across the thickness of the placed layer of concrete. Whereas the averaged ultrasonically determined strength of the concrete in the same samples does not vary across the thickness of the analyzed slab. Therefore, it was concluded that the slight increase in concrete compressive strength with depth below the top surface is a natural thing and need not be taken into account in the assessment of the strength of concrete in the structure.

Keywords: concrete slabs and floorings; horizontal casting; compressive strength; ultrasonic tests

---

1. Introduction

The view that the compressive strength of concrete varies across the thickness of horizontally cast elements (concrete slabs, floorings, etc.) is seldom expressed in the literature on the subject. Opinions as to the significance of this variation are widely divided, as the following survey of literature indicates.

The research published by Stawiski [1–3] provided the direct incentive for this study of the distribution of concrete compressive strength along the height of horizontally cast elements. On the basis of ultrasonic tests of core samples taken from concrete, Stawiski found the compressive strength of the concrete to be lower in the top zone than in the bottom zone by as much as 40–50% [1–3]. The variation in compressive strength along the height of the cross section was approximately linear. The fall in ultrasonic wave velocity at the sample's top surface is ascribed by Stawiski [1] to the surface weakening effect connected with concrete consolidation resulting in the segregation of concrete components. The main factor responsible for the decrease in concrete compressive strength is considered to be porosity, which very strongly affects ultrasonic wave velocity. Also the inadequate curing of fresh concrete, damage to the structure of concrete caused by corrosion, and mechanical damage to the top surface of the concrete which can arise in the course of the service life of the element are also possible factors.

Stawiski [3] proposed to introduce (besides the grade of concrete) strength gradient ∇fc into the evaluation of concrete in horizontally cast elements (e.g., floor toppings). On the basis of his research [3] Stawiski pointed out that in, e.g., an approximately 15 cm thick element the strength gradient of the concrete at the depth of 10 cm from the bottom amounted to 0.7 MPa/cm, whereas in the layers situated closer to the top surface it varied markedly (−3.0, −4.5, −8.0 MPa/cm). Therefore, Stawiski calls for [3] defining allowable variations in concrete compressive strength, e.g., $\nabla f_c \leq 1.0$ MPa/cm. The increase of 1.0 MPa/cm in the strength of concrete in the lower situated layers relative to the top layer suggested by Stawiski [3] seems to be very large.

On the basis of ultrasonic tests of the compressive strength across the thickness of samples taken from cut out pieces of 40, 45 and 60 mm thick floorings made of cement mortars, Hoła, Sadowski and Hoła A. [4] found the strength was not identical and varied across the thickness. The lowest strength was in the top zone, the highest in the bottom zone, while in the middle zone, it was close to the destructively determined compressive strength. In the considered case, the strength gradient of the mortar across the thickness of the flooring amounted to 6–7 MPa/cm.

Petersons in [5] found the compressive strength of the lower situated layers to be higher than that of the top layer, but only by 10–20%. No further increase in concrete strength was observed in the layers situated below 300 mm. The difference in compressive strength between the top surface and the bottom surface in slabs was ascribed to the inadequate curing of the concrete [5].

In monograph [6], Dąbrowski, Stachurski and Zieliński found that the deeper situated layers of concrete had higher strength than the surface layer. Below 80 cm, this increase in strength stabilized at the level of approximately 10%. In the authors' opinion [6], this is due to the well-known property of concrete—it reaches higher strength when hardening under a moderate pressure—and that is why this phenomenon does not occur in samples of low height.

Yuan, Ragab, Hill and Cook [7] found that the compressive strength of concrete along the height of the placed layer did not vary significantly. Suprenant [8] found that the compressive strength of concrete in slabs varied minimally, and only in a small upper part of the element. The most marked variation in concrete strength along element height has been observed in walls and beams. This is mainly due to the greater static pressure exerted by the concrete situated above.

Neville [9] found that the slight increase in concrete compressive strength below the top surface was a natural thing, but need not be taken into account. When testing a reinforced concrete wall and beam by means of the ultrasonic method, Watanabe, Hishikawa, Kamae and Namiki [10] found the compressive strength of concrete in samples taken from the lower part of the element was slightly higher than in samples taken from its upper part. They treated this as a natural thing which did not need to be taken into account.

Neville [11] ascribed the variation in the compressive strength of concrete along the height of the sample to the presence of retained water, occurring during concrete bleeding.

According to standard [12], the compressive strength of concrete in a structure can be lower in the top layer than in the bottom layer by as much as 25%. Concrete characterized by lower compressive strength usually occurs to a depth of 300 mm or to 20% of the height of the cross section, depending on which of the values is lower. According to standard [13], the range of variation in the compressive strength of concrete in a structure can differ between the particular portions of the structure. The variation is random and often forced (by, e.g., the relative density, the degree of compaction, the curing conditions, etc.).

Therefore, the questions arises: Is the variation in the compressive strength of concrete across the thickness of horizontally cast elements negligibly small or rather needs to be taken into account at the design stage?

2. Description of Author's Investigations

2.1. Ultrasonic Tests of Concrete

In order to verify the phenomenon of concrete compressive strength variation across the thickness of horizontally cast elements, samples with diameter d = 100 mm height h = 350 mm (Table 1), taken from a specially cast slab made of concrete C25/30 (the grade of the concrete was determined using concrete cubes cast when casting the slab) were subjected to ultrasonic tests. CEMII/BS-32.5 cement

(270 kg/m³), fly ash additive (60 kg/m³), plasticizer (2.43 kg/m³), water (170 kg/m³) and 1879 kg/m³ of natural aggregate (sand 0/2 mm −40%, gravel 2/8 mm −26%, gravel 8/16 mm −34%) were used in the concrete mix for the slab construction. The latter had been compacted by means of an immersion vibrator and cured for 28 in the laboratory conditions defined in standard [14]. The slab had been exposed to variable weather conditions for two years. Samples (01−06 in Figure 1) were drilled out of the slab perpendicularly to its top surface. For reference purposes specimens (07−12) were drilled out of the slab parallel with its top surface. Prior to the tests, the actual dimensions of the samples and their weight were determined (Table 1).

Table 1. The mean dimensions of core samples and their weight.

Sample Number	$d_{śr}$	$h_{śr}$	m	V	ρ
	mm		g	cm³	g/cm³
01	98.7	351.7	6158.5	2691	2.29
02	98.5	351.4	6167.5	2680	2.30
03	98.6	351.5	6138.0	2682	2.29
04	98.6	351.4	6176.5	2682	2.30
05	98.5	350.0	6121.5	2669	2.29
06	98.6	350.2	6121.0	2675	2.29
07	98.6	350.0	6122.0	2672	2.29
08	98.7	351.3	6172.5	2688	2.30
09	98.5	351.4	6168.2	2678	2.30
10	98.5	350.0	6122.4	2667	2.30
11	98.6	351.5	6140.0	2684	2.29
12	98.5	351.4	6137.6	2678	2.29

Figure 1. The samples to be tested using ultrasonic device.

The measuring points spaced at every 1 cm (Figures 1 and 2) were marked on the sides of the core samples. Velocity C_L of ultrasonic wave passage through concrete was measured in two perpendicular directions. No concrete/probe coupling material was used. The probes were set perpendicularly to the tested side surface of the sample (Figure 2). The distributions of velocity C_L of longitudinal ultrasonic wave passage through the concrete were obtained from the tests (Tables 2 and 3, and Figures 3 and 4).

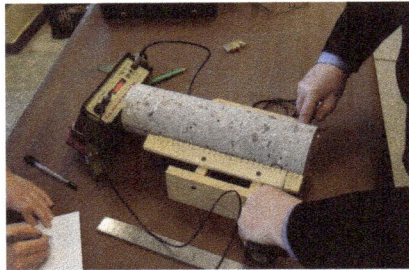

Figure 2. The measurement of velocity C_L of longitudinal ultrasonic wave.

Table 2. Velocities C_L ultrasonic wave propagation, measured in two perpendicular directions in core samples taken perpendicularly to top surface of slab.

Measuring Place Distance from Sample Top	Sample 01		Sample 02		Sample 03		Sample 04		Sample 05		Sample 06	
	C_{L1}	C_{L2}	C_{L1}	C_{L2}	C_{L1}	C_{L2}	C_{L1}	C_{L2}	C_{L1}	C_{L2}	C_{L1}	C_{L2}
cm							km/s					
0.5	3.19	3.39	3.49	3.18	3.32	3.44	3.40	3.31	3.36	3.36	3.23	3.29
1.5	3.18	3.35	3.48	3.02	3.28	3.38	3.29	3.26	3.35	3.30	3.22	3.33
2.5	3.30	3.41	3.49	3.48	3.31	3.37	3.40	3.36	3.28	3.33	3.27	3.36
3.5	3.58	3.63	3.62	3.57	3.43	3.54	3.57	3.62	3.33	3.54	3.55	3.65
4.5	3.60	3.68	3.60	3.61	3.49	3.52	3.53	3.62	3.53	3.52	3.61	3.63
5.5	3.51	3.66	3.64	3.54	3.66	3.62	3.66	3.62	3.62	3.67	3.57	3.61
6.5	3.60	3.68	3.70	3.52	3.57	3.67	3.68	3.44	3.47	3.66	3.60	3.63
7.5	3.53	3.59	3.62	3.48	3.64	3.58	3.58	3.60	3.53	3.62	3.62	3.58
8.5	3.46	3.60	3.72	3.61	3.57	3.50	3.70	3.60	3.54	3.61	3.56	3.58
9.5	3.53	3.57	3.74	3.57	3.26	3.62	3.71	3.58	3.44	3.59	3.60	3.65
10.5	3.61	3.59	3.70	3.57	3.57	3.63	3.54	3.53	3.47	3.58	3.67	3.79
11.5	3.58	3.59	3.62	3.66	3.60	3.62	3.62	3.56	3.50	3.58	3.58	3.62
12.5	3.56	3.65	3.62	3.65	3.55	3.65	3.60	3.58	3.52	3.62	3.41	3.62
13.5	3.32	3.63	3.61	3.60	3.57	3.49	3.65	3.61	3.57	3.57	3.57	3.70
14.5	3.38	3.65	3.72	3.66	3.55	3.63	3.70	3.70	3.63	3.62	3.66	3.76
15.5	3.45	3.67	3.73	3.57	3.65	3.63	3.44	3.53	3.61	3.65	3.73	3.68
16.5	3.60	3.66	3.73	3.58	3.62	3.51	3.72	3.59	3.50	3.61	3.62	3.65
17.5	3.53	3.68	3.72	3.53	3.53	3.62	3.66	3.53	3.54	3.61	3.65	3.70
18.5	3.68	3.67	3.64	3.49	3.62	3.53	3.70	3.59	3.54	3.66	3.61	3.70
19.5	3.58	3.72	3.66	3.59	3.47	3.48	3.65	3.57	3.62	3.66	3.63	3.62
20.5	3.59	3.68	3.63	3.47	3.57	3.39	3.64	3.70	3.43	3.70	3.62	3.66
21.5	3.62	3.57	3.57	3.58	3.53	3.56	3.62	3.54	3.54	3.62	3.48	3.65
22.5	3.40	3.61	3.62	3.61	3.55	3.45	3.70	3.50	3.54	3.61	3.58	3.70
23.5	3.45	3.64	3.62	3.65	3.68	3.59	3.59	3.61	3.40	3.37	3.58	3.65
24.5	3.35	3.62	3.51	3.62	3.52	3.58	3.64	3.70	3.48	3.53	3.62	3.65
25.5	3.55	3.66	3.65	3.48	3.53	3.50	3.53	3.65	3.46	3.57	3.64	3.57
26.5	3.38	3.71	3.69	3.61	3.57	3.32	3.48	3.65	3.44	3.61	3.58	3.58
27.5	3.54	3.55	3.62	3.55	3.54	3.43	3.58	3.63	3.37	3.58	3.54	3.49
28.5	3.54	3.61	3.52	3.57	3.56	3.59	3.62	3.66	3.57	3.66	3.62	3.65
29.5	3.61	3.65	3.63	3.58	3.60	3.59	3.68	3.69	3.48	3.54	3.57	3.57
30.5	3.57	3.62	3.64	3.57	3.57	3.57	3.55	3.62	3.57	3.53	3.61	3.54
31.5	3.45	3.61	3.58	3.52	3.45	3.48	3.57	3.29	3.57	3.50	3.27	3.51
32.5	3.24	3.20	3.52	3.25	3.13	3.34	3.30	3.42	3.48	3.26	3.43	3.34
33.5	3.33	3.25	3.57	3.22	3.28	3.20	3.51	3.42	3.54	3.41	3.42	3.42
34.5	3.39	3.29	3.59	3.23	3.47	3.04	3.62	3.49	3.49	3.49	3.40	3.49

Table 3. Velocities C_L of longitudinal ultrasonic wave propagation, measured in two perpendicular directions in core samples taken parallel with the top surface of the slab.

Measuring Place Distance from Sample Top	Sample 07		Sample 08		Sample 09		Sample 10		Sample 11		Sample 12	
	C_{L1}	C_{L2}	C_{L1}	C_{L2}	C_{L1}	C_{L2}	C_{L1}	C_{L2}	C_{L1}	C_{L2}	C_{L1}	C_{L2}
cm							km/s					
0.5	3.31	3.42	3.60	3.40	3.34	3.60	3.60	3.35	3.30	3.28	3.50	3.38
1.5	3.27	3.33	3.50	3.28	3.25	3.30	3.32	3.34	3.29	3.26	3.40	3.31
2.5	3.55	3.47	3.31	3.60	3.36	3.56	3.41	3.51	3.32	3.41	3.45	3.33
3.5	3.56	3.45	3.25	3.46	3.59	3.42	3.62	3.53	3.52	3.54	3.43	3.59
4.5	3.54	3.49	3.54	3.42	3.45	3.55	3.50	3.51	3.40	3.41	3.46	3.56
5.5	3.55	3.46	3.46	3.60	3.40	3.50	3.48	3.44	3.41	3.55	3.42	3.36
6.5	3.57	3.50	3.41	3.45	3.51	3.44	3.41	3.44	3.28	3.38	3.46	3.37
7.5	3.49	3.50	3.57	3.42	3.51	3.53	3.44	3.45	3.39	3.50	3.56	3.43
8.5	3.45	3.43	3.56	3.60	3.47	3.54	3.55	3.41	3.49	3.51	3.51	3.49
9.5	3.55	3.43	3.60	3.52	3.44	3.57	3.48	3.51	3.50	3.44	3.45	3.50
10.5	3.40	3.47	3.49	3.50	3.45	3.50	3.40	3.45	3.41	3.49	3.39	3.39
11.5	3.56	3.43	3.46	3.51	3.54	3.46	3.41	3.55	3.36	3.46	3.44	3.50
12.5	3.55	3.38	3.42	3.50	3.54	3.46	3.42	3.47	3.52	3.55	3.43	3.43
13.5	3.57	3.43	3.38	3.46	3.51	3.56	3.37	3.40	3.39	3.50	3.55	3.51
14.5	3.48	3.47	3.41	3.55	3.57	3.51	3.38	3.40	3.51	3.55	3.46	3.52
15.5	3.59	3.47	3.55	3.51	3.43	3.59	3.42	3.41	3.48	3.58	3.44	3.45
16.5	3.54	3.45	3.56	3.46	3.42	3.46	3.38	3.38	3.41	3.45	3.56	3.40
17.5	3.50	3.40	3.53	3.35	3.57	3.56	3.49	3.44	3.40	3.38	3.41	3.39
18.5	3.47	3.44	3.46	3.47	3.49	3.45	3.40	3.36	3.42	3.39	3.38	3.40
19.5	3.42	3.43	3.41	3.55	3.45	3.43	3.40	3.46	3.42	3.50	3.43	3.54
20.5	3.40	3.48	3.47	3.46	3.44	3.50	3.44	3.43	3.38	3.40	3.38	3.55
21.5	3.45	3.50	3.50	3.39	3.42	3.42	3.44	3.56	3.48	3.44	3.44	3.49
22.5	3.40	3.41	3.41	3.42	3.46	3.42	3.40	3.44	3.50	3.43	3.42	3.43
23.5	3.43	3.43	3.41	3.45	3.54	3.41	3.41	3.49	3.33	3.50	3.37	3.41
24.5	3.50	3.38	3.53	3.50	3.38	3.42	3.49	3.42	3.52	3.38	3.51	3.50
25.5	3.42	3.50	3.51	3.38	3.40	3.57	3.46	3.40	3.46	3.41	3.45	3.53
26.5	3.56	3.38	3.45	3.36	3.44	3.37	3.45	3.25	3.55	3.45	3.49	3.44
27.5	3.54	3.45	3.51	3.47	3.39	3.31	3.30	3.36	3.36	3.43	3.41	3.46
28.5	3.42	3.37	3.55	3.46	3.40	3.38	3.48	3.36	3.37	3.37	3.50	3.55
29.5	3.50	3.43	3.46	3.34	3.43	3.36	3.44	3.35	3.43	3.35	3.53	3.50
30.5	3.56	3.50	3.42	3.43	3.45	3.45	3.44	3.53	3.50	3.37	3.37	3.32
31.5	3.57	3.40	3.55	3.33	3.44	3.43	3.46	3.50	3.40	3.37	3.37	3.45
32.5	3.40	3.34	3.40	3.53	3.40	3.41	3.50	3.56	3.40	3.45	3.45	3.44
33.5	3.50	3.40	3.59	3.51	3.47	3.45	3.42	3.49	3.51	3.40	3.44	3.47

A Unipan Materials Tester Type 543 with point-contact exponential probes [15] and a frequency of 40 kHz (Figure 2) was used to measure the time taken for the longitudinal ultrasonic wave to pass through the tested sample. Prior to the tests, the instrument had been calibrated to determine the time taken for the ultrasonic wave to pass through the probes alone (t_0 = 36.6 µs). The details of the operation of exponential heads with point-to-point contact with the examined surface are described in detail in the paper [15].

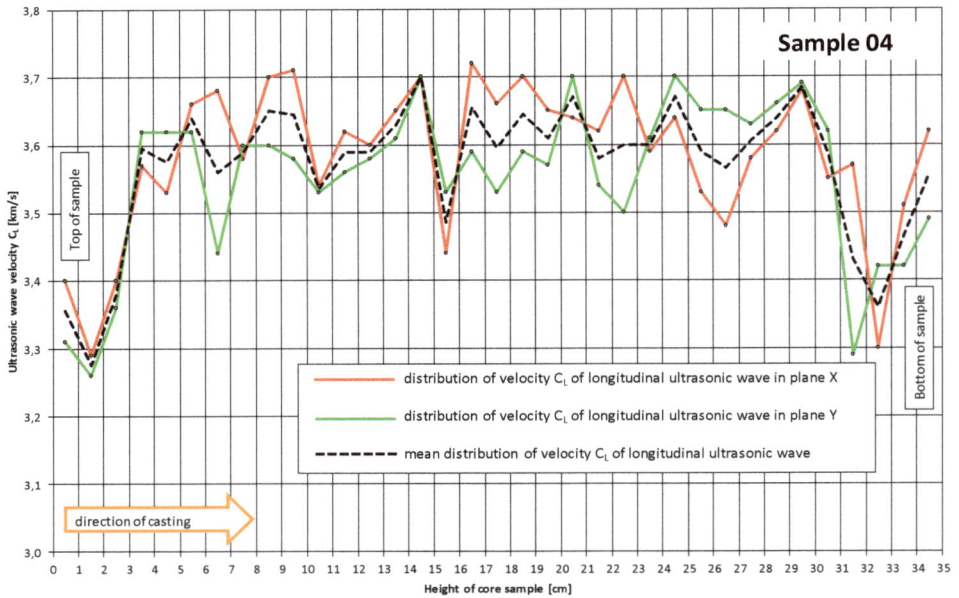

Figure 3. Exemplary distribution of velocity C_L of longitudinal ultrasonic wave along the height of sample 04 taken perpendicularly to the top surface of the slab.

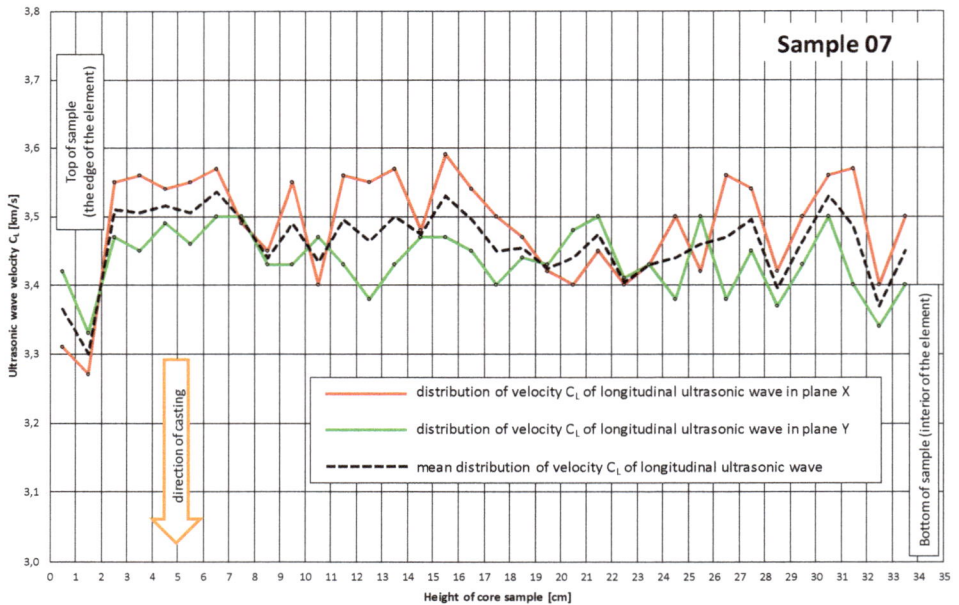

Figure 4. Exemplary distribution of velocity C_L of longitudinal ultrasonic wave along height of sample 07 taken parallel with the top surface of the slab.

Tables 2 and 3 and Figures 3 and 4 showed that the distributions of velocity C_L of the longitudinal ultrasonic wave along the height of the core samples with d $\cong$ 100 mm were quite constant (except for the top layers). The mean velocity C_L of ultrasonic wave propagation in the samples (01–06) taken perpendicularly to the top surface of the slab was $C_{L1} = 3.54$ km/s, a standard deviation $s_{CL1} = 0.13$ km/s and a variation coefficient $v_{CL1} = 3.67\%$. In the case of the samples (07–12) taken parallel with the top of the slab, the following were obtained: $C_{L2} = 3.45$ km/s, $s_{CL2} = 0.072$ km/s and $v_{CL2} = 2.09\%$. On the basis of the obtained velocities C_L of the longitudinal ultrasonic wave in the range: $C_L = 3.5$–4.5 km/s, it can be assessed [16] whether the quality of the concrete as good, whereas the velocities in the range of 3.0–3.5 km/s indicated dubious quality.

The observed slight fluctuations of velocity C_L of the longitudinal ultrasonic wave in the inner layers of the concrete are due to local concrete defects (e.g., air voids) or local strengthening with larger aggregate grains. Lower velocities C_L of the longitudinal ultrasonic wave were registered in the samples (01–06) taken perpendicularly to the top surface of the slab (Figures 3 and 5). In the samples (07–12) taken parallel with the top of the slab, a decline in velocity CL of the longitudinal ultrasonic wave was observed only at the edge constituting the side edge of the slab (Figures 4 and 6).

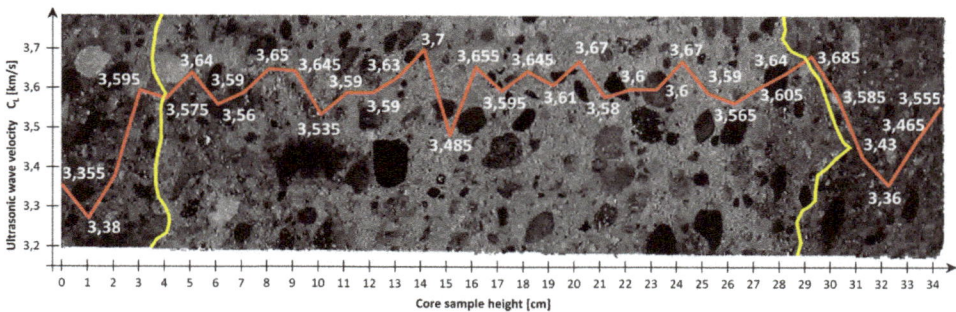

Figure 5. The structure of concrete of the sample taken perpendicularly to the top surface of the slab (slab top surface on left). The image also shows distribution of the mean velocity C_L of longitudinal ultrasonic wave (red line) and disturbance zone at the top and bottom surface of the slab (yellow line).

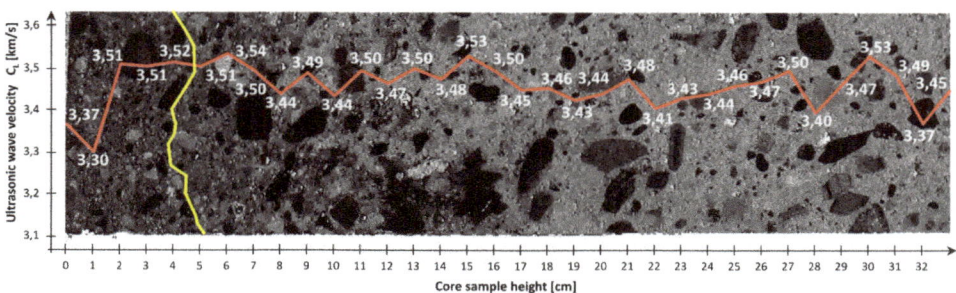

Figure 6. The structure of concrete of the sample taken parallel with the top surface of the slab (slab top surface on left). The image also shows distribution of the mean velocity C_L of longitudinal ultrasonic wave (red line) and zone of disturbance at the lateral surface of the slab (yellow line).

Initially, it was though that the falls in velocity C_L of the longitudinal ultrasonic wave were due to the sample end effect. Ultimately, it was decided that this phenomenon in samples 01–06 was caused from the top by bleeding [17] and concrete sedimentation, and from the bottom by the improper vibration of the concrete by means of the immersion vibrator (the vibrator was not fully immersed in the freshly placed concrete). In samples 07–12, the fall in velocity C_L of the longitudinal ultrasonic wave can be caused by the wall effect [9,10]. Figures 5 and 6 show the structure of the concrete along the height of the samples taken perpendicularly to and parallel with the top surface of the slab. The image of the structure of the samples in Figures 5 and 6 was prepared in GIMP 2.10.4 using the filter: LCHH(ab) component with a contrast of 50%. In Figure 5 the altered structure of the concrete is visible in the sample's upper part (an approximately 30–40 mm thick layer) and lower part (an approximately 30–50 mm thick layer). In these places, reduced velocities of the longitudinal wave velocity were observed. In Figure 6, the altered structure of the concrete can be seen in an approximately 20–80 mm thick layer located at the side wall of the slab. Also in this layer, falls in the velocity of the longitudinal ultrasonic wave were recorded.

2.2. Ultrasonic Tests of Concrete

For further investigations the core samples (Table 1) were cut into smaller samples (Figure 7) with height h = 100 mm (h/d = 1). In the terminal samples (e.g., 01/1—the top of the sample, 01/3—the bottom of the sample) the actual end faces were left unchanged (or were slightly trimmed to make them level). The middle sample (e.g., 01/2) was cut to the required size of 100 mm. Then, the end faces were prepared by grinding for compressive strength tests (Figure 8).

Figure 7. Core samples with height/diameter ratio h/d = 1, obtained from samples 01–06.

Figure 8. One of the core samples during grinding of its end surface.

The compressive strength tests were carried out in conformance with standard [13] in the ZD100 strength testing machine (Figure 9a) satisfying the requirements of standard [18]. All the samples showed the same type of failure (Figure 9b). The parameters of the samples and the test results are presented in Table 2.

(a)   (b)

Figure 9. The sample in strength testing machine: (**a**) during loading, (**b**) after failure.

The concrete strength values (Table 4) yielded by the tests carried out on the 100 mm high samples were used to evaluate the class of the concrete and the variation of strength along the core sample height (h = 350 mm) and were correlated with the results obtained using the ultrasonic method. According to standard [19], the result of concrete compressive strength tests carried out on cylindrical specimens with diameter d = 100 mm and height h = 100 mm, cut out of a structure directly corresponded to the strength of concrete determined on 150 × 150 × 150 mm standard cubes ($f_{ck,is} = f_{ck,is,cube}$).

Table 4. The results of concrete compressive strength tests.

Samp. No.	d_m	h_m	m	A_c	V_c	ρ	F_{is}	f_{is}	Samp. No.	d_m	h_m	m	A_c	V_c	ρ	F_{is}	f_{is}
	mm		g	cm²	cm³	g/cm³	kN	MPa		mm		g	cm²	cm³	g/cm³	kN	MPa
01/1	98.7	99.9	1725	76.51	764	2.26	236	30.85	07/1	98.6	100.0	1760	76.36	764	2.30	226	29.60
01/2	98.7	99.8	1743	76.51	764	2.28	240	31.37	07/2	98.6	99.9	1751	76.36	763	2.30	231	30.25
01/3	98.7	99.8	1765	76.51	764	2.31	242	31.63	07/3	98.6	99.8	1746	76.36	762	2.29	232	30.38
02/1	98.5	100.0	1724	76.20	762	2.26	236	30.97	08/1	98.7	99.9	1755	76.51	764	2.30	230	30.06
02/2	98.5	99.8	1761	76.20	760	2.32	238	31.23	08/2	98.7	99.9	1761	76.51	764	2.30	235	30.71
02/3	98.5	99.9	1762	76.20	761	2.31	243	31.89	08/3	98.7	99.9	1757	76.51	764	2.30	237	30.98
03/1	98.5	99.8	1711	76.20	760	2.25	233	30.58	09/1	98.5	99.7	1744	76.20	760	2.30	234	30.71
03/2	98.5	99.9	1741	76.20	761	2.29	241	31.63	09/2	98.5	99.9	1740	76.20	761	2.29	238	31.23
03/3	98.6	99.6	1768	76.36	761	2.32	248	32.48	09/3	98.5	100.0	1760	76.20	762	2.31	243	31.89
04/1	98.5	99.9	1731	76.20	761	2.27	237	31.10	10/1	98.5	99.6	1754	76.20	759	2.31	239	31.36
04/2	98.5	99.9	1755	76.20	761	2.31	240	31.50	10/2	98.5	99.8	1742	76.20	760	2.29	245	32.15
04/3	98.5	99.9	1779	76.20	760	2.34	240	31.50	10/3	98.5	99.8	1748	76.20	760	2.30	248	32.55
05/1	98.5	99.9	1717	76.20	761	2.26	234	30.71	11/1	98.6	99.7	1741	76.36	761	2.29	237	31.04
05/2	98.5	99.9	1745	76.20	761	2.29	240	31.50	11/2	98.6	99.7	1747	76.36	761	2.29	232	30.38
05/3	98.5	99.9	1769	76.20	761	2.32	242	31.76	11/3	98.6	99.9	1750	76.36	763	2.29	238	31.17
06/1	98.5	99.9	1716	76.20	761	2.25	240	31.50	12/1	98.5	99.9	1766	76.20	761	2.32	240	31.50
06/2	98.5	99.9	1741	76.20	761	2.29	242	31.76	12/2	98.5	99.8	1754	76.20	760	2.31	243	31.89
06/3	98.5	99.9	1762	76.20	761	2.31	245	32.15	12/3	98.5	100.1	1750	76.20	763	2.29	245	32.15

Standard [19] states that due to drilling, which undoubtedly can slightly damage the core's material, the strengths of core samples determined in-situ are usually lower than the strengths of the standard samples. For this reason, it is allowed to use a correction factor of 0.85, understood as a ratio of the in-situ characteristic compressive strength to the characteristic compressive strength determined on the standard samples. As a result, the concrete compressive strength values coming directly from strength tests are increased.

The mean compressive strength of the concrete, determined on the 18 samples taken perpendicularly to the top surface of the slab, amounted to $f_{m(18),is}$ = 31.45 MPa (the minimum value $f_{is,lowest}$ = 30.58 MPa). The mean standard deviation amounted to s = 0.49 MPa. The coefficient k_1 = 1.48 was assumed. The characteristic compressive strength of the concrete in the structure ($f_{ck,is}$) was determined on the basis of standard [13], from the condition: $f_{ck,is}$ = min($f_{m(18),is}$ − k_1 × s; $f_{is,lowest}$ + 4) = (30.72 MPa, 34.58 MPa). Thus the characteristic cube compressive strength of the concrete determined on samples taken perpendicularly to the top surface of the slab amounted to $f_{ck,is}$ = $f_{ck,is,cube}$ = 30.72 MPa.

The mean compressive strength of the concrete determined on the 18 samples taken parallel with the top surface of the slab amounted to $f_{m(18),is}$ = 31.11 MPa (the minimum value $f_{is,lowest}$ = 29.60 MPa). The mean standard deviation amounted to s = 0.81 MPa. The characteristic cube compressive strength of the concrete, determined on the samples taken parallel with the top surface of the slab, amounted to $f_{ck,is}$ = (29.91 MPa, 33.60 MPa) = 29.91 MPa. The cube strength determined on the samples taken parallel to the top surface of the slab was 3% lower than the strength determined on the samples taken perpendicularly to the top surface of the slab. This confirms the observation that the strength of core samples drilled out horizontally is lower (on average by 8% [12,13]) than that of core samples drilled out vertically.

On the basis of the obtained $f_{ck,is,cube}$ values, the actual strength class of the concrete in the structure is estimated to be $f_{ck,is,cube}$ = 30.72 MPa and 29.91 MPa, respectively. When the correction factor of 0.85 is applied, this gives the concrete strength class respectively $f_{ck,is,cube}$ = 36.1 MPa and 35.2 MPa, which corresponds to at least concrete class C25/30.

2.3. Scaling of Correlation Curve

On the basis of the measurements, the mean longitudinal ultrasonic wave passage velocities C_L were correlated with the mean compressive concrete strengths f_{is}. When calculating the mean longitudinal ultrasonic wave passage velocity C_L, the values from the areas of disturbances near the ends of the samples were rejected. The correlation curve was scaled according to the procedure described in standard [13] (version 2). In accordance with [16], a hypothetical base regression curve for ordinary concrete, i.e., $f_{CL,b}$ = $2.39C_L^2$ − $7.06C_L$ + 4.2 for C_L = 2.4–5.0 km/s, was adopted. Then, the differences δf between experimental compressive strength f_{is} and the strength obtained from base curve f_{CL}, as well as the mean value $δf_{m(n)}$ of the differences and standard deviation s were determined. The shift of the base correlation curve was calculated from the relation $Δf$ = $δf_{m(n)}$ − $k_1·s$ for coefficient k_1 = 1.48 [13]. Ultimately, the corrected correlation curve f_{CL} = $f_{CL,b}$ + $Δf$ has the form f_{CL} = $2.39C_L^2$ − $7.06C_L$ + 25.09. The obtained curve only slightly differs from the curves determined separately for samples 01 ÷ 06 and 07–12.

The obtained correlation was evaluated using two accuracy characteristics, i.e., the correlation coefficient η > 0.75 and the mean square relative deviation v_k ≤ 12 ≤ %. The correlation coefficient amounted to:

$$\eta = [0.25 \times \sum(f_{CL,I} - f_{CL(36),v})^2]^{1/2} \div [0.25 \times \sum(f_{is} - f_{m(36),is})^2]^{1/2}$$
$$\eta = [0.25 \times 15.34]^{1/2} \div [0.25 \times 16.33]^{1/2} = 0.97 > 0.75 \tag{1}$$

and the mean square relative deviation to:

$$v_k = 100 \times \{[1/(n-1)] \times \sum[(f_{CL,i} - f_{is})/f_{CL,i}]^2\}^{1/2}$$
$$v_k = 100 \times [(1/35)] \times 0.12555]^{1/2} = 5.99\% < 12\%. \tag{2}$$

Thus, it can be said that a good correlation between the mean longitudinal ultrasonic wave passage velocities C_L and the mean concrete compressive strengths f_{is} was obtained.

The class of the concrete in the structure was determined on the basis of the concrete compressive strength values obtained from the correlation curve f_{CL} = 2.39 × C_L^2 − $7.06C_L$ + 25.09 for the mean longitudinal ultrasonic wave passage velocities C_L. The mean compressive strength of the concrete determined using the ultrasonic method amounted to $f_{CL(36),is}$ = 29.80 MPa (minimal $f_{CL,is,lowest}$ = 28.80 MPa) and the mean standard deviation to s = 0.66 MPa. Hence, the characteristic compressive strength of the concrete amounted to $f_{ck,is}$ = $f_{ck,is,cube}$ ≤ (29,80 − 1.48 × 0.66, 28.80 + 4) = (28.82 MPa, 32.80 MPa) = 28.82 MPa. When the correction factor of 0.85 was taken into account, concrete class C25/30 was obtained. The concrete class determined on the basis of the compressive strength values obtained from the correlation curve confirmed the destructively determined class of the concrete.

3. Analysis of Test Results

With the corrected correlation curve: f_{CL} = 2.39·C_L^2 − 7.06·C_L + 25.09, it was possible to trace the variation of the compressive strength of the concrete along the height of the analyzed samples. Figures 10 and 11 show the variations of concrete compressive strength along the height of the core samples respectively, and perpendicular to (Figure 10) and parallel with (Figure 11) the top surface of the slab. Further, the results of the destructive tests and the averaged results of the ultrasonic tests for the particular samples with a height h = 100 mm are included in the diagrams.

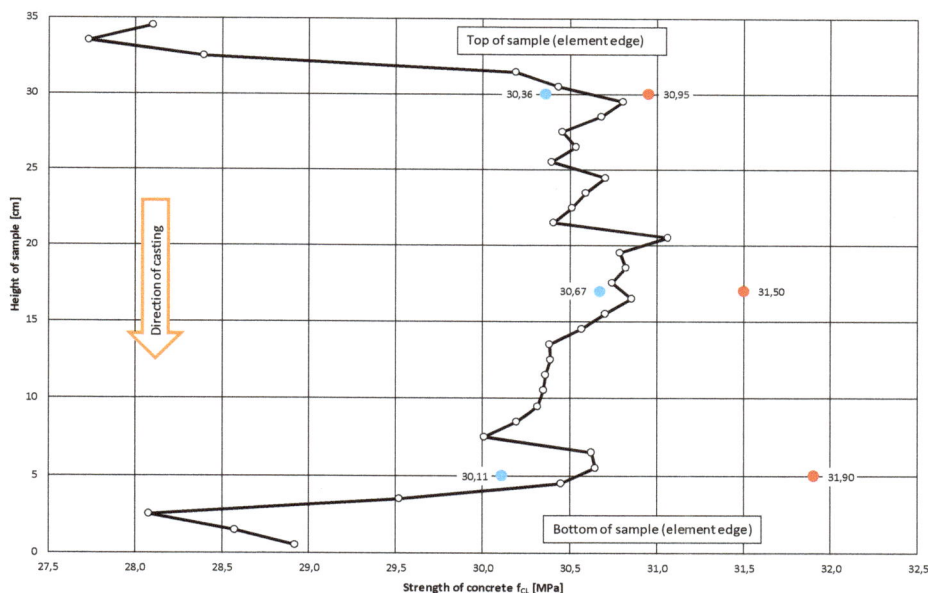

Figure 10. The mean distribution of compressive strength f_{CL} of concrete along the height of the sample taken perpendicularly to top surface of slab. The diagram includes the mean concrete strengths for the top, middle and bottom samples, obtained from ultrasonic tests (marked blue) and destructive tests (marked red).

An analysis of the concrete compressive strength values for the particular samples 01–06 (Figure 10) taken perpendicularly to the top plane of the slab indeed showed a slight increase (by 3%) in this strength in the sample's lower part relative to its upper part. A similar phenomenon (also an increase by approximately 2.6%) was observed for samples 07–12 (Figure 11) taken parallel with the top plane of the slab. However, it should be noted that the compressive strength values were strongly averaged for the samples and included the effect of various factors connected with the destructive test itself.

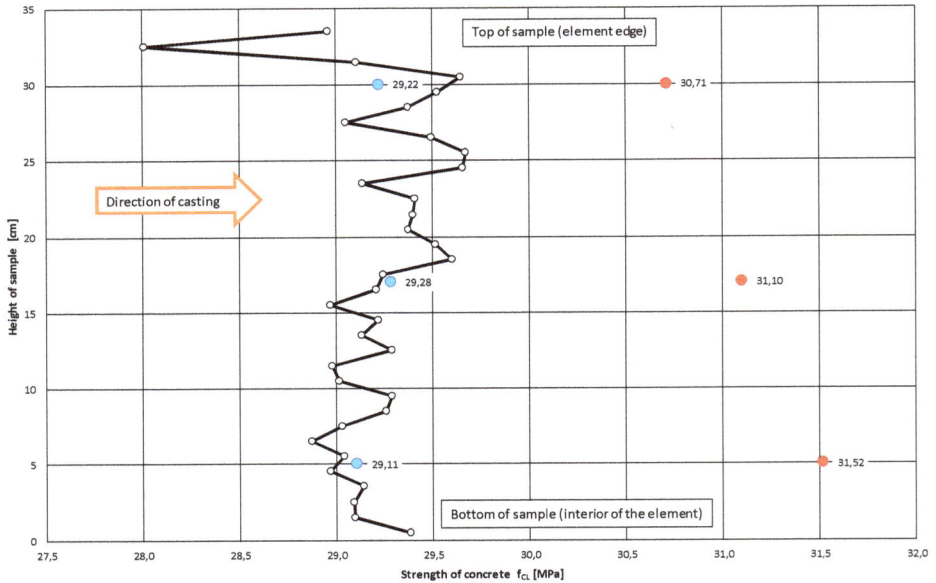

Figure 11. The mean distribution of compressive strength f_{CL} of concrete along the height of the sample taken parallel with the top surface of the slab. The diagram includes the mean concrete strengths for the top, middle and bottom samples obtained from ultrasonic tests (marked blue) and destructive tests (marked red).

The averaged compressive strength results obtained from the ultrasonic measurements showed (Figures 10 and 11), however, that there was no increase in the compressive strength of the concrete along the height of the sample. This applies to the samples taken both perpendicularly to and parallel with the top plane of the slab.

The ultrasonic tests indicate that the variation in the compressive strength of concrete along the height of the sample is minimal and random. It can even be considered as negligible. The obtained results do not corroborate Stawiski's theses [1–3], but confirm the results reported in [7–10].

The decreases in the compressive strength of the concrete occurring at the ends of the samples taken perpendicularly to the top plane of the slab (samples 01–06) and at the edge constituting the side edge of the slab for the samples taken parallel with the top plane of the slab (samples 07–12) were found to be interesting.

4. Conclusions

The following conclusions emerge from the investigations of the compressive strength of concrete carried out on core samples taken perpendicularly to and parallel with the top surface of the approximately 35 cm thick element, using different testing methods (the ultrasonic method and the destructive method):

The concrete compressive strength destructively determined along the height of the placed layer of concrete changed slightly (by 3%—samples 01–06) with a depth below the top surface of the element. The averaged concrete strength determined on the basis of the ultrasonic tests of the same samples did not vary across the thickness of the analyzed slab.

1. The obtained compressive strength increments across the thickness of the placed layer of concrete do not corroborate Stawiski's theses [1–3], but confirm the results reported in, [7–9]. Therefore, there can be agreement with Neville's statement [10] that the slight increase in concrete compressive

strength with depth below the top surface is a natural thing and need not to be taken into account in the evaluation of the strength of concrete in the structure.

2.   The concrete compressive strength determined on core samples only slightly depends on the depth of where the sample came from (provided the ingredients of the concrete do not segregate as it is being placed and compacted).

3.   The use of the ultrasonic method for testing concrete with point-contact exponential probes showed the variation in concrete strength along the height of the core sample could be quite accurately evaluated and areas of lower quality concrete could be indicated. This was mainly from the thick layer (approximately 30–40 mm) extending from the top edge and the thick layer (approximately 30–50 mm) extending from the bottom edge of the samples 01–06 taken perpendicularly to the upper plane of the element. Further, from the thick layer (approximately 20–80 mm) extending from the edge constituting the side plane of the slab for the samples taken parallel with the top plane of the element (samples 07–12). The decline in the strength of the concrete in the upper part of samples 01–06 is caused by the bleeding of water from the concrete mixture (the bleeding phenomenon [17]) and the sedimentation of the latter. While in the lower part of the samples, it is due to the improper vibration of the placed layer of concrete mixture. The decrease in concrete strength at the side edge of the slab in the case of samples 07–12 can be due to the wall effect [9,10].

4.   From the point of view of the assessment of the concrete structure, supplementary tests on the slab in the future should be carried out using ultrasonic tomography [20,21].

5.   The ultrasonic method of testing concrete by means of point-contact exponential probes enables the accurate assessment of the quality of concrete (the segregation of concrete components, porosity, density, strength, etc.) along the height of a core sample drilled out perpendicularly to the placed layer.

Funding: This research received no external funding.

Conflicts of Interest: The authors declare no conflicts of interest.

References

1.   Stawiski, B. *Ultrasonic Testing of Concrete and Mortar Using Point Probes*; Wroclaw University of Technology: Wroclaw, Poland, 2009; p. 154. (In Polish)
2.   Stawiski, B. The heterogeneity of mechanical properties of concrete in formed constructions horizontally. *Arch. Civ. Mech. Eng.* **2012**, *12*, 90–94. [CrossRef]
3.   Stawiski, B. Concrete strength gradients in industrial floors. *Mater. Bud.* **2017**, *543*, 22–24. (In Polish)
4.   Hoła, J.; Sadowski, L.; Hoła, A.M. The effect of failure to comply with technological and technical requirements on the condition of newly built cement mortar floors. *Proc. Inst. Mech. Eng. Part L J. Mater. Des. Appl.* **2019**, *233*, 268–275. [CrossRef]
5.   Petersons, N. Should standard cube test specimens be replaced by test specimens taken from structures? *Mater. Struct.* **1968**, *1*, 425–435. [CrossRef]
6.   Dąbrowski, K.; Stachurski, W.; Zieliński, J.L. *Concrete Constructions*; Wydawnictwo Arkady: Warsaw, Poland, 1982. (In Polish)
7.   Yuan, R.L.; Ragab, M.; Hill, R.E.; Cook, J.E. Evaluation of core strength in high-strength concrete. *Concr. Int.* **1991**, *13*, 30–34.
8.   Suprenant, B.A. *Core Strength Variation of in-Place Concrete*; The Aberdeen Group: Boston, MA, USA, 1995.
9.   Neville, A. Core tests: Easy to perform, not easy to interpret. *Concr. Int.* **2001**, *11*, 59–68.
10. Watanabe, S.; Hishikawa, K.; Kamae, K.; Namiki, S. Study on estimation of compressive strength of concrete in structure using ultrasonic method. *J. Struct. Constr. Eng.* **2016**, *81*, 191–198. [CrossRef]
11. Neville, A.M. *Properties of Concrete*; Polski Cement Sp. z o.o.: Cracow, Poland, 2000. (In Polish)
12. PN-EN 12390-2:2011. *Testing Hardened Concrete. Part 2. Making and Curing Specimens for Strength Tests*; Polish Committee for Standardization: Warsaw, Poland, 2011.

13. PN-EN 13791:2008. *Assessment of In-Situ Compressive Strength in Structures and Precast Concrete Components*; Polish Committee for Standardization: Warsaw, Poland, 2008.
14. PN-EN 12390-3:2011. *Testing Hardened Concrete. Part 3: Compressive Strength of Test Specimens*; Polish Committee for Standardization: Warsaw, Poland, 2008.
15. Gudra, T.; Stawiski, B. Non-destructive strength characterization of concrete using surface waves. *NDT E Int.* **2000**, *33*, 1–6. [CrossRef]
16. Drobiec, L.; Jasiński, R.; Piekarczyk, A. *Diagnosis of Reinforced Concrete Structures. Methodology, Field Tests, Laboratory Tests of Concrete and Steel*; Wydawnictwo Naukowe PWN: Warsaw, Poland, 2010. (In Polish)
17. Soshiroda, T. Effects of bleeding and segregation on the internal structure of hardened concrete. In *Properties of Fresh Concrete*; Taylor & Francis: Mild Park, UK, 1990; pp. 225–232.
18. PN-EN 12390-4:2001. *Testing Hardened Concrete. Part 4: Compressive Strength. Specification for Testing Machines*; Polish Committee for Standardization: Warsaw, Poland, 2001.
19. PN-EN 12504-1:2011. *Testing Concrete in Structures. Part 1: Cored Specimens. Taking, Examining and Testing in Compression*; Polish Committee for Standardization: Warsaw, Poland, 2011.
20. Schabowicz, K. Ultrasonic tomography—The latest nondestructive technique for testing concrete members—Description, test methodology, application example. *Arch. Civ. Mech. Eng.* **2014**, *14*, 295–303. [CrossRef]
21. Schabowicz, K.; Suvorov, V.A. Nondestructive testing of a bottom surface and construction of its profile by ultrasonic tomography. *Russ. J. Nondestruct. Test.* **2014**, *50*, 109–119. [CrossRef]

materials

MDPI

Article

Viscoelastic Parameters of Asphalt Mixtures Identified in Static and Dynamic Tests

Piotr Mackiewicz * and Antoni Szydło

Faculty of Civil Engineering, Wrocław University of Science and Technology, 50-370 Wrocław, Poland
* Correspondence: piotr.mackiewicz@pwr.edu.pl; Tel.: +48-71-320-45-57

Received: 10 June 2019; Accepted: 26 June 2019; Published: 28 June 2019

check for
updates

Abstract: We present two methods used in the identification of viscoelastic parameters of asphalt mixtures used in pavements. The static creep test and the dynamic test, with a frequency of 10 Hz, were carried out based on the four-point bending beam (4BP). In the method identifying viscoelastic parameters for the Brugers' model, we included the course of a creeping curve (for static creep) and fatigue hysteresis (for dynamic test). It was shown that these parameters depend significantly on the load time, method used, and temperature and asphalt content. A similar variation of parameters depending on temperature was found for the two tests, but different absolute values were obtained. Additionally, the share of viscous deformations in relation to total deformations is presented, on the basis of back calculations and finite element methods. We obtained a significant contribution of viscous deformations (about 93% for the static test and 25% for the dynamic test) for the temperature 25 °C. The received rheological parameters from both methods appeared to be sensitive to a change in asphalt content, which means that these methods can be used to design an optimal asphalt mixture composition—e.g., due to the permanent deformation of pavement. We also found that the parameters should be determined using the creep curve for the static analyses with persistent load, whereas in the case of the dynamic studies, the hysteresis is more appropriate. The 4BP static creep and dynamic tests are sufficient methods for determining the rheological parameters for materials designed for flexible pavements. In the 4BP dynamic test, we determined relationships between damping and viscosity coefficients, showing material variability depending on the test temperature.

Keywords: viscoelastic parameters; creep test; fatigue tests; asphalt mixtures; Burgers model; four point bending beam

1. Introduction

The selection of correct material parameters is very important, both in engineering practice and scientific study. The determination of reliable material properties is also essential in further structural analyses. The appropriate material parameters and the model enable the use of efficient numerical methods, and determine the state of stresses and deformations in the construction model. It is especially important for asphalt mixtures used as the main material in vulnerable road pavements. Such mixtures are thermo-rheological, changing their properties under thermal conditions and load time. In various conditions, both under static and dynamic loads, they reveal their rheological characteristics. These properties are much more important in the description of the material in higher temperatures than in lower ones, in which linear–elastic models are sufficient to model the material parameters. The asphalt layers in the road pavements show both elastic and viscous features. The elastic properties dominate at the lower temperatures, and are responsible for irreversible deformations of the asphalt pavement, whereas the viscous features are typical of the higher temperatures. Therefore, proper identification of the rheological parameters of asphalt mixtures based on the results of laboratory tests is not easy. The typical static tests in which these parameters are defined include the static creep, testing under the constant load when the cylindrical specimens are compressed and the beams are bent [1]. Dynamic tests are analogous, testing with compressed cylindrical specimens [2] and bend fatigue beams.

Different rheological parameters can be obtained in various mounting schemes and load conditions, characterized by duration and frequency. Therefore, the choice of proper research method is important. This method is used to determine these parameters and the models describing the behaviour of the structure. In the case of road pavements, both static (parking lots, crossroads, etc.) and repetitive loads with short-term impact are analysed. As mentioned earlier, the asphalt mixtures become viscous over time in high temperatures for a long-term static load, whereas the accumulation of permanent deformations resulting in permanent deformation (i.e., ruts) occurs under dynamic loading. However, changing the viscoelastic dissipative energy is also important in fatigue tests. This change significantly affects the fatigue destruction of the material. This publication analyses the behaviour of the asphalt mixture under the static and dynamic loading for a four-point bending beam (4BP).

2. Identification of Rheological Parameters

Many rheological models are used in the common road practice. As already mentioned, the asphalt mixtures expose their rheological properties at high temperatures. The viscoelastic models are used to describe these properties. The viscoelasticity theory is increasingly being used in the analysis of asphalt pavements, due to its good description of flow and deformation of road materials.

According to Reiner and Ward [3], the first papers about rheology come from the thirties of the previous century. However, this discipline has intensively developed since the 1950s [4], and deals with materials and constructions of buildings as well as road pavements. In the fundamental work edited by Reiner and Ward [3], there is a chapter devoted to the rheology of materials and asphalt pavements written by Van der Poela [5]. Regarding asphalt mixtures, one of the first works by Monismith et al. [6] deserves special attention. The authors found that asphalt pavement mixtures also exhibit linear viscoelastic properties at very low deformations. By studying the creep of the asphalt pavement mixtures, Vakili [7] draws the same conclusions. Goodrich [8] studied asphalt mixtures with mineral fillers, as well as the large aggregate under oscillations with small amplitudes, and found again that these materials show linear viscoelastic features at very small deformations.

In theoretical considerations, Kisiel and Lysik [9], Nowacki [10], and Jakowluk [11] contributed significantly to development the rheology in construction. The use of rheological models in the description of asphalt mixtures can be found, among others, in the works [12–15]. The identified rheological parameters were also studied under different static [16–19] and dynamic load conditions [6,13,20]. However, no comprehensive comparisons have been made to the four-point study of static and dynamic conditions, although the 4BP is commonly used. There are also no comparisons to other various laboratory studies.

Currently, there are many analytic methods [21–23] and numerical models, including micromechanical models [24] and anisotropic models [25,26], in which the material parameters of asphalt mixtures are used in the assessment of the behaviour of flexible pavement.

Currently, due to the high availability of software for numerical calculations, no attention is paid to the selection of the appropriate research method, the application of the appropriate model, or the use of valid parameters in the models of surfaces. Moreover, the entire creep curve is not included with the load curve in the determination of parameters. Both simple and complex rheological models were analysed. For example, the complex constitutive models, with and without damage, can be found in [27–31]. Other studies have addressed advanced pavement structural models with and without dynamic effects [32–34].

It has been found that the Burgers model, among many other viscoelastic models, reliably describes asphalt concrete behaviour [12,17,35,36]. The model diagram along with its parameters is shown in Figure 1.

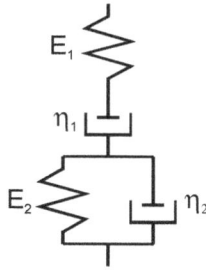

Figure 1. Burgers rheological model. E_1: immediate elastic modulus in the Burgers model (Pa); E_2: delayed elastic modulus in the Burgers model (Pa); η_1: viscosity coefficient in the Burgers model (Pa·s), η_2: viscosity coefficient of elastic delay in the Burgers model (Pa·s).

The study of static creeping was performed under the 4BP bending conditions. The creep curve in the Burgers model has its graphic interpretation, shown in Figure 2. Parameters can be determined by immediate deformations, maximum deformations (elastic moduli), and the rate of deformations (viscosity coefficients). However, such interpretation is not very accurate, because large errors can occur when the immediate deformations are registered during elastic recurrence. Therefore, we proposed to determine these parameters using numerical methods, taking into account the overall creep curve at loading and the curve at unloading.

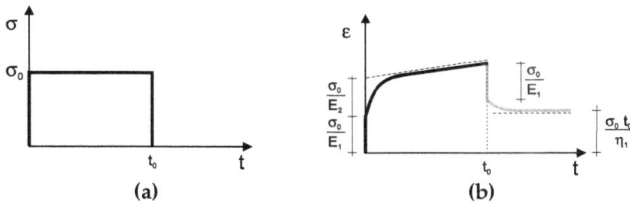

Figure 2. The creep curve in the Burgers model. The graphical finding of the parameters is presented. (a)—stress vs. time curve, (b)—strain vs. time curve.

A conjugated gradient method was used to approximate the laboratory creep curve using the theoretical curve. This is an effective method for solving optimization problems. The minimum of function was determined from each point in a given search direction. The rheological parameters of the Burgers model were the sought variables: E_1, E_2, η_1, η_2 (see Figure 1). The target function is described by Equation (1):

$$\Delta p = \sqrt{\frac{\sum\limits_{i=1}^{i=l} (\varepsilon_{ti} - \varepsilon_{li})^2}{l}} \cdot 100\% \tag{1}$$

where ε_{li} is the deformation measured on the sample, ε_{ti} is the theoretical deformation calculated for the model, and l is the number of measured points.

The theoretical deformations were determined from the equations of the Burgers model, starting from the differential constitutive relationship between stress σ and deformation ε:

$$\sigma + a\dot{\sigma} + b\ddot{\sigma} = c\dot{\varepsilon} + d\ddot{\varepsilon} \tag{2}$$

$$\sigma + \left(\frac{\eta_1}{E_1} + \frac{\eta_1}{E_2} + \frac{\eta_2}{E_2}\right) \cdot \dot{\sigma} + \frac{\eta_1 \eta_2}{E_1 E_2}\ddot{\sigma} = \eta_1 \dot{\varepsilon} + \frac{\eta_1 \eta_2}{E_2}\ddot{\varepsilon} \tag{3}$$

After the solution of these equations, the relationship between the deformation $\varepsilon(t)$ and the time t was obtained Equations (4) and (5):

$$\text{for the load } t < t_0 \quad \varepsilon(t) = \sigma_0 \left[\frac{1}{E_1} + \frac{t}{\eta_1} + \frac{1}{E_2}\left(1 - e^{\frac{-tE_2}{\eta_2}}\right)\right], \tag{4}$$

$$\text{or the unload } t > t_0 \quad \varepsilon(t) = \sigma_0 \left[\frac{t_0}{\eta_1} - \frac{1}{E_2} e^{\frac{-tE_2}{\eta_2}}\left(1 - e^{\frac{t_0 E_2}{\eta_2}}\right)\right] \tag{5}$$

The identification of rheological parameters can effectively contribute to the optimization of mixture composition also under fatigue conditions, when there exists energy dissipation due to microcracks. The procedure for determining the rheological parameters under repetitive stress conditions was performed. In this test, the parameters were determined at the 10 Hz load frequency. This frequency was applied according to the European Standard EN 12697-24:2012 [37], in order to evaluate the fatigue characteristics of asphalt mixtures. The identification of the parameters was performed by the selection of parameters in the Burgers model for hysteresis, describing the relationship between stress σ and deformation ε (Figure 3), and using the conjugate gradient method to minimize the function in Equation (1).

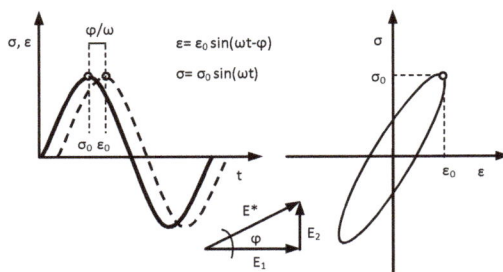

Figure 3. Fatigue hysteresis used for the identification of parameters. ϕ: phase angle ($°$); ω: angular frequency $= 2\pi f$ (1/s); t: time (s); σ_0: the amplitude of stress (MPa); ε_0: the amplitude of deformation (-); E^*: composite modulus (MPa); E_1: the real element of the composite modulus (MPa); E_2: the imaginary element of composite modulus (MPa).

The controlled amplitude of displacement and the time of its delay in relation to the acting force were recorded directly on the basis of the fatigue strength test, using the variable force. Based on the basic dependencies for the 4BP beam (Equations (6)–(8)), it is possible to determine the required stress value σ, deformation ε, and phase angle ϕ in any cycle and at any time of the load, as follows:

$$\sigma = \frac{3Pa}{bh^2} \tag{6}$$

$$\varepsilon = \frac{12\Delta h}{3L^2 - 4a^2} \tag{7}$$

$$\varphi = 360 fs \tag{8}$$

where P is the force (N); b and h are the beam width and height, respectively (m); a is the distance between the support and the force (m), and $a = L/2$; Δ is the displacement (m); L is the spacing of the supports (m); f is the frequency (Hz), and $f = \omega/2\pi$; and s is the delay time between the force P and the displacement Δ (s).

According to Figure 3, it is possible to determine the complex stiffness modulus $E*$ and phase angle φ between deformation and stress:

$$E* = \frac{\sigma_0 \sin(\omega t)}{\varepsilon_0 \sin(\omega t - \varphi)} \tag{9}$$

$$tg\varphi = \frac{E_2}{E_1} \tag{10}$$

Additional conditions for the agreement between the phase angle and the composite modulus found in the test and model were introduced in the search criteria for the most accurate matching of the laboratory results, with the hysteresis determined by the Burgers model. The dependence of the phase angle and the complex modulus on the parameters in the Burgers model are described by Equations (11) and (12):

$$E* = \omega \left[\frac{c^2 + (d\omega)^2}{(b\omega^2 - 1)^2 + (a\omega)^2} \right]^{1/2} \tag{11}$$

$$tg\varphi = \frac{ad\omega^2 - (b\omega^2 - 1)c}{(b\omega^2 - 1)d\omega + ac\omega} \tag{12}$$

Using Equation (2), for the cyclic symmetrical sinusoidal deformation $\varepsilon = \varepsilon_0 \sin(\omega t - \phi)$, we obtain the relationship

$$\sigma + a\dot{\sigma} + b\ddot{\sigma} = -\varepsilon_0 \omega [d\omega \sin(\omega t) - c \cos(\omega t)] \tag{13}$$

The variables a, b, c, and d present in the constitutive Equation (13) are described by Equations (14)–(17):

$$a = \frac{\eta_1}{E_1} + \frac{\eta_1}{E_2} + \frac{\eta_2}{E_2} \tag{14}$$

$$b = \frac{\eta_1 \eta_2}{E_1 E_2} \tag{15}$$

$$c = \eta_1 \tag{16}$$

$$d = \frac{\eta_1 \eta_2}{E_2} \tag{17}$$

3. Materials and Methods

The static creeping test with 4BP bending was performed on the NAT (Nottingham Asphalt Tester, University of Nottingham, Nottingham, UK) device, which enables the efficient testing of many asphalt mixtures under various mounting and loading patterns. This device is characterized by the good reproducibility of results. The technical conditions for the study were adopted according to manual (Cooper Research Technology [38]). The following conditions were applied: a constant load with 0.30 MPa (15% of the bending strength at 25 °C), a load time of 1800 s, and an unload time of 510 s. In order to determine rheological parameters, we applied four temperatures: −5 °C, 0 °C, 10 °C, and 25 °C. The dimensions of the samples were as follows: the width was 60 mm, the height was 50 mm, and the length was 384 mm. In Figure 4, a schematic diagram of the static 4BP creep test is shown.

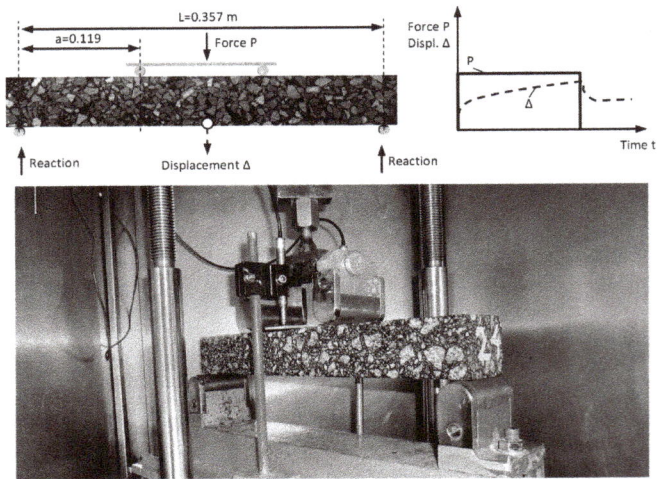

Figure 4. The scheme of the static four-point bending beam (4BP) creep test.

The 4BP dynamic test consists of the cyclic bending of the beam supported in four points, as shown in Figure 5 (accordance with EN 12697-24:2012 [37]). Due to common research practice, the study was conducted under the sinusoidal kinematic constraints, with the controlled deformation. The amplitude of deformation was 100×10^{-6}. This method allows us to compare the received results to the known fatigue criteria in the design practice [35,39]. The dimensions of the bending beams and the temperature conditions were assumed as for the static testing. The basic parameter that was determined during the test was the fatigue hysteresis, which depends on the number of load cycles. The tests with the fixed peak-to-peak strain allowed us to record the change in stresses in relation to the load cycles.

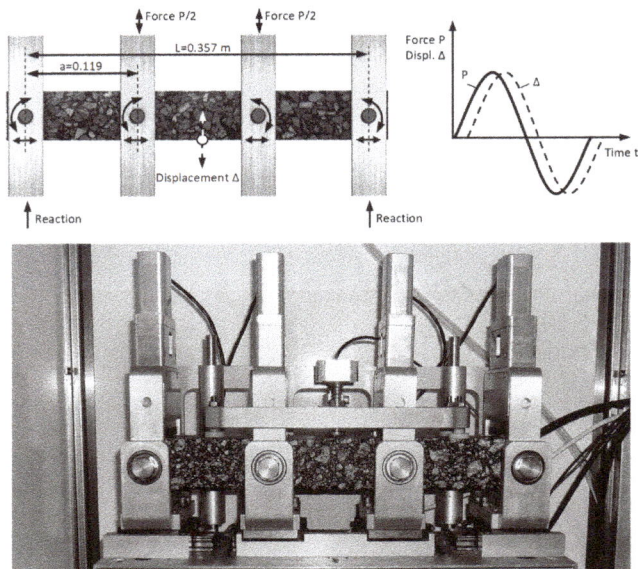

Figure 5. The scheme of the dynamic 4BP test.

The research was carried out with Cooper Research Technology Ltd. Beam-Flex, on typical asphalt mixture commonly used in building road pavements, which was laid on the binding surfaces AC16W with asphalt 35/50. Mixtures with different asphalt content were analyzed—i.e., 4.0%, 4.5%, and 5.3%. The formulas of the mixtures were previously designed in accordance with the current technical requirements.

4. Results of the Parameters' Identification

Based on the presented procedure for the identification of the viscoelastic parameters and the tests performed under various temperature conditions and asphalt content, we derived the parameters of the Burgers model for the creep study at static (Table 1) and dynamic (Table 2) loading.

Table 1. Rheological parameters of the Burger's model obtained for the creep 4BP study.

Temperature	−5 °C	0 °C	10 °C	25 °C
	asphalt content: 4.0%			
E_1 (MPa)	5470	3619	1498	607
η_1 (MPa·s)	1,497,854	769,973	308,374	121,832
E_2 (MPa)	4311	3250	2235	1213
η_2 (MPa·s)	1,881,548	821,849	34,011	28,569
	asphalt content: 4.5%			
E_1 (MPa)	5733	3902	1762	742
η_1 (MPa·s)	1,570,013	830,165	362,676	148,891
E_2 (MPa)	4519	3504	2628	1483
η_2 (MPa·s)	1,972,192	886,096	40,000	34,914
	asphalt content: 5.3%			
E_1 (MPa)	5135	3474	1601	485
η_1 (MPa·s)	1,406,134	739,012	329,545	97,405
E_2 (MPa)	4047	3119	2388	970
η_2 (MPa·s)	1,766,333	788,801	36,346	22,841

Table 2. Rheological parameters of the Burger's model obtained for the dynamic 4BP study.

Temperature	−5 °C	0 °C	10 °C	25 °C
	asphalt content: 4.0%			
E_1 (MPa)	26,494	20,164	11,703	8001
η_1 (MPa·s)	23,454	17,520	9441	4327
E_2 (MPa)	31,980	23,912	12,865	5797
η_2 (MPa·s)	5165	3399	870	132
	asphalt content: 4.5%			
E_1 (MPa)	27,770	21,740	13,764	9778
η_1 (MPa·s)	24,584	18,890	11,103	5288
E_2 (MPa)	33,521	25,781	15,130	7084
η_2 (MPa·s)	5414	3665	1023	161
	asphalt content: 5.3%			
E_1 (MPa)	24,871	19,353	12,507	6397
η_1 (MPa·s)	22,018	16,816	10,089	3459
E_2 (MPa)	30,022	22,950	13,748	4634
η_2 (MPa·s)	4849	3263	930	105

For a mixture with asphalt content 4.5%, the results of study and the approximation of curves using the Burgers creep model in the static test for various temperatures are shown in Figure 6, whereas the results of the dynamic test, as well as the approximation of curves $\sigma-\varepsilon$ using the Burgers model, are presented in Figures 7–10.

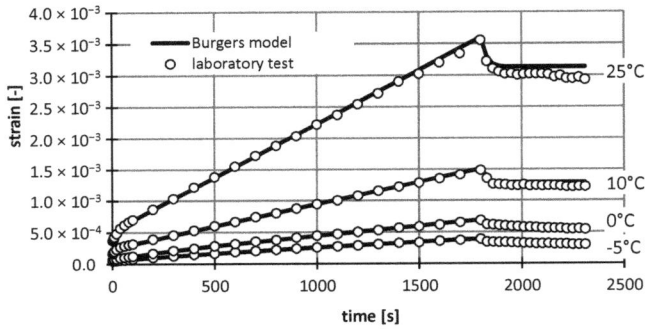

Figure 6. The results of laboratory study and the approximation of creep curves using the Burgers model in the static test (asphalt mixtures with an asphalt content of 4.5%).

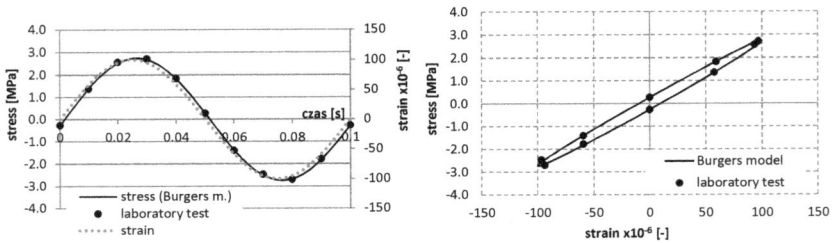

Figure 7. The results of laboratory study and the approximation of curves σ–ε using the Burgers model in the dynamic test at the temperature −5 °C (asphalt mixtures with an asphalt content of 4.5%).

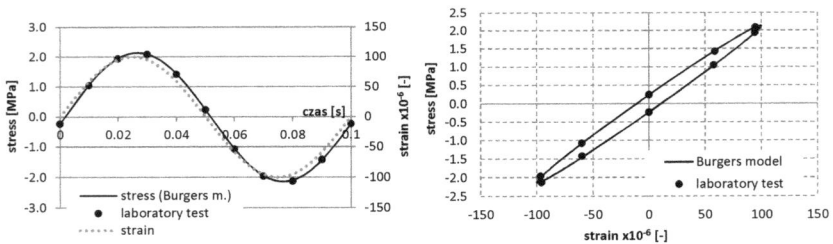

Figure 8. The results of laboratory study and the approximation of curves σ–ε using the Burgers model in the dynamic test at the temperature 0 °C (asphalt mixtures with an asphalt content of 4.5%).

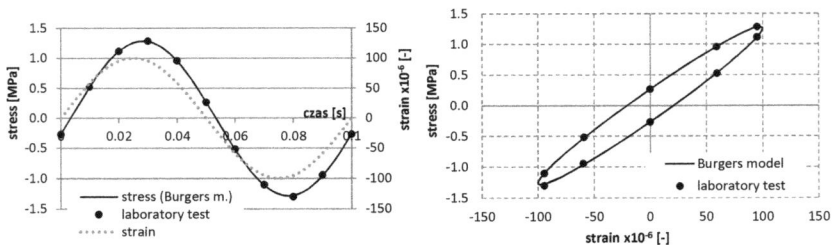

Figure 9. The results of laboratory study and the approximation of curves σ–ε using the Burgers model in the dynamic test at the temperature 10 °C (asphalt mixtures with an asphalt content of 4.5%).

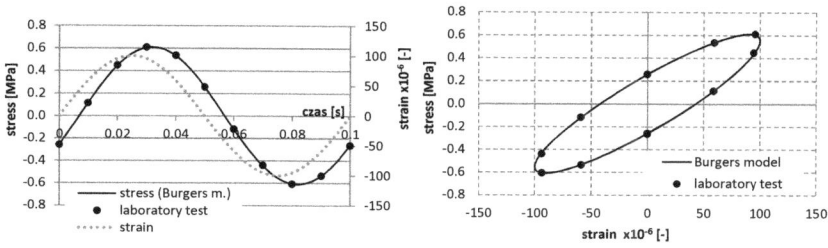

Figure 10. The results of laboratory study and the approximation of curves σ–ε using the Burgers model in the dynamic test at the temperature 25 °C (asphalt mixtures with an asphalt content of 4.5%).

It may be noticed that the obtained parameters differ. Moduli of instant elasticity E_1 obtained in the dynamic 4BP test are about 5 to 13 times larger than those received in the static test. Similarly, moduli of delayed elasticity E_2 are about 5 to 7 times greater in the dynamic than in the static tests. On the other hand, viscosity coefficients η_1 and η_2 are about two and three times smaller in the dynamic test than in the static test, respectively. This results from the very short time of variable loading, and consequently, of the short time of the material deformation response. A comprehensive comparison of changes in parameter values for different temperatures and asphalt content is shown in Figure 11 (a static test) and Figure 12 (a dynamic test).

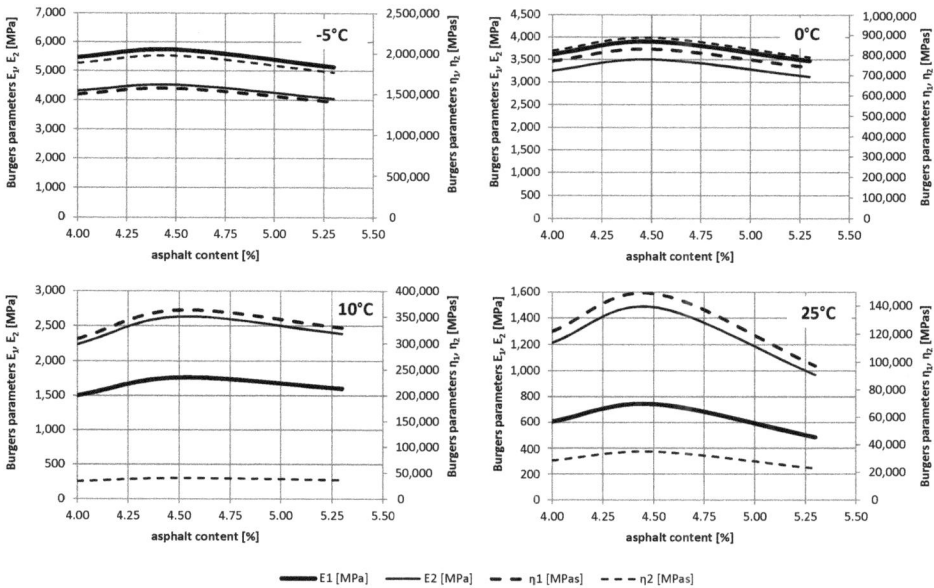

Figure 11. The relationship of Burgers parameters on the temperature in the static test.

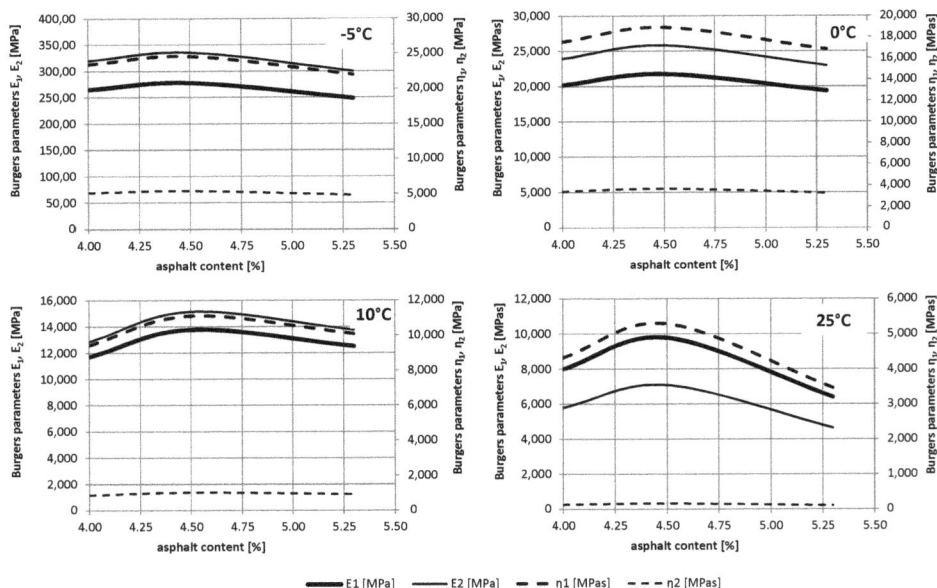

Figure 12. The relationship of Burgers parameters on the temperature in the dynamic test.

For the temperature and asphalt content range analysed, changes in the parameters were observed. For lower temperatures, a smaller change in parameters is observed depending on the asphalt content. For temperature 25 °C, the largest parameter values (except for η_2) were obtained for the optimal asphalt content 4.5%. For other temperatures, there are less pronounced extremes associated with the composition of the mixture.

Moreover, it is worth noting as the load time increased in the creep static test, smaller values of the elastic parameters E_1 and E_2 were obtained, whereas the viscosity parameters η_1 and η_2 were higher. In the dynamic test, the response of material influenced by the short-variable load was more elastic. However, the rheological features were visible over the entire range of analysed temperatures. It is worth noting that the values of parameters, mainly related to the viscosities η_1 and η_2, were correlated with the phase angle ϕ. Its value increases with increasing temperature. The angle change in the low temperature range is practically linear (Figure 13). At higher temperatures and higher angles, the material will have a larger contribution of viscous rather than elastic characters, while the lower the angle, the more elastic the material. Moreover, at higher temperatures, there is a greater variation in the angle value depending on the asphalt content in the mixture. For 5.3% asphalt content, the highest values of the phase angle ϕ were obtained.

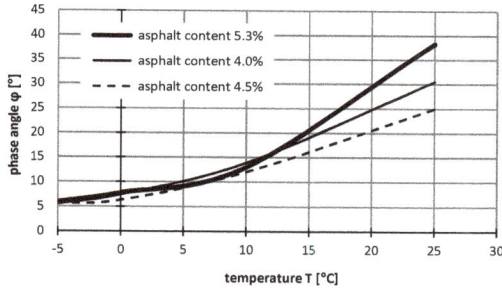

Figure 13. The relationship between the phase angle and temperature.

5. Numerical Verification of Rheological Parameters

For a selected mixture with the optimum asphalt content of 4.5%, the numerical verification of rheological parameters was performed using the finite element method. Three-dimensional static and dynamic models were developed. They included the appropriate sample geometry and load conditions that are consistent with the previously described test procedure (Figure 14). Previously, we analysed the division of the model into finite elements. The dimensions of the model were in agreement with those of the laboratory tested sample. To build the model, we used 410,000 eight-node volume elements. In the middle part of the beam, the density of the element grid was greater. Such discretization allowed for the convergence of results for displacements and deformations. The calculation of the model was carried out in the SOLIDWORKS-COSMOS/M software, ver. 2010, Structural Research and Analysis Corporation, Santa Monica, CA, USA.

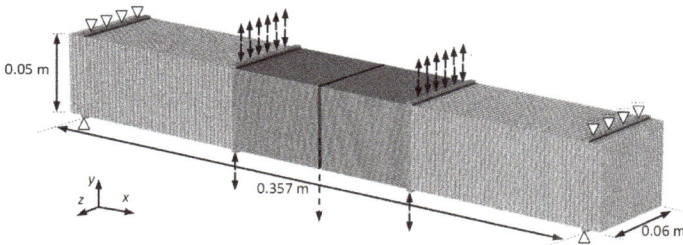

Figure 14. The relationship between the phase angle and temperature.

Rheological parameters were appropriately applied for the static and dynamic testing, according to the date presented in Table 1; Table 2. In the dynamic study, we assumed the density to be 2400 kg/m^3, with appropriate damping parameters associated with the dynamic analysis included in the static testing. The dynamic problem of discretization in Finite Element Method (FEM) is described by the classical equation [40]

$$[M]\{\ddot{u}(t)\} + [C]\{\dot{u}(t)\} + [K]\{u(t)\} = \{F(t)\} \tag{18}$$

where: $[M]$ is the matrix mass, $[C]$ is the damping matrix, $[K]$ is the stiffness matrix, $\{F(t)\}$ is the load vector variable in time t, $\{\ddot{u}(t)\}$ is the acceleration vector in the time t, $\{\dot{u}(t)\}$ is the velocity vector in time t, and $\{u(t)\}$ is the displacement vector in time t.

Selecting appropriate damping for the material is an important issue in the dynamic model. This is a complex phenomenon that involves the dissipation of energy through a variety of mechanisms, such as internal friction, cyclic thermal effects, microscopic material deformation, and micro- and macrocracks. The damping process consists of damping material, structural damping, and viscous damping associates with energy dissipation. To realistically simulate the material behavior under a short-term load, the damping factor is included as an important parameter. This is difficult to model, but the existing damping models are available in numerical calculation programs. The damping models can depend on the frequency or viscosity. The Rayleigh's damping model is quite often used in the structural dynamic analysis. To include damping effects, the damping coefficients α and β should be calculated. They are present in Rayleigh's damping matrix [40]:

$$[C] = \alpha[M] + \beta[K] \tag{19}$$

The damping coefficients are related to the angular frequency, in the form of Rayleigh's damping coefficient:

$$\xi = \frac{\alpha}{2\omega} + \frac{\beta\omega}{2} \tag{20}$$

At present, no effective experimental methods have been developed to identify damping parameters for asphalt mixtures. There is no simple correlation between the damping and static or dynamic deflections. In practice, it is possible to use FEM, and it is enough to associate only the damping with the rigidity of the system [41]. The damping was also applied for this beam model, but with only the stiffness obtaining satisfactory results. The damping parameter β was identified in the model using iterative back-calculations maintaining the agreement between the deformations from the laboratory studies. The following damping parameters β were obtained: 0.002 1/s for -5 °C, 0.0025 1/s for 0 °C, 0.004 1/s for 10 °C, and 0.005 1/s for 25 °C. Correlation between damping parameters and viscosity coefficients was determined (Figure 15). We have described the relationships between viscosity parameters and the damping coefficient using linear regression functions. High correlation coefficients (close to 1) were obtained for this function. As the value of viscosity parameters increases, the value of damping coefficients decreases.

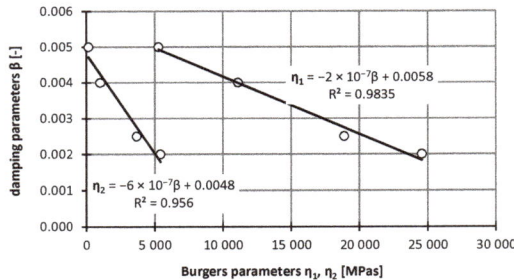

Figure 15. The relationship between the phase angle and the damping parameters (asphalt mixtures with an asphalt content of 4.5%).

Figure 16 shows the results of deformation for the static creep tests for different temperatures, and Figure 17 presents the results for the fatigue dynamic test.

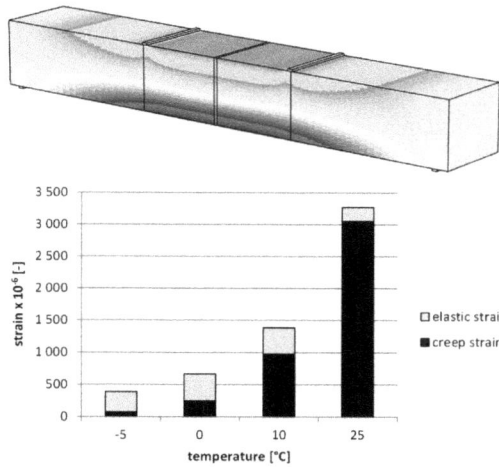

Figure 16. The deformation results for the static creep test (asphalt mixtures with an asphalt content of 4.5%).

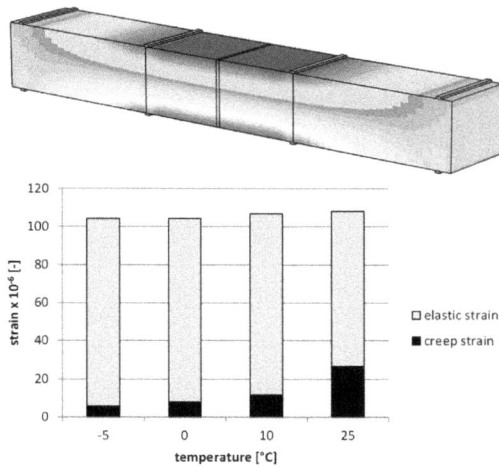

Figure 17. The deformation results for the fatigue dynamic test (asphalt mixtures with an asphalt content of 4.5%).

The numerical analysis allows us to indicate the proportion of elastic and viscous deformations depending on the test temperature. It has been found that with increasing temperature, the proportion of viscous deformations also increases, and in a dynamic temperature test is 6% at −5 °C, 8% at 0 °C, 11% at 10 °C, and 25% at 25 °C. The results for the creeping test are 19% at −5 °C, 37% at 0 °C, 71% at 10 °C, and 93% at 25 °C. The differentiation of the contribution of weak deformations results from different time intervals of the load in both studies. It is worth noting, however, that in the case of the static creep testing, the increase in temperature results in a significant nonlinear growth in the value of viscous deformations.

6. Conclusions

The different values of viscoelastic parameters in a Burgers model were determined. The variability of the parameters in temperature was obtained for the static and dynamic tests.

1. These parameters depended significantly on the duration of the load. Therefore, appropriate parameters should be chosen depending on the load time when the behaviour of asphalt mixtures in the pavement is modelled.
2. For the static long-term load tests, the parameters should be derived from creep curves, and for dynamic tests, they should be determined from the hysteresis.
3. It was found that the use of the Burgers viscoelastic model is justified for dynamic loads with the frequency of 10 Hz. For higher frequencies and at lower temperatures, the determination of the parameters may be of lesser importance, because the material has parameters similar to the elastic model, due to its low phase angle.
4. The creep test using static and dynamic 4BP loading is an effective method for determining rheological parameters under the assumed load time, the number of cycles, and temperatures. The linear viscoelastic Burgers model is helpful in this regard, because interprets the thermoplastic features of the road pavement material, such as the asphalt mixtures, well.
5. The numerical analysis using the finite element method allows us to identify the contribution of viscous deformations relative to the total, and show the significant variation of these deformations for two tests, according to the temperature.
6. The rheological parameters also depend on the composition of the bituminous mixture. For the optimal asphalt content (4.5%), the highest values of rheological parameters were obtained, demonstrating the best mechanical features and resistance to permanent deformations. For the increased asphalt content, viscosity coefficients clearly decrease, which corresponds to the increase in the value of phase angle ϕ and material damping values.
7. The obtained rheological parameters from both methods proved to be sensitive to a change in asphalt content, which means that the methods can be used to design the optimal asphalt mixtures composition—e.g., due to permanent deformation of road surfaces.

In further publications, the calculations using the finite element method for both tests, taking into account the Burgers model, will be verified. In addition, the Burgers parameters will be analysed in the dynamic fatigue test. These parameters change due to the dissipation processes and structural variation in the material.

Author Contributions: Conceptualization, P.M. and A.S.; methodology, P.M.; software, P.M.; validation, P.M., and A.S.; formal analysis, A.S.; investigation, P.M.; resources, P.M.; data curation, P.M.; writing—original draft preparation, P.M.; writing—review and editing, P.M., and A.S.; visualization, P.M.; supervision, P.M., and A.S.; project administration, P.M.; funding acquisition, A.S.

Funding: This research received no external funding.

Conflicts of Interest: The authors declare no conflict of interest.

References

1. Judycki, J. Bending Test of Asphaltic Concrete Mixtures Under Statical Loading. In Proceedings of the IV International RILEM Symposium Mechanical Tests for Bituminous Mixes—Characterization, Design and Quality Control, Budapest, Hungry, 20 September 1990; pp. 207–233.
2. Bonaquist, R. *Refining the Simple Performance Tester for Use in Routine Practice*; Report 614; NCHRP, Transportation Research Board: Washington, WA, USA, 2008.
3. Reiner, M.; Ward, A.G. Building Materials Their Elasticity and Inelasticity. North Holland Publishing Company: Amsterdam, The Netherlands, 1954.
4. Brodnyan, J.G.; Gaskin, F.H.; Philppoff, W.; Lendart, E.G. The rheology of asphalt I. *Trans. Soc. Rheol.* **1958**, *2*, 285–306. [CrossRef]

5. Van der Poel, C. Road Asphalt of the chapter IX. In *Building Materials Their Elasticity and Inelasticity*; Reiner, M., Ward, A.G., Eds.; North Holland Publishing Company: Amsterdam, The Netherlands, 1954.

6. Monismith, C.L.; Alexander, R.L.; Secor, K.E. Rheologic behavior of asphalt concrete. *Assoc. Asph. Paving Technol. Proc.* **1966**, *35*, 400–450.

7. Vakili, J. Creep behavior of asphalt-concrete under tension. *J. Rheol.* **1984**, *25*, 573–580. [CrossRef]

8. Goodrich, J.L. Asphalt binder rheology, asphalt concrete rheology and asphalt concrete mix properties. *J. AAPT.* **1991**, *60*, 80–119.

9. Kisiel, I.; Lysik, B. *An Outline of Soil Rheology, the Effect of Static Load on the Ground*; Wydawnictwo Arkady: Warszawa, Poland, 1966. (In Polish)

10. Nowacki, W. *Creep Theory*; Wydawnictwo Arkady: Warszawa, Poland, 1963. (In Polish)

11. Jakowluk, A. *Creep and Fatigue Processes in Materials*; WKŁ: Warszawa, Poland, 1993. (In Polish)

12. Hopman, P.C.; Nilsson, R.N.; Pronk, A.C. Theory, Validation and Application of the Visco-Elastic Multilayer Program VEROAD. In Proceedings of the eight International Conference on Asphalt Pavements, Seattle, WA, USA, 10–14 August 1997.

13. Collop, A.C.; Cebon, D.; Hardy, M.S. Viscoelastic approach to rutting in flexible pavements. *J. Transp. Eng.* **1995**, *121*, 82–93. [CrossRef]

14. Rowe, G.M.; Brown, S.F.; Sharrock, M.J.; Bouldin, M.G. Visco-elastic analysis of hot mix asphalt pavement structures. *Transp. Res. Rec.* **1995**, *1482*, 44–51.

15. Sousa, J.B.; Weissman, S.L.; Sackman, J.L.; Monismith, C.L. A non-linear elastic viscous with damage model to predict permanent deformation of asphalt concrete mixes. In Proceedings of the 72nd Annual Meeting of the Transportation Research Board, Washington, WA, USA, 10–14 January 1993.

16. Mackiewicz, P.; Szydło, A. Effect of load repeatability on deformation resistance of bituminous mixtures in creep and rutting tests. *Arch. Civ. Eng.* **2003**, *49*, 35–51.

17. Szydło, A.; Mackiewicz, P. Verification of bituminous mixtures' rheological parameters through rutting test. *Road Mater. Pavement Des.* **2003**, *4*, 423–428. [CrossRef]

18. Szydło, A.; Mackiewicz, P. Asphalt mixes deformation sensitivity to change in rheological parameters. *J. Mat. Civ. Eng.* **2005**, *17*, 1–9. [CrossRef]

19. Jahangiri, B.; Karimi, M.; Tabatabaee, N. Relaxation of hardening in asphalt concrete under cyclic compression loading. *J. Mat. Civ. Eng.* **2017**, *29*, 1–8. [CrossRef]

20. Blab, R.; Harvey, J.T. Viscoelastic rutting model with improved loading assumtions. In Proceedings of the Ninth International Conference on Asphalt Pavements, Copenhagen, Denmark, 17–22 August 2002.

21. You, L.; Yan, K.; Hu, Y.; Ma, W. Impact of interlayer on the anisotropic multi-layered medium overlaying viscoelastic layer under axisymmetric loading. *Appl. Math. Model.* **2018**, *61*, 726–743. [CrossRef]

22. Chen, E.Y.; Pan, E.; Green, R. Surface Loading of a Multilayered Viscoelastic Pavement: Semianalytical Solution. *J. Eng. Mech.* **2009**, *135*, 517–528. [CrossRef]

23. Karimi, M.; Tabatabaee, N.; Jahangiri, B.; Darabi, M.K. Constitutive modeling of hardening-relaxation response of asphalt concrete in cyclic compressive loading. *Constr. Build. Mat.* **2017**, *137*, 169–184. [CrossRef]

24. Yin, H.M.; Buttlar, W.G.; Paulino, G.H.; Di Benedetto, H. Assessment of Existing Micro-mechanical Models for Asphalt Mastics Considering Viscoelastic Effects. *Road Mater. Pavement Des.* **2008**, *9*, 31–57. [CrossRef]

25. You, L.; You, Z.; Yan, K. Effect of anisotropic characteristics on the mechanical behavior of asphalt concrete overlay. *Front. Struct. Civ. Eng.* **2019**, *13*, 110–122. [CrossRef]

26. Karimi, M.; Tabatabaee, N.; Jahanbakhsh, H.; Jahangiri, B. Development of a stress-mode sensitive viscoelastic constitutive relationship for asphalt concrete: Experimental and numerical modeling. *Mech. Time-Depend. Mat.* **2017**, *21*, 383–417. [CrossRef]

27. Kim, Y.R.; Little, D.N. One-dimensional constitutive modeling of asphalt concrete. *J. Eng. Mech.* **1990**, *116*, 751–772. [CrossRef]

28. Lee, H.J.; Kim, Y.R. A viscoelastic constitutive model for asphalt concrete under cyclic loading. *J. Eng. Mech.* **1998**, *124*, 32–40. [CrossRef]

29. Lee, H.J.; Kim, Y.R. Viscoelastic continuum damage model of asphalt concrete with healing. *J. Eng. Mech.* **1998**, *124*, 1224–1232. [CrossRef]

30. Souza, F.V.; Soares, J.B.; Allen, D.H.; Evangelista, F., Jr. Model for predicting damage evolution in heterogeneous viscoelastic asphaltic mixtures. *Transp. Res. Rec.* **2004**, *1891*, 131–139. [CrossRef]

31. Tashman, L.; Masad, E.; Zbib, H.; Little, D.; Kaloush, K. Microstructural viscoplastic continuum model for permanent deformation in asphalt pavements. *J. Eng. Mech.* **2005**, *131*, 47–57. [CrossRef]

32. Zafir, Z.; Siddharthan, R.; Sebaaly, P.E. Dynamic pavement strain histories from moving traffic load. *J. Transp. Eng.* **1994**, *120*, 821–842. [CrossRef]

33. Siddharthan, R.; Yao, J.; Sebaaly, P.E. Pavement strain from moving dynamic 3D load distribution. *J. Transp. Eng.* **1998**, *124*, 557–566. [CrossRef]

34. Collop, C.; Scarpas, A.T.; Kasbergen, C.; de Bondt, A. Development and finite element implementation of a stress dependent elasto-visco-plastic constitutive model with damage for asphalt. *Transp. Res. Rec.* **2003**, *1832*, 96–104. [CrossRef]

35. Judycki, J. Comparison of fatigue criteria for design of flexible and semi-rigid road pavements. In Proceedings of the Eight International Conference on Asphalt Pavements, Seattle, WA, USA, 10–14 August 1997; pp. 919–937.

36. Jaczewski, M.; Judycki, J. Effects of deviations from thermo-rheologically simple behaviour of asphalt mixes in creep on developing of master curves of their stiffness modulus. In Proceedings of the 9th International Conference Environmental Engineering, Vilnius, Lithuania, 22–23 May 2014.

37. *European Standard EN 12697–24:2012. Bituminous Mixtures—Test Methods for Hot Mix Asphalt—Part 24: Resistance to Fatigue*; PKN: Warsaw, Poland, 2012.

38. Manual of Nottingham Asphalt Tester. Available online: https://www.cooper.co.uk (accessed on 28 June 2019).

39. ARA, Inc. ERES Consultants Division. Guide for Mechanistic-Empirical Design of New and Rehabilitated Pavement Structures, Final Report, Part 3—Design and Analysis. Transportation Research Board (TRB), National Cooperative Highway Research Program (NCHRP): Washington, WA, USA, 2004.

40. Mario, P. *Structural Dynamics*, 1st ed.; Springer: Boston, MA, USA, 1991. [CrossRef]

41. Saouma, V.; Miura, F.; Lebon, G.; Yagome, Y. A simplified 3D model for soil-structure interaction with radiation damping and free field input. *Bull. Earthq. Eng.* **2011**, *9*, 1387–1402. [CrossRef]

![materials logo]

materials

MDPI

Article

Waste Brick Dust as Potential Sorbent of Lead and Cesium from Contaminated Water

Barbora Doušová [1,*], **David Koloušek** [1], **Miloslav Lhotka** [1], **Martin Keppert** [2], **Martina Urbanová** [3], **Libor Kobera** [3] and **Jiří Brus** [3]

[1]   University of Chemistry and Technology Prague, Faculty of Chemical Technology, Technická 5, 166 28 Praha 6, Czech Republic; koloused@vscht.cz (D.K.); lhotkam@vscht.cz (M.L.)
[2]   Faculty of Civil Engineering, Czech Technical University in Prague, Thákurova 7, 166 29 Praha 6, Czech Republic; martin.keppert@fsv.cvut.cz
[3]   Institute of Macromolecular Chemistry AS CR, Heyrovského nám. 2, 162 06 Praha 6, Czech Republic; urbanova@imc.cas.cz (M.U.); kobera@imc.cas.cz (L.K.); brus@imc.cas.cz (J.B.)
*   Correspondence: dousovab@vscht.cz; Tel.: +420-22044-4381

Received: 26 April 2019; Accepted: 19 May 2019; Published: 20 May 2019

check for updates

Abstract: Adsorption properties of waste brick dust (WBD) were studied by the removing of Pb[II] and Cs[I] from an aqueous system. For adsorption experiments, 0.1 M and 0.5 M aqueous solutions of Cs^+ and Pb^{2+} and two WBD (Libochovice—LB, and Tyn nad Vltavou—TN) in the fraction below 125 μm were used. The structural and surface properties of WBD were characterized by X-ray diffraction (XRD) in combination with solid-state nuclear magnetic resonance (NMR), supplemented by scanning electron microscopy (SEM), specific surface area (S_{BET}), total pore volume and zero point of charge (pH_{ZPC}). LB was a more amorphous material showing a better adsorption condition than that of TN. The adsorption process indicated better results for Pb^{2+}, due to the inner-sphere surface complexation in all Pb^{2+} systems, supported by the formation of insoluble $Pb(OH)_2$ precipitation on the sorbent surface. A weak adsorption of Cs^+ on WBD corresponded to the non-Langmuir adsorption run followed by the outer-sphere surface complexation. The leachability of Pb^{2+} from saturated WBDs varied from 0.001% to 0.3%, while in the case of Cs^+, 4% to 12% of the initial amount was leached. Both LB and TN met the standards for Pb[II] adsorption, yet completely failed for any Cs[I] removal from water systems.

Keywords: waste brick dust; adsorption; lead; cesium; surface complexation; precipitation; solid-state NMR spectroscopy

---

1. Introduction

The contamination of waters and soils by toxic carcinogens and radioactive nuclides is a pressing environmental problem. There is particular concern for accumulative poisonous effects in local environments.

Lead (Pb) is one such element that has well known chronic influence on the central nervous system, where it replaces the residing zinc in neurons. This process can cause mental retardation, behavioral changes, paralysis and anemia. Pb is easily accumulated in body over the lifetime, yet its expulsion is very difficult. It is generally believed that the anthropogenic sources of Pb are limited to the use of tetraethyl lead, however, other Pb origins, such as battery manufacturers, lead smelters and ammunition industries are all major Pb polluters of the environment [1,2]. Cesium as ^{137}Cs is an important source of radioactivity from nuclear waste, as it is the major radionuclide of spent nuclear fuel [3]. Due to nuclear accidents over the last decades (e.g., Fukushima in 2011, Chernobyl in 1986), ^{137}Cs has infiltrated soils and groundwater and even further into the biosphere thanks to its high solubility [4]. A potential risk of both Pb and ^{137}Cs originates from their toxicity and harmful concentrations to organisms in various part of the biosphere [5]. While Pb poisons the soil microorganisms, which limits heterotrophic breakdown of organic matter [6], ^{137}Cs is toxic due to its long radio-isotopic half-life (30 years) [7]. Furthermore, ^{137}Cs is metabolically similar to potassium, causing it to accumulate in plants and therefore it is incorporated into many food chains [8].

Despite a number of modern cleaning technologies (ultrafiltration, reverse osmosis, biological methods, etc.), adsorption remains an effective technique due to its simple application in wastewater treatment. The demand for environmentally and economically friendly technologies has resulted in the development and testing of new natural and synthetic adsorbents, preferably low-cost [5,9–12].

Waste building materials have attained increased professional concern due to their availability and properties analogous to aluminosilicates, such as composition, chemical stability, fine structure and environmentally safe nature. Generally, they are recycled either as concrete admixtures [13] or when fine waste particles have pozzolanic activity, they can be added to cement-based materials to declining the Portland cement consumption [14].

Dousova et al. [15] followed out the pilot study on waste brick dusts (WBD) as potential adsorbents of a series of risk cations, anions and radioactive residues. As has been found, WBD showed to be selective cation-active sorbents. The anionic As form (AsO_4^{3-}) was also adsorbed onto WBD, but with more than four times higher sorbent dosage. In terms of possible technological applications, the long-term stability of toxic ions in saturated WBD, as well as its successful incorporation into cement building materials, are crucial.

The aim of this work was to study the characteristics of WBD as a cation-active adsorbent of Pb^{II} and Cs^{I} particles from aqueous solutions. A deep structural characterization of the materials of interest was conducted for potential technological applications based on the combination of X-ray diffraction (XRD) and solid-state NMR spectroscopy; these complementary methods are especially informative. While XRD analysis can easily describe crystalline fractions and changes in mineralogical composition, solid-state NMR spectroscopy gives information about the structure of both amorphous and crystalline phases, local framework defects and extra framework components. Specifically, [133]Cs and [207]Pb solid-state NMR spectra can solely provide insight into the chemical nature of adsorbed and solidified contaminants. The information on the structural and surface changes related to the adsorption process may lead to an adsorption mechanism, in which the adsorption rate can be characterized through important parameters such as adsorption capacity, adsorption efficiency, surface complexation and binding energy.

2. Materials and Methods

2.1. Waste Brick Dust

Waste brick dusts (WBD) arose as a waste (grinding powder) during the production of vertically perforated ceramic blocks intended for thin joint masonry. The amount of recyclable WBD is limited, as surplus ceramic waste is typically dumped.

The fraction below 125 μm of two WBD samples from different brick factories in the Czech Republic (Libochovice—LB and Tyn nad Vltavou—TN) were tested as selective cation-active adsorbents for Pb/Cs removal from model aqueous systems. The chemical characteristics of both materials (Table 1) indicates more than 10 times alkalinity, slightly less Fe content and almost double S_{BET} of LB compare to TN. Both WBD were mineralogically similar, with dominant quartz content (ca. 25%), and the presence of illite, albite, anorthite and hematite. The presence of calcite was detected in LB only, while sillimanite was found in TN.

Table 1. Chemical composition, pH of the zero point of charge (pH$_{ZPC}$) and specific surface area (S$_{BET}$) of waste brick dust (WBD) from Libochovice (LB) and Tyn nad Vltavou (TN).

Composition (wt%)	LB	TN
Na$_2$O/K$_2$O	1.1/4.3	2.5/2.8
MgO/CaO	1.5/14.4	1.3/1.6
Al$_2$O$_3$	19.7	32.6
SiO$_2$	46.4	60.3
P$_2$O$_5$	0.4	0.3
SO$_3$	1.5	0.6
Fe$_2$O$_3$	3.7	5.4
TiO$_2$	0.4	0.8
MnO$_2$	0.03	0.04
S$_{BET}$ (m^2 g^{-1})	4.1	2.5
V$_{total}$ (cm^3 g^{-1})$^{*)}$	0.01	0.006
pH $^{**)}$	~9–10	~5–6
pH$_{ZPC}$	4.3	5.5

*) total pore volume of mesopores; **) water leachate at 20 °C, solid-liquid ratio 1:30

2.2. Model Solutions

Model solutions of Pb^{2+} and Cs$^+$ were prepared from inorganic salts (PbCl$_2$ and CsCl, respectively) of analytical grade and distilled water, with concentrations of 0.1 and 0.5 mM L^{-1} and their natural pH values ($\approx$3.5). The concentration ranges were selected as appropriate to simulate a slightly increased amount of the contaminant in a water system to a heavily contaminated solution.

2.3. Sorption Experiments

The suspension of model solution (50 mL) and defined dosage (1–15 g L^{-1} for Pb and 1–30 g L^{-1} for Cs) of WBD was agitated in a batch manner at laboratory temperature (20 °C) for 24 h [16]. The product was filtered, and the filtrate was analyzed for residual concentration of cations, while the solid residue (saturated WBD) was kept for subsequent solid-state analyses (NMR, SEM, S$_{BET}$).

All adsorption data were fitted to the Langmuir isotherm [17,18], according to the equations (1) and (2), which was verified in many papers [9,15–17] as a suitable adsorption model for natural sorbents, including aluminosilicates and soils. The obvious adsorption parameters (q_{max}—maximum equilibrium adsorption capacity; Q_t—theoretical adsorption capacity; R^2—correlation factor; K_L—Langmuir adsorption constant) were then used to evaluate the effectiveness of adsorption systems.

$$q = \frac{Q_t \cdot K_L \cdot c}{1 + K_L \cdot c},$$ (1)

and its linearized form:

$$\frac{1}{q} = \frac{1}{Q_t} + \frac{1}{Q_t \cdot K_L \cdot c},$$ (2)

where q is an equilibrium concentration of adsorbed ion (adsorbate) in solid phase [mol g^{-1}], c is an equilibrium concentration of adsorbate in solution [mol L^{-1}], Q_t indicates the theoretical adsorption capacity [mol g^{-1}], and K_L is a Langmuir adsorption constant [L mol^{-1}].

The equilibrium amount (q) of adsorbate caught in the solid phase was calculated from experimental data by Equation (3):

$$q = \frac{V_0(c_0 - c)}{m},$$ (3)

where V_0 is the volume of solution [L], c_0 is the initial concentration of adsorbate in solution [mol L^{-1}], and m is the mass of solid phase [g].

2.4. Leaching Test

The stability of cations in the saturated WBDs was tested by a standard procedure EN 12457–2 [19]. First, a suspension of dry sorbent and distilled H_2O at a ratio of 1:10 was agitated in a sealed polyethylene bottle at 20 °C for 24 h. The suspension was filtered out and the filtrate was analyzed for residual Pb^{2+}/Cs^+ concentration.

2.5. Analytical Methods

Powder XRD was measured with a Seifert XRD 3000P diffractometer (Seifert, Ahrensburg, Germany) with CoK_α radiation (λ= 0.179026 nm, graphite monochromator, goniometer with Bragg–Brentano geometry) in the 2θ range of 5–60° with a step size of 0.05° 2θ.

The X-ray Fluorescence (XRF) analyses of the solid phase were carried out using an ARL 9400 XP+ spectrometer (ARL, Ecublens, Switzerland) at a voltage of 20–60 kV, probe current of 40–80 mA, and with an effective area of 490.6 mm^2. UniQuant 4 software was used to evaluate the data (Thermo ARL, Ecublens, Switzerland).

The S_{BET} was measured on a Micromeritics ASAP 2020 (accelerated surface area and porosimetry) analyzer using gas sorption. The ASAP 2020 model (Micromeritics, Norcross, GA, USA) assesses single and multipoint BET surface area, Langmuir surface area, Temkin and Freundlich isotherm analysis, pore volume and pore area distributions in the micro- and macro-pore ranges. The macro-pore and micro-pore samples were analyzed by the Horvath-Kavazoe method (BJH method), respectively. The BHJ method used N_2 as the analysis adsorptive and an analysis bath temperature of −195.8 °C. Samples were degassed at 313 K for 1000 min.

Solid-state NMR spectra were measured at 11.7 T using a Bruker AVANCE III HD 500 WB/US NMR spectrometer (Bruker, Karlsruhe, Germany). The ^{27}Al Magic Angle Spinning Nuclear Magnetic Resonance Spectroscopy (MAS NMR)spectra were acquired at a spinning frequency of 11 kHz, Larmor frequency of 130.287 MHz and recycle delay of 2 s. The spectra were referenced to the external standard $Al(NO_3)_3$ (0 ppm). The ^{29}Si MAS NMR spectra were acquired at a spinning frequency of 11 kHz, Larmor frequency of 99.325 MHz and recycle delay of 10 s. The number of scans for the acquisition of a single ^{29}Si MAS NMR spectrum was 6144. The spectra were referenced to the external standard M_8Q_8 (−109.8 ppm). ^{133}Cs MAS NMR spectra were acquired at a spinning frequency of 11 kHz, Larmor frequency of 65.601 MHz, recycle delay of 2 s and the number of scans of 73,000. The spectra were referenced to the external standard $CsOH \cdot H_2O$ (124.1 ppm). Due to the large chemical shift anisotropy the ^{207}Pb NMR spectra were measured using the ^{207}Pb MAS NMR and WURST-QCPMG NMR experiments combined with variable offset cumulative spectroscopy (VOCS) technique [20]. The Larmor frequency was 104.640 MHz and the recycle delay was 2 s for all ^{207}Pb NMR spectra. The number of scans was set to 30000 for each sub-spectrum. The applied experimental parameters were optimized using a mixture of PbO and $Pb(OH)_2$ the resonance frequency. The ^{207}Pb NMR spectra were referenced to the external standard $Pb(NO_3)_2$ (−3473.6 ppm). High-power ^{1}H decoupling (SPINAL64 for MAS experiments and Continuous Wave (CW) for static experiments) was used to eliminate heteronuclear dipolar couplings in all measurements. The NMR experiments were performed at a temperature of 303 K, and temperature calibration was done to compensate for the frictional heating of the samples [21].

The structure of samples was determined using scanning electron microscopy (SEM) on the Tescan Vega 3 (Brno-Kohoutovice, Czech Republic). Energy dispersion spectrometry (EDS) was conducted on the Inca 350 spectrometer (Oxford Instruments, Abingdon, UK).

The concentration of Pb and Cs in aqueous solutions was measured by atomic absorption spectrometry (AAS) using a SpectrAA-880 VGA 77 unit (Varian, Palo Alto, CA, USA) in flame mode. An accuracy of AAS analyses was guaranteed by the Laboratory of Atomic Absorption Spectrometry of UCT Prague, CR, with the detection limit of 0.5 µg L^{-1}, with a standard deviation ranging from 5%–10% of the mean).

3. Results and Discussion

3.1. Specification of WBD Adsorbents in Relation to Pb/Cs Chemistry and pH Value

The chemical and mineralogical composition of both WBD are similar apart from the high content of alkalis in LB as compared to TN, which results in the former exhibiting a significant higher pH (see Table 1). Whereas the pH/pH$_{ZPC}$ values (Table 1) are responsible for the selectivity of adsorption, the points of pH$_{ZPC}$ in relation with pH ranges of adsorption solution are shown in Figure 1. At the pH$_{ZPC}$ of LB (the crosses in both phase diagrams) lower than the pH range of adsorption solution (the dashed areas in the phase diagrams), when pH/pH$_{ZPC}$ > 2, the sorbent surface primarily deprotonates to balance the pH gradient and the negatively charged surface of LB then attracts cations from the solution [22]. In the case of TN, the negligible pH/pH$_{ZPC}$ gradient (pH/pH$_{ZPC}$ ≈ 1, spotted areas vs. the empty rings in the phase diagrams) caused a weak attraction at the solid-liquid interface. In terms of WBD with the average pH$_{ZPC}$ of 5, a higher pH of adsorption has been essential for selective removal of Pb^{2+}/Cs^{+}. The optimal pH resulting from adsorption experiments ranged from 7.5 to 10.

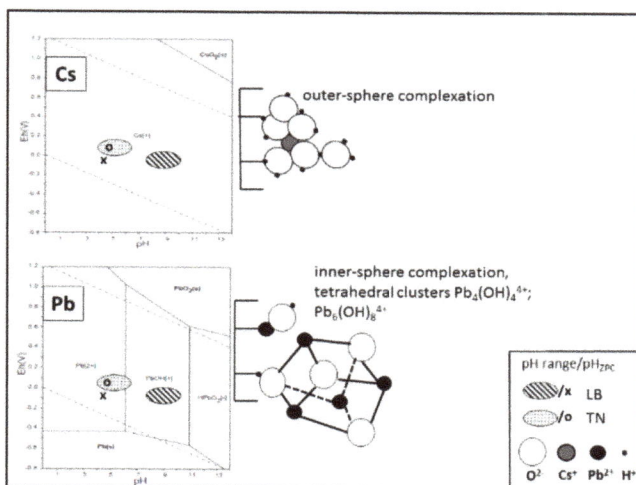

Figure 1. Surface chemistry and expected complexation of Cs^{+}/Pb^{2+} in aqueous system.

As shown in Table 1, the S$_{BET}$ and total porosity values also favored the adsorption selectivity of LB to TN, because a more porous material with a larger surface area would generally provide more sorption sites for surface binding.

The adsorption behavior of the investigated systems was also based on the different aqueous chemistry of Pb and Cs, respectively (Table 2, Figure 1). In an aqueous environment, Cs^{+} was expected to form a small hydrated ion, because of its low charge and large crystallographic radius (1.69 Å). While the adsorption of Cs was minimally pH dependent, Cs^{+} ions were bound primarily to cation exchange sites, forming only outer-sphere complexes. On the other hand, Pb^{2+} (1.32 Å) was subjected to high hydrolysis, easily losing the part of its primary hydration shell, especially at high pH [23]. The hydrolyzed ions initially formed strong inner-sphere complexes with sorbent surfaces and eventually led to the formation of polynuclear complexes or surface precipitates [24].

Table 2. Ionic radius and thermodynamic properties of hydrated Cs^+/Pb^{2+} [25].

Cation	r(Å)	r'(Å)	n	ΔG_{hydr} (kJ mol^{-1})
Cs^+	1.69	3.28	2.1	-250
Pb^{2+}	1.32	4.03	6.1	-1425

r—ionic radius; r'—hydrated radius; n—number of water molecules in hydration shell; ΔG_{hydr}—molar Gibbs energy of ion hydration.

3.2. Structural Changes Following Adsorption

The SEM micro-observation of initial and Cs^+/Pb^{2+} saturated WBD (Figure 2) showed completely different morphology of surface particles, confirming the above described unequal hydrolyses and binding strategies of Cs^+ and Pb^{2+} ions. Separately hydrated Cs^+ ions, that formed outer-sphere complexes with active surface sites, did not change the surface morphology remarkably (Figure 2), while the predicted formation of surface precipitates of $Pb_4(OH)_4^{4+}/Pb_6(OH)_8^{4+}$ (Figure 1) led to the homogeneous covering of sorbent surface.

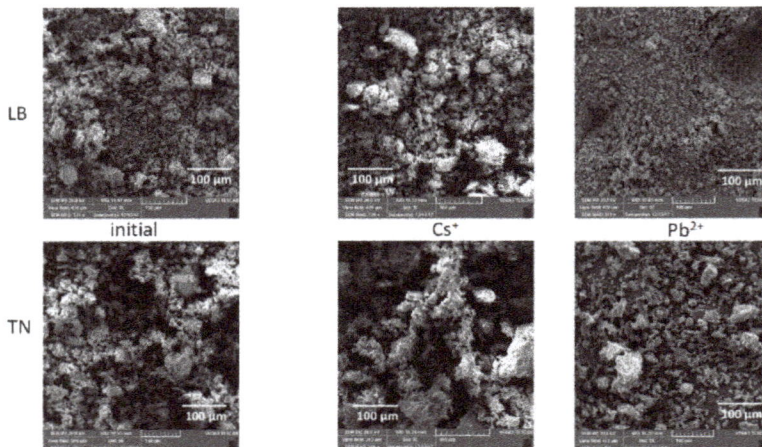

Figure 2. SEM images of WBD before and after Cs^+/Pb^{2+} adsorption on LB and TN (magnification of 300×).

The MAS NMR spectra of both materials recorded before and after Cs^+/Pb^{2+} adsorption indicate structural changes, crystallinity and framework defects resulting from the adsorption process. The ^{27}Al MAS NMR spectra (Figure 3) demonstrate that aluminum sites in both systems occupy tetrahedral coordination Al^{IV}, as reflected by the most intense signals at 60 and 58 ppm for LB and TN systems, respectively. The low-intensity signal at approximately 5 ppm reflects a small amount of hexagonally coordinated aluminum species present in the LB system. Additional information relating to the crystallinity of investigated systems can be deduced from the line-widths of the recorded signals. While the LB system is represented by a relatively broad signal with the half-width of about 1.6 kHz, the TN system is reflected by considerably narrower signal (1.3 kHz). This finding, which agrees with the ^{29}Si MAS NMR spectra (discussed later), indicates a more disordered and amorphous character of aluminosilicate framework of the LB system.

The ^{29}Si MAS NMR spectra (Figure 4) provide more detailed insight into the framework of investigated systems and further highlight their distinct structures. The main signal at -91 ppm is clearly attributed to the Q^3(0Al) sites in dehydroxylated illite, whereas the shoulder at -85 ppm is assigned to a mixture of Q^3(1Al) and Q^3(2Al) usually present in natural illite and related systems. The signal resonating at approximately -75 ppm likely reflects the incompletely condensed Q^1 and Q^2

units of mineral defects and the amorphous SiO_2 fraction. The narrow signal at -108 ppm therefore reflects the Q^4 units of crystalline SiO_2 (quartz, approx. 20%).

On the other hand, the ^{29}Si MAS NMR spectrum of neat brick dust from TN reflects a much narrower distribution of aluminosilicate building blocks. The signal is dominated by the narrow resonance at -108 ppm, reflecting the Q^4 units of crystalline SiO_2 (quartz, approx. 70%). Deconvolution of the recorded spectrum further revealed additional much broader resonance centered at -92 ppm, which would reflect the $Q^3(0Al)$ units in dehydroxylated illite (approx. 30%).

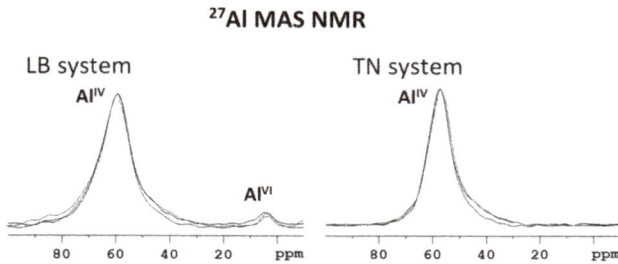

Figure 3. Comparison of ^{27}Al MAS NMR spectra of neat LB and TN WBD systems, before and after Cs^+/Pb^{2+} adsorption.

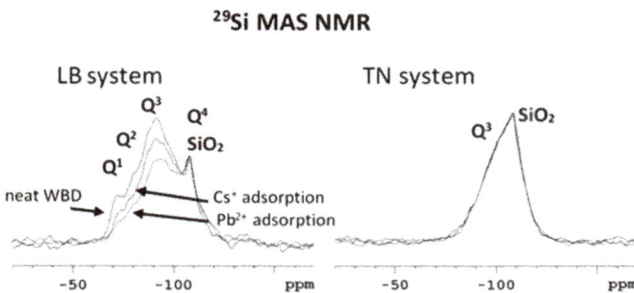

Figure 4. Comparison of ^{29}Si MAS NMR spectra of neat LB and TN WBD systems, before and after Cs^+/Pb^{2+} adsorption.

The structural changes associated with the sorption process were mostly observed at the ^{29}Si MAS NMR spectra of LB system (Figure 4). The intensities of broad signals of Q^1, Q^2 and Q^3 units considerably decreased relative to the signal intensity of Q^4 units of crystalline SiO_2. In the case of Pb^{2+} adsorption, the observed decrease is more intense in comparison with the decrease observed during the sorption of Cs^+ ions, which indicates the partial dissolution of disordered and amorphous fractions rich in the SiO_2 phase. In contrast, there were no considerable changes in the ^{29}Si MAS NMR spectra of the TN system, thus exhibiting structural stability of this material during the adsorption of Pb^{2+} and Cs^+ ions.

3.3. Adsorption of Pb^{2+}/Cs^+ on WBD

The adsorption series were performed under the same conditions (described in Section 2.3) in 'order to evaluate the applicability of WBD for the removal of toxic cations from aqueous systems.

The Langmuir adsorption isotherms (Figure 5) illustrate improved adsorption properties of LB for both cations and a considerably higher affinity of Pb^{2+} for both WBD. As shown by the adsorption parameters calculated from the linearized Langmuir equation (Section 2.3, Table 3), only Pb^{2+} adsorption corresponded to the Langmuir model indicating an inner-sphere surface complexation [26,27] for all Pb^{2+} adsorptions. The differences in theoretical adsorption capacities (Q_{theor}) and binding energies (K_L) of LB/Pb^{2+} as compared to the TN/Pb^{2+} systems resulted from the different pH/pH_{ZPC} ratios (Section 3.1), which could be considered the prior adsorption power [15,16]. Accordingly, LB/Pb^{2+} adsorption system showed much more robust process in all aspects (Q_{theor}, K_L, R^2) (Table 3).

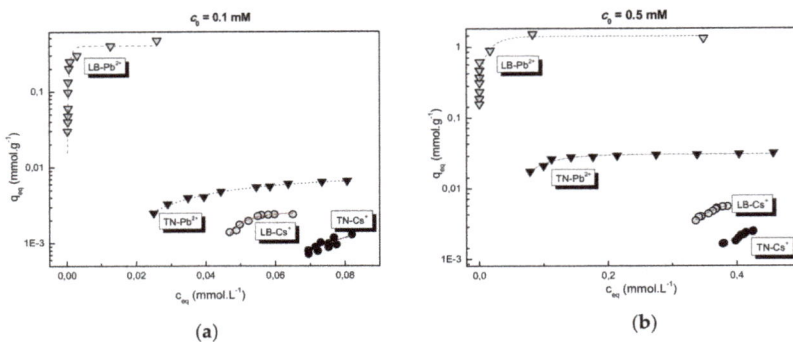

Figure 5. Langmuir adsorption isotherms of Pb^{2+}/Cs^+ adsorption on LB and TN at different initial concentrations; (**a**) c_0 = 0.1 mmol L^{-1}, (**b**) c_0 = 0.5 mmol L^{-1}. Sorbent dosage ≈ 1–15 g L^{-1} for Pb^{2+} and 1–30 g L^{-1} for Cs^+.

Table 3. Adsorption parameters for Pb and Cs adsorption from model solutions on LB and TN.

Sorbent/ c_0 (mM)	Pb^{2+}				Cs^+			
	q_{max} (mmol g^{-1})	Q_{theor} (mmol g^{-1})	K_L (L mmol^{-1})	R^2	q_{max} (mmol g^{-1})	Q_{theor} (mmol g^{-1})	K_L (L mmol^{-1})	R^2
LB/0.5	1.32	1.6	776.4	0.973	5.5×10^{-3}	-	-	-
LB/0.1	0.85	1.1	171.5	0.930	2.4×10^{-3}	-	-	-
TN/0.5	3.4×10^{-2}	4.5×10^{-2}	16.7	0.827	2.9×10^{-3}	-	-	-
TN/0.1	5.5×10^{-3}	6.8×10^{-3}	39.8	0.838	1.1×10^{-3}	-	-	-

On the other hand, the adsorption of hydrated Cs^+ ions did not follow Langmuir run, but typical physical adsorption with a weak electrostatic surface binding (outer-sphere complexation) was observed. Maximum equilibrium sorption capacities (q_{max}) of both LB and TN for Cs^+ were one to three orders of magnitude lower than for Pb^{2+}.

An almost 100% adsorption efficiency at low sorbent consumption for Pb^{2+} (Figure 6) supported the abovementioned formation of poorly soluble $Pb(OH)_2$ clusters on the sorbent surface, which markedly increased the adsorption yields.

The fate of adsorbed Cs^+/Pb^{2+} ions was also observed in the ^{113}Cs and ^{207}Pb solid-state NMR spectra. The spectra were recorded only for the LB-system, which in both cases exhibited considerably higher sorption capacities. The ^{133}Cs MAS NMR spectrum (Figure 7) shows relatively broad signal covering the frequency range from −7 to approx. −20 ppm. According to literature data, these resonances can be assigned to surface species, in which more shielded (more negative) resonances reflect Cs^+ sites more tightly bound to the surface, whereas the less shielded resonances correspond to less tightly bound species. This phenomenon can be attributed to the formation of multilayer deposits of Cs^+ ions.

When measuring ^{207}Pb solid-state NMR spectra, there is relatively low natural abundance of ^{207}Pb spins (approx. 22.6 %, I=1/2), which is combined with usually very large chemical shift anisotropy (CSA). Consequently, under the standard spinning speeds (10–30 kHz) manifolds of the central signals and spinning sidebands in ^{207}Pb MAS NMR spectra can cover extremely broad spectral windows 2000–4000 ppm, rendering them nearly undetectable. Therefore, the attempt here to record

[207]Pb solid-state NMR spectra was nearly unsuccessful. This effect was observed even in the model system, which consisted of a mixture of neat powdered PbO and Pb(OH)$_2$ (Figure 8a). However, the application of an alternative experimental technique, [207]Pb Wideband Uniform-Rate Smooth Truncation-Quadrupolar Carr-Purcell Meiboom-Gill (WURST-QCPMG) NMR, seems to be more suitable for the measurement of broad spectral lines. As demonstrated in Figure 8b, the intensities of the recorded [207]Pb WURST-QCPMG NMR signals are considerably higher than that recorded by the standard [207]Pb MAS NMR experiment.

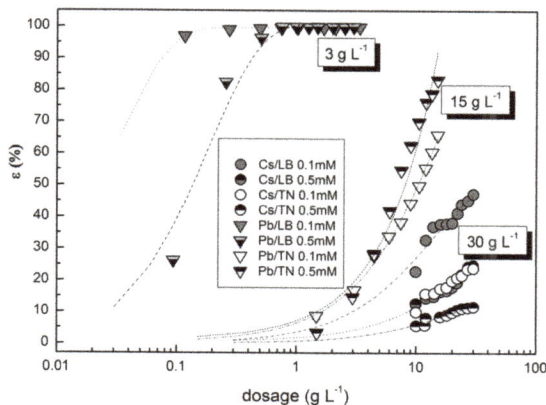

Figure 6. Adsorption efficiency of Cs$^+$/Pb^{2+} adsorption on LB and TN in relation to sorbent dosage.

[133]Cs MAS NMR

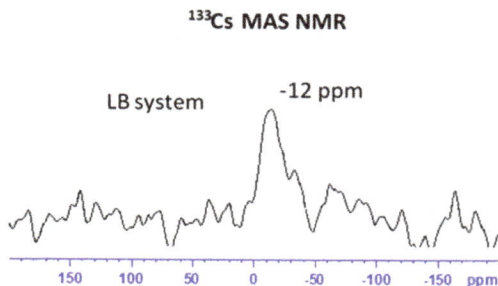

Figure 7. [133]Cs MAS NMR spectrum of LB system after Cs$^+$ adsorption.

In spite of a long accumulation time (48 h), the resulting spectrum is very poor, indicating very low amounts of the surface Pb^{2+} species (Figure 8c). Nevertheless, two spectral regions with [207]Pb NMR resonances were identified in the recorded spectrum. One of them covers frequencies from 1200 to 2200 ppm, whereas the second one ranges from −400 to −1000 ppm. These results imply that Pb^{2+} ions on the surface of LB-system form two chemically distinct fractions. One of these fractions is likely structurally related to precipitated hydroxide species, whereas the second one can be rather attributed to Pb^{2+} binding to a silicate tetrahedral [28].

Consistent with the previously mentioned results, the adsorption of Cs$^+$ on WBDs was worse in all aspects when compared to Pb^{2+}. A low effect of Cs$^+$ adsorption was mainly given by steric properties. According to Rajec et al. [29], the majority of active sorption sites for Cs$^+$ binding have been associated with permanent (pH independent) charge cation exchange sites on zeolites and smectite clays rather than with surface hydroxyl sites on quartz and feldspars. Therefore, zeolites have been frequently used to remove radionuclides and metal cations because of their molecular sieve structure and high cation exchange capacity [30]. Contrastingly, WBDs consist mostly of quartz, feldspars and hematite and exhibit rather poor Cs adsorbing behavior [31].

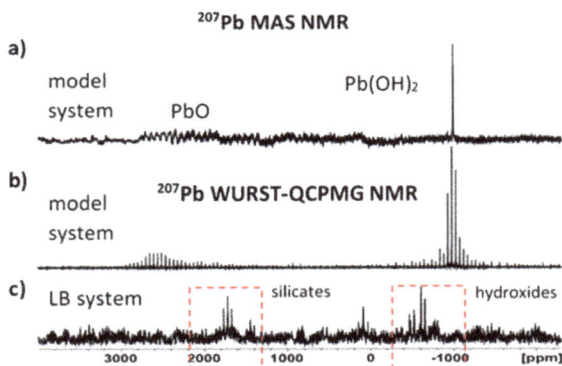

Figure 8. ^{207}Pb solid-state NMR spectra. (**a**) Standard 207 Pb MAS NMR spectrum of a reference mixture of neat powdered PbO and Pb(OH)$_2$ recorded at 12 kHz (NS = 50,000); (**b**) the ^{207}Pb WURST-QCPMG NMR spectrum of the reference system (NS = 8000); and (**c**) the corresponding ^{207}Pb WURST-QCPMG NMR spectrum of LB system after Pb^{2+} adsorption (NS = 50,000).

3.4. Leachability of Pb^{2+}/Cs$^+$ from Saturated Sorbents

The stability of cations in saturated WBD was tested by the leaching test (Section 2.4). Not surprisingly, the stability of Pb^{2+}/Cs$^+$ in saturated WBD was closely related to the selectivity of a particular adsorption (Table 4); Pb^{2+} adsorption on LB carried out by high sorption capacities and efficiencies, linking the formation of very stable inner-sphere surface complexes at both initial concentrations (only about 0.001% of Pb^{2+} was leached). The favorable stability (less than 0.3% of the original amount leached) was recorded for Pb^{2+} in TN. The non-selective Cs$^+$ adsorption resulted in a low binding stability during the leaching test, where 4 to 12% of the initial amount was leached. According to the standards (CSN EN 12457), LB and TN saturated with Pb^{2+} fall into the non-hazardous waste. Although the regulation has not specified the limits for a Cs hazard, a high leachability of Cs$^+$ from both saturated WBDs ($\approx$ 10% wt.) indicates its environmental risk. In line with the previous results, LB proved better adsorption selectivity and leaching stability to cations than TN. The stability of testified adsorption systems decreased in the following order: Pb^{2+}/LB >> Pb^{2+}/TN >> Cs$^+$/LB > Cs$^+$/TN.

Table 4. Leachability of Cs/Pb from WBDs.

Element/Initial Concentration	Initial Amount in Saturated WBD (mg·g^{-1})	Leached in H$_2$O (%)	Class of Leachability*
	Libochovice (LB)		
Cs$^+$/0.1 mM	0.32	3.7	not limited
Cs$^+$/0.5 mM	0.74	7.4	not limited
Pb^{2+}/0.1 mM	146.3	1.2×10^{-3}	IIa—other waste
Pb^{2+}/0.5 mM	212.8	6.6×10^{-4}	I—inert waste
	Tyn nad Vltavou (TN)		
Cs$^+$/0.1 mM	0.15	12.3	not limited
Cs$^+$/0.5 mM	0.33	11.4	not limited
Pb^{2+}/0.1 mM	0.93	0.25	IIa—other waste
Pb^{2+}/0.5 mM	5.98	0.18	IIa—other waste

* Regulation No. 294/2005 (CSN EN 12457).

4. Conclusions

The adsorption selectivity of WDBs to Pb(II) and Cs(I) was initially affected by the structural properties and morphology of WBDs being tested as Pb^{2+}/Cs^+ adsorbents. The unequal binding forms of both cations presented the next important aspect of adsorption process; while Cs^+ ions formed separate particles weakly bound by outer-sphere surface complexes, Pb^{2+} appeared in two co-existed phases, namely as the hydrolyzed particles Pb^{2+} bound by strong inner-sphere complexes, and the poorly soluble $Pb(OH)_2$ clusters that catch on the sorbent surface. Therefore, the adsorption of Pb^{2+} was much more effective, strongly supported by the almost insoluble surface precipitation ($Pb(OH)_2$). In connection with that process, the adsorption of Pb^{2+} was pH dependent, with an optimal pH range of 8 to 11. On the other hand, Cs^+ ions were equally adsorbed within a wide range of pHs.

The stability of Cs^+/Pb^{2+} saturated sorbents subjected to the standard leaching tests corresponded well with the quality and effectiveness of adsorption in particular; the better parameters of Pb^{2+} adsorption, following the Langmuir adsorption model, indicated a high binding energy of surface complexation that resulted in a considerably higher stability of WBD that was saturated with Pb^{2+} when compared to Cs^+. The stability of Pb^{2+} saturated WBDs was about three orders in magnitude higher than that of Cs^+. In all aspects, LB proved to be a much better sorbent than TN. Both WBDs (LB and TN) are promising prospective adsorbents for Pb^{II}.

Author Contributions: B.D. wrote the paper, carried out adsorption experiments, managed the methodology of research steps; D.K. performed XRD analyses; M.L. measured and evaluated the S_{BET} and porosity of WBDs; M.K. provided suitable material and its characterization; M.U. and L.K. performed NMR study, while J.B. supervised and evaluated the MMR experiments.

Funding: This research was funded by Czech Science Foundation under the project 19-11027S "Concrete slurry-hazardous waste or secondary raw material?".

Conflicts of Interest: The authors declare no conflict of interest.

References

1. Wong, C.S.C.; Li, X.D. Pb contamination and isotopic composition of urban soils in Hong Kong. *Sci. Total Environ.* **2004**, *319*, 185–195. [CrossRef]
2. Jiang, M.; Wang, Q.; Jin, X.; Chen, Z. Removal of Pb(II) from aqueous solution using modified and unmodified kaolinite clay. *J. Hazard. Mater.* **2009**, *170*, 332–339. [CrossRef] [PubMed]
3. Benedicto, A.; Missana, T.; Fernández, A.M. Interlayer Collapse Affects on Cesium Adsorption onto Illite. *Environ. Sci. Technol.* **2014**, *48*, 4909–4915. [CrossRef] [PubMed]
4. Cornell, R.M. Adsorption of cesium on minerals: A review. *J. Radioanal. Nuclear Chem.* **1993**, *171*, 483–500. [CrossRef]
5. Dahiya, S.; Tripathi, R.M.; Hegde, A.G. Biosorption of heavy metals and radionuclide from aqueous solutions by pre-treated arcashell biomass. *J. Hazard. Mater.* **2008**, *150*, 376–386. [CrossRef]
6. Honeyman, B.D.; Santsch, P.H. Metals in aquatic environment. *Environ. Sci. Technol.* **1988**, *22*, 862–871. [CrossRef] [PubMed]
7. Song, K.C.; Lee, H.K.; Moon, H.; Lee, K.J. Simultaneous removal of the radiotoxic nuclides ^{137}Cs & ^{129}I from aqueous solution. *Sep. Purif. Technol.* **1997**, *12*, 215–217.
8. Jalali, R.R.; Ghafourian, H.; Asef, Y.; Dalir, S.T.; Sahafipour, M.H.; Gharanjik, B.M. Biosorption of cesium by native and chemically modified biomass of marine algae: Introduce the new biosorbent for biotechnology applications. *J. Hazard. Mater.* **2004**, *B116*, 125–134. [CrossRef] [PubMed]
9. Han, R.; Lu, Z.; Zou, W.; Daotong, W.; Shi, J.; Jiujun, Y. Removal of copper(II) and lead(II) from aqueous solution by manganese oxide coated sand. II. Equilibrium study and competitive adsorption. *J. Hazard. Mater.* **2006**, *B137*, 480–488. [CrossRef]
10. Mousavi, H.Z.; Seyedi, S.R. Kinetic and equilibrium studies on the removal of Pb(II) from aqueous solution using nettle ash. *J. Chil. Chem. Soc.* **2010**, *55*, 307–311. [CrossRef]

11. Ding, D.; Lei, Z.; Yang, Y.; Feng, C.; Zhang, Z. Selective removal of cesium from aqueous solutions with nickel (II) hexacyanoferrate (III) functionalized agricultural residue–walnut shell. *J. Hazard. Mater.* **2014**, *270*, 187–195. [CrossRef]

12. Badescu, I.S.; Bulgariu, D.; Ahmad, I.; Bulgariu, L. Valorisation possibilities of exhausted biosorbents loaded with metal ions—A review. *J. Environ. Manag.* **2018**, *224*, 288–297. [CrossRef]

13. Silva, R.V.; de Brito, J.; Dhir, R.K. Properties and composition of recycled aggregates from construction and demolition waste suitable for concrete production. *Con. Build. Mat.* **2014**, *65*, 201–217. [CrossRef]

14. Vejmelková, E.; Keppert, M.; Rovnaníková, P.; Ondráček, M.; Keršner, Z.; Černý, R. Properties of high performance concrete containing fine-ground ceramics as supplementary cementitious material. *Cement Concrete Compos.* **2012**, *34*, 55–61. [CrossRef]

15. Dousova, B.; Kolousek, D.; Keppert, M.; Machovic, V.; Lhotka, M.; Urbanova, M.; Brus, J.; Holcova, L. Use of waste ceramics in adsorption technologies. *Appl. Clay Sci.* **2016**, *134*, 145–152. [CrossRef]

16. Doušová, B.; Grygar, T.; Martaus, A.; Fuitová, L.; Koloušek, D.; Machovič, V. Sorption of As^V on aluminosilicates treated with Fe^{II} nanoparticles. *J. Colloid Interface Sci.* **2006**, *302*, 424–431. [CrossRef]

17. Jeong, Y.; Fan, M.; Singh, S.; Chuang, C.L.; Saha, B.; van Leeuwen, J.H. Evaluation of iron oxide and aluminium oxide as potential arsenic(V) adsorbents. *Chem. Eng. Process.* **2007**, *46*, 1030–1039. [CrossRef]

18. Misak, N.Z. Langmuir isotherm and its application in ion-exchange reactions. *React. Polym.* **1993**, *21*, 53–64. [CrossRef]

19. CEN 2002. CEN 2002. Characterization of Waste-Leaching-Compliance Test for Leaching of Granular Waste Material and Sludge Part 2. Comite Europeen de Normalisation 2002, EN 12457-2. Available online: https://www.sis.se/en/produkter/environment-health-protection-safety/wastes/solid-wastes/ssen124574/ (accessed on 1 October 2003).

20. Massiot, D.; Farnan, I.; Gautier, N.; Trumeau, D.; Trokiner, A.; Coutures, J.P. Ga-71 and Ga-69 Nuclear-Magnetic-Resonance Study of Beta-Ga$_2$O$_3$-Resolution of 4-Fold and 6-Fold Coordinated Ga Sites in Static Conditions. *Solid State Nucl. Magn. Reson.* **1995**, *4*, 241–248. [CrossRef]

21. Brus, J. Heating of samples induced by fast magic-angle spinning. *Solid State Nucl. Magn. Reson.* **2000**, *16*, 151–160. [CrossRef]

22. Walther, J.V. *Essentials of Geochemistry*, 2nd ed.; Jones and Bartlett Publishers Inc.: Sudbury, MA, USA, 2009; 486p.

23. Bohn, H.L.; Mc Neal, B.L.; O'Conner, G.A. *Soil Chemistry*, 2nd ed.; John Wiley and Sons: New York, NY, USA, 1985; 341p.

24. Um, W.; Papelis, C. Sorption mechanisms of Sr(II) and Pb(II) on zeolitized tuffs from the Nevada Test Site as a function of pH and ionic strength. *Am. Mineral.* **2003**, *88*, 2028–2039. [CrossRef]

25. Marcus, Y. Thermodynamics of Solvation of Ions. *J. Chem. Soc. Faraday Trans.* **1991**, *87*, 2995–2999. [CrossRef]

26. Ding, M.; de Jong, B.H.W.S.; Roosendaal, S.J.; Vredenberg, A. XPS studies on the electronic structure of bonding between solid and solutes: Adsorption of arsenate, chromate, phosphate, Pb^{2+}, and Zn^{2+} ions on amorphous black ferric oxyhydroxide. *Geochim. Cosmochim. Acta* **2000**, *64*, 1209–1219. [CrossRef]

27. Wen, X.; Du, Q.; Tang, H. Surface complexation model for the heavy metal adsorption on natural sediment. *Environ. Sci. Technol.* **1998**, *32*, 870–875. [CrossRef]

28. Fayon, F.; Farnan, I.; Bessada, C.; Coutures, J.; Massiot, D.; Coutures, J.P. Empirical Correlations between [207]Pb NMR Chemical Shifts and Structure in Solids. *J. Am. Chem. Soc.* **1997**, *119*, 6837–6843. [CrossRef]

29. Rajec, P.; Macasek, F.; Feder, M.; Misaelides, P.; Samajova, E. Sorption of caesium and strontium on clinoptilolite- and mordenite-containing sedimentary rocks. *J. Radioanal. Nucl. Chem.* **1998**, *229*, 49–56. [CrossRef]

30. Faghihian, H.; Maragheh, M.G.; Kazemian, H. The use of clinoptilolite and its sodium form for removal of radioactive cesium and strontium from nuclear wastewater and Pb^{2+}, Ni^{2+}, Cd^{2+}, Ba^{2+} from municipal wastewater. *Appl. Radiat. Isotopes* **1999**, *50*, 655–660. [CrossRef]

31. Bergaoui, L.; Lambert, J.F.; Prost, R. Caesium adsorption on soil clay: Macroscopic and spectroscopic measurements. *Appl. Clay Sci.* **2005**, *29*, 23–29. [CrossRef]

materials

MDPI

Article

Wave Frequency Effects on Damage Imaging in Adhesive Joints Using Lamb Waves and RMS

Erwin Wojtczak * and **Magdalena Rucka**

Department of Mechanics of Materials and Structures, Faculty of Civil and Environmental Engineering, Gdansk University of Technology, Narutowicza 11/12, 80-233 Gdansk, Poland; magdalena.rucka@pg.edu.pl or mrucka@pg.edu.pl
* Correspondence: erwin.wojtczak@pg.edu.pl; Tel.: +48-58-347-2497

Received: 7 May 2019; Accepted: 5 June 2019; Published: 6 June 2019

check for
updates

Abstract: Structural adhesive joints have numerous applications in many fields of industry. The gradual deterioration of adhesive material over time causes a possibility of unexpected failure and the need for non-destructive testing of existing joints. The Lamb wave propagation method is one of the most promising techniques for the damage identification of such connections. The aim of this study was experimental and numerical research on the effects of the wave frequency on damage identification in a single-lap adhesive joint of steel plates. The ultrasonic waves were excited at one point of an analyzed specimen and then measured in a certain area of the joint. The recorded wave velocity signals were processed by the way of a root mean square (RMS) calculation, giving the actual position and geometry of defects. In addition to the visual assessment of damage maps, a statistical analysis was conducted. The influence of an excitation frequency value on the obtained visualizations was considered experimentally and numerically in the wide range for a single defect. Supplementary finite element method (FEM) calculations were performed for three additional damage variants. The results revealed some limitations of the proposed method. The main conclusion was that the effectiveness of measurements strongly depends on the chosen wave frequency value.

Keywords: Lamb waves; scanning laser vibrometry; adhesive joints; non-destructive testing; damage detection; excitation frequency

1. Introduction

Adhesive bonding is one of the effective methods for joining elements in metallic structures, besides welding, riveting, and bolting [1]. It has dozens of applications in the aerospace, machine, automotive, military, and electronics industries [2]. Structural adhesive joints have numerous advantages in comparison to other joining techniques. Firstly, adhesives do not interfere with the structure of adherends (joined elements), which is what happens in bolted joints (openings weakening joined parts) or welded joints (internal stresses after welding). Moreover, gluing enables the creation of heterogenic connections, especially when welding or hole drilling is forbidden. There are also some disadvantages; from these, among the most significant is high vulnerability to accuracy in the processes of preparation and manufacturing. Particularly, the most important issues for the strength of the joint are the accurate surface treatment [3] and the protection against any contamination [4]. Any inaccuracy may lead to the formation of kissing defects or voids [5,6]. Their presence can cause a significant decrease in the strength of the joint and, as a result, its failure. The problematic issue is that kissing defects are not detectable in the visual assessment, because of their existence in the internal structure of the joint. This creates the necessity of application of non-destructive testing (NDT). There are a number of promising methods that have also been successfully applied for damage identification in adhesive joints, using ultrasounds [7–9], thermography [10], radiography [11], laser-induced breakdown spectroscopy [12], or electric time-domain reflectometry [13]. These methods are the basis for structural health monitoring (SHM) systems that provide a real-time evaluation of analyzed structures of different types, such as bridges [14–16], tunnels [17], or marine structures [18]. Nowadays, SHM strategies are becoming more and more popular for composite materials also [19–21].

The guided wave propagation phenomenon is commonly used for damage identification in structures of different types (e.g., [22–27]). Lamb waves are a specific type of guided waves that propagate in plate-like elements. It is worth noticing that they are multimodal; i.e., in general, an infinite number of different modes (symmetric and antisymmetric) can propagate in each medium. Another significant feature is the dispersive nature, which means that wave characteristics such as the wavenumber and propagation velocities of each mode are frequency-dependent. These properties make the question of wave propagation a complex problem. For certain frequency ranges, some modes do not propagate, whereas for a different range, the same modes can travel with certain velocities; thus, they influence the wave propagation. For this reason, the appropriate choice of the excitation frequency is an essential issue for the effectiveness of obtained results. High sensitivity to any disruption of geometry and changes in material properties create many applications of guided waves in non-destructive diagnostics of existing structures. Previous studies prove their usefulness for the identification of damages of different types, such as cracks in metallic beams and plates [28,29] or delamination and flaws in composites [30–32]. With regard to the adhesive joints, guided waves were efficiently used for the identification of disbond areas in the single-lap joints of plates made from different materials. Ren and Lissenden [7] detected damaged areas in the adhesive film in a CFRP (carbon fiber reinforced polymer) plate stiffened with a stringer using the adjustable angle beam transducers. Nicassio et al. [33] analyzed debonding in an adhesive joint of aluminum plates using piezo sensors. Sunarsa et al. [34] used air-coupled ultrasonic transducers to detect debonding and weakened bonding areas of different shapes in adhesively bonded aluminum plates. The time of flight of the measured signals was estimated with the support of the wavelet transform. Parodi et al. [35] analyzed numerically and experimentally wave propagation in a wall of a composite pressure vessel with flaws in the interface between the aluminum layer and CFRP coating, considering the excitation frequency in a range of 20 to 100 kHz. Ultrasonic waves are also successfully used for the evaluation of adhesion levels between adhesive and adherends. Gauthier et al. [36] analyzed the influence of different adherend surface treatment methods on the guided wave propagation in a single aluminum plate covered with epoxy-based adhesive. Castaings [37] considered a contamination of the overlap surface by an oil pollutant in the single-lap adhesive joints of aluminum plates.

The guided wave propagation method usually consists of the excitation of waves in one point of an analyzed structure and a collection of signals in some other points. If the number of measurement points (fitted with ultrasonic or piezoelectric transducers) is relatively small, the actual state of the considered structure is determined by the analysis of registered time histories. For a greater number of measurements, the non-contact methods are beneficial, allowing to sense the guided wave field in a considered area. The scanning laser Doppler vibrometry (SLDV) is one of the methods that provide a more accurate analysis [38–41]. As the effect, the plane representation of propagating waves (the so-called SLDV map) can be obtained. The existence of any defect in the scanned area results in the disturbance of the wave front shape, but its actual position and shape are indeterminable. Therefore, further signal processing is required to obtain a useful defect image. For example, Sohn et al. [42] detected delamination and disbond in composite plates based on the SLDV maps processed with the use of different techniques such as Laplacian image filtering. Another quite simple but effective method of damage imaging is based on the vibration energy distribution, and requires root mean square (RMS) calculations or its alternative weighted variant (WRMS). Recently, it has been successfully applied for the damage identification of different structures [43–49]. Saravanan et al. [43] detected missing bolts, attached masses, and openings in aluminum specimens assuming the excitation frequency as 50 kHz. Radzieński et al. [44] analyzed the detection of additional mass in aluminum and composite plates for frequencies of 35 and 10 kHz, respectively. In another work [45], they examined aluminum plates strengthened with riveted L-shape stiffeners, considering different excitation frequencies (5, 35, and 100 kHz). Aluminum plates with notches of different directions were studied by Lee and Park [46], who proved that the orientation of defects to the incident wave front was significant. In another research study, Lee et al. [47] investigated the notches and corrosion defects of different areas using the weighted root mean square and edge detection algorithms. Rucka et al. [48] studied the influence of a weighting factor on the efficiency of WRMS maps. Aryan et al. [49] visualized defects in the form of corrosion, surface cracks, and dents in aluminum plates and the delamination in a composite beam using scanning laser Doppler vibrometer and RMS calculations. The excitation frequencies were chosen from a range of 100 to 300 kHz. To sum up, the above-mentioned works present the application of root mean square calculations of registered guided wave signals without extensive consideration of the influence of the excitation frequency. This parameter was usually arbitrary assumed; notwithstanding, it can significantly affect the legibility of obtained RMS maps.

The aim of the study is damage imaging in a Lamb-wave based inspection of adhesively bonded joints. Particular attention was paid to the influence of the excitation frequency on the efficiency of obtained results. The guided wave signals were collected by the scanning laser Doppler vibrometer and further processed using root mean square calculations. The experimental research was conducted on a real-scale physical model of a single-lap joint of metal plates bonded with the epoxy-based adhesive. The verification of measurements was provided by numerical analyses carried out on finite element method (FEM) models. A novel element of the study is the proposition of choosing the adequate excitation frequency by the qualitative measure of the effectiveness of RMS damage imaging. The hypothesis is that an efficient frequency range for experimental measurements can be determined in the way of initial FEM calculations for artificial defects. The relative difference between RMS values in the damaged and intact areas of the joint can be assumed as the measure of the efficiency.

2. Materials and Methods

2.1. Specimen Description

The investigations were conducted on the single-lap adhesive joints of steel plates. The geometry of the specimens is presented in Figure 1. The dimensions of each plate were 270 mm × 120 mm × 3 mm. The overlap surface was 120 mm × 60 mm. The internal defect in the form of partial debonding was designed in the adhesive film in four variants (#1 to #4, as shown in Figure 2), from which the first one was chosen for experimental measurements. The defect (#1) was obtained by sticking a PTFE (polytetrafluoroethylene) tape of 0.2-mm thickness in the middle of the overlap before manufacturing the connection. To avoid creating unintended debonding areas, the overlap surface of each adherend was treated with fine sandpaper (grit size 120) and degreased with Loctite-7063 cleaner just before joining. The epoxy-based adhesive Loctite Hysol 9461 (Henkel, Düsseldorf, Germany) was used to join the plates. The measured bondline thickness was equal to approximately 0.2 mm. To control the expected geometry of the defect, the adherends were disconnected after the experiments. The separated plates are presented in Figure 3. The failure occurred in the interfaces between the glue layer and the steel plates (mainly the lower one); it has a purely adhesive character. There are visible leaks of adhesive into the area of the intended defect (on the upper plate). Moreover, the defect edges are irregular (visible mainly on the lower plate).

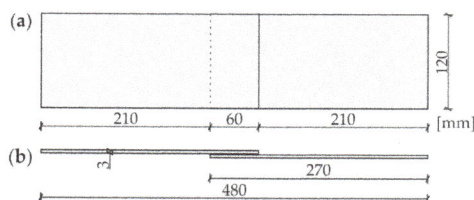

Figure 1. Geometry of investigated specimen: (**a**) plane view; (**b**) side view.

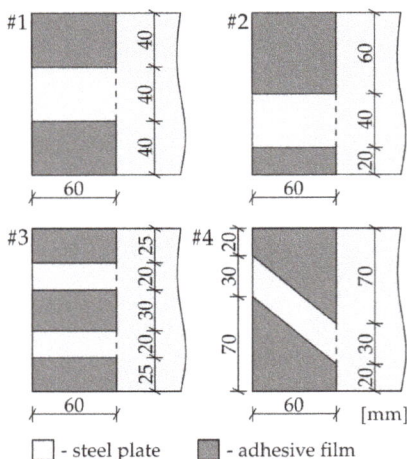

Figure 2. Variants of defects (#1 to #4).

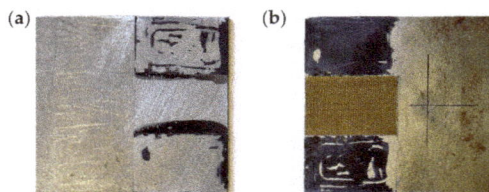

Figure 3. Photograph of experimental specimen after separation: (**a**) upper plate; (**b**) lower plate.

2.2. Experimental Setup

The experimental examination of prepared specimen #1 consisted of the excitation and the acquisition of the Lamb wave propagation signals in the specified area of the joint by the scanning laser Doppler vibrometry method. The experimental setup is presented in Figure 4a. The generation of the input wave signal was provided by the arbitrary function generator AFG 3022 (Tektronix, Inc., Beaverton, OR, USA) with the support of the high-voltage amplifier PPA 2000 (EC Electronics, Krakow, Poland). The plate piezoelectric actuator NAC2024 (Noliac, Kvistgaard, Denmark) with dimensions of 3 mm × 3 mm × 3 mm was used for the excitation of the guided wave field in one of the adherends. The actuator was attached to the top surface of the specimen by the petro wax 080A109 (PCB Piezotronics, Inc., Depew, NY, USA). The input signal was a wave packet obtained from the five periods of the sinusoidal function by the Hanning window modulation. The excitation frequency was individual for each measurement and varied from 20 to 350 kHz. The signals of the guided wave field were recorded by the scanning head of the laser vibrometer PSV-3D-400-M (Polytec GmbH, Berlin, Germany) equipped with a VD-07 velocity decoder. The sampling frequency was assumed to be 2.56 MHz. The improvement of light backscatter was provided by covering the scanned surface with a retro-reflective sheeting. The out-of-plane components of velocity values were acquired in the time domain in 3721 points distributed over the area featuring the overlap surface and the part of the plate after it (at the top side of the specimen); see Figure 4b,c. The scanning was performed point by point in the quadratic mesh of 61 rows and 61 columns, resulting in the resolution of about 1.93 mm. The representations of acquired signals for specific time instances show the propagation of the full guided wave field (SLDV maps).

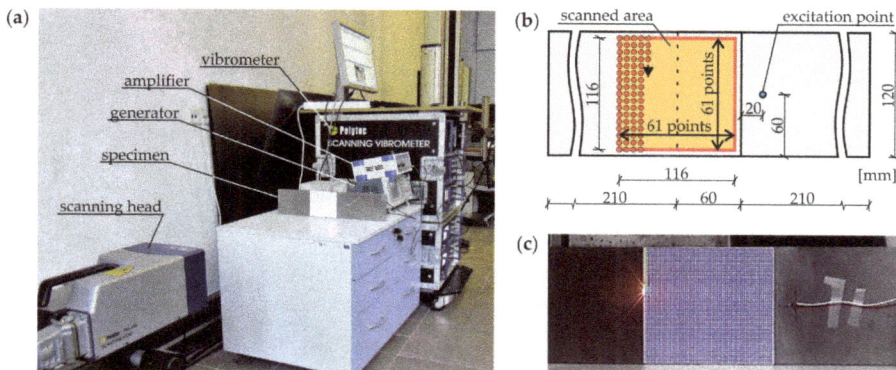

Figure 4. Experimental measurements: (**a**) experimental setup for the generation and acquisition of Lamb waves; (**b**) view of a specimen with the position of a scanned area and excitation point; (**c**) investigated specimen with indicated scanning points.

2.3. FEM Modeling

The numerical modeling of elastic wave propagation in composite structures, such as adhesive joints, is a complex problem, mainly because of material inhomogeneity and an uncertainty of contact at the interfaces between different materials. An effective contribution to this issue was made by Chronopoulos [50] and Apalowo and Chronopoulos [51]. In the present paper, numerical analysis of the guided wave propagation in the considered adhesive joints (#1 to #4) was conducted using the finite element method in Abaqus/Explicit software. Some assumptions were made to simplify the modeling process and shorten the calculations. Three-dimensional FEM models were prepared for a transient dynamic analysis. Each structure was discretized by eight-node solid elements with reduced integration (C3D8R) from the explicit element library. The appropriate mapping of the wave behavior requires at least 20 nodes for the shortest wavelength of interest [52]. According to this limitation, the mesh was initially assumed to be regular and consisted of cube-shaped elements with a global size of 1 mm, which was reduced to 0.2 mm for the thickness of the adhesive layer (see Figure 5). The mesh convergence test was conducted taking into account a few refined meshes. The out-of-plane velocity values in some randomly chosen points at specific time instances were assumed as the measure of the convergence. The relative differences between results were negligible; thus, the exact calculations were conducted with the use of the above-mentioned mesh with the global element size of 1 mm. The boundary conditions were free at all the edges. The materials were adopted to meet the assumptions of a homogenous, isotropic material model. The material parameters were: for steel $E_s = 195.2$ GPa, $v_s = 0.30$, $\rho_s = 7741.7$ kg/m^3, and for adhesive $E_a = 5$ GPa, $v_a = 0.35$, and $\rho_a = 1330$ kg/m^3. The material damping was neglected because of its marginal influence on the RMS damage imaging. Both adherends and the adhesive film were assumed to be independent structures combined rigidly at the part of their surfaces by means of a tie connection (compatibility of translational degrees of freedom at all the contacting nodes). The excitation of guided waves was applied at the lower adherend in the form of the concentrated force surface load with the amplitude varying in time in accordance with the wave packet signal. The excitation frequency range was extended in comparison to experimental measurements (20 to 500 kHz). The dynamic analysis was conducted with the use of the central difference method with a fixed time step of 10^{-7} s. This value meets the recommendation of at least 20 points per each cycle of the wave with the higher frequency [52]. The results of the analysis were out-of-plane velocity signals collected at 3721 points spread over the area of the joint corresponding with experimental measurements.

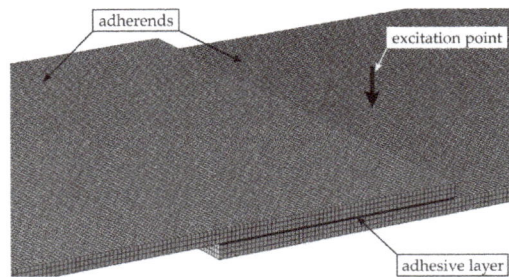

Figure 5. View of a discretized model for numerical calculations with an indicated excitation point.

2.4. RMS Damage Imaging

The signals of propagating waves acquired during scanning with a laser vibrometer need further processing techniques that allow detecting damaged areas and defining their actual shape. The essential point of damage imaging is to show the differences between the undamaged and damaged part of an analyzed structure. One of the simplest method consists of the calculation of the root mean square (RMS) for each recorded signal. The RMS value for the continuous time signal $s(t)$ can be calculated with the following formula:

$$RMS = \sqrt{\frac{1}{t_2 - t_1} \int_{t_1}^{t_2} s(t)^2 dt} \tag{1}$$

where t_1 is the beginning and t_2 is the end of the time window, which is defined as the difference between these two values. For a discrete signal $s_k = s(t_k)$ recorded with the time interval Δt, the RMS value can be calculated as follows:

$$RMS = \sqrt{\frac{1}{n} \sum_{k=1}^{n} s_k^2} \tag{2}$$

where n is the number of samples, and the time window is defined as $T = \Delta t(n - 1) = t_2 - t_1$. The map prepared from the calculated RMS values allows identifying and determining the geometry of any possible defects existing in the scanned area. Overall, for damaged areas of an analyzed element (e.g., delamination, crack, opening), different RMS values are attained because of different characteristics of Lamb wave propagation (changes due to material stiffness or geometry disturbance).

3. Results and Discussion

3.1. Dispersion Curves

The initial step in the damage detection of adhesive joints was the comparison of Lamb wave characteristics in a three-layer medium (steel–adhesive–steel, simulating a properly prepared adhesive joint) and in a single-layer medium (single steel plate or disbonded area of the adhesive joint). For this purpose, dispersion curves were prepared experimentally and numerically for a steel plate with dimensions of 240 mm × 300 mm × 3 mm (sample D1) and for two plates with an adhesive film with a thickness of 0.2 mm bonding them together (sample D2). In each specimen, a wave packet in the form of a single-cycle Hanning windowed sinusoidal function was excited. The carrier frequency was changing in the range from 50 to 300 kHz with a step of 50 kHz. Additionally, for each frequency, symmetric and antisymmetric Lamb modes were excited independently. The velocity signals (out-of-plane components) were acquired in 101 points distributed along the straight line with a total length of 100 mm. The dispersion curves in the form of the maps representing wavenumber–frequency relations were obtained in the way of 2D-FFT (two-dimensional fast Fourier transform) calculations for each of 12 measurements (cf. [6,36,53]). The final result was the superposition of all the compound maps (Figure 6).

A comparison of experimental and numerical curves led to the conclusion that both approaches gave consistent results. This also proved the appropriateness of the assumption of adhesive material parameters. To track the theoretical dispersion curves of Lamb waves in the investigated media, our own code was developed in the Matlab® software (9.3.0.713597, The MathWorks, Inc., Natick, MA, USA), implementing the transfer matrix method [54,55]. Figure 7a shows wavenumber–frequency relations for both media (samples D1 and D2). The shape of the curves is approximately the same as that shown in the maps in Figure 6. However, the possibility of the effective excitation of certain modes was not the same in the results of the measurements (in both experimental and numerical curves for the single plate and the joint). In sample D1 (Figures 6a and 7, black curves), only fundamental S_0 and A_0 modes can propagate in the considered frequency range, notwithstanding that the S_0 curve is not as strongly exposed as A_0 in the maps, which may be the result of the acquisition of only out-of-plane components on the upper surface that are related mainly to antisymmetric modes. In sample D2 (Figures 6b and 7, red curves), in addition to the fundamental pair, A_1 and S_1 modes are present starting from the frequencies of about 130 and 260 kHz, respectively. Moreover, the shape of the S_0 curve changes meaningfully compared with the single-layer plate. The shape of the A_0 curve does not change significantly in comparison to sample D1. The differences between the two types of media are also clearly visible on group velocity–frequency relations (Figure 7b), which will be useful in further considerations.

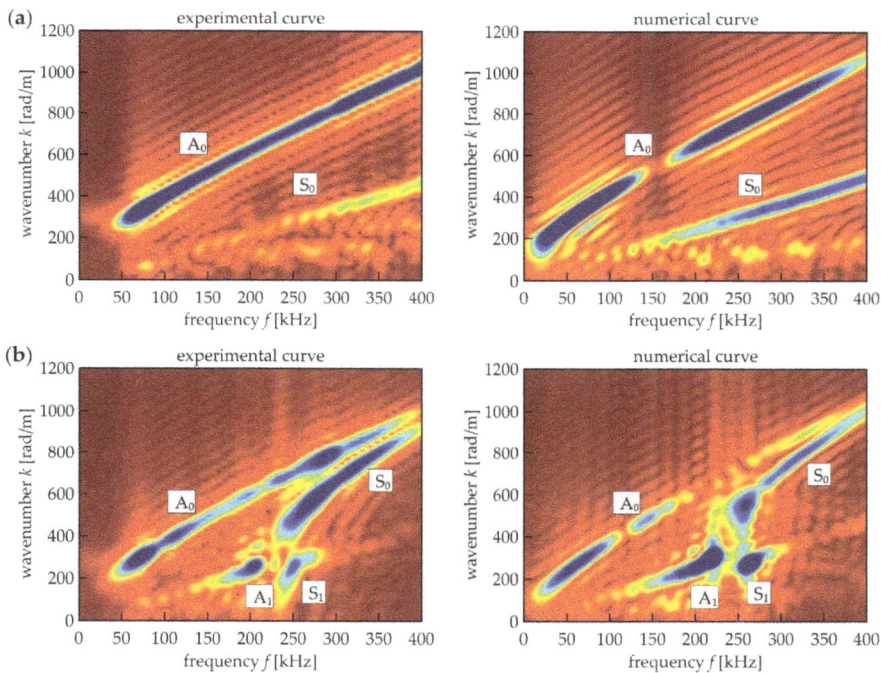

Figure 6. Experimental and numerical dispersion curves: (**a**) steel plate (sample D1)—single layer medium (d_s = 3 mm, E_s = 195.2 GPa, ν = 0.3, ρ_s = 7741.7 kg/m³); (**b**) adhesive joint (sample D2)—three-layer medium consisted of two steel plates (parameters same as in (**a**)) and adhesive film (d_a = 0.2 mm, E_a = 5 GPa, ν_a = 0.35, ρ_a = 1330 kg/m³).

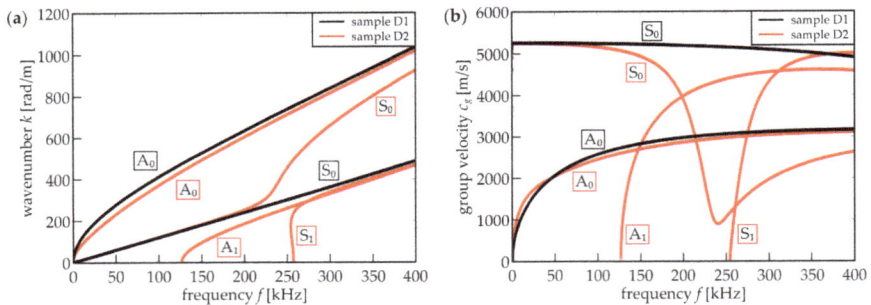

Figure 7. Theoretical dispersion curves for a steel plate (D1, black curves)–single layer medium (d_s = 3 mm, E_s = 195.2 GPa, vs. = 0.3, ρ_s = 7741.7 kg/m^3) and adhesive joint (D2, red curves)–three-layer medium consisting of two steel plates (parameters same as above) and adhesive film (d_a = 0.2 mm, E_a = 5 GPa, v_a = 0.35, ρ_a = 1330 kg/m^3): (**a**) wavenumber–frequency relations; (**b**) group velocity–frequency relations.

3.2. Influence of Excitation Frequency on the RMS Damage Imaging

The analysis of the influence of the excitation frequency on the effectiveness of RMS damage imaging was performed for specimen #1. Experimental and numerical approaches were applied for an analysis of guided Lamb wave fields and RMS maps.

3.2.1. Guided Wave Fields

Guided wave fields representing out-of-plane velocity values were prepared for a specific time instance t = 30 µs. Certain frequencies (50, 100, 150, 210, 300, and 350 kHz) were chosen for a comparative analysis of experimental and numerical results. The maps of propagating waves are presented in Figure 8. The comparison of presented snapshots revealed the variability of group velocity in relation to the excitation frequency. For the lowest frequency (50 kHz, Figure 8a), the wavefront is moderately visible (disturbance only at the initial part of the overlap). In the case of higher frequencies, the wavefront moved to the left side of the overlap, which suggests the greater speed of the excited wave packet. In fact, the individual selection of the time instance for each measurement can reduce these differences. Minor differences were observed between higher frequencies, because the group velocity was similar. Knowing that the excitation has an antisymmetric character, the A_0 mode is expected to be dominant. These observations agree with the dispersion curves (Figure 7b). The A_0 curve for the three-layer plate indicates significant growth in the group velocity value in the initial frequency range, and almost no variations in the further range. This explains why there are meaningful differences between snapshots for lower frequencies (50, 100, and 150 kHz), but wave fields are comparable for higher ones (210, 300, and 350 kHz), neglecting considerable changes in periods of wave packets.

Comparing the experimental and numerical results, there are some slight differences. Firstly, the numerical maps are symmetric, whereas the symmetry of the experimental wave fields is vaguely disturbed, probably by an imperfect preparation of the specimen and an inaccurate assumption of the scanning area for measurements. Moreover, the wavefronts are disturbed sharply at the edges of the defect in the numerical snapshots, but this effect is not that demonstrable in the experimental results because of the irregularities in the shape of defect edges (cf. Figure 3). Additionally, the group velocity is slightly higher for the experimental maps. The reason might lie in the differences between the mechanical properties of both materials (steel, adhesive) or the geometry of plates and the adhesive film (especially thickness). The observation of each snapshot allows identifying the defect. Significant disturbances of the wavefront indicate the intended lack of the adhesive in the middle of the overlap. Nonetheless, the determination of the actual geometry of the damaged area is not possible, and additional signal processing is required.

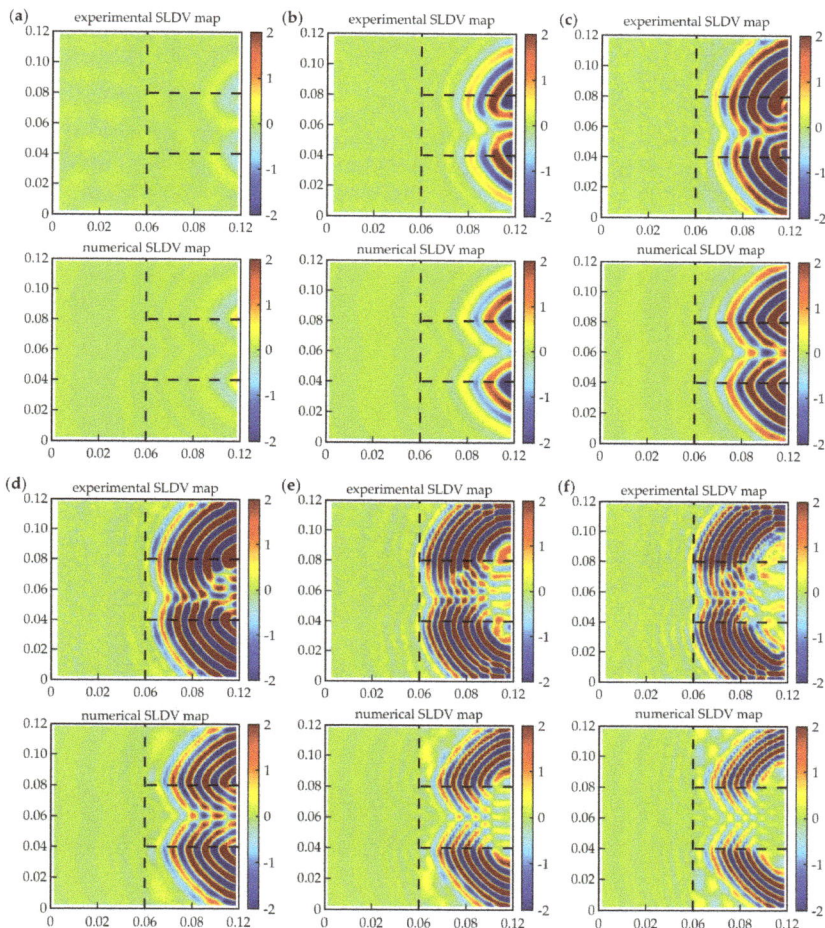

Figure 8. Experimental and numerical guided wave fields (values in m/s·10^{-3}) for a specific time, 30 μs, and different excitation frequencies: (**a**) f = 50 kHz; (**b**) f = 100 kHz; (**c**) f = 150 kHz; (**d**) f = 210 kHz; (**e**) f = 300 kHz; and (**f**) f = 350 kHz.

3.2.2. RMS Imaging

Figure 9 shows the RMS maps normalized to unity for experimental and numerical signals collected for specimen #1. The chosen frequencies were the same as those for the SLDV maps. Each RMS value was calculated with respect to Equation (2). The time window covered the whole time of signal acquisition, i.e., T = 3.2 ms. The individual characterization of a single-layer medium (steel plate, such as the damage area) and a three-layer medium (properly prepared adhesive joint) should result in the clear difference of the calculated RMS values. However, it is undeniable that excitation frequency is an essential factor affecting the effectiveness of RMS damage imaging. For the lower frequencies (50 kHz, Figure 9a; 100 kHz, Figure 9b; 150 kHz, Figure 9c), the damaged area is characterized by highest RMS values rather than an appropriately prepared joint (similar to the single plate after the joint). The difference between these two areas is clear. Some differences between experimental and numerical maps result from irregularities in defect geometry (cf. Figure 3). It is worth noticing that at the lower excitation frequency, the lower resolution can be obtained in the map and, as a result, the larger defects can be omitted. This may be very important in the case of small defects; however, for the considered damage area, it is not essential. The frequency 210 kHz (Figure 9d) give an ineffectual result: there is almost no difference between the defect and intact joint, especially in the numerical map. In the experimental RMS, some boundary effects (intensification of the wave energy on the irregular edge) led to higher RMS values. This example shows that the invalid choice of the excitation frequency can make the measurement results useless. For the frequencies higher than 210 kHz (300 kHz, Figure 9e; 350 kHz, Figure 9f) the RMS values are lower in the damaged area than in the intact joint. The correlation between RMS values for these two areas is inverted. What is important is that the visual assessment of RMS maps shows that the distinction between the damaged and intact area of the joint is much more pronounced in the lower frequency range, especially for experimental maps, where the whole area of the adhesive layer does not have the same value. This may be the effect of the limitations of the used experimental setup. What is more, the increase in the excitation frequency is related to the increase in the wave attenuation.

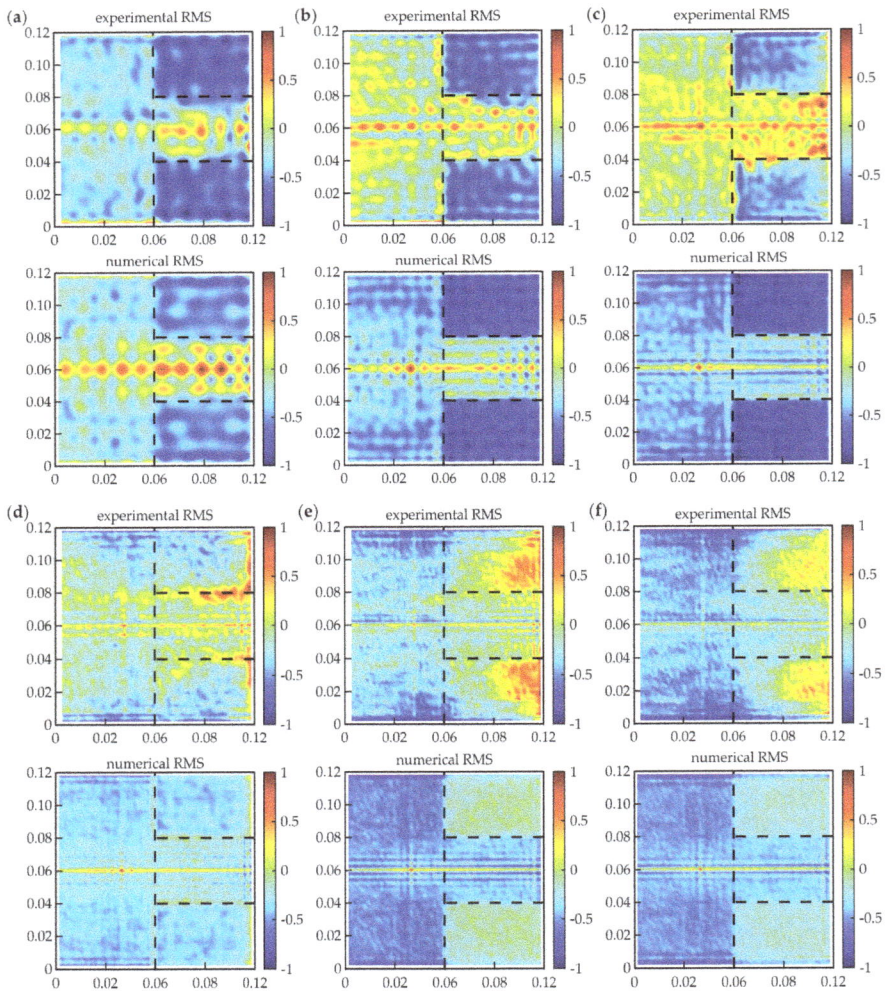

Figure 9. Experimental and numerical root mean square (RMS) maps for different excitation frequencies: (a) $f = 50$ kHz; (b) $f = 100$ kHz; (c) $f = 150$ kHz; (d) $f = 210$ kHz; (e) $f = 300$ kHz; and (f) $f = 350$ kHz.

The effectiveness of RMS damage identification is an important issue, so there is the need for a qualitative measure of contradistinction between the damaged and intact areas of the joint. The proposition is the relative difference between the level of the RMS in these two areas, which can be expressed by the relation:

$$R_d = \frac{l_d - l_i}{l_i} \qquad (3)$$

where l_i and l_d denote the mean RMS value in the properly prepared area of the overlap and in the damaged area, respectively. The definition of the R_d value induces that the damaged area is defined, so it cannot be used if the joint has any unknown defects. Nevertheless, the aim of R_d calculations is only the demonstration of changes in the effectiveness of RMS imaging in relation to the excitation frequency. The area of the overlap was divided before the calculations into two parts (damaged and intact) with a rejection of points localized on the edges of the defect and on the longitudinal axis of the joint, because for these points, the RMS values are distinctly high (intensification of energy evoked by the symmetry and boundaries, cf. Figure 9). The mean RMS values were calculated for the points of both areas, and the R_d value was calculated for measurements over the whole considered range of frequency.

Figure 10 shows the relation between R_d and excitation frequency for experimental and numerical results. The curves are slightly different, but both have some characteristic points. The first one is a local maximum for 50 kHz. For this value, the A_0 modes for single and three-layer plates are crossing on dispersion curves (cf. Figure 7b). Then, there are some fluctuations that differ between the two curves. Another peak repeating for both experimental and numerical curves is for about 120 kHz, when the A_1 mode for the three-layer plate appears. Further, the curves are falling monotonically. The numerical curve has the root equal to approximately 215 kHz, and this is the frequency value for which the damaged area and intact joint are not distinguishable (cf. Figure 9). For the experimental curve, the root is translated to approximately 250 kHz. This may be the result of the energy intensification on the edges of the defect (cf. Figure 9d). The global minimum for the numerical curve is attained for approximately 260 kHz (the appearance of the S_1 mode for the three-layer plate). This is the frequency for which the joint and the defect can be distinguished with maximal efficiency in the frequency range above 210 kHz. Further, the curve is slightly rising until obtaining another root (about 500 kHz) at the end of the frequency range. The experimental curve does not obtain the local minimum above 210 kHz; instead, it is constantly falling to the end of the assumed frequency range up to 350 kHz. Generally, the positive values of R_d are obtained when the damaged area is characterized by higher RMS values than the intact joint. Negative values indicate on the inverted relation. If R_d equals zero, there is no possibility of identifying the defects. What is important is that the absolute values of R_d are higher below the first root (about 215 kHz), which suggests that the lower frequencies allow obtaining a better differentiation of three-layer and single layer media. However, the above-mentioned decrease in map resolution cannot be neglected. To compromise both of these factors (the differentiation between defect and intact joint and the map resolution), the excitation frequency should be chosen from an approximate range of 120 to 180 kHz. The determination of an optimal frequency value requires mathematical optimization and a proposition of an objective function containing components linked with the image resolution and the R_d value.

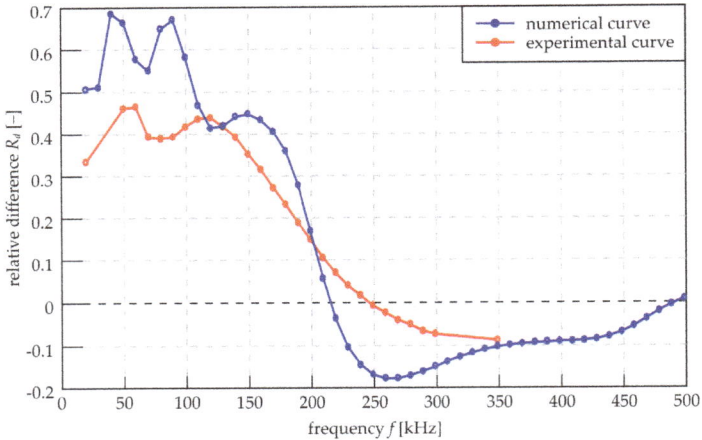

Figure 10. Relative difference between mean RMS values for damaged and intact areas of overlap for experimental (frequency range from 20 to 350 kHz) and numerical results (frequency range from 20 to 500 kHz).

3.2.3. Statistical Analysis of RMS Values

In addition to the foregoing considerations, the statistical analysis was conducted. All the RMS values calculated for the whole overlap surface, rejecting points at the edges of the defect and on the longitudinal axis (as for R_d calculations), were treated as the single series of values. Histograms were calculated for each dataset. If there are no defects in the analyzed area, the purely unimodal distribution would be obtained, because all the RMS values should accumulate over a single value, which is symbolized above by l_i. The presence of a defect in the adhesive layer should result in the bimodal distribution caused by the existence of two dominant values for the intact joint l_i and for defect l_d.

Figure 11 presents RMS histograms that have been prepared for certain frequencies (the same as for the RMS maps). The results are normalized, both for the RMS value and the quantity axes. For the lower frequencies (50, 100, and 150 kHz), bimodal distributions were obtained for the experimental and numerical data. The first mode is characterized by the lower RMS values and related to the intact joint area (cf. Figure 9a–c). The second mode indicates the defect existence (higher RMS values), and it obtains less quantity than the first mode, because the defect surface is twice as small as the intact joint surface. For 210 kHz (Figure 9d), the histograms are unimodal, which results from the equality of mean RMS values calculated for the defect and intact joint (R_d = 0, not efficient damage imaging). The experimental histogram is not as narrow as the numerical one, which is related to the translation of root of curves from Figure 10. For higher frequencies (300 kHz, 350 kHz), the distributions are not unimodal; the damage can be identified, and its mode is related to the lower RMS values. The dissociation of modes is not as clear as for lower frequencies—this effect is related to the lower absolute values of R_d for higher frequencies. In numerical histograms, the defect mode is characterized by the lower intensity than the intact joint mode (compatibly with the relation of surfaces of damaged and intact areas). The experimental histograms do not cover the same rule; the intact joint mode has a lower intensity because higher RMS values are not obtained in the whole area of the properly prepared joint (cf. Figure 9e,f). To sum up, the histogram analysis can reveal the existence of damaged areas, but it has some limitations. Firstly, the geometry of the defects cannot be determined. Moreover, the method is not efficient for small defects, because the damage modes would be of small quantity, which makes it impossible to identify them on the histograms.

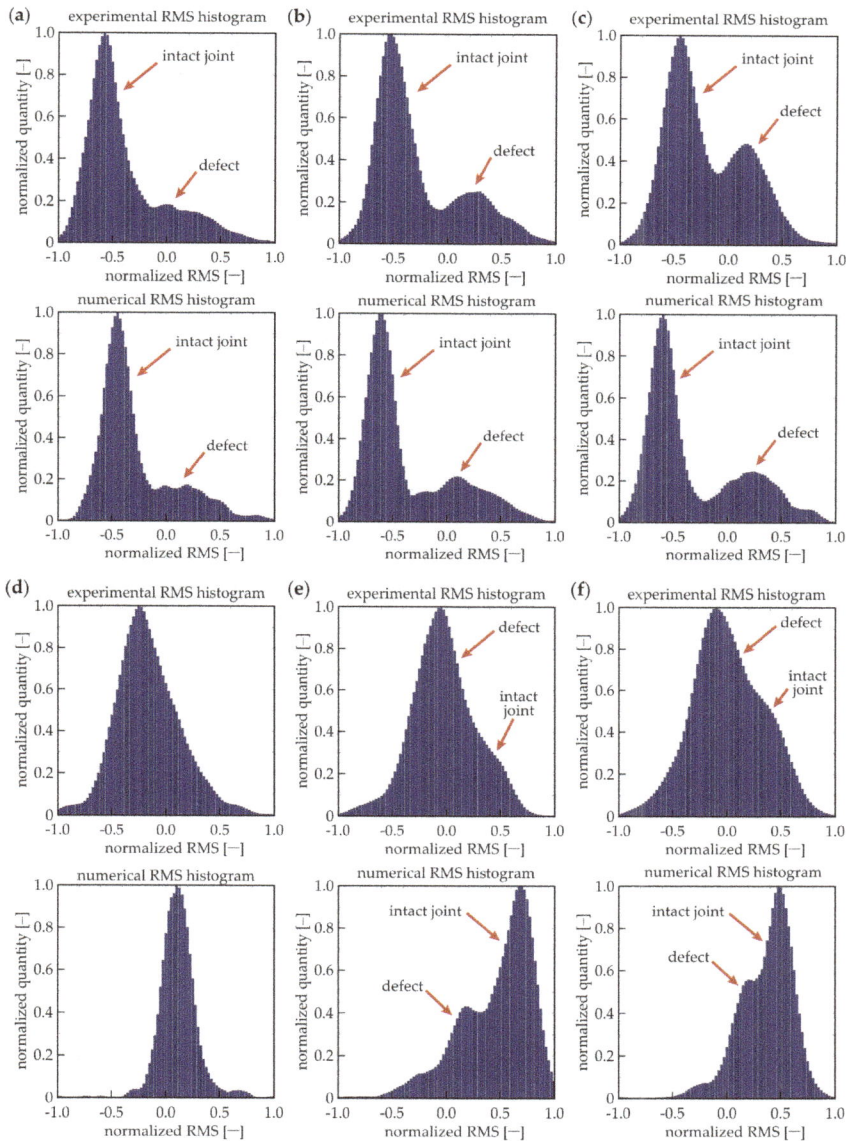

Figure 11. Experimental and numerical RMS histograms for different excitation frequencies: (**a**) $f = 50$ kHz; (**b**) $f = 100$ kHz; (**c**) $f = 150$ kHz; (**d**) $f = 210$ kHz; (**e**) $f = 300$ kHz; and (**f**) $f = 350$ kHz.

3.3. Influence of Different Defect Geometry

The above considerations were conducted only for a single joint, #1. Next, the observed effects were verified on specimens #2 to #4 by the way of numerical calculations for three certain frequencies (100 kHz, 210 kHz, and 300 kHz). The normalized RMS maps are presented in Figure 12. It is visible that for the frequency of 100 kHz, higher RMS values were obtained for the damaged areas than for the intact joint (for all the analyzed specimens). The frequency of 210 kHz appeared to be inefficient for RMS damage imaging, i.e., the difference between the damaged and properly prepared area was not significant. The frequency of 300 kHz resulted in lower RMS values in the damaged area. Moving to the histograms (Figure 13), the frequency 210 kHz gave the unimodal distribution (no difference between the defect and intact joint). The histograms for 100 and 300 kHz gave bimodal distributions, but for lower frequencies, the defect mode was related to higher RMS values, whereas for higher frequencies, it was related to lower RMS values. The quantity for the defect mode was always smaller than that for an intact joint. The considerations for joints #2 to #4 provided the same results as for specimen #1. Summarizing, the efficiency of a measurement with a specific excitation frequency does not change with the geometry of a damaged area. It is a satisfying conclusion, because generally, the geometry of the damaged area is unknown. This means that an additional preliminary study consisting of numerical calculations can reveal the appropriate excitation frequency value and reduce the number of measurements. However, the possibility of the damage identification due to a frequency value depends strongly on the characteristics of the considered media, so they need to be determined.

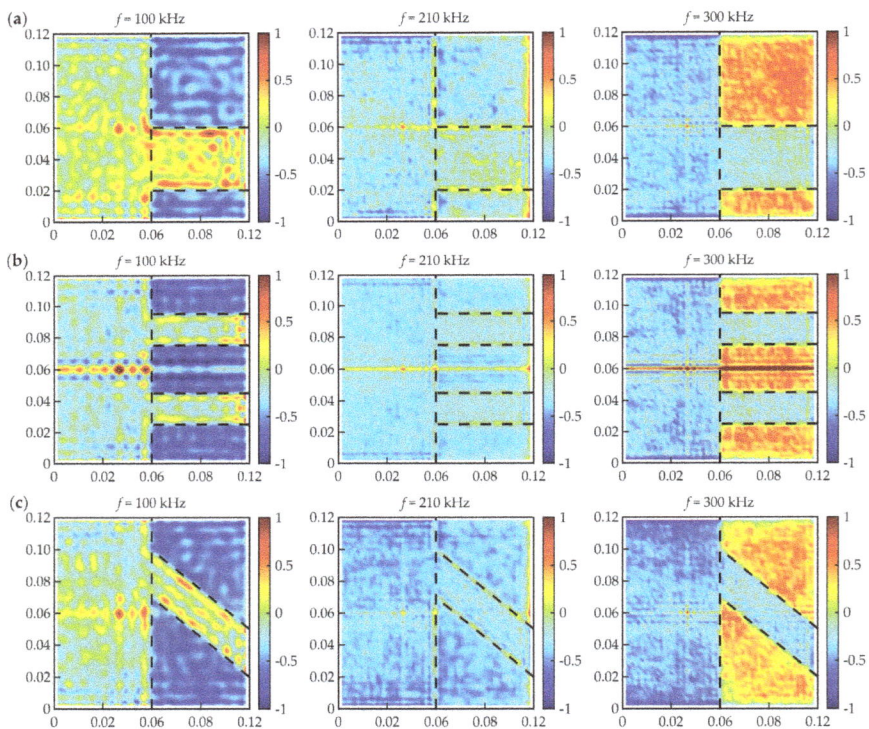

Figure 12. Numerical RMS maps for different excitation frequencies (100, 210, and 300 kHz) and different defects: (**a**) #2; (**b**) #3; and (**c**) #4.

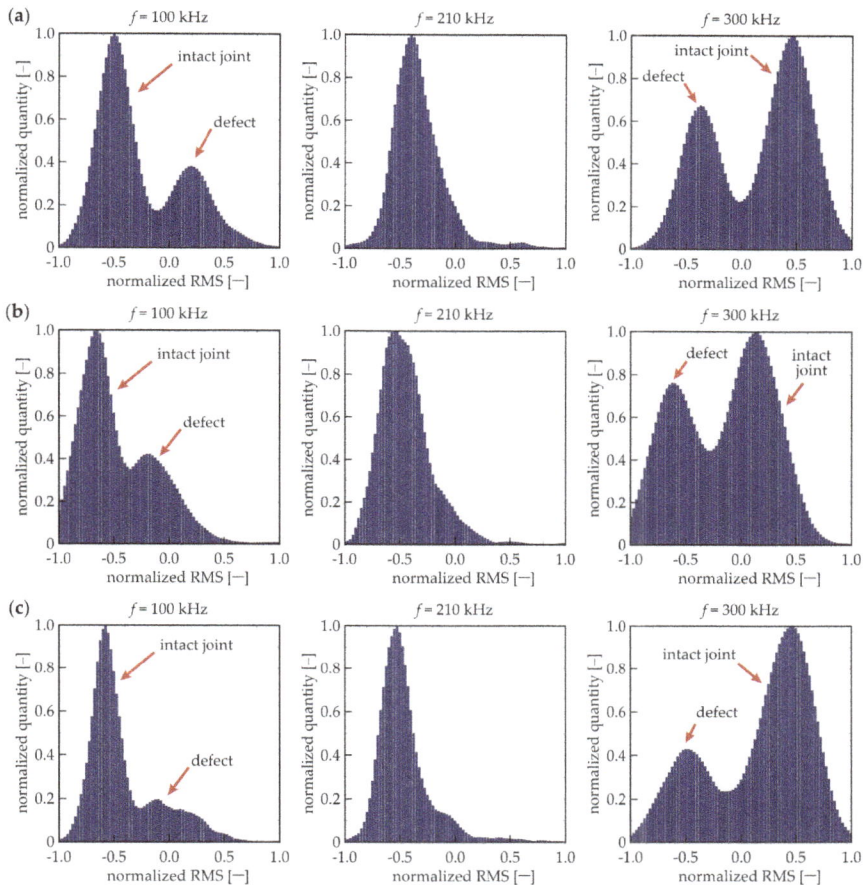

Figure 13. Numerical RMS histograms for different excitation frequencies (100, 210, and 300 kHz) and different defects: (**a**) #2; (**b**) #3; and (**c**) #4.

4. Conclusions

The paper discussed the effects of the wave frequency on the efficiency of damage detection in adhesive joints of steel plates using Lamb wave propagation and RMS imaging. Experimental and numerical approaches were applied. The research comprised the visual appreciation of obtained RMS maps and statistical analysis of calculated values. The study resulted in the conclusions presented below.

- The guided wave fields enabled identifying the occurrence of the defect regardless of the excitation frequency. However, the actual location and shape are indeterminable; thus, guided wave field measurements can only be an initial step for further analyses.
- The RMS maps allowed determining the geometry of the damaged areas. The effectiveness of damage visualization was strongly dependent on the excitation frequency.
- The variability of the relative difference between the mean RMS values for the intact joint and the damage was fully compatible with the clarity of the RMS maps. Some analogies between the relative RMS difference and dispersion curves were observed.

- The statistical analysis was successfully used to determine the effectiveness of the results obtained for different excitation frequencies based on the RMS histograms. The important advantage of this approach is the independence of the defect geometry.
- The statistical analysis in a certain frequency range on the single numerical model with a random defect can be sufficient for the determination of the adequate frequency for the further experimental testing of samples with an unknown state.

The guiding conclusion was that the Lamb wave-based inspection of adhesive joints with the use of scanning laser Doppler vibrometry and signal processing, such as root mean square calculations, provides a successful method for damage imaging. To obtain valuable results, some initial analyses need to be conducted before the exact measurements. The main factor is the choice of an appropriate excitation frequency, which can be conducted using numerical calculations supported by statistical analysis.

Author Contributions: Conceptualization and Methodology, E.W. and M.R.; Experimental Investigations, E.W. and M.R.; FEM Calculations, E.W.; Formal Analysis, E.W.; Visualization, E.W.; Writing—Original Draft Preparation, E.W.; Writing—Review and Editing, M.R.; Supervision, Project Administration and Funding Acquisition, M.R.

Funding: The research work was carried out within project No. 2015/19/B/ST8/00779, financed by the National Science Centre, Poland.

Acknowledgments: Abaqus calculations were carried out at the Academic Computer Centre in Gdańsk.

Conflicts of Interest: The authors declare no conflict of interest.

References

1. Adams, R.D.; Wake, W.C. *Structural Adhesive Joints in Engineering*; Elsevier Applied Science Publishers: London, UK, 1986; ISBN 978-94-010-8977-7.
2. Dillard, D.A. *Advances in Structural Adhesive Bonding*, 1st ed.; Woodhead Publishing: Cambridge, UK, 2010; ISBN 9781845694357.
3. Martínez-Landeros, V.H.; Vargas-Islas, S.Y.; Cruz-González, C.E.; Barrera, S.; Mourtazov, K.; Ramírez-Bon, R. Studies on the influence of surface treatment type, in the effectiveness of structural adhesive bonding, for carbon fiber reinforced composites. *J. Manuf. Process.* **2019**, *39*, 160–166. [CrossRef]
4. Jeenjitkaew, C.; Guild, F.J. The analysis of kissing bonds in adhesive joints. *Int. J. Adhes. Adhes.* **2017**, *75*, 101–107. [CrossRef]
5. Sengab, A.; Talreja, R. A numerical study of failure of an adhesive joint influenced by a void in the adhesive. *Compos. Struct.* **2016**, *156*, 165–170. [CrossRef]
6. Ong, W.H.; Rajic, N.; Chiu, W.K.; Rosalie, C. Lamb wave–based detection of a controlled disbond in a lap joint. *Struct. Health Monit.* **2018**, *17*, 668–683. [CrossRef]
7. Ren, B.; Lissenden, C.J. Ultrasonic guided wave inspection of adhesive bonds between composite laminates. *Int. J. Adhes. Adhes.* **2013**, *45*, 59–68. [CrossRef]
8. Puthillath, P.K.; Yan, F.; Kannajosyula, H. Inspection of Adhesively Bonded Joints Using Ultrasonic Guided Waves. In Proceedings of the 17th World Conference on Nondestructive Testing, Shanghai, China, 25–28 October 2008; pp. 25–28.
9. Korzeniowski, M.; Piwowarczyk, T.; Maev, R.G. Application of ultrasonic method for quality evaluation of adhesive layers. *Arch. Civ. Mech. Eng.* **2014**, *14*, 661–670. [CrossRef]
10. Tighe, R.C.; Dulieu-Barton, J.M.; Quinn, S. Identification of kissing defects in adhesive bonds using infrared thermography. *Int. J. Adhes. Adhes.* **2016**, *64*, 168–178. [CrossRef]
11. Opdam, N.J.M.; Roeters, F.J.M.; Verdonschot, E.H. Adaptation and radiographic evaluation of four adhesive systems. *J. Dent.* **1997**, *25*, 391–397. [CrossRef]
12. Sato, T.; Tashiro, K.; Kawaguchi, Y.; Ohmura, H.; Akiyama, H. Pre-bond surface inspection using laser-induced breakdown spectroscopy for the adhesive bonding of multiple materials. *Int. J. Adhes. Adhes.* **2019**, 1–9. [CrossRef]
13. Steinbild, P.J.; Höhne, R.; Füßel, R.; Modler, N. A sensor detecting kissing bonds in adhesively bonded joints using electric time domain reflectometry. *NDT E Int.* **2019**, *102*, 114–119. [CrossRef]

14. Malik, H.; Zatar, W. Software Agents to Support Structural Health Monitoring (SHM)-Informed Intelligent Transportation System (ITS) for Bridge Condition Assessment. *Procedia Comput. Sci.* **2018**, *130*, 675–682. [CrossRef]
15. dos Reis, J.; Oliveira Costa, C.; Sá da Costa, J. Local validation of structural health monitoring strain measurements. *Meas. J. Int. Meas. Confed.* **2019**, *136*, 143–153. [CrossRef]
16. Comisu, C.C.; Taranu, N.; Boaca, G.; Scutaru, M.C. Structural health monitoring system of bridges. *Procedia Eng.* **2017**, *199*, 2054–2059. [CrossRef]
17. Yang, J.P.; Chen, W.Z.; Li, M.; Tan, X.J.; Yu, J.X. Structural health monitoring and analysis of an underwater TBM tunnel. *Tunn. Undergr. Space Technol.* **2018**, *82*, 235–247. [CrossRef]
18. Miśkiewicz, M.; Pyrzowski, Ł.; Wilde, K.; Mitrosz, O. Technical Monitoring System for a New Part of Gdańsk Deepwater Container Terminal. *Polish Marit. Res.* **2017**, *24*, 149–155. [CrossRef]
19. Gomes, G.F.; Mendéz, Y.A.D.; da Silva Lopes Alexandrino, P.; da Cunha, S.S.; Ancelotti, A.C. The use of intelligent computational tools for damage detection and identification with an emphasis on composites—A review. *Compos. Struct.* **2018**, *196*, 44–54. [CrossRef]
20. Martins, A.T.; Aboura, Z.; Harizi, W.; Laksimi, A.; Khellil, K. Structural health monitoring for GFRP composite by the piezoresistive response in the tufted reinforcements. *Compos. Struct.* **2019**, *209*, 103–111. [CrossRef]
21. Chroscielewski, J.; Miskiewicz, M.; Pyrzowski, L.; Rucka, M.; Sobczyk, B.; Wilde, K. Dynamic Tests and Technical Monitoring of a Novel Sandwich Footbridge. In *Dynamics of Civil Structures, Volume 2*; Pakzad, S., Ed.; Conference Proceedings of the Society for Experimental Mechanics Series; Springer: Berlin, Germany, 2019; pp. 55–60.
22. Ostachowicz, W.; Kudela, P.; Krawczuk, M.; Zak, A. *Guided Waves in Structures for SHM: The Time-Domain Spectral Element Method*; Wiley: Hoboken, NJ, USA, 2012; ISBN 9781119965855.
23. Rose, J.L. *Ultrasonic Guided Waves in Solid Media*; Cambridge University Press: New York, NY, USA, 2014; ISBN 9781107273610.
24. Yu, X.; Zuo, P.; Xiao, J.; Fan, Z. Detection of damage in welded joints using high order feature guided ultrasonic waves. *Mech. Syst. Signal Process.* **2019**, *126*, 176–192. [CrossRef]
25. Zhang, W.; Hao, H.; Wu, J.; Li, J.; Ma, H.; Li, C. Detection of minor damage in structures with guided wave signals and nonlinear oscillator. *Meas. J. Int. Meas. Confed.* **2018**, *122*, 532–544. [CrossRef]
26. Pan, W.; Sun, X.; Wu, L.; Yang, K.; Tang, N. Damage Detection of Asphalt Concrete Using Piezo-Ultrasonic Wave Technology. *Materials (Basel)* **2019**, *12*, 443. [CrossRef]
27. Schabowicz, K. Ultrasonic tomography - The latest nondestructive technique for testing concrete members - Description, test methodology, application example. *Arch. Civ. Mech. Eng.* **2014**, *14*, 295–303. [CrossRef]
28. He, S.; Ng, C.T. Guided wave-based identification of multiple cracks in beams using a Bayesian approach. *Mech. Syst. Signal Process.* **2017**, *84*, 324–345. [CrossRef]
29. Pahlavan, L.; Blacquière, G. Fatigue crack sizing in steel bridge decks using ultrasonic guided waves. *NDT E Int.* **2016**, *77*, 49–62. [CrossRef]
30. Munian, R.K.; Mahapatra, D.R.; Gopalkrishnan, S. Lamb wave interaction with composite delamination. *Compos. Struct.* **2018**, *206*, 484–498. [CrossRef]
31. Shoja, S.; Berbyuk, V.; Boström, A. Delamination detection in composite laminates using low frequency guided waves: Numerical simulations. *Compos. Struct.* **2018**, *203*, 826–834. [CrossRef]
32. Xiao, H.; Shen, Y.; Xiao, L.; Qu, W.; Lu, Y. Damage detection in composite structures with high-damping materials using time reversal method. *Nondestruct. Test. Eval.* **2018**, *33*, 329–345. [CrossRef]
33. Nicassio, F.; Carrino, S.; Scarselli, G. Elastic waves interference for the analysis of disbonds in single lap joints. *Mech. Syst. Signal Process.* **2019**, *128*, 340–351. [CrossRef]
34. Sunarsa, T.Y.; Aryan, P.; Jeon, I.; Park, B.; Liu, P.; Sohn, H. A reference-free and non-contact method for detecting and imaging damage in adhesive-bonded structures using air-coupled ultrasonic transducers. *Materials (Basel)* **2017**, *10*, 1402. [CrossRef]
35. Parodi, M.; Fiaschi, C.; Memmolo, V.; Ricci, F.; Maio, L. Interaction of Guided Waves with Delamination in a Bilayered Aluminum-Composite Pressure Vessel. *J. Mater. Eng. Perform.* **2019**, 1–11. [CrossRef]
36. Gauthier, C.; Ech-Cherif El-Kettani, M.; Galy, J.; Predoi, M.; Leduc, D.; Izbicki, J.L. Lamb waves characterization of adhesion levels in aluminum/epoxy bi-layers with different cohesive and adhesive properties. *Int. J. Adhes. Adhes.* **2017**, *74*, 15–20. [CrossRef]

37. Castaings, M. SH ultrasonic guided waves for the evaluation of interfacial adhesion. *Ultrasonics* **2014**, *54*, 1760–1775. [CrossRef] [PubMed]

38. Kudela, P.; Wandowski, T.; Malinowski, P.; Ostachowicz, W. Application of scanning laser Doppler vibrometry for delamination detection in composite structures. *Opt. Lasers Eng.* **2016**, *99*, 46–57. [CrossRef]

39. Rothberg, S.J.; Allen, M.S.; Castellini, P.; Di Maio, D.; Dirckx, J.J.J.; Ewins, D.J.; Halkon, B.J.; Muyshondt, P.; Paone, N.; Ryan, T.; et al. An international review of laser Doppler vibrometry: Making light work of vibration measurement. *Opt. Lasers Eng.* **2017**, *99*, 11–22. [CrossRef]

40. Derusova, D.; Vavilov, V.; Sfarra, S.; Sarasini, F.; Krasnoveikin, V.; Chulkov, A.; Pawar, S. Ultrasonic spectroscopic analysis of impact damage in composites by using laser vibrometry. *Compos. Struct.* **2019**, *211*, 221–228. [CrossRef]

41. Pieczonka, Ł.; Ambroziński, Ł.; Staszewski, W.J.; Barnoncel, D.; Pérès, P. Damage detection in composite panels based on mode-converted Lamb waves sensed using 3D laser scanning vibrometer. *Opt. Lasers Eng.* **2017**, *99*, 80–87. [CrossRef]

42. Sohn, H.; Dutta, D.; Yang, J.Y.; Desimio, M.; Olson, S.; Swenson, E. Automated detection of delamination and disbond from wavefield images obtained using a scanning laser vibrometer. *Smart Mater. Struct.* **2011**, *20*, 045017. [CrossRef]

43. Saravanan, T.J.; Gopalakrishnan, N.; Rao, N.P. Damage detection in structural element through propagating waves using radially weighted and factored RMS. *Measurement* **2015**, *73*, 520–538. [CrossRef]

44. Radzieński, M.; Doliński, L.; Krawczuk, M.; Zak, A.; Ostachowicz, W. Application of RMS for damage detection by guided elastic waves. *J. Phys. Conf. Ser.* **2011**, *305*, 1–10. [CrossRef]

45. Radzieński, M.; Doliński, Ł.; Krawczuk, M.; Palacz, M. Damage localisation in a stiffened plate structure using a propagating wave. *Mech. Syst. Signal Process.* **2013**, *39*, 388–395. [CrossRef]

46. Lee, C.; Park, S. Flaw Imaging Technique for Plate-Like Structures Using Scanning Laser Source Actuation. *Shock Vib.* **2014**, *2014*, 725030. [CrossRef]

47. Lee, C.; Zhang, A.; Yu, B.; Park, S. Comparison study between RMS and edge detection image processing algorithms for a pulsed laser UWPI (Ultrasonic wave propagation imaging)-based NDT technique. *Sensors (Switzerland)* **2017**, *17*, 1224. [CrossRef] [PubMed]

48. Rucka, M.; Wojtczak, E.; Lachowicz, J. Damage imaging in Lamb wave-based inspection of adhesive joints. *Appl. Sci.* **2018**, *8*, 522. [CrossRef]

49. Aryan, P.; Kotousov, A.; Ng, C.T.; Cazzolato, B.S. A baseline-free and non-contact method for detection and imaging of structural damage using 3D laser vibrometry. *Struct. Control Health Monit.* **2017**, *24*, 1–13. [CrossRef]

50. Chronopoulos, D. Calculation of guided wave interaction with nonlinearities and generation of harmonics in composite structures through a wave finite element method. *Compos. Struct.* **2018**, *186*, 375–384. [CrossRef]

51. Apalowo, R.K.; Chronopoulos, D. A wave-based numerical scheme for damage detection and identification in two-dimensional composite structures. *Compos. Struct.* **2019**, *214*, 164–182. [CrossRef]

52. Moser, F.; Jacobs, L.J.; Qu, J. Modeling elastic wave propagation in waveguides with the finite element method. *NDT E Int.* **1999**, *32*, 225–234. [CrossRef]

53. Gauthier, C.; Galy, J.; Ech-Cherif El-Kettani, M.; Leduc, D.; Izbicki, J.L. Evaluation of epoxy crosslinking using ultrasonic Lamb waves. *Int. J. Adhes. Adhes.* **2018**, *80*, 1–6. [CrossRef]

54. Lowe, M.J.S. Matrix Techniques for Modeling Ultrasonic-Waves in Multilayered Media. *IEEE Trans. Ultrason. Ferroelectr. Freq. Control* **1995**, *42*, 525–542. [CrossRef]

55. Maghsoodi, A.; Ohadi, A.; Sadighi, M. Calculation of Wave Dispersion Curves in Multilayered Composite-Metal Plates. *Shock Vib.* **2014**, *2014*, 1–6. [CrossRef]

MDPI

St. Alban-Anlage 66

4052 Basel

Switzerland

Tel. +41 61 683 77 34

Fax +41 61 302 89 18

www.mdpi.com

Materials Editorial Office

E-mail: materials@mdpi.com

www.mdpi.com/journal/materials

www.ingramcontent.com/pod-product-compliance
Lightning Source LLC
Chambersburg PA
CBHW051704210326
41597CB00032B/5360